Kutner, Marc
Leslie.

Astronomy

$26.50

DATE			

ASTRONOMY

ASTRONOMY
A Physical Perspective

Marc L. Kutner

Rensselaer Polytechnic Institute

WILEY

JOHN WILEY & SONS, NEW YORK
Chichester, Brisbane, Toronto, Singapore

On the cover: Locations of infrared sources found by the Infrared Astronomy Satellite (IRAS). The bright band across the middle is the plane of the Milky Way galaxy.

Astronomy: A Physical Perspective
Copyright © 1987 by John Wiley & Sons, Inc.

Library of Congress Cataloging-in-Publication Data

Kutner, Marc Leslie.
  Astronomy: a physical perspective.

  Includes index.
  1. Astronomy. I. Title.
QB45.K88 1987   520   86-19561
ISBN 0-471-60499-2

9 8 7 6 5 4 3 2

BRIEF CONTENTS

Detailed Contents   vii
Preface   xvii
  1. **Introduction**   1

PART I   PROPERTIES OF ORDINARY STARS   5

  2. **Continuous Radiation from Stars**   7
  3. **Spectral Lines in Stars**   27
  4. **Telescopes**   49
  5. **Binary Stars and Stellar Masses**   85
  6. **The Sun: A Typical Star**   110

PART II   STELLAR EVOLUTION   137

  7. **Relativity**   139
  8. **Protostars**   176
  9. **The Main Sequence**   199
10. **Stellar Old Age**   229
11. **The Fate of High-Mass Stars**   248
12. **Evolution in Close Binaries**   272
13. **Clusters of Stars**   287

PART III   THE MILKY WAY GALAXY   303

14. Contents of the Interstellar Medium   305
15. Star Formation   335
16. The Milky Way Galaxy   335
17. Normal Galaxies   375

PART IV   THE UNIVERSE AT LARGE   399

18. Clusters of Galaxies   401
19. Active Galaxies and Quasars   419
20. Cosmology   440
21. The Big Bang   460

PART V   THE SOLAR SYSTEM   491

22. Overview of the Solar System   493
23. The Earth and Moon   514
24. The Inner Planets   548
25. The Outer Planets   565
26. Minor Bodies in the Solar System   595

APPENDIX A   Glossary of Symbols   607
APPENDIX B   Physical and Astronomical Constants   611
APPENDIX C   Units and Conversions   613
APPENDIX D   Planet and Satellite Properties   614
APPENDIX E   Properties of Main Sequence Stars   617
APPENDIX F   Astronomical Coordinates
                     and Timekeeping   618
APPENDIX G   Abundances of the Elements   622

Photo Credits   625

Index   628

DETAILED CONTENTS

Preface   **xvii**
1. Introduction   **1**

PART I   PROPERTIES OF ORDINARY STARS   **5**

2. Continuous Radiation from Stars   **7**
  2.1 Brightness of Starlight   7
  2.2 The Electromagnetic Spectrum   9
  2.3 Colors of Stars   11
      2.3.1 Quantifying Color   11
      2.3.2 Blackbodies   12
  2.4 Planck's Law and Photons   15
      2.4.1 Planck's Law   15
      2.4.2 Photons   17
  2.5 Stellar Colors   18
  2.6 Stellar Distances   19
  2.7 Absolute Magnitudes   23
  Chapter Summary   24
  Questions and Problems   25

3. Spectral Lines in Stars   **27**
  3.1 Spectral Lines   27
  3.2 Spectral Types   29
  3.3 The Origin of Spectral Lines   30
      3.3.1 Bohr Atom   31
      3.3.2 Quantum Mechanics   37

3.4 Formation of Spectral Lines   38
    3.4.1 Excitation   38
    3.4.2. Ionization   39
    3.4.3 Intensities of Spectral Lines   41
3.5 The Hertzsprung–Russell Diagram   43
Chapter Summary   47
Questions and Problems   47

4. Telescopes   **49**
4.1 What a Telescope Does   49
    4.1.1 Light Gathering   49
    4.1.2 Angular Resolution   50
    4.1.3 Image Formation in a Camera   52
4.2 Refracting Telescopes   55
4.3 Reflecting Telescopes   57
4.4 Data Handling   62
    4.4.1 Detection   62
    4.4.2 Spectroscopy   64
4.5 Observatories   66
    4.5.1 Ground-Based Observing   66
    4.5.2 Observations From Space   68
4.6 Observing in the Ultraviolet   69
4.7 Observing in the Infrared   71
4.8 Radio Astronomy   75
4.9 X-ray Astronomy   81
Chapter Summary   83
Questions and Problems   83

5. Binary Stars and Stellar Masses   **85**
5.1 Binary Stars   85
5.2 Doppler Shift   86
    5.2.1 Moving Sources and Observers   87
    5.2.2 Circular Orbits   89
5.3 Binary Stars and Circular Orbits   91
5.4 Elliptical Orbits   97
    5.4.1 Geometry of Ellipses   97
    5.4.2 Angular Momentum in Elliptical Orbits   98
    5.4.3 Energy in Elliptical Orbits   99
    5.4.4 Observing Elliptical Orbits   101
5.5 Stellar Masses   102
5.6 Stellar Sizes   104
Chapter Summary   108
Questions and Problems   108

6. The Sun: A Typical Star   **110**
    6.1   Basic Structure   111
    6.2   Elements of Radiation Transfer Theory   113
    6.3   The Photosphere   116
       6.3.1   Appearance of the Photosphere   117
       6.3.2   Temperature Distribution   118
       6.3.3   Doppler Broadening of Lines   121
    6.4   The Chromosphere   123
    6.5   The Corona   124
       6.5.1   Parts of the Corona   124
       6.5.2   Temperature of the Corona   126
    6.6   Solar Activity   128
       6.6.1   Sunspots   128
       6.6.2   Other Activity   132
    Chapter Summary   *134*
    Questions and Problems   *134*

PART II   **STELLAR EVOLUTION**   **137**

7. Relativity   **139**
    7.1   Foundations of Special Relativity   139
       7.1.1   Problems With Electromagnetic Radiation   139
       7.1.2   Problems With Simultaneity   141
    7.2   Time Dilation   142
    7.3   Length Contraction   145
    7.4   The Doppler Shift   148
       7.4.1   Moving Source   148
       7.4.2   Moving Observer   149
       7.4.3   General Result   150
    7.5   Space-Time   151
       7.5.1   Four-Vectors and Lorentz Transformation   151
       7.5.2   Energy and Momentum   153
    7.6   General Relativity   155
       7.6.1   Curved Space-Time   155
       7.6.2   Principle of Equivalence   157
    7.7   Tests of General Relativity   159
       7.7.1   Orbiting Bodies   161
       7.7.2   Bending Electromagnetic Radiation   162
       7.7.3   Gravitational Red Shift   163
       7.7.4   Gravitational Radiation   166
       7.7.5   Competing Theories   166
    7.8   Black Holes   166
       7.8.1   The Schwarzschild Radius   166
       7.8.2   Approaching a Black Hole   169

7.8.3 Stellar Black Holes   171
7.8.4 Nonstellar Black Holes   172
Chapter Summary   173
Questions and Problems   174

8. Protostars   **176**
  8.1 Star Formation   176
    8.1.1 Gravitational Potential and Kinetic Energy of
               a Sphere   177
    8.1.2 Collapsing Clouds   178
  8.2 The Virial Theorem   184
  8.3 Luminosity of Collapsing Clouds   187
  8.4 Evolutionary Tracks for Protostars   190
  8.4 T Tauri Stars and Related Objects   192
  Chapter Summary   197
  Questions and Problems   197

9. The Main Sequence   **199**
  9.1 Stellar Energy Sources   199
  9.2 Nuclear Physics   201
    9.2.1 Nuclear Building Blocks   201
    9.2.2 Binding Energy   202
    9.2.3 Nuclear Reactions   204
    9.2.4 Overcoming the Fusion Barrier   206
  9.3 Nuclear Energy for Stars   209
  9.4 Stellar Structure   215
    9.4.1 Hydrostatic Equilibrium   216
    9.4.2 Energy Transport   220
  9.5 Stellar Models   222
  9.6 Solar Neutrinos   223
  Chapter Summary   227
  Questions and Problems   227

10. Stellar Old Age   **229**
  10.1 Evolution Off the Main Sequence   229
    10.1.1 Low-Mass Stars   229
    10.1.2 High-Mass Stars   232
  10.2 Cepheid Variables   232
    10.2.1 Variable Stars   232
    10.2.2 Cepheid Mechanism   234
    10.2.3 Period–Luminosity Relation   235
  10.3 Planetary Nebulae   237

10.4 White Dwarfs   240
    10.4.1 Electron Degeneracy   240
    10.4.2 Properties of White Dwarfs   242
    10.4.3 Relativistic Effects   245
Chapter Summary   246
Questions and Problems   247

11. The Fate of High-Mass Stars   **248**
  11.1 Supernovae   248
    11.1.1 Core Evolution of High Mass Stars   248
    11.1.2 Supernova Remnants   250
  11.2 Neutron Stars   252
    11.2.1 Neutron Degeneracy Pressure   254
    11.2.2 Rotation of Neutron Stars   256
    11.2.3 Magnetic Fields of Neutron Stars   257
  11.3 Pulsars   258
    11.3.1 Discovery   258
    11.3.2 What are Pulsars?   260
    11.3.3 Period Changes   263
  11.4 Pulsars and Dispersion   268
  11.5 Stellar Black Holes   269
Chapter Summary   270
Questions and Problems   270

12. Evolution in Close Binaries   **272**
  12.1 Close Binaries   272
  12.2 Systems with White Dwarfs   275
  12.3 Neutron Stars in Close Binary Systems   277
  12.4 Systems with Black Holes   280
  12.5 SS433   283
Chapter Summary   285
Questions and Problems   286

13. Clusters of Stars   **287**
  13.1 Types of Clusters   287
  13.2 Distances to Moving Clusters   289
  13.3 Clusters as Dynamical Entities   291
    13.3.1 Energies   292
    13.3.2 Cluster Dynamics   294
  13.4 *H–R* Diagrams for Clusters   297
  13.5 The Concept of Populations   299
Chapter Summary   300
Questions and Problems   300

PART III   THE MILKY WAY GALAXY   303

14. Contents of the Interstellar Medium   305
   14.1 Overview   305
   14.2 Interstellar Extinction   306
      14.2.1 The Effect of Extinction   306
      14.2.2 Star Counting   308
      14.2.3 Reddening   309
      14.2.4 Extinction Curves   311
      14.2.5 Polarization   312
      14.2.6 Scattering versus Absorption   312
   14.3 Physics of Dust Grains   313
      14.3.1 Size and Shape   313
      14.3.2 Composition   314
      14.3.3 Electric Charge   314
      14.3.4 Temperature   317
      14.3.5 Evolution   318
   14.4 Interstellar Gas   319
      14.4.1 Optical Studies   319
      14.4.2 Radio Studies of Atomic Hydrogen   320
   14.5 Interstellar Molecules   324
      14.5.1 Discovery   324
      14.5.2 Chemistry   327
      14.5.3 Observing Molecules   328
   14.6 Thermodynamics of the Interstellar Medium 331
   Chapter Summary   333
   Questions and Problems   334

15. Star Formation   335
   15.1 Problems in Star Formation   335
   15.2 Magnetic Effects and Star Formation   336
   15.3 Molecular Clouds and Star Formation   339
   15.4 Regions of Recent Star Formation   342
      15.4.1 HII Regions   342
      15.4.2 Masers   346
      15.4.3 Energetic Flows   349
   15.5 Picture of a Star-Forming Region: Orion   352
   Chapter Summary   353
   Questions and Problems   353

16. The Milky Way Galaxy   355
   16.1 Overview   355
   16.2 Differential Galactic Rotation   356
      16.2.1 Rotation and Mass Distribution   357

16.2.2 Rotation Curve and Doppler Shift 358
16.2.3 Oort Formulae 360
16.3 Determination of the Rotation Curve 364
16.4 Average Gas Distribution 367
16.5 Spiral Structure in the Milky Way 369
16.3.1 Optical Tracers of Spiral Structure 369
16.5.2 Radio Tracers of Spiral Structure 370
16.6 The Galactic Center 371
Chapter Summary 373
Questions and Problems 373

17. Normal Galaxies **375**
17.1 Types of Galaxies 375
17.1.1 Elliptical Galaxies 376
17.1.2 Spiral Galaxies 379
17.1.3 Other Types of Galaxies 380
17.2 Star Formation in Spiral Galaxies 382
17.3 Explanations of Spiral Structure 388
17.4 Mass-to-Light Ratios in Galaxies 394
Chapter Summary 397
Questions and Problems 397

PART IV   THE UNIVERSE AT LARGE **399**

18. Clusters of Galaxies **401**
18.1 Distribution of Galaxies 401
18.2 Cluster Dynamics 403
18.3 Expansion of the Universe 409
18.3.1 Hubble's Law 409
18.3.2 Determining the Hubble Constant 413
18.4 Superclusters 416
Chapter Summary 418
Questions and Problems 418

19. Active Galaxies and Quasars **419**
19.1 Radio Galaxies 419
19.1.1 Properties of Radio Galaxies 419
19.1.2 Model For Radio Galaxies 421
19.1.3 The Problem of Superluminal Expansion 424
19.2 Seyfert and N Galaxies 427
19.3 Quasars 429
19.3.1 Discovery of Quasars 429
19.3.2 Properties of Quasars 432
19.3.3 Energy–Redshift Problem 434

19.3.4 Gravitationally Lensed Quasar   437
19.3.5 BL Lacertae Objects   438
Chapter Summary   438
Questions and Problems   439

20. Cosmology   **440**
  20.1 The Scale of the Universe   440
  20.2 Expansion of the Universe   443
    20.2.1 Olbers's Paradox   443
    20.2.2 Keeping Track of Expansion   444
  20.3 Cosmology with Newtonian Gravitation   446
  20.4 Cosmology and General Relativity   450
  20.5 Is the Universe Open or Closed?   455
  Chapter Summary   458
  Questions and Problems   458

21. The Big Bang   **460**
  21.1 The Cosmic Background Radiation   460
    21.1.1 Origin of the Cosmic Background Radiation   460
    21.1.2 Observations of the Cosmic Background Radiation   463
    21.1.3 Isotropy of the Cosmic Background Radiation   467
  21.2 Big Bang Nucleosynthesis   473
  21.3 Fundamental Particles and Forces   475
    21.3.1 Fundamental Particles   475
    21.3.2 Fundamental Forces   477
    21.3.3 The Role of Symmetries   478
    21.3.4 Color   479
    21.3.5 Unification of Forces   481
  21.4 Merging of Physics of the Big and Small   483
    21.4.1 Back to the Earliest Times   483
    21.4.2 Galaxy Formation   485
    21.4.3 Inflation   486
  Chapter Summary   488
  Questions and Problems   488

PART V   THE SOLAR SYSTEM   **491**

22. Overview of the Solar System   **493**
  22.1 Motions of the Planets   493
  22.2 The Motion of the Moon   497
  22.3 Studying the Solar System   501
  22.4 Traveling through the Solar System   503

22.5 Formation of the Solar System   507
Chapter Summary   512
Questions and Problems   512

23. The Earth and Moon   **514**
  23.1 History of the Earth   514
    23.1.1 Early History   514
    23.1.2 Radioactive Dating   516
    23.1.3 Plate Tectonics   518
  23.2 Temperature of a Planet   520
  23.3 The Atmosphere   524
    23.3.1 Pressure Distribution   524
    23.3.2 Retention of an Atmosphere   524
    23.3.3 Temperature Distribution   526
    23.3.4 General Circulation   532
  23.4 The Magnetosphere   534
  23.5 Tides   537
  23.6 The Moon   541
    23.6.1 The Lunar Surface   543
    23.6.2 The Lunar Interior   545
    23.6.3 Lunar Origin   545
Chapter Summary   546
Questions and Problems   546

24. The Inner Planets   **548**
  24.1 Basic Features   548
    24.1.1 Mercury   548
    24.1.2 Venus   549
    24.1.3 Mars   549
    24.1.4 Radar Mapping of Planets   550
  24.2 Surfaces   552
    24.2.1 Mercury   552
    24.2.2 Venus   553
    24.2.3 Mars   554
  24.3 Interiors   557
    24.3.1 Basic Considerations   557
    24.3.2 Mercury   559
    24.3.3 Venus   559
    24.3.4 Mars   559
  24.4 Atmospheres   560
    24.4.1 Mercury   560
    24.4.2 Venus   561
    24.4.3 Mars   562
  24.5 Moons   563
Chapter Summary   564
Questions and Problems   564

25. The Outer Planets   **565**
        25.1 Basic Information   565
                  25.1.1 Jupiter   565
                  25.1.2 Saturn   566
                  25.1.3 Uranus   567
                  25.1.4 Neptune   568
        25.2 Atmospheres   569
                  25.2.1 Jupiter   569
                  25.2.2 Saturn   573
                  25.2.3 Uranus and Neptune   574
        25.3 Interiors   574
                  25.3.1 Jupiter   574
                  25.3.2 Saturn   575
                  25.3.3 Uranus and Neptune   576
        25.4 Rings   576
                  25.4.1 Basic Properties   576
                  25.4.2 Ring Dynamics   580
        25.5 Moons   586
                  25.5.1 Jupiter   587
                  25.5.2 Saturn's Moons   590
                  25.5.3 Moons of Uranus   591
        25.6 Pluto   592
        Chapter Summary   594
        Questions and Problems   594

26. Minor Bodies in the Solar System   **595**
        26.1 Asteroids   595
        26.2 Comets   597
        26.3 Meteroids   603
        Chapter Summary   606
        Questions and Problems   606

APPENDIX A   Glossary of Symbols   **607**
APPENDIX B   Physical and Astronomical Constants   **611**
APPENDIX C   Units and Conversions   **613**
APPENDIX D   Planet and Satellite Properties   **614**
APPENDIX E   Properties of Main Sequence Stars   **617**
**APPENDIX F   Astronomical Coordinates
                and Timekeeping**   **618**
APPENDIX G   Abundances of the Elements   **622**

Photo Credits   **625**

Index   **628**

PREFACE

The study of astronomy has blossomed in a variety of ways in the past decade. Some of these advances have been the result of careful planning, but many have been quite unexpected. Much of this information has resulted from new ground-based and space-based instruments. We have also taken advantage of computers that allow us to analyze vast quantities of data and apply theories to increasingly complex situations.

The most amazing aspect of all of this progress is that we can still provide reasonable answers to the naive question, "How does it all work?" As our astronomical horizon expands, we can still use familiar physics to explain a wealth of phenomena. Even when the explanation at the research level requires a complex application of certain physical laws, there is usually still a way of understanding the phenomena based on introductory-level physics. Perhaps this is just a realization that the laws of physics are small in number, and apply universally. There are a few exceptions where the astronomical problems are helping to drive back the frontiers of physics, but even these can be explained in more familiar terms.

This book is dedicated to the student who would like more out of even a brief study of astronomy than a list of what there is. It is for the student who wants to understand why certain phenomena occur, and how astronomical objects work. In addition, it addresses the question of how we collect and interpret information about such remote objects.

Like many texts, this book started as a set of notes for a course. In this case, it was our introductory astronomy course at RPI. The two-semester sequence is taken primarily by sophomores. We take advantage of the fact that almost all RPI freshmen take calculus and a calculus-based physics course. We therefore presume that the students have seen the classical physics needed for the astronomy course, but we do not presume any knowledge of "modern" physics. The notes that led to this book were developed after the course had been taught for many years with no suitable text. In my travels to other schools, I found many professors facing the same problem, and I was encouraged to write this book.

My thanks go to the RPI students who took the course while I was using first the notes and then the manuscript for the text. It is not easy for students to give up the security of a normal text. Their feedback on the book's strengths and weaknesses was an invaluable part of the development process. Of all the people who have reviewed the book, the students were the least inhibited in pointing out parts they didn't like. I also appreciate the patience and feedback of Drs. Frances Verter and George Spagna, who taught the course, using the manuscript for the text.

I am grateful for the support and patience of the entire staff at Harper & Row throughout the process of producing this book. As the book entered production, Lisa Berger was the editor who kept the project moving. Through all of the editorial changes, I appreciated the constant support of Logan Campbell. I am particularly indebted to Steven Pisano for his patience with me and for bringing all the material together in the production phase.

Reviews by other astronomers have been very important in refining the book. I am particularly indebted to Professors Steven Gottesman (University of Florida), Michael Jura (UCLA), Craig Sarazin (University of Virginia), Dennis Hegyi (University of Michigan), David Helfand (Columbia University) and Thomas Jones (University of Minnesota), who read the whole manuscript (with the first three reading it in two different drafts). There are as many ways to write an astronomy text as there are astronomers. However, these reviewers were able to go beyond telling me how they would write a particular section, to evaluating what I had written in terms of the goals I had set for the book. In addition, sections of the book were read by Professors Joseph Veverka (Cornell University; solar system), R. Bruce Partridge (Haverford College; relativity and cosmology), and Dr. R. S. Harrington (U. S. Naval Observatory, solar system). I also appreciate many helpful comments by Dr. Kathryn N. Mead (RPI).

Many astronomers and physicists have contributed data and illustrations which I have used directly. They are too numerous to mention here, but they are listed in figure captions and illustration credits. My special thanks go to those who were anxious for me to have the most recent data or best pictures.

I have benefited from the special atmosphere at RPI, which encourages research, but also rewards contributions to education. The book was completed while I was on sabbatical in the Astronomy Department of the University of Texas at Austin. I am grateful to the numerous colleagues who helped me clear up various last-minute questions.

Marc L. Kutner

Introduction

Our curiosity about the world around us is most naturally manifested when we look up into the night sky. We don't need any special instruments to tell us that something interesting is going on. However, only under the scrutiny afforded by a variety of instruments can these patches of light, and the dark regions between them, offer clues about their nature. We have to be clever to collect these clues, and just as clever to interpret them. It is the total of these studies that we call astronomy.

We are fortunate to be living in an era of extraordinary astronomical discovery. For centuries, astronomers were restricted to making visual observations from the surface of the Earth. We can now detect virtually any type of radiation given off by an astronomical object, from radio waves to gamma rays. Where necessary, we can put our observatories in space. For the solar system, we can even visit the objects of our study.

However, for all of these capabilities, there is one drawback. We cannot do traditional experiments on remote astronomical objects. We cannot change their environment and see how they respond. We must passively study the radiation they give off. For this reason, we refer to astronomy as an *observational* science, rather than an experimental one. It is because of this difference that we must be clever in using the information that we do receive. In this book we will see what information we can obtain, and how the clues are processed.

One of the most fascinating aspects of astronomy is that many phenomena can be understood in terms of relatively simple physics. This doesn't mean that we can explain every detail. However, we can explain the basic phenomena. In this book, we emphasize the application of a few physical principles to a variety of situations. For this purpose some background in physics is needed. We assume that the reader has had a course in "classical" physics (mechanics, electricity and magnetism, thermodynamics). We also use quite a bit of "modern" physics (relativity, atomic and nuclear physics). However, the modern physics will be developed as we need it. In addition, a familiarity with the concepts of calculus is assumed. While most of the material can be mastered without actually taking derivatives and working out integrals, the concepts of derivatives representing changes and integrals representing sums are used. The reader may also notice variation in the mathematical and physics level from subject to subject. This is because the goal in writing this book has been to present each astronomical subject at the simplest level that still provides a reasonable understanding.

In organizing an astronomy text, one important question is where to put the material on the solar system. The ''traditional'' approach has been to place the solar system first. This allows the student to start with familiar, nearby, objects, and work out from there. Also, if a nonquantitative description is all that is needed, the material on the solar system can be presented as a collection of facts, thereby not challenging the student at the very beginning of the book. In this book the solar system is placed last for a couple of reasons. First, the methods by which we study the solar system are not typical of those that we use for the remainder of the astronomical objects. Among astronomical objects, the members of the solar system are unique in that they are within our reach. The reader may gain a better appreciation of how astronomy is done by starting with any other subject but the solar system. Second, at the level of this book, we can apply the physical ideas developed for understanding stars and galaxies to gain a deeper understanding of processes on other planets.

We start with the stars, those points of light in the night sky. This allows us to develop physical ideas we can use for the rest of the book. We will see how information is obtained on the properties of stars. The Sun will be studied, in detail, as an example of a typical star. We will then put these stellar properties together, and describe a theoretical picture of how stars work. In Part II, we will see how stars evolve. We will look at the cycle of stellar birth, life, and death. In stellar death, we will encounter a variety of exotic objects, such as neutron stars and black holes.

In Part III, we will see how stars are organized into galaxies. We will study our own galaxy as well as others. We will also study the nonstellar part of galaxies, including the interstellar medium, the material out of which stars form, and ''dark'' matter, which makes its mass felt, but which is invisible. We will also see that there are galaxies known as active galaxies, which are giving off much more energy than our own. We will follow the trail of active galaxies to quasars, objects that have puzzled astronomers for two decades.

In Part IV, we will look at the overall structure of the universe, including the arrangement of galaxies and their motions. The early history of the universe will be described, and we will see how we can look for clues to both its past history and eventual fate. It is in Part IV that we will encounter one of the most fascinating recent developments, the merging of physics on the smallest and largest scales. This is the blending of theories on the ultimate structure of matter with our theories on the overall structure of the universe.

In the final part, we will come to the solar system. We will see how the formation of the solar system can be fit into the ideas already developed about star formation. We will encounter a variety of surfaces, atmospheres, and rings that can be explained using the physical ideas already developed. We will also see how certain parts of the solar system contain clues to its history.

While the organization of the book is around the astronomical topics, the presentation of the topics emphasizes the application of the underlying physics. Almost all of the physical tools will apply to several topics. (A great strength of physical theories is the generality of their applicability.) For example, orbital mechanics can tell us about the masses of binary stars and help us plan a probe to Mars. Radiative transfer helps us understand the appearance of the Sun, the physical conditions in interstellar clouds, and the temperatures of planetary atmospheres. Tidal effects may explain the visual appearance of certain galaxies, the rings around planets, and the internal heating of Io, one of Jupiter's moons.

Though an understanding of the physics is our goal, astronomy's foundation is in observation. We will see how observations often define a problem—the discovery of new phenomena, for example. Observations usually provide crucial information for solving a problem. Finally, observations provide a check on theories that are developed. In this book, we will therefore emphasize the interplay between observation and an understanding of the physics. We will see how some observations may yield numbers with great precision, while others may only give an order-of-magnitude determination, but both types can be equally important in deciding between theories.

With the current pace of astronomical discovery, there is an important caution to keep in mind. When you read an introductory text on classical physics, you are reading about theories that were worked out and tested over a century ago. No question is raised about the correctness of those theories. In astronomy, new ideas or new observations are constantly changing the thinking on various problems. Many of the topics discussed in this book are far from being settled. Sometimes more than one explanation is presented for the same phenomenon. This is done either because we don't know which is correct, or to show how one theory is eliminated in favor of another. Just because this is a ''text,'' it doesn't mean that it has the final word. If you understand where the problems lie, and the reasoning behind the explanations, then you should be able to follow future explanations as they appear in newspapers and scientific magazines and journals.

This, then, is the plan. As you study the material that follows, see how far we can go with a little bit of physics and a lot of curiosity and ingenuity.

PART I
PROPERTIES OF ORDINARY STARS

Stars provide a good starting point for our astronomical investigations. When we look at the dark night sky, the existence of a multitude of stars is obvious. In this part, and in the next part, we will see the central role played by stars. They seem to be the building blocks of our galaxy. The light given off by our galaxy is really the sum of the light given off by all of the 10^{11} stars in the galaxy.

Besides being important sources of energy, the stars are the source of most of the elements. We think that our galaxy was formed with only hydrogen and helium. All of the other elements have been made in the interiors of stars and dispersed into interstellar space, to be incorporated into the next generation of stars. We owe our very existence to the stars.

What would we like to learn about stars? We would like to understand how they work—what makes them shine. We would like to know what determines their size, temperature, and brightness. We would like to know how they are born, evolve, and die.

In our study of stars we will apply the laws of physics to astronomical situations. This is the essence of what we mean by *astrophysics*. The astrophysicist is like a detective who must piece fragments of evidence into a whole picture. A detective will use some of the evidence to posit a scenario, or model, for the situation, and then see how other aspects of the case fit into the scenario. The astrophysicist does something similar. Given some information, the astrophysicist uses the laws of physics to propose a model for how stars work. Those models are then used to make predictions about additional stellar properties. If those predicted properties are observed, more predictions are made; if not, a new model is needed.

To apply physical laws we must have some idea of the masses, sizes, temperatures, and composition of stars. These quantities can only be determined from observations of stars. This points out a very basic difference between astrophysics and other branches of physics. In solid-state physics, we can study the properties of a given material by taking a sample of that material to our laboratory to subject it to a variety of tests. In particular, we can isolate its response to various conditions by varying those conditions in a controlled way. The objects we study in astronomy are so remote we cannot produce any changes in their environment. We must observe them as they are. This is the essential difference between an *experimental* science and an *observational* one.

In this part we will see how observations, along with some clever reasoning, can be used to determine the basic properties of stars. Then we will see how those properties can be explained by using the laws of physics to arrive at models for stars. These models can even be used to tell us how stars change with time, and how they eventually die. We will first see how we gather and interpret the light from stars. We will then look at how temperatures, sizes, and masses are determined. Finally, we will take a close look at the Sun, the one star that we can study in great detail. In Part II, we will combine these basic properties with physical laws to see how stars work and evolve.

Continuous Radiation from Stars

2.1 BRIGHTNESS OF STARLIGHT

When we look at the sky, we note that some stars appear brighter than others. At this point we are not concerned with the causes of these brightness differences. (They may result from the stars actually having different power outputs, or they may result from the stars being at different distances.) All we know at first glance is that the stars *appear* to have different brightnesses.

We would like to have some way of quantifying the observed brightnesses of stars. When we speak, loosely, of brightness, we are really talking about the *energy flux f*, which is the energy per unit area per unit time, received from the star. This can be measured with current instruments (as we will discuss in Chapter 4). However, the study of stellar brightness started long before such instruments, or even telescopes, were available. Ancient astronomers made naked eye estimates of brightness. Hipparchus, the Greek astronomer, and later Ptolemy, a Greek living in Alexandria, Egypt, around A.D. 150, divided the stars visible to the naked eye into six classes of brightness. These classes were called *magnitudes*. This was an ordinal arrangement, with first-magnitude stars being the brightest and sixth-magnitude stars the faintest.

When quantitive measurements were made, it was found that each jump of one magnitude corresponded to a fixed *flux ratio,* not a flux difference. Because of this, the magnitude scale is essentially a logarithmic scale. This is not too surprising, since the eye is logarithmic in its response to brightness. This type of response allows us to see in very low and very high light levels. (We say that the eye has a large dynamic range. This range is achieved at a sacrifice in our ability to discriminate small brightness differences.)

The next step was to make the scale continuous, so that a star that is somewhere between second and third magnitude could be placed accurately, and to extend the scale, so that the brightness of stars we can see only through telescopes can be included. It was found that a *difference of five magnitudes corresponds to a factor of 100 in energy flux*. In setting up the magnitude scale, this relation is defined to be exact. Let f_1 and f_2 be the observed energy fluxes of two stars, and m_1 and m_2 be the corresponding magnitudes. The statement that a five-magnitude difference gives a flux ratio of 100 corresponds to

$$f_1/f_2 = 100^{(m_2 - m_1)/5} \tag{2.1}$$

We can see that this equation guarantees that each time $m_2 - m_1$ increases by five, f_1/f_2 decreases by a factor of 100. Remember, *increasing* the energy flux *decreases* the magnitude. This point sometimes confuses astronomers, too. That is why you will often hear astronomers talk about something being so many magnitudes "brighter" or "fainter" than something else, without worrying about whether that makes m larger or smaller.

Equation (2.1) gives flux ratios in powers of 100, but we usually work in powers of ten. To convert this we write 100 as 10^2, so equation (2.1) becomes

$$f_1/f_2 = 10^{(m_2 - m_1)/2.5} \tag{2.2}$$

This equation can be used to calculate a brightness ratio for a given magnitude difference. If we want to calculate a magnitude difference for a given brightness ratio, we take the logarithm (base 10) of both sides, giving

$$m_2 - m_1 = 2.5 \log(f_1/f_2) \tag{2.3}$$

To see how this works, let's look at a few simple examples. The magnitude range for stars visible to the naked eye is 1–6 mag. This corresponds to a flux ratio

$$f_1/f_2 = 10^{(6-1)/2.5} = 10^2$$

The largest ground-based telescopes extend our range from 6 mag to 26 mag. This corresponds to a flux ratio

$$b_1/b_2 = 10^{(26-6)/2.5} = 10^8$$

We can also find the magnitude difference Δm corresponding to a factor of 10^6 in flux:

$$\Delta m = 2.5 \log_{10}(10^6) = (2.5)(6) \text{ mag} = 15 \text{ mag}$$

We have taken the original six magnitude groups and come up with a continuous scale that can be extended to fainter or brighter objects. Brighter objects than magnitude 1 can have magnitude 0, or even negative magnitudes.

2.2   THE ELECTROMAGNETIC SPECTRUM

Thomas Young first demonstrated interference effects in light, showing that light is a wave phenomenon. If we pass white light through a prism (Fig. 2.1), we can see that the light is spread out into different colors. We call this range of colors the *visible spectrum*. These colors have different wavelengths (Fig. 2.2). For example, the red light has a wavelength around 650 nm (= 650×10^{-9} m = 6.5×10^{-7} cm). We still express this in terms of angstrom units, after the Swedish physicist, A. J. Ångstrom, as 6500 Å. Even though it is not part of the official metric system, the angstrom is a convenient unit to use since it is about the size of a typical atom. At the opposite end of the visible spectrum from red is violet, with a wavelength of about 400 nm (4000 Å).

In a vacuum all wavelengths of light have the same speed, $c = 3 \times 10^{10}$ cm/s (3×10^8 m/s, 3×10^5 km/s). At this speed, light can travel a distance equal to the Earth's circumference 7.5 times per second. A light pulse takes 1.3 s to reach the Moon. The speed of light is so large that measuring it requires the accurate measurement of time over short intervals or the passage of light over long distances. Until late in the nineteenth century, the large distances between astronomical objects were used to provide reasonably long travel times. More recently, accurate timing devices have made laboratory measurements feasible.

All waves have a frequency associated with them. The frequency tells us the number of oscillations per second, or the number of crests that pass per second. The product of the wavelength λ and the frequency v gives us the speed of the wave. That is,

$$\lambda v = c \tag{2.4}$$

The higher the frequency is, the shorter is the wavelength. For example, we can find the frequency for light of wavelength 600 nm:

$$v = c/\lambda$$
$$= \frac{3.0 \times 10^8 \text{ m/s}}{600 \times 10^{-9} \text{ m}}$$
$$= 5.0 \times 10^{14} \text{ cps}$$

For 1 cps we use the unit 1 hertz (Hz).

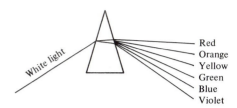

Figure 2.1   The colors of visible light. When white light passes through a prism, the rays of different colors are deflected by different amounts. The colors are listed, from top to bottom, in order of decreasing wavelength.

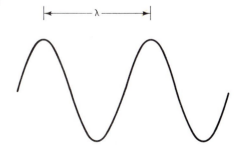

Figure 2.2   The wavelength λ is the distance between the corresponding points of a wave in successive cycles. For example, it can be from peak to peak.

When we talk about light waves with the above frequency, what is actually varying 5×10^{14} times per second? This question was answered by James Clerk Maxwell, who pointed out the unity between electric and magnetic fields. The behavior of these fields, and their relationship to charged matter, is described by four equations known as *Maxwell's equations*.

In these equations, Maxwell was mostly summarizing the work of others, but it was he who put the whole picture together. For example, one of Maxwell's equations is Faraday's law of induction, which describes how a changing magnetic field can produce an electric field. (This is the basis for the production of electricity in a generator.) Maxwell realized that if there is a symmetry between electric and magnetic fields, then a varying electric field should also be able to produce a magnetic field.

This realization serves as the basis for our understanding of *electromagnetic waves*. An electric field that varies sinusoidally (as a sine wave) produces a varying magnetic field, which then produces a varying electric field, and so on. These varying fields can propogate through space, even empty space. All wavelengths are possible. The speed of these waves can be predicted from Maxwell's equations. The speed in a vacuum is the same for all wavelengths, and is numerically equal to c, the speed of light. Light is just one form of electromagnetic waves. Other forms have wavelengths that fall in different ranges.

The full set of electromagnetic waves is called the *electromagnetic spectrum* (see Table 2.1). The visible spectrum is just a small part of the electromagnetic spectrum. At longer wavelengths are infrared and radio waves. At shorter wavelengths are the ultraviolet, x-ray, and gamma-ray parts of the spectrum. Even though there is no real difference between the waves in the various parts of the spectrum, we use the divisions because different techniques are used to detect electromagnetic waves in the various wavelength ranges. For example, our eyes are sensitive to waves with wavelengths between 400 and 700 nm. This is not too surprising, since this is where the Sun gives off most of its energy. It makes sense that we would have evolved with eyes best able to make use of the illuminating sunlight.

TABLE 2.1   The Electromagnetic Spectrum

Region	Wavelength	Frequency (Hz)
Radio	>1 mm	$<3 \times 10^{11}$
Infrared	700 nm–1 mm	3×10^{11}–4.3×10^{14}
Visible	400–700 nm	4.3×10^{14}–7.5×10^{14}
Ultraviolet	10–400 nm	7.5×10^{14}–3×10^{16}
X-ray	0.1–10 nm	3×10^{16}–3×10^{18}
Gamma-ray	<0.1 nm	$>3 \times 10^{18}$

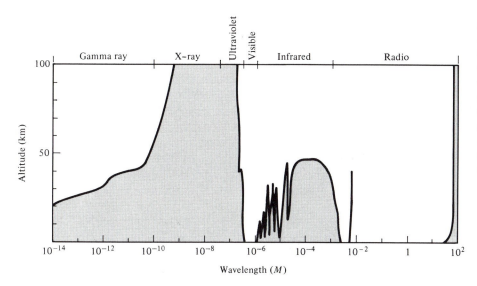

Figure 2.3 Atmospheric transmission as a function of wavelength. The curve gives the altitude from which you would have to observe to get 50% of the signal. Note the high transmission in the visible and radio parts of the spectrum. Also note a few narrow ranges, or "windows," of relatively good transmission in the infrared.

We now know that astronomical objects give off radiation in all parts of the spectrum. However, the Earth's atmosphere limits what we can actually detect (Fig. 2.3). Ultraviolet and shorter wavelengths are blocked by the atmosphere. Visible light passes through the clear atmosphere (but is blocked by clouds). Most infrared wavelengths are blocked by the atmosphere, but some wavelengths get through. For the most part, radio waves pass through the atmosphere with little absorption. We speak of *visible* and *radio windows* in the atmosphere, as well as of some narrow windows in the infrared. A window is simply a wavelength range in which the atmosphere is at least partially transparent.

Until relatively recently, astronomers could only gather information in the visible part of the spectrum, because of the lack of equipment. Much of the development of astronomy was biased by this handicap. Some 50 years ago, the radio part of the spectrum was opened for astronomical observations. Even more recently, the other parts of the spectrum have become available to us. Observing in various parts of the spectrum will be discussed throughout this book.

2.3   COLORS OF STARS

2.3.1   Quantifying Color

When we look at a star, we would like to know how much energy it gives off at various wavelengths. We sometimes refer to a graph, or some equivalent representation, showing intensity as a function of wavelength (or frequency), as a *spectrum*. It is really not proper to talk about the energy given off at a particular wavelength. If we can specify a wavelength to an arbitrary number of decimal places, then even a small wavelength range has an infinite number of wavelengths. If an object gives off even a little energy at each of those infinite number of wavelengths, it will give off an infinite amount of energy.

Instead, we talk about the energy given off over some wavelength range. For example, we define the intensity function $I(\lambda)$ such that $I(\lambda)\,d\lambda$ is the (energy/unit time)/unit surface area given off by an object in the wavelength range λ to $\lambda + d\lambda$. Similarly, $I(\nu)\,d\nu$ gives the (energy/unit time)/unit surface area in the frequency range ν to $\nu + d\nu$.

When we make a plot of $I(\lambda)$ vs. λ for a star we find that the graph varies smoothly over most wavelengths. There are some wavelengths at which we see sharp increases or decreases in $I(\lambda)$ over a very narrow wavelength range. These sharp increases and decreases are called *spectral lines* and will be discussed in the next chapter. In this chapter, we will concern ourselves with the smooth, or *continuous*, part of the spectrum. This is also called the *continuum*.

When we look at stars we see that they have different colors. Stars with different colors have different continuous spectra. Stars that appear red have continuous spectra that peak in the red. Stars that appear blue have continuous spectra that peak at shorter wavelengths. The color of a star depends on its temperature. We know that as we heat an object, it first glows red, it then turns yellow/green, and, finally, it turns blue as it gets hotter.

We can therefore measure the temperature of a star by measuring its continuum. In fact, it is not necessary to measure the whole spectrum in detail. We can measure the amount of radiation received in certain wavelength ranges. These ranges are defined by *filters* that let a given wavelength range pass through. By comparing the intensity of radiation received in various filters, we can come up with a quantitative way of determining the color of a star, and therefore its temperature.

2.3.2 **Blackbodies**

We can understand the relationship between color and temperature by considering objects called *blackbodies*. This is a theoretical idea that closely approximates many real objects in thermodynamic equilibrium. (We say that an object is in thermodynamic equilibrium with its surroundings when energy is freely interchanged and a steady state is reached in which there is no *net* energy flow.) A blackbody is an object that absorbs all the radiation that strikes it. One way to construct an object that is very close to this idealization is to make a cavity with a very small hole in one wall (Fig. 2.4). Any radiation that strikes the hole from outside ends up in the cavity. It then bounces around until it is absorbed in one of its collisions with the walls or until it leaves

Figure 2.4 Blackbody radiation. The cavity has reflecting walls. However, the reflections are not perfect, and some radiation is absorbed at each reflection. Since the hole is small, almost all radiation getting into the box bounces around until it is absorbed. In this way, the hole acts as a perfect absorber. The radiation coming out of the hole is blackbody radiation at the temperature of the cavity.

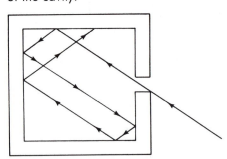

via the hole. However, since the hole is so small, it will strike the walls many times before it finds the hole. Therefore, it is likely to get absorbed by the walls before it finds the hole. In this sense, we can think of the hole as the blackbody. All radiation striking the hole gets absorbed. When physicists were first trying to understand blackbody radiation, they thought in terms of such cavities.

A blackbody can also emit radiation. In fact, if a blackbody is to maintain a constant temperature, it must radiate energy at the same rate it absorbs energy. If it radiates less than it absorbs, it will heat up. If it radiates more than it absorbs, it will cool. However, this doesn't mean that the spectrum of emitted radiation must match the spectrum of the absorbed radiation. Only the total energies must balance. The spectrum of emitted radiation is determined by the temperature of the blackbody. As the temperature changes, the spectrum changes. The blackbody will adjust its temperature so that its emitted spectrum contains just enough energy to balance the absorbed energy. When the temperature is reached to give this balance, we say that the blackbody is in *equilibrium.*

Figure 2.5 shows some sample blackbody spectra. We note that *at any wavelength, a hotter blackbody gives off more energy than a cooler blackbody of the same size. We also see that, as the temperature increases, the peak of the spectrum shifts to shorter wavelengths.* The relationship between the wavelength at which the peak occurs, λ_{max}, and the temperature T is very simple. It is given by *Wien's displacement law:*

$$\lambda_{max} T = 2.9 \times 10^7 \text{ Å K} \tag{2.5}$$

In this law, we must use temperature on an absolute (Kelvin) scale. (The temperature on the Kelvin scale is the temperature on the Celsius scale plus 273.1.)

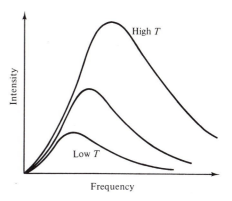

Figure 2.5 Blackbody spectra. Note the shift of the peak to higher frequency (shorter wavelength) at higher temperature. Note, also, that at any frequency, a hotter blackbody gives off more radiation than a cooler one.

EXAMPLE 2.1 Using Wien's Displacement Law

(a) Find the temperature of an object whose blackbody spectrum peaks in the middle of the visible part of the spectrum, $\lambda = 5500$ Å. (b) The Earth has an average temperature of about 300 K. At what wavelength does the Earth's blackbody spectrum peak?

Solution

(a) Given the wavelength, we solve equation (2.5) for the temperature

$$T = \frac{2.9 \times 10^7 \text{ Å K}}{5500 \text{ Å}} = 5270 \text{ K}$$

This is close to the temperature of the Sun.

(b) Given the temperature, we solve equation (2.5) for the wavelength

$$\lambda = \frac{2.9 \times 10^7 \text{ Å K}}{300 \text{ K}}$$

$$= 1 \times 10^5 \text{ Å}$$

$$= 10 \times 10^{-6} \text{ m}$$

$$= 10 \text{ } \mu\text{m}$$

This is in the infrared part of the spectrum. Even though the Earth is giving off radiation, we don't see it glowing in the visible part of the spectrum. Similarly, objects around us that are at essentially the same temperature as the Earth give off radiation in the infrared part of the spectrum, but very little visible light. The visible light we see from surrounding objects is partially reflected sunlight or room light.

Suppose we are interested in the total energy given off by a blackbody (per unit time per unit surface area) over the whole electromagnetic spectrum. We must add the contributions at all wavelengths. This amounts to taking an integral over the blackbody curves such as those in Fig. 2.5. Since a hotter blackbody gives off more energy at all wavelengths than a cooler one, and is particularly dominant at shorter wavelengths, we would expect a hotter blackbody to give off much more total energy than a cooler one. Indeed, this is the case. The total energy per unit time per unit surface area, E, given off by a blackbody, is proportional to the fourth power of the temperature. That is,

$$E = \sigma T^4 \tag{2.6}$$

This relationship is called the *Stefan–Boltzmann law*. The constant of proportionality σ is called the Stefan–Boltzmann constant, and has the value 5.7×10^{-5} erg/(cm^2 K^4 s). This law was first determined experimentally, but, as we shall see below, it can also be derived theoretically. The T^4 dependence means that E depends strongly on T. If we double the temperature of an object, the rate at which it gives off energy is increased by a factor of 16. If we change the temperature by a factor of 10 (say, from 300 K to 3000 K), the energy radiated goes up by a factor of 10^4.

For a star, we are interested in the total *luminosity*. The luminosity is the total energy per second (or power) given off by the star. The quantity σT^4 is only the energy per second per unit surface area. Therefore, it must be multiplied by the surface area of the star. If the spherical star has a radius R, the surface area is $4\pi R^2$; so the luminosity is

$$L = (4\pi R^2)(\sigma T^4) \tag{2.7}$$

EXAMPLE 2.2   Luminosity of the Sun

The surface temperature of the Sun is about 5800 K and its radius is 7×10^5 km (7×10^{10} cm). What is the luminosity of the Sun?

Solution

We use equation (2.7) to find the luminosity:

$$L = 4\pi(7 \times 10^{10} \text{ cm})^2[5.7 \times 10^{-5} \text{ erg}/(\text{cm}^2 \text{ K}^4 \text{ s})](5.8 \times 10^3 \text{ K})^4$$

$$= 4 \times 10^{33} \text{ erg/s}$$

This quantity is called a *solar luminosity* $L_\odot$, and serves as a convenient unit for expressing the luminosities of other stars.

2.4 PLANCK'S LAW AND PHOTONS

2.4.1   Planck's Law

The study of blackbody radiation plays an important role in the development of what we refer to as ''modern'' physics (even though these developments took place early in this century). When physicists tried to apply classical ideas of radiation, they could not derive blackbody spectra that agreed with the experimental results. The classical calculations yielded an intensity $T(\nu, T)$ given by

$$I(\nu, T) = 2kT\nu^2/c^2 \tag{2.8}$$

This is known as the *Rayleigh–Jeans law*. The constant k that appears in this law is the Boltzmann constant (not to be confused with the Stefan–Boltzmann constant). Its value is 1.38×10^{-16} erg/K. (The quantity kT is proportional to the kinetic energy per particle in a gas.) The Rayleigh–Jeans law agrees with experimental results at low frequencies (long wavelengths), but disagrees at high frequencies. In fact, you can see from equation (2.8), that as we go to higher and higher frequencies, the energy given off gets arbitrarily large, implying that a blackbody gives off an infinite amount of energy. This is clearly not the case. The classical prediction of arbitrarily large energies at short wavelengths is called the *ultraviolet catastrophe*.

The first step in solving the problem is to deduce an *empirical* formula for the observed spectra. By ''empirical'' we mean a formula that is put together to describe the observations, but which is not derived from theory.

In 1900, Max Planck, a German physicist, produced an empirical formula that accurately describes the experimental blackbody spectra:

$$I(\nu, T) = \frac{2h\nu^3/c^2}{\exp(h\nu/kT) - 1} \tag{2.9}$$

In this equation the constant h is called *Planck's constant,* and has the numerical value 6.625×10^{-27} erg/s. This value was determined to provide the best agreement with observed blackbody spectra.

Since the Rayleigh–Jeans law adequately describes blackbody spectra at low frequencies, the Planck law must reduce to the Rayleigh–Jeans law at low frequencies. We can see this if we take low frequencies to mean that $h\nu \ll kT$ (or, equivalently, $h\nu/kT \ll 1$). In this case we can take advantage of the fact that for $x \ll 1$, $e^x \cong 1 + x$. The Planck function then becomes

$$I(\nu, T) \cong \frac{2h\nu^3}{c^2} \frac{kT}{h\nu}$$

$$= \frac{2kT\nu^2}{c^2}$$

which is the Rayleigh–Jeans law.

Equation (2.9) gives the Planck function in terms of frequency. How do we find it as a function of wavelength? Your first guess might be simply to substitute c/λ for each occurrence of ν in equation (2.9). However, we must remember that $I(\nu, T)$ gives the energy per second per frequency interval, and $I(\lambda, T)$ gives the energy per second per wavelength interval. The functions must reflect that difference (especially since they will need different units). We therefore require that

$$I(\lambda, T)\, d\lambda = I(\nu, T)\, d\nu$$

Solving for $I(\lambda, T)$ gives

$$I(\lambda, T) = I(\nu, T)(d\nu/d\lambda)$$

To find $I(\lambda, T)$, we must be able to evaluate $d\nu/d\lambda$. We do this by remembering that $\nu = c/\lambda$, so that

$$d\nu/d\lambda = -c/\lambda^2 \tag{2.10a}$$

We don't care about the minus sign, which just tells us that frequency increases when wavelength decreases. Using this result gives

$$I(\lambda, T) = I(\nu, T)(c/\lambda^2)$$

Now we can substitute c/λ for ν to get the final result:

$$I(\lambda, T) = \frac{2hc^2/\lambda^5}{\exp(hc/\lambda kT) - 1} \qquad (2.10b)$$

Remember, the Planck function accurately describes blackbody spectra, but it was originally presented as an empirical formula. There was still no theoretical understanding of the origin of the formula. Planck continued his work in an effort to derive the formula from some theory. Planck found that he could derive the formula from classical physics if he inserted a mathematical trick. This trick amounted to taking a sum rather than an integral. The trick corresponded to the physical statement that a blackbody can only radiate energy at frequency ν in integer multiples of $h\nu$. That is, the energy could only be emitted in small bundles, or *quanta* (singular *quantum*). The quanta have energy $h\nu$. Even though Planck was able to derive the blackbody formula correctly, he was still not satisfied. There was no justification for the restriction that energy must be quantized.

2.4.2 Photons

An explanation for why energy must be quantized was proposed by Albert Einstein, in 1905. (It was for this explanation that Einstein was later awarded the Nobel Prize in physics.) Einstein was trying to explain a phenomenon known as the *photoelectric effect*, in which electrons can be ejected from metal surfaces if light falls on the surface. Laboratory studies had shown that increasing the intensity of the light falling on the surface increased the number of electrons ejected from the surface, but not their energy. Einstein said that all radiation (whether from a blackbody or otherwise) must come in small bundles, called *photons*. The energy E of a photon with frequency ν is given by

$$E = h\nu \qquad (2.11)$$

This explains the observed properties of the photoelectric effect by stating that each electron is ejected by a single photon striking the surface. Increasing the intensity of the light increases the number of photons striking the surface per second, and therefore increases the *rate* at which electrons are ejected. Increasing the frequency of the light increases the *energy* at which the electrons are ejected. (This latter prediction was finally tested by Robert Millikan in 1916.) A further test of this hypothesis came in an analysis of collisions between light (photons) and electrons, by A. H. Compton. This is why Planck had to assume that energy is quantized in deriving the formula for blackbody spectra.

The assertion that light is essentially a particle went against the then accepted ideas about light. The question of whether light is a particle or a wave had been going on for centuries. For example, Newton believed that it was a particle, and he worked out a theory of refraction—the bending of light when it passes, for example, from air to glass—on the basis of light speeding up when it enters the glass. (We now know, however, that, as a wave, it slows down.) The wave theory became dominant following the demonstration of interference effects by Young and the explanation of electromagnetic waves by Maxwell. In explaining the photoelectric effect, Einstein was saying that the particle picture must be revived. The explanation was that, somehow, light can exhibit both particle and wave properties. This is referred to as the wave–particle duality. This concept is the foundation of what we refer to as the *quantum revolution,* since it was such a radical departure from previous theories. We will discuss this point further in the next chapter.

2.5   STELLAR COLORS

We have seen that the color of a star can tell us the star's temperature. However, we now need a way of quantifying a color, rather than just saying something is red, green, or blue. For example, if we compare two blue stars, how do we decide which one is bluer?

We define two standard wavelength ranges, centered at λ_1 and λ_2, and take the ratio of the observed energy fluxes, $f(\lambda_1)/f(\lambda_2)$. We then convert this brightness ratio into a magnitude difference [using equation (2.3)], giving

$$m_1 - m_2 = -2.5 \log_{10}[f(\lambda_1)/f(\lambda_2)] \tag{2.12}$$

We define the quantity $m_1 - m_2$ as the color, measured in magnitudes, corresponding to the wavelength pair λ_1, λ_2. For definiteness, let's assume that $\lambda_2 > \lambda_1$. As we increase the temperature, $f(\lambda_1)/f(\lambda_2)$ increases. This means that the quantity $m_1 - m_2$ decreases, since the magnitude scale runs backward. If we know that an object is radiating exactly like a blackbody, we need only take the ratio of brightnesses at any two wavelengths to determine the temperature.

As we have said, we don't really measure the intensity of a wavelength. Instead we measure the amount of energy received in some wavelength interval. We can control that wavelength interval by using a filter that only passes light in that wavelength range. When we use a filter, we are actually measuring the integral of $I(\lambda, T)$ over that wavelength range. Actually, the situation is more complicated. The transmission of any real filter is not 100% over the selected range, and this must be factored in.

Another complication is that continuous spectra of stars do not exactly

TABLE 2.2 Filter Systems

Filter	Peak wavelength (Å)	Width (Å)[a]
U	3550	700
B	4150	950
V	5350	800

[a]Width is full-width at half-maximum.

follow blackbody curves. Therefore, observations through two filters are not generally sufficient to tell us the temperature of the star. Over the years, a system of standard filters has been developed, so that astronomers at various observatories can compare their results. The wavelength ranges of various filters are given in Table 2.2. The most commonly discussed filters are the U (for *u*ltraviolet), B (for *b*lue), and V (for *v*isible). For example, the B − V color is defined by

$$B - V = 2.5 \log_{10}[I(\lambda_V)/I(\lambda_B)] + \text{constant}$$

where $I(\lambda_V)$ and $I(\lambda_B)$ are the intensities averaged over the filter ranges. (The constant is adjusted so that B − V is zero for a particular temperature star, designated A0. These designations will be discussed in the next chapter.) As the temperature of an object increases, the ratio of blue to visible intensity increases. This means that the B − V color decreases (again because the magnitude scale runs backward).

2.6 STELLAR DISTANCES

So far we have discussed how bright stars appear as seen from Earth. However, the apparent brightness depends on two quantities, the intrinsic luminosity of the star and its distance from us. (As we will see in Section 14.2, starlight is also dimmed when it passes through clouds of interstellar dust.) Two identical stars at different distances will have different apparent brightnesses. If we want to understand how stars work, we must know their total luminosities. This requires correcting the apparent brightness for the distance to the star.

If we have a star of luminosity L, we can calculate its observed energy flux at a distance d. If no radiation is absorbed along the way, all the energy per second leaving the surface of the star will cross a sphere at a distance d in the same time. It will just be spread over a larger area. Therefore, the energy per second reaching d is still L, but it is spread over an area of $4\pi d^2$, so the energy flux is

$$f = L/4\pi d^2 \tag{2.13}$$

The flux falls off inversely as the square of the distance.

Unfortunately, distances to astronomical objects are generally hard to determine. There is a direct method for determining the distances to nearby stars. It is called *trigonometric parallax*. It amounts to triangulation from two different observing points. You can demonstrate parallax for yourself by holding out a finger at arm's length and viewing it against a distant background. Look at the finger alternately using your left and right eye. The finger appears to shift against the distant background. Bring the finger closer and repeat the experiment. The shift now appears larger. If you could move your eyes farther apart, the effect would be even greater.

EXAMPLE 2.3   Angular Measure

The natural unit for measuring angles is the *radian*. If we have a circle of radius R and two lines from the center making an angle θ with each other, then the length of the arc bounded by the two lines is

$$L = \theta(\text{rad})R$$

where $\theta(\text{rad})$ is the value of θ, measured in radians. Since the full circumference of a circle is $2\pi R$, the angle corresponding to a full circle must be 2π rad. This tells us that a full circle, that is, $360°$, is equal to 2π rad, or $180°$ is equal to π rad. In astronomy we often deal with very small angles, and measurements in arc seconds ($''$) are convenient. We convert measurements by saying that

$$\theta('') = \theta(\text{rad}) \frac{180°}{\pi \text{ rad}} \frac{60'}{1°} \frac{60''}{1'}$$

$$= 2.06 \times 10^5 \, \theta(\text{rad})$$

When we take the derivatives of trigonometric functions [for example, $d(\sin \theta)/d\theta = \cos \theta$], it is assumed that the angles are in radians. If not, a conversion factor must be carried through the differentiation.

When angles expressed in radians are much less than one, we can use a Taylor series to approximate them:

$$\sin \theta \cong \sin 0 + \theta \left. \frac{d(\sin \theta')}{d\theta'} \right|_{\theta=0}$$

$$= \theta \cos 0$$

$$\cong \theta$$

$$\tan \theta \cong \tan 0 + \theta \left. \frac{d(\tan \theta')}{d\theta'} \right|_{\theta=0}$$

$$= \theta \sec^2 0$$

$$\cong \theta$$

$$\cos \theta \cong \cos 0 + \theta \left. \frac{d(\cos \theta')}{d\theta'} \right|_{\theta=0}$$

$$= 1 - \theta \sin 0$$

$$\cong 1$$

Note that for small θ, $\sin \theta$ and $\tan \theta$ are both approximately equal to θ, so they must equal each other. (Remember, in each of the above expressions, θ must be expressed in radians.)

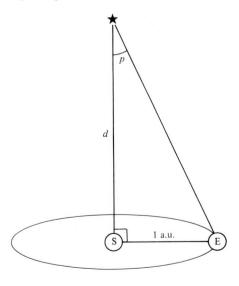

Figure 2.6 Geometry for parallax measurements. No matter where a star is, there will always be two times a year when the Earth, the Sun, and the star form a right triangle, with the Sun at the right angle.

Even the closest stars are too far to demonstrate parallax when you just use your eyes. However, we can take advantage of the fact that the Earth orbits the Sun at a distance defined to be 1 astronomical unit (AU). Therefore, six months apart we have viewing points that are located 2 AU apart. The situation is illustrated in Fig. 2.6. We note the position of the star against the background star, and then six months later we note the angle by which it has shifted. If we take half the value of this angle, we have the *parallax angle p*.

Once we know the value of p, we can construct a right triangle, with a base of 1 AU and the other leg being of length d, the unknown distance to the star. From the right triangle, we see that

$$\tan p = \text{AU}/d \qquad (2.14)$$

Since p is small, $\tan(p) \cong p(\text{rad})$, where $p(\text{rad})$ is the value of p, measured in radians. Equation (2.14) then gives us

$$p(\text{rad}) = 1 \text{ AU}/d \qquad (2.15)$$

It is not very convenient measuring such small angles in radians, so we convert to arc seconds (see Example 2.3).

$$p('') = 2.06 \times 10^5 \, p(\text{rad})$$

where $p('')$ is the parallax angle, measured in arc seconds. Substituting this into equation (2.15) gives

$$d/1 \text{ AU} = 2.06 \times 10^5 \, p('') \qquad (2.16)$$

This gives us the distance in AU (1 AU $= 1.5 \times 10^8$ km).

This method suggests a convenient unit for measuring distances. We define the *parsec* (abbreviated as pc) as the distance of a star that produces a parallax angle p of $1''$. From equation (2.16) we see that 1 pc $= 2.06 \times 10^5$ AU (or 3.09×10^{13} km, or 3.26 light years). We rewrite equation (2.16) as

$$d(\text{pc}) = 1/p('') \qquad (2.17)$$

Remember, as an object gets farther away, the parallax angle decreases. Therefore, a star at a distance of 2 pc will have a parallax angle of $0.5''$.

EXAMPLE 2.4   Distance to the Nearest Star

The nearest star (Proxima Centauri) has a parallax $p = 0.76''$. Find its distance.

Solution

We use equation (2.17) to give

$$d(\text{pc}) = 1/0.76$$

$$= 1.32 \text{ pc}$$

With current ground-based equipment, we can measure parallax to within a few thousandths of an arc second. Parallax measurements are therefore useful for the few thousand nearest stars. They are the starting point for a very complex system of determining distances to astronomical objects. We will encounter a variety of distance determination methods throughout this book. The trigonometric parallax method is the only that is direct and free of any assumptions. For this reason, astronomers would like to extend their capability for measuring parallax. We will come back to this point when we discuss observing techniques in Chapter 4.

2.7 ABSOLUTE MAGNITUDES

The magnitudes discussed in Section 2.1, based on observed energy fluxes, are called *apparent magnitudes*. In order to compare the intrinsic luminosities of stars we define a system of *absolute magnitudes*. The absolute magnitude of a star is the magnitude that it would appear to have as viewed from a standard distance d_0. This standard distance is chosen to be 10 pc. From this definition, you can see that for a star actually at a distance of 10 pc, the absolute and apparent magnitudes will be the same.

To see how this works, consider two identical stars, one at a distance d and the other at the standard distance d_0. We let m be the apparent magnitude of the star at distance d and M be that of the star at d_0. (Of course, M will also be the absolute magnitude for both stars.) The energy flux of a star falls off inversely as the square of the distance. Therefore, the ratio of the flux of the star at d to that from the star at d_0 is $(d_0/d)^2$. Equation (2.3) then gives us

$$m - M = 2.5 \log_{10}(d/d_0)^2$$

Using the fact that $\log(x^2) = 2 \log(x)$, gives

$$m = M + 5 \log_{10}(d/10 \text{ pc}) \tag{2.18}$$

The quantity $5 \log_{10}(d/10 \text{ pc})$, which is equal to $m - M$, is called the *distance modulus* of the star. It indicates the amount by which distance has dimmed the starlight. If you know any two of the three quantities (m, M, or d), you can use equation (2.18) to find the third. For any star that we can observe, we can always measure m, its apparent magnitude. Therefore, we are generally faced with knowing M and finding d, or knowing d and finding M.

EXAMPLE 2.5 Absolute Magnitude

A star is at a distance of 100 pc, and its apparent magnitude is $+5$. What is its absolute magnitude?

Solution

We use equation (2.18) to find

$$M = m - 5 \log(d/10 \text{ pc})$$
$$= 5 - 5 \log(100 \text{ pc}/10 \text{ pc})$$
$$= 5 - 5 \log(10)$$
$$= 0$$

We should note that changing the distance of a star changes its apparent magnitude, but it does not change any of its colors. Because colors are defined to be differences in magnitudes, each is changed by the distance modulus. For example, using equation (2.18),

$$m_B = M_B + 5 \log_{10}(d/10 \text{ pc})$$
$$m_V = M_V + 5 \log_{10}(d/10 \text{ pc})$$

Taking the difference gives

$$m_B - m_V = M_B - M_V$$

Therefore, the distance modulus never appears in any colors.

When we talk about determining an absolute magnitude, we are really only determining it over some wavelength range, corresponding to the wavelength range of the observations. We would like to have an absolute magnitude that corresponds to the total luminosity of the star. This magnitude is called the *bolometric magnitude* of the star. (A bolometer is a device for measuring the total energy received from an object, and will be discussed further in Chapter 4.) For any type of star we can define a number, called the *bolometric correction*, abbreviated BC, which relates the bolometric magnitude to the absolute visual magnitude M_V. Therefore,

$$M_{\text{bol}} = M_V + \text{BC} \tag{2.19}$$

so

$$\text{BC} = 2.5 \log(L_V/L_{\text{bol}}) \tag{2.20}$$

CHAPTER SUMMARY

We saw in this chapter what can be learned from the brightness and spectrum of the continuous radiation from stars.

We introduced a logarithmic scale, the magnitude scale, for keeping track of brightness. Apparent magnitude is related to the observed energy flux from the star, and absolute magnitude is related to the intrinsic luminosity of the star.

We saw how, even though stars are obvious to us in the visible part of the spectrum, they, and other astronomical objects, give off radiation in other parts of the spectrum. The richness of information in other parts of the spectrum is a theme that we will come back to throughout the book.

We introduced the concept of a blackbody, which is useful because the continuous spectrum of a star closely resembles that of a blackbody. Hotter blackbodies give off more power per unit surface area than cooler ones (Stefan–Boltzmann law), and also have their spectra peaking at shorter wavelengths (Wien's displacement law). We saw how attempts to understand the details of blackbody spectra (Planck's law) contributed to the idea of light coming in bundles, called photons, with specific energies. With a knowledge of blackbody spectra, we saw how stellar colors can be used to deduce stellar temperatures.

We saw how finding distances to astronomical objects is very important, but can be quite difficult. If we don't know the distance to an object, we can't convert its apparent brightness into an intrinsic luminosity. We introduced one method of measuring distances, trigonometric parallax. It is the most direct method, but only works for nearby stars. The problem of distance determination will come up throughout this book.

QUESTIONS AND PROBLEMS

2.1 What magnitude difference corresponds to a factor of 10 change in energy flux?

2.2 The apparent magnitude of the Sun is -26.8. How much brighter does the Sun appear than the brightest star?

2.3 (a) What is the distance modulus for the Sun? (b) What is the Sun's absolute magnitude?

2.4 Suppose two objects have energy fluxes, f and $f + \Delta f$, where $\Delta f << f$. Derive an approximate expression for the magnitude difference Δm between these objects. Your expression should have Δm proportional to Δf. (*Hint:* Use the fact that $\ln(1 + x) \cong x$, where $x << 1$.)

2.5 Show that our definition of magnitudes has the following property: If we have three stars with energy fluxes, f_1, f_2, and f_3, and we define

$$m_2 - m_1 = 2.5 \log_{10}(f_1/f_2)$$

$$m_3 - m_2 = 2.5 \log_{10}(f_2/f_3)$$

then

$$m_3 - m_1 = 2.5 \log_{10}(f_1/f_3)$$

2.6 Describe how observations of the timing of eclipses of Jupiter's moons can be used to measure the speed of light.

2.7 Suppose we measure the speeed of light in a laboratory, with the light travelling a path of 10 m. How accurately do you have to time the light travel to measure c to eight significant figures?

2.8 Let λ_1, λ_2 (ν_1, ν_2) be the wavelength (frequency) limits of the visible part of the spectrum. Compare $(\lambda_1 - \lambda_2)/(\lambda_1 + \lambda_2)$ with $(\nu_1 - \nu_2)/(\nu_1 + \nu_2)$. Comment on the significance.

2.9 Why was Maxwell's realization that a varying electric field can create a magnetic field important in understanding electromagnetic waves?

2.10 (a) Estimate the number of people on the Earth who are *exactly* 2 m tall. (By ''exactly'' we mean to an arbitrary number of decimal places.) (b) How does this relate to the way we define the intensity function $I(\lambda)$?

2.11 Suppose we receive light from a star for which the received energy flux is given by the function $f(\lambda)$. Suppose we observe the star through a filter for which the fraction of light transmitted is $t(\lambda)$. Derive an expression for the total energy detected from the star.

2.12 Suppose we wish to construct a cubic blackbody cavity with sides of length L. The wall material is such that every time a beam of light strikes the material, 10^{-3} of the photons get absorbed. How small would we have to make the hole to guarantee that 99.9% of the photons entering the hole will never get out?

2.13 (a) Calculate the frequencies corresponding to the wavelengths 5000.0 and 5001.0 Å. Use these to check the accuracy of equation (2.10a). Repeat the process for the second wavelength being 5010.0 Å, and again for 5100.0 Å. What do you conclude?

2.14 (a) Use equation (2.9) to derive ν_{max}, the frequency a which $I(\nu, T)$ peaks. Convert this ν_{max} into a wavelength λ_{max}. (b) Use equation (2.10b) to find the wavelength at which it peaks. (c) How do the results in (a) and (b) compare?

2.15 For a 300-K blackbody, over what wavelength range would you expect the Rayleigh–Jeans law to be a good approximation?

2.16 Derive an approximation to the Planck function valid at high frequencies ($h\nu >> kT$).

2.17 Derive an expression for the shift in peak wavelength $\Delta\lambda_{max}$ corresponding to a small shift in temperature ΔT.

2.18 Calculate the energy per second reaching the Earth from the Sun.

2.19 (a) How does the absolute magnitude of a star vary with the size of the star? (b) Does this result depend on wavelength?

2.20 (a) How does the absolute bolometric magnitude of a star vary with the temperature of the star? (b) Does the absolute visual magnitude vary in the same way?

2.21 For a star whose radiation follows a blackbody spectrum at temperature T, derive an expression for the bolometric correction.

2.22 If we determine $I(\nu, T)$ for a blackbody by a measurement of $I(\nu, T)\,d\nu$ over some narrow frequency range, find an expression for the temperature of that blackbody (a) in the Rayleigh–Jeans limit; and (b) in the general case.

2.23 (a) Consider two wavelengths, λ_1 and λ_2. For an object emitting radiation with a blackbody spectrum, derive an expression for the magnitude difference, $m_1 - m_2$, at the two wavelengths, as a function of temperature T. (b) For the specific case of the two wavelengths corresponding to B and V, graph your result over a reasonable range of T.

2.24 (a) What is the energy of a photon in the middle of the visible spectrum ($\lambda = 550$ nm)? (b) How many photons per second are emitted by (2) a 100-W light bulb and (b) the Sun?

2.25 If we double the surface area of a blackbody, by how much must we change the temperature to keep the luminosity constant?

2.26 Derive an expression for the distance of an object as a function of the parallax angle seen by your eyes.

2.27 (a) If we can measure parallaxes as small as 0.1″, what is the greatest distance that can be measured using the method of trigonometric parallax? (b) By what factor will the volume of space over which we can measure parallax change if we can measure to 0.01″? (c) Why is the volume of space important?

2.28 How would parallax measurements improve if we could do our observations from Mars?

2.29 As we determine the astronomical unit more accurately, how does the relationship between the astronomical unit and the parsec change?

2.30 If we lived on Mars instead of on the Earth, how large would the parsec be?

2.31 Suppose we discover a planet orbiting a nearby star. The distance to the star is 3 pc. We observe the radius of the planet's orbit to be 0.1″. How many astronomical units is the planet from the star?

2.32 Derive an expression for the distance to a star in terms of its distance modulus.

2.33 If we make a 0.05-mag error in measuring the apparent magnitude of a star, what error does this introduce in our distance determination (assuming its absolute magnitude is known exactly)?

Spectral Lines in Stars

3.1 SPECTRAL LINES

In the last chapter we discussed the continuous spectra of stars and saw that they could be closely described by blackbody spectra. In this chapter we will discuss the situations in which the spectrum shows an increase or decrease in intensity over a very narrow wavelength range.

We know that if we pass white light through a prism, light of different colors (wavelengths) will emerge at different angles with respect to the initial beam of light. If we pass the white light through a slit before it strikes the prism (Fig. 3.1), and then let the spread-out light fall on a screen, at each position on the screen we will get the image of the slit at a particular wavelength.

Both William Hyde Wollaston (1804) and Josef von Fraunhofer (1811) used this method to examine sunlight. They found that the normal spectrum was crossed by dark lines, as shown in Fig. 3.2. These lines represent wavelengths at which there is less radiation than at nearby wavelengths. (The lines are only dark in comparison to the nearby bright regions.) The linelike appearance comes from the fact that at each wavelength we are seeing the image of the slit. It is the linelike appearance that leads to calling these features *spectral lines*. If we were to make a graph of intensity vs. wavelength, we would find narrow dips superimposed on the continuum. The solar spectrum with dark lines is sometimes referred to as the *Fraunhofer spectrum*.

The origin of these spectral lines was a mystery for some time. In 1859, the German chemist Gustav Robert Kirchhoff noticed a similar phenomenon in the laboratory. He found that when a beam of white light passed through a tube containing some gas, the spectrum showed dark lines. The gas was absorbing energy in a few specific narrow wavelength bands. In this situation, we refer to the lines as *absorption lines*. When the white light was removed, the spectrum of gas showed bright lines, or *emission lines*, at the wavelengths

Figure 3.1 If we allow white light to fall all over a prism, the red from one part of the prism will overlap the blue from another part, and we won't see a clear spectrum. Instead, we pass the white light through a slit first. The beam of light is then spread out. On the screen, we are really seeing a succession of images of the slit in different colors. If there is a color missing from the white light, this will appear as a gap on the screen in the shape of the slit.

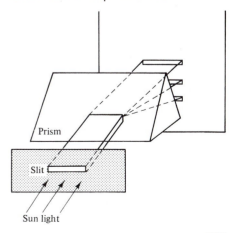

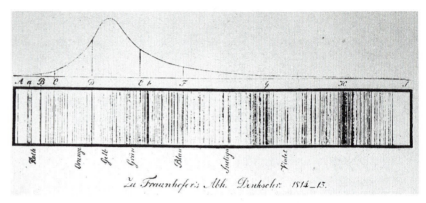

Figure 3.2   The Fraunhofer solar spectrum. The strongest lines still carry Fraunhofer's letter designations.

where absorption lines had previously appeared. The gas could emit or absorb energy only in certain wavelength bands.

Kirchhoff found that the wavelengths of the emission or absorption lines depend only on the type of gas that is used. Each element has its own set of special wavelengths. If two elements which don't react chemically are mixed, the spectrum shows the lines of both elements. Thus, the emission or absorption spectrum of an element identifies it as uniquely as a fingerprint identifies a person. This identification can be carried out without understanding why it works.

Whether we see absorption or emission lines depends in part on whether or not there is a strong enough background source providing energy to be absorbed (Fig. 3.3). The strength of the spectral lines also depends on the quantity of gas and on the temperature of the gas. Sample emission and absorption spectra are shown in Fig. 3.4.

Figure 3.3   Conditions for the formation of emission and absorption lines. (a) We look only at a cloud of gas with the atoms or molecules capable of producing spectral lines. Since there is no continuum radiation to absorb, we can only have emission. (b) We now look at a background continuum source through the cloud of gas. This can produce absorption lines.

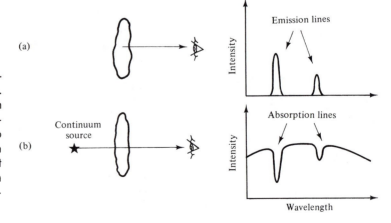

3.2   SPECTRAL TYPES

When spectra were taken of stars other than the Sun, they also showed absorption spectra. Presumably, the continuous radiation produced in a star passes through an atmosphere in which the absorption lines are produced. Not all of the stars have absorptions at the same wavelength.

Astronomers began to classify and catalogue the spectra, even though they still did not understand the mechanism for producing the lines. This points out an important general technique in astronomy—studying large numbers of objects to find general trends. In one very important study, over 200,000 stars were classified by Annie Jump Cannon at the Harvard College Observatory. The benefactor of this study was Henry Draper, and the catalogue of stellar spectra was named after him. The stars in this catalogue are still known by their HD numbers.

One set of spectral lines common to many stars was recognized as belonging to the element hydrogen. The stars were classified according to the strength of the hydrogen absorption lines. In this system, A stars have the strongest hydrogen lines, B stars the next strongest, and so on. These letter designations were called *spectral classes* or *spectral types*. We now know that the different spectral type stars have different surface temperatures. However, the sequence A, B, . . . , is not a temperature-ordered sequence. For reasons we will discuss below, the hydrogen lines are strongest in intermediate-temperature stars.

The spectral classes we use, in order of decreasing temperature, are O, B, A, F, G, K, M. We break each of these classes into ten subclasses, identified by a number from zero to nine, for example, the sequence O7, O8, O9, B0, B1, B2, . . . , B9, A0, A1, ... (For O stars, the few hottest subclasses are not used.) For some of the hotter spectral types, we even use half-subclasses, for example, B1.5. It was originally thought that stars got cooler as they evolved, so that the temperature sequence was really an evolutionary sequence. Therefore, the hotter spectral types were called *early* and

Figure 3.4   Samples of stellar spectra, in a negative image. Wavelength increases from left to right. The light band marked LiI, at 6707 Å, is in absorption. The dark band marked [OI] at 6300 Å is in emission. The band marked Hα is in emission, but has absorption to the short wavelength side. (The unmarked spectrum on top is a laboratory emission spectrum.)

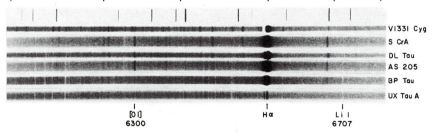

the cooler spectral types were called *late*. We now know that these evolutionary ideas are not correct. However, the nomenclature still remains. We even talk about a B0 or B1 star as being an ''early B'' and a B8 or B9 star as being a ''late B.''

3.3   THE ORIGIN OF SPECTRAL LINES

The processes that result in atoms being able to emit or absorb radiation only at certain wavelengths are tied to the nature of matter and light. In the last chapter, we saw the beginnings of the quantum revolution with the realization that light exhibits both particle and wave properties. We now see how the ideas of quantization apply to the structure of the atom.

The modern picture of the atom begins with the experiments of Ernest Rutherford, who studied the scattering of alpha particles (helium nuclei) off gold atoms. Most of the alpha particles passed through the atoms without being deflected, suggesting that the atom is mostly empty space! Some alpha particles were deflected through large angles, suggesting a concentration of positive charge in the center of the atom. This concentration is called the *nucleus*. A sufficient number of electrons orbit the nucleus to keep the whole atom neutral.

There were still some problems with this picture. It did not explain why the electron orbits were stable. Classical electricity and magnetism tells us that an accelerating charge gives off radiation. An electron going in a circular orbit is accelerating, since the direction of its motion is always changing. Therefore, the electrons orbiting the nucleus should lose energy via radiation and spiral into the nucleus. The second problem concerns the origin of spectral lines. There is nothing in Rutherford's atom that allows for spectral lines.

The arrangement of spectral lines in a particular element is not random. For example, in 1885, Johann Jakob Balmer, a Swiss teacher, realized that there was a regularity in the wavelengths of the spectral lines of hydrogen. (The hydrogen spectrum is shown in Fig. 3.5.) They obeyed the simple relationship known as the *Balmer formula:*

$$1/\lambda = R(1/2^2 - 1/n^2) \tag{3.1}$$

The constant R is called the *Rydberg constant,* and its value is given by

Figure 3.5   The Balmer spectrum of atomic hydrogen. In each row the top and bottom spectra are for comparison.

$1/R = 911.7636$ Å. The quantity n is any integer greater than two. By setting n to 3, 4, . . . , we get the wavelengths for the visible hydrogen lines (also known as the *Balmer series*.) Of course, this was just an empirical formula, with no theoretical justification.

EXAMPLE 3.1   First Balmer Line

Calculate the wavelength of the longest-wavelength Balmer line. This line is known as Balmer-alpha, or, simply, H-alpha.

Solution

We set $n = 3$ in equation (3.1) to get

$$1/\lambda = R(1/2^2 - 1/3^2)$$

Inverting gives

$$\lambda = 6564.7 \text{ Å}$$

This is the wavelength as measured in a vacuum. We generally refer to the wavelength in air, since that is how we measure it at a telescope. The wavelength in air is that in vacuum, divided by the index of refraction of air, 1.00029, giving 6562.8 Å. (It is interesting to note that, when spectroscopists tabulate wavelengths, those longer than 2000 Å are given as they would be in air, since that is how they will usually be measured. Radiation with wavelengths less than 2000 Å doesn't penetrate through air, and its wavelengths are usually measured in a vacuum, so the vacuum values are tabulated.)

3.3.1   Bohr Atom

The next advancement was by the Danish physicist Niels Bohr who tried to understand hydrogen (the simplest atom), illustrated in Fig. 3.6. He postulated the existence of certain *stationary states*. If the electron is orbiting in one of these states, the atom is stable. Each of these states has a particular energy. We can let the energy of the nth state be E_n, and the energy of the mth state be E_m. For definiteness, let $E_n > E_m$.

Under the right conditions, transitions between states can take place. If the electron is in the higher-energy state, it can drop down to the lower energy state, as long as a photon is emitted with an energy equal to the energy difference between the two states. If the frequency of the photon is ν, this condition means that $h\nu = E_m - E_n$. If the electron is in the lower state, it can make a transition to the upper state if the atom absorbs a photon

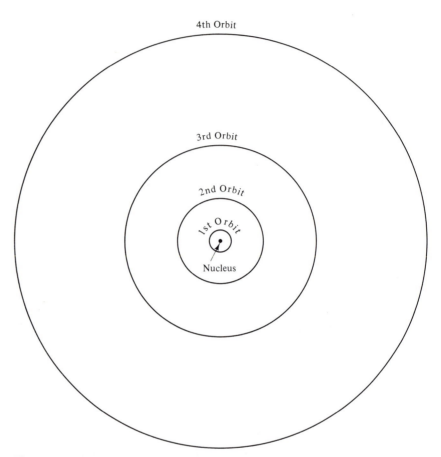

Figure 3.6 The Bohr atom. Electrons orbit the nucleus in allowed orbits. The relative sizes of the orbits is correct, but, on that scale, the nucleus should be much smaller.

with exactly the right energy. This explanation incorporated Einstein's idea of photons.

Bohr pointed out that one could get the energies of the allowed states by assuming that the angular momentum J of the orbiting electron is quantized in integer multiples of $h/2\pi$. The combination $h/2\pi$ appears so often that we give it its own symbol, $\hbar$ (spoken as ''h-bar''). We apply this to a hydrogen atom, with an electron, with charge $-e$, orbiting a nucleus, with charge $+e$, at a distance r. We assume that the nucleus is much more massive than the electron, so we can ignore the small motion of the nucleus, since both the nucleus and electron orbit their common center of mass.

We first look at the energy of the orbiting electron. The kinetic energy is

$$\mathrm{KE} = \tfrac{1}{2}mv^2$$

The potential energy, relative to the potential energy begin zero when the electron is infinitely far from the nucleus, is given by

$$PE = -e^2/r$$

(By writing the potential energy in this form, rather than with a factor of $1/4\pi\epsilon_0$, we are using cgs units. This means that charges are expressed in electrostatic units, esu, with the charge on the proton being 4.8×10^{-10} esu.) The total energy is therefore

$$E = \tfrac{1}{2}mv^2 - e^2/r \tag{3.2}$$

We can relate v and r by noting that the electrical force between the electron and the nucleus e^2/r^2 must provide the acceleration to keep the electron in a circular orbit v^2/r. This tells us that

$$mv^2/r = e^2/r^2$$

Multiplying both sides by r gives

$$mv^2 = e^2/r \tag{3.3}$$

We put this into equation (3.2) to get the total energy in terms of r:

$$E = -\tfrac{1}{2}e^2/r \tag{3.4}$$

The minus sign indicates that the total energy is negative, which means that the system is *bound*. We would have to add an energy of $\tfrac{1}{2}e^2/r$ to tear the electron away from the nucleus.

We still have to find the allowed values of r. The angular momentum is $J = mvr$. The quantization condition becomes

$$mvr = n\hbar$$

Solving for v gives

$$v = n\hbar/mr$$

Squaring both sides and multiplying by m gives

$$mv^2 = n^2\hbar^2/mr^2$$

By equation (3.2), we have

$$e^2/r = n^2\hbar^2/mr^2$$

We now solve for r, giving r, the radius of the nth orbit:

$$r = n^2\hbar^2/me^2 \tag{3.5}$$

Substituting into equation (3.4) gives the energy of the nth state:

$$E = -\tfrac{1}{2}e^4m/n^2\hbar^2 \tag{3.6}$$

Note that this has the $1/n^2$ dependence we would expect from the Balmer formula.

One modification that we should make is to account for the motion of the nucleus (since it is not infinitely massive). We should replace the mass of the electron m in equation (3.6) by the *reduced mass* of the electron and the proton. The reduced mass m_r is defined such that the motion of the electron, as viewed from the (moving) proton, is as if the proton were fixed and the electron's mass were reduced to m_r. An expression for m_r (see Problem 3.4) is

$$m_r = \frac{m_e m_p}{m_e + m_p}$$

$$= 0.9995\, m_e$$

EXAMPLE 3.2 H Atom Energy

Compute the energy of the lowest (ground) energy level in a hydrogen atom. Also, find the radius of the orbit of the electron in this state.

Solution

We use equation (3.6), with $n = 1$, to give

$$E_1 = \frac{-\tfrac{1}{2}(4.8 \times 10^{-10}\text{ esu})^4(9.11 \times 10^{-28}\text{ g})}{(1)^2(1.05 \times 10^{-27}\text{ erg s})^2}$$

$$= -2.2 \times 10^{-11}\text{ erg}$$

An erg is not a convenient unit for keeping track of such small energies, so we convert to electron volts (1 eV = 1.6×10^{-12} erg is the energy acquired by an electron while being accelerated through a potential difference of 1 V), giving

$$E = -13.6\text{ eV}$$

The radius is given by equation (3.5):

$$r_1 = \frac{(1.05 \times 10^{-27} \text{ erg s})^2}{(9.11 \times 10^{-28} \text{ g})(4.8 \times 10^{-10} \text{ esu})^2}$$

$$= 5.25 \times 10^{-9} \text{ cm}$$

$$= 0.525 \text{ Å}$$

Note that if we take $n = \infty$ in equation (3.6), we get $E = 0$. However, $n = \infty$ corresponds to a free electron. Therefore, to get the electron far from the nucleus, we must add 13.6 eV. The energy that we must add to an atom to break it apart is called the *binding energy*. The energy does the work against the electric attraction between the electron and the nucleus as you try to pull the electron away.

Now that we have evaluated E_n, we can rewrite equation (3.6) as

$$E_n = -13.6 \text{ eV}/n^2 \qquad (3.6a)$$

It is then very easy to calculate the energies of other levels. For example, $E_2 = -3.4$ eV. Therefore, the energy difference, $E_2 - E_1$, is equal to 10.2 eV. The energy levels are shown in Fig. 3.7. Note that they get closer together as one goes to higher values of n.

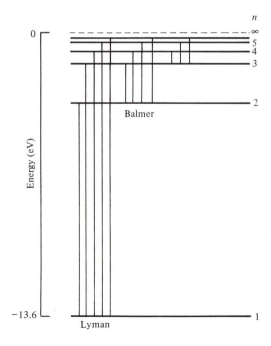

Figure 3.7 Hydrogen energy levels. The right-hand column gives the principle quantum number n. The energies are relative to the state in which the electron and proton are infinitely far apart, so the ground state energy is −13.6 eV. Transitions (which can be either emission or absorption) are grouped according to the lower level of the transition. For example, the Balmer series consists of al emissions ending with the electron in the $n = 2$ state and all absorptions starting with the electron in the $n = 2$ state.

Since the zero of potential energy is arbitrarily defined, we sometimes choose to shift our energy scale by 13.6 eV. This would make the energy of the ground state ($n = 1$) zero, and for the $n = 2$ state, $+10.2$ eV. A free electron would have an energy of $+13.6$ eV, or greater. The energy *differences* between states are unaffected by this shift in the zero point of energy.

We can use equation (3.6) to derive the Balmer formula. First, we rewrite the equation as

$$E_n = -hcR/n^2 \tag{3.6b}$$

where

$$R = -e^4 m/(4\pi)c\hbar^3 \tag{3.6c}$$

The energy of a photon emitted or absorbed must equal the energy difference between the two states:

$$E_n - E_m = h\nu$$

$$= hc/\lambda$$

Taking the energies from equation (3.6b) gives

$$1/\lambda = R(1/m^2 - 1/n^2) \tag{3.7}$$

This looks very similar to the Balmer formula, except that the Balmer formula has a 2 instead of the *m*. This means that the Balmer series all have the lowest energy level as the second level.

We can use equation (3.7) to divide the hydrogen spectrum into different series. A given series is characterized by having the same lower-energy state. For example, the Balmer series consists of absorptions accompanying transitions from level 2 to any higher levels, and emissions accompanying transitions from higher levels down to level 2. The first Balmer transition (involving levels 2 and 3) has the smallest energy difference of the series. (Clearly, the energy difference between levels 2 and 3 is less than the energy difference between levels 2 and 4 or between levels 2 and 5, and so on.) The Balmer series is important because the first few transitions fall in the visible part of the spectrum. The series with the lower level being level 1 is called the *Lyman series*. Even the lowest transition in the Lyman series is in the ultraviolet.

NOTE: We have developed a labeling system for the various transitions. First we give the chemical symbol for the element (e.g., H for hydrogen). Then we give the *m* for the lowest level that characterizes the series (1 for Lyman, 2 for Balmer, etc.). Finally, we give a Greek letter denoting the number of levels jumped. For example, if $n = m + 1$, we have an alpha (α) transition; if $n = m + 2$, we have a beta (β) transition. The first Balmer line would then be designated as H2α. (Note that for the Balmer series of hydrogen only, we sometimes drop the 2 and just say Hα, Hβ, etc.)

3.3.2 Quantum Mechanics

The Bohr model for the atom allowed physicists to understand the organization of energy levels. However, it was far from a complete theory. One shortcoming was that it did not explain why some spectral lines were stronger than others. More fundamentally, it was still an *ad hoc* theory. Bohr had no explanation for why stationary states exist, or why angular momentum must be quantized in some particular way. These were just postulates. A much deeper understanding was needed.

An important step was made by Louis de Broglie, who proposed the revolutionary idea that if light could exhibit a particle–wave duality, then maybe all matter could. That is, an electron orbiting a nucleus has certain wavelike properties, and it is those properties that determine the states that are stable. One could think of the electron as having a certain wavelength. Stationary states could be those whose circumference contained an integral number of wavelengths, producing a pattern that is reinforced during each orbit (like a standing wave). It was necessary to have some expressions for the wavelength and frequency of a particle, and de Broglie noted that if the wavelength was taken as h/p (where p is the momentum of the particle) and the frequency as $E/h,$ then the orbits that were allowed by the standing wave idea were the same as the orbits that Bohr found from his postulates (see Problem 3.11).

This is clearly a departure from our normal way of looking at matter around us, and we cannot go into all of the ramifications here. To this point we have gone far enough to understand stellar spectra. The picture, as described by Bohr and de Broglie, is the quantum theory in its most naive form. It was realized that if particles behave, in some fashion, like waves, then the description of particle motions (mechanics) must be changed from the Newtonian laws of motion to laws of motion involving waves. Theories that do this are called *wave mechanics* or *quantum mechanics*. One such theory was presented in 1925 by the German physicist Erwin Schrödinger. In his theory, the information about the motion of a particle is contained in a function, called a *wave function*. Schrödinger's interpretation of the wave function was that it is related to the probability of finding a particle in a particular place with a particular momentum. This replaced the absolute determinism of classical physics with the statement that we can only predict where a particle is *likely* to be, but not *exactly* where it will be. However, we can predict the average locations and momenta of a large group of particles, and it is these average properties that we see in our everyday world. Many physicists (including Einstein) were not comfortable with this probabilistic interpretation, but quantum theory has been quite successful in predicting the outcome of a variety of experiments. We will pick up some more threads of the quantum revolution later in this book.

3.4 FORMATION OF SPECTRAL LINES

Now that we have some idea of how emission and absorption of radiation by atoms takes place, we can return to stellar spectra. The first point to realize is that in a star we are not talking about radiation from a single hydrogen atom but from a large number of them. We see a strong Hα absorption line in stars because many photons are removed from the continuum by this process. It is clear, however, that having a lot of hydrogen does not assure us of a strong Hα absorption. In order for such an absorption to take place, a significant number of atoms must be in level 2, ready to absorb a photon. If all of the hydrogen is in level 1, you will not see the Balmer series, no matter how much hydrogen is present.

3.4.1 Excitation

In general, the strength of a particular transition (emission or absorption) will depend on the number of atoms in the initial state for that transition. The number of atoms per unit volume in a particular state is called the *population* of that state. In this section, we will look at the factors that determine the populations of various states. We refer to the processes that alter the populations as *excitation* processes. We have already seen one type of excitation process—the emission and absorption of photons. Electrons can jump to a lower level when a photon is emitted, or they can jump to a higher level when a photon is absorbed.

Populations can also be governed by collisions with other atoms. For example, atom 1 can be in state j. It then undergoes a collision with atom 2, and makes a transition to a higher state, i. In the process, the kinetic energy of atom 2 is reduced by the energy difference between the two states in atom 1, $E_i - E_j$. The reverse process is also possible, with atom 2 gaining kinetic energy while atom 1 goes from a higher- to a lower-energy state.

The collisional excitation rates will depend on the kinetic temperature of the gas. For atoms at kinetic temeprature T_k the average kinetic energy per atom is $\frac{3}{2}kT_k$. As the temperature increases, more energy is available for collisions. This makes higher-energy states easier to reach. Also, as we increase the energy, the particles are moving faster and collisions take place more frequently. This is illustrated in Fig. 3.8.

When a gas is in thermodynamic equilibrium, with a kinetic temperature

Figure 3.8 Level populations as a function of temperature T for a two-level system. At $T = 0$ K, all of the atoms are in the ground state. At high temperature, the ratio of the populations approaches the ratio of the statistical weights of the levels.

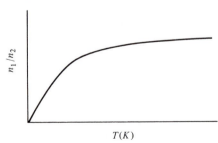

T_k, the ratios of the level populations are given by a *Boltzmann distribution*. If we let n_i and n_j be the populations of levels i and j, respectively, their ratio is given by

$$\frac{n_i}{n_j} = \frac{g_i}{g_j} \exp \left[\frac{-(E_i - E_j)}{kT_k} \right] \tag{3.8}$$

In this equation g_i and g_j are called *statistical weights*. They are needed because certain energy levels are actually groupings of sublevels that have the same energy. The statistical weight of a level is just a count of the number of sublevels in that level.

To see some properties of this distribution, we assume that level i is higher than level j. As we let E_i get larger, keeping T_k constant, the number in the square brackets becomes more negative, meaning that the population ratio gets smaller. This makes sense, since it should be harder to excite a state if its energy is very high. Now we look at what happens to the population ratio if we change the temperature. As we increase T_k, the number in the square brackets becomes less negative, making n_i/n_j larger. Note that for $T_k = 0$, this ratio is 0, which makes sense. If there is no energy in collisions, no excitation will take place. If we let T_k go to infinity, the population ratio approaches the ratio of the statistical weights (not infinity).

The Boltzmann distribution provides us with a convenient reference point even for a system that is not in thermodynamic equilibrium. For any given population ratio n_i/n_j, we can always find some value of T to plug into equation (3.8) to make the equation correct. We call such a temperature an *excitation temperature*. When we are not in equilibrium, the excitation temperature can be different for any pair of levels. In thermodynamic equilibrium, all excitation temperatures are equal to each other and to the kinetic temperature.

3.4.2 Ionization

If we know the temperature in the atmosphere of a star, we can use the Boltzmann equation to predict how many atoms will be in each state i and predict the strengths of the various spectral lines. However, there is still an effect that we have not taken into account—ionization. If the temperature is very high, some of the colliding particles will have kinetic energies greater than the ionization energy of the atom, so the electron will be torn away in the collision. Once a hydrogen atom is ionized, it can no longer participate in line emission or absorption.

When the gas is ionized, electrons and positive ions will sometimes collide and recombine. When the total rate of ionizations is equal to the total rate of recombinations, the gas is in *ionization equilibrium*. If the gas is in thermal equilibrium, then the *Saha equation* tells us the relative abundances of various ions. We let $n(X_r)$ and $n(X_{r+1})$ be the densities of the r and $r +$ 1 ionization states, respectively, of the element X. (For example, if $r = 0$, we are talking about the neutral species and the first ionized state.) The ionization energy to go from r to $r + 1$ is E_i. The electron density is n_e and the kinetic temperature is T_k. Finally, g_r and g_{r+1} are the statistical weights for the ground states of X_r and X_{r+1} (assuming that most of the species are in the ground state). The Saha equation tells us that

$$\frac{n_e n(X_{r+1})}{n(X_r)} = \frac{2g_r}{g_{r+1}} \left(\frac{2\pi m_r kT_k}{h^2}\right)^{3/2} \exp\left[\frac{-E_i}{kT_k}\right] \tag{3.9}$$

The Saha equation has the same exponential energy dependence as the Boltzmann distribution. However, there is the additional factor of $T_k^{3/2}$. This comes from the fact that a free electron has more states available to it at a higher T_k than at a lower T_k. In addition, there is a factor of n_e on the left. This is because a high abundance of electrons leads to a higher rate of recombinations, driving down the fraction of atoms that are ionized. The number of electrons must be the number of electrons from all sources, since an electron that came from a hydrogen atom can recombine with a helium ion. Just as we did with an excitation temperature in the Boltzmann equation, we can always find an *ionization temperature* T_i which makes the Saha equation correct, even for a gas not in thermodynamic equilibrium.

The ionization energies of some common atoms and ions are given in Table 3.1. This is useful in deciding which ions you are likely to encounter at various temperatures. There is a shorthand representation of various ions. A roman numeral I is used to designate the neutral species, II to designate the singly ionized species, III the doubly ionized species, and so on. For example, neutral hydrogen is H(I), ionized hydrogen (H^+) is H(II), doubly ionized carbon is C(III).

TABLE 3.1   Ionization Energies (eV)

Atom	Singly ionized	Doubly ionized
H	13.6	
He	24.6	54.4
C	11.3	24.4
N	14.5	29.6
O	13.6	35.1
Na	5.1	47.3
K	4.3	31.8
Ca	6.1	11.9
Fe	7.9	16.2

3.4.3   Intensities of Spectral Lines

We are now in a position to discuss the intensities of various absorption lines in stars. We will take Hα as an example to see the combined effects of excitation and ionization. At low temperatures, essentially all the hydrogen is neutral, and most of it is in the ground state. Since little H will be in the second state, there will be few chances for Hα absorption. The Hα line will be weak.

As we go to moderate temperatures, most of the hydrogen is still neutral. However, more of the hydrogen is in excited states, meaning that a reasonable amount will be in level 2. Hα absorption is possible. As the temperature increases, the Hα absorption is seen to get stronger.

At very high temperatures, the hydrogen becomes ionized. Since there is less neutral hydrogen, the Hα line becomes weaker. This explains why the Hα line is strongest in middle-temperature stars, and why the original scheme of classifying by hydrogen line strengths did not produce a sequence ordered in temperature.

We can apply a similar analysis to other elements. The details will differ because of the different energy level structures and the different ionization energies. It should be noted that after hydrogen and helium, the abundances of the elements fall off drastically. In fact, astronomers often refer to hydrogen, helium, and ''everything else.'' The ''everything else'' are collectively called *metals,* even though many of the elements don't fit our common definition of a metal.

We now look at the properties of different spectral types, in order of increasing temperature. Sample spectra are shown in Fig. 3.9, and the behaviors of a few lines are shown in Fig. 3.10.

M   Temperatures in M stars are below 3500 K, explaining their red color. The temperature is not high enough to produce strong Hα absorption, but some lines from neutral metals are seen. The stars are cool enough for simple molecules to form, and many lines are seen from molecules such as CN and TiO (titanium oxide). If cool stars show strong CH lines, we designate them as C-type, or ''carbon stars.'' If an M star has strong ZrO (zirconium oxide) lines as opposed to TiO lines, we call it an S-type.

K   Temperatures range from 3500 to 5000 K. There are many lines from neutral metals. The H lines are stronger than in M stars, but much of the H is still in the ground state.

G   Temperatures are in the range 5000–6000 K. The sun is a G2 star. The H lines are stronger than in K stars, as more atoms are in excited states. The temperature is high enough for metals with low ionization energies to be partially ionized. Two prominent lines are from Ca(II). When Fraunhofer studied the solar spectrum, he gave the strongest lines letter designations. These Ca(II) lines are the H and K lines in his sequence.

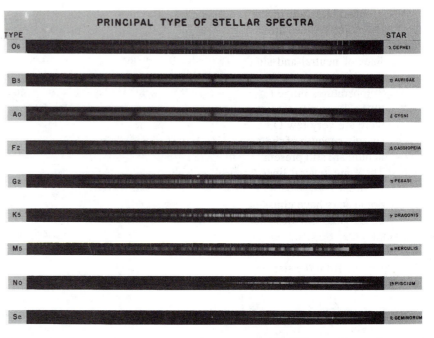

PRINCIPAL TYPE OF STELLAR SPECTRA

TYPE		STAR
O6		λ CEPHEI
B3		η AURIGAE
A0		δ CYGNI
F2		β CASSIOPEIA
G2		η PEGASI
K5		γ DRACONIS
M5		α HERCULIS
No		19 PISCIUM
Se		R GEMINORUM

Figure 3.9 Samples of spectra from stars with different spectral types. This is a positive printing, so the continuum is bright, and the absorption lines appear dark. The name of the star appears on the right of each spectrum and the spectral type appears on the left. In each spectrum, wavelength increases from left or right.

F   Temperatures range from 6000 to 7500 K. The H lines are a little stronger than in G stars. The ionized metal lines are also stronger.

A   Temperatures range from 7500 to 10,000 K. These stars are white–blue in color. They have the strongest H lines. Lines of ionized metals are still present.

Figure 3.10 The relative strength of spectral lines from important species as a function of spectral type. Each species shows the effects of excitation and ionization. For example, the increase in H line strengths from K to A stars is because the increasing temperature results in more hydrogen in the $n = 2$ levels and higher. However, the higher tempertures of the B and O stars ionize the hydrogen, and the lines get weaker.

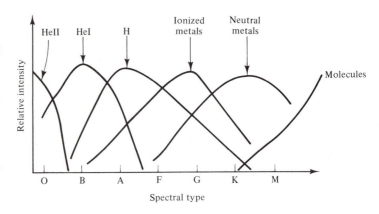

B   Temperatures are in the range 10,000–30,000 K, and the stars appear blue. The H lines are beginning to get weaker because the temperatures are high enough to ionize a significant fraction of the hydrogen. The lines of neutral and singly ionized helium begin to appear. Otherwise, there are relatively few lines in the spectrum.

O   Temperatures range from 30,000 to over 60,000 K, and the stars appear blue. The earliest spectral types that have been seen are O3 stars, and there are very few O3 and O4 stars. The hydrogen line strengths fall off sharply because of the high rate of ionization. The lines of singly ionized helium are still present, but there are very few lines overall in the visible. There are several lines in the ultraviolet.

Some stars have emission lines as well as absorption lines in their spectra. These stars are designated with an ''e'' after the spectral class, for example, Oe, Be, Ae, etc. O stars with very broad emission lines are called *Wolf–Rayet* stars. These stars probably have circumstellar material that has been ejected from the star. (Wolf–Rayet stars are not the only stars with such outflowing material.)

3.5   THE HERTZSPRUNG–RUSSELL DIAGRAM

Even though we cannot study any one star (except for the Sun) in great detail, we can compensate somewhat by having a large number of stars to study. From statistical studies we learn about general trends. For example, we think that any property that is common to many stars must be telling us about the laws of physics that are important in understanding the structure of stars.

One of the early statistical studies was carried out in 1910 independently by the Danish astronomer Ejnar Hertzsprung, and by the American astronomer Henry Norris Russell. They plotted the properties of stars on a diagram in which the horizontal axis is some measure of the temperature (spectral type or color) and the vertical axis is some measure of the luminosity. We call such a diagram a *Hertzsprung–Russell diagram,* or, simply, an *H–R diagram*.

If a random group of stars is chosen, all at different distances, a comparison of apparent magnitudes is not very meaningful. The apparent magnitude must be corrected to give the absolute magnitude. However, if we find a group of stars at the same distance, we can plot their apparent magnitude, since the distance modulus will be the same for all the stars in the group. For this purpose we use clusters of stars.

When this was done, the result was something like that shown in Fig. 3.11(a). The first thing we notice is that stars appear only in certain parts of the diagram. Arbitrary combinations of temperature and luminosity are not allowed. Remember, for a given temperature, the luminosity still depends on

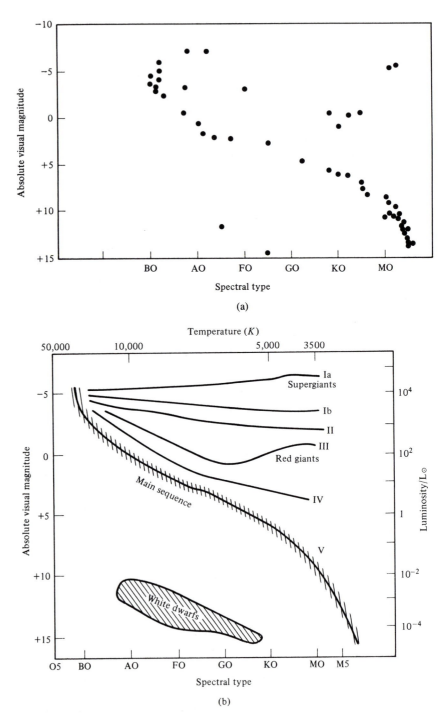

Figure 3.11 (a) *H–R* diagram. (b) Schematic *H–R* diagram. Luminosity classes are indicated by Roman numerals.

the radius of the star, so the H–R diagram is telling us that arbitrary combinations of radius and temperature are not possible.

Most of the stars are found in a narrow band called the *main sequence*. The significance of the main sequence is that most stars of the same temperature have essentially the same luminosity; hence, equation (2.7) tells us that most stars of the same temperature have essentially the same radius. This close relationship between size and temperature must be imposed by the laws of physics, and gives us hope that we can understand stellar structure by applying the known laws. It also gives us a crucial test: Any theory of stellar structure must predict the existence of the main sequence.

Not all the stars appear on the main sequence. Some appear above the main sequence. This means that they are more luminous than main sequence stars of the same temperature. If two stars have the same temperature, but one is more luminous, it must be larger than the other. Stars appearing above the main sequence are therefore larger than main sequence stars. We call these stars *giants*. By contrast, main sequence stars are called *dwarfs*. We actually subdivide the giants into three groups: subgiants, giants, and supergiants.

To keep track of the size of a star of a given spectral type, we append a *luminosity class* to the spectral type. The luminosity class is denoted by a roman numeral. Main sequence stars are luminosity class V. The Sun, for example, is a G2V star. Subgiants are luminosity class IV, and giants are class III. Luminosity class II stars are somewhere between giants and supergiants. Supergiants are luminosity class I. We further divide supergiants into Ia and Ib, with Ia being the larger. When we look at the spectral lines from a star, we can actually tell something about the size of the star. Stars of different size will have different accelerations of gravity at their surface. The surface gravity affects the detailed appearance of certain spectral lines.

There are also stars that appear below the main sequence. These stars are typically 10 mag fainter than main sequence stars. They are clearly much smaller than main sequence stars. Since most of them appear at middle spectral types, and appear white, we refer to them as *white dwarfs*. (Do not confuse dwarfs, which are main sequence stars, with white dwarfs, which are much smaller then ordinary dwarfs.)

EXAMPLE 3.3   Size of White Dwarfs

Suppose that some white dwarf has the same spectral type as the Sun, but has an absolute magnitude that is 10 mag fainter than the Sun. What is the ratio of the radius of the white dwarf, R_{wd}, to the radius of the Sun, $R_\odot$?

Solution

The luminosity is proportional to the square of the radius, so

$$L_{wd}/L_\odot = (R_{wd}/R_\odot)^2$$

We use equation (2.2) to find the luminosity ratio for a 10 mag difference.

$$L_{wd}/L_\odot = 10^{(M_\odot - M_{wd})/2.5}$$

$$= 10^{-4}$$

Combining these two results to find the ratio of the radii,

$$R_{wd}/R_\odot = (L_{wd}/L_\odot)^{1/2}$$

$$= (10^{-4})^{1/2}$$

$$= 10^{-2}$$

The radius of the white dwarf is only 1% of the radius of the Sun!

For any cluster for which we plot an H–R diagram, we only know the apparent magnitudes, not the absolute magnitudes. If we know the absolute magnitude for some spectral type, we can find the distance modulus for stars of that spectral type in the cluster. The distance modulus is the same for all of the stars in the cluster, so we can calibrate the whole H–R diagram in terms of absolute magnitudes. To obtain a reliable calibration, we would really like to carry it out for many stars. We have already seen that there is a small group of nearby stars for which trigonometric parallax give us a good distance measurement. However, this group of stars does not contain a good representation of spectral types. In Chapter 13, we'll see how we can improve on this sample.

Once we know the absolute magnitude for a given spectral type, we have a very useful way for determining distances. For any given star, we measure m, its apparent magnitude. We take a spectrum of the star and determine its spectral type. From the spectral type, we know the absolute magnitude M. Since we know both m and M, we know the distance modulus $m - M$, and therefore the distance. This procedure is called *spectroscopic parallax*. The word ''spectroscopic'' refers to the fact that we use the star's spectrum to determine its absolute magnitude.

EXAMPLE 3.4 Spectroscopic Parallax

For a B0 star ($M = 3$), we observe an apparent magnitude, $m = 10$. What is the distance to the star, d?

Solution

The distance modulus is

$$m - M = 10 - (-3) = 13 \text{ mag}$$

We use this in equation (2.17) to find the distance:

$$5 \log(d/10 \text{ pc}) = 13 \text{ mag}$$

$$\log(d/10 \text{ pc}) = 2.6$$

Solving for d gives

$$d/10 \text{ pc} = 400$$

$$d = 4000 \text{ pc}$$

CHAPTER SUMMARY

In this chapter we looked at how spectral lines are formed, and how spectral lines can tell us about the physical conditions in the atmosphere of a star.

We saw that stars were originally classified into spectral types before the nature of the temperature sequence was understood.

We saw how an explanation of spectral lines, in general, requires an atomic theory in which the electrons can occupy only certain energy states. An atom can go from one state to another by emitting or absorbing a photon with the appropriate energy. We saw that a relatively simple theory could explain the spectrum of the hydrogen atom.

In a star, the strength of a spectral line depends on the abundance of the particular atom and on the relative number in the appropriate ionization and energy states. The distribution among energy states is described by the Boltzmann equation, and the distribution among ionization states is described by the Saha equation.

Finally, we saw what could be learned from a Hertzsprung–Russell diagram, in which the horizontal axis is some measure of the temperature, and the vertical axis is some measure of luminosity. Most of the points representing stars on the H–R diagram fall along a narrow band, called the main sequence. This tells us that, for most stars, there must be a simple relationship between size and temperature. Stars that do not lie along the main sequence are identified as being various classes of giants, for the brighter ones, and white dwarfs, for the fainter ones.

QUESTIONS AND PROBLEMS

3.1 Arrange the standard spectral sequence—O, B, A, F, G, K, M—in order of decreasing Hα strength.

3.2 What were (a) the strong points and (b) the weak points of (i) Rutherford's atom and (ii) Bohr's atom?

3.3 What evidence supports the idea of photons?

3.4 Rederive equation (3.6) without making the assumption of an infinitely massive nucleus, and show that

one obtains the same expression except for the reduced mass replacing the electron mass.

3.5 Tabulate the electron–nucleus reduced mass for an interesting range of nuclear masses.

3.6 Consider a carbon atom that has six electrons orbiting a nucleus with six protons and six neutrons. Suppose five of the electrons are in their lowest possible states, but the sixth is in a very high state. The energy levels of this sixth electron will be given approximately by equation (3.6) with the appropriate reduced mass. Explain why this should be, since the charge of the C nucleus is six times the charge of the H nucleus. (*Hint:* Think of the electric field seen by the sixth electron.)

3.7 Find all the $Hn\alpha$ transitions that fall in the visible part of the spectrum.

3.8 Find the wavelength of $H1\alpha$ (Lyman-alpha).

3.9 How much energy is required to ionize hydrogen already in the $n = 2$ state?

3.10 An electron in a hydrogen atom is in a high n state. It drops down one state at a time. What is the first transition to give a visible photon?

3.11 Using the de Broglie wavelength h/p, show that orbits whose angular momentum is quantized according to the Bohr condition ($J = n\hbar$) correspond to orbits whose circumference is an integer number of wavelengths.

3.12 If the populations of two levels with energies E_i and E_j and statistical weights g_i and g_j ($E_i > E_j$) are found to be n_i and n_j, respectively, derive an expression for the excitation temperature of this transition.

3.13 Assuming that all the level populations are given by equation (3.8), derive an expression for f_i, the fractional population of the ith level, defined as

$$f_i = n_i \bigg/ \sum_{j=1}^{N} n_j$$

where N is the highest populated level.

3.14 Consider only the three lowest levels in hydrogen, with $g_1 = 2$, $g_2 = 6$, $g_3 = 10$. Plot the fraction of hydrogen in level 2 $[n_2/(n_1 + n_2 + n_3)]$ vs. T for an interesting temperature range.

3.15 For an atom whose populations are given by equation (3.8), (a) in what temperature limit is the ratio of the populations equal to the ratio of the statistical weights, and (b) in what temperature range would the ratio of the populations be greater than the ratio of the statistical weights?

3.16 Suppose for a given sample of hydrogen that 50% is ionized at a temperature of 5000 K. (a) What fraction would be ionized at 10^4 K? (b) What fraction would be ionized with twice as much hydrogen in the same volume at 5000 K? (Assume that $N_e = n_+$ and that the hydrogen that isn't ionized is in the ground state.)

3.17 Explain how the $H\alpha$ absorption strength changes as we raise the temperature of a star.

3.18 Explain the advantage of studying an H–R diagram for a cluster.

3.19 What is the significance of the main sequence?

3.20 How much larger is an M0Ia star than an M0V star? (See Appendix E for stellar properties.)

3.21 For an A3 star, we measure an apparent magnitude, $m = 12$. How far away is the star (assuming that it is a main sequence star)? (See Appendix E for stellar properties.)

3.22 If we are limited to $m = 6$ or brighter for naked eye observations, make a graph of the maximum distance we can see a star versus its spectral type (for main sequence stars).

3.23 Assume that at a certain electron pressure P_e half of the hydrogen is ionized at a temperature of 8000 K. Make a graph of n_e/n vs. T for T in the range of temperatures for main sequence stars, where $n = n_+ + n_0$.

Telescopes

The past decades have seen dramatic improvements in our observing capabilities. There have been improvements in our ability to detect visible radiation, and there have also been exciting extensions into other parts of the spectrum. These improved observing capabilities have had a major impact on astronomy and astrophysics. In this chapter we will first discuss the basic concepts behind optical observations. We will then discuss observations in other parts of the spectrum.

4.1 WHAT A TELESCOPE DOES

An optical telescope provides us with two important capabilities:

1. It provides us with *light-gathering* power. This means that we can see fainter objects with a telescope than we can see with our naked eye.
2. It can provide us with improved *angular resolution*. This means that we can see greater detail with a telescope than without.

For ground-based optical telescopes, the light-gathering power is usually the most important feature.

4.1.1 Light Gathering

We can think of the light from a star as a steady stream of photons, striking the ground with a certain number of photons per unit area per second. If we look straight up at the star, we will only see the photons that directly strike our eye. If we can somehow collect photons over an area much larger than our eye, and concentrate them on the eye, then the eye will receive more photons per second than the unaided eye. A telescope provides us with a large collecting area, to intercept as much of the beam of incoming photons as possible, and then has the optics to focus those photons on the eye, or a camera, or onto some detector.

EXAMPLE 4.1   Light-Gathering Power

Compare the light-gathering power of the naked eye, with a pupil diameter of 5 mm, to that of a 1-m-diam telescope. Make the comparison both in terms of luminosity and magnitude.

Solution

Let d_1 be the diameter of the pupil and d_2 be the diameter of the telescope. The collecting area is proportional to the square of the diameter. The ratio of the areas is

$$\left(\frac{d_2}{d_1}\right)^2 = \left(\frac{1.0 \text{ m}}{5.0 \times 10^{-3} \text{ m}}\right)^2$$

$$= 4.0 \times 10^4$$

This is the ratio of luminosities that we can see with the naked eye and with the telescope. We can express this in magnitudes,

$$m_1 - m_2 = 2.5 \log(4.0 \times 10^4)$$

$$= 11.5 \text{ mag}$$

This means that the faintest objects we can see with the telescope are 11.5 mag fainter than the faintest objects we can see with the naked eye. If the naked eye can see down to 6 mag, the telescope-aided eye can see down to 17.5 mag. This illustrates the great improvement in light-gathering power.

A major advantage of the camera or photoelectric detector over the eye is its ability to collect light for a long time. In the eye, "exposures" are fixed at about $\frac{1}{30}$ s. With modern detectors, exposures of several hours are possible. Therefore, the limiting magnitude for direct visual observing is not as faint as for photography or photoelectric detectors.

4.1.2   Angular Resolution

We now look at resolving power. Resolution is the ability to separate the images of stars that are close together. It also allows us to discern details in an extended object.

One phenomenon that limits resolution is *diffraction*. Diffraction is the

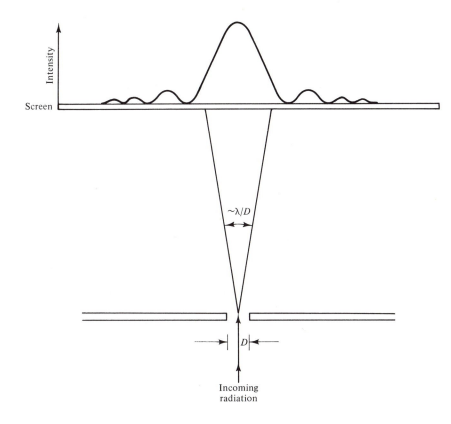

Figure 4.1 Diffraction. A light ray enters from the bottom, and passes through a slit of diameter D. Diffraction spreads the beam out and it falls on a screen. The intensity as a function of position on the screen is shown at the top. Most of the energy is in a main peak, whose angular width is approximately λ/D (in radians). Smaller peaks occur at larger angles.

bending or spreading of waves when they strike a barrier or pass through an aperture. As they spread out, waves from different parts of the aperture interfere with one another, producing maxima and minima, as shown in Fig. 4.1. Most of the power is in the central maximum, whose angular width $\Delta\theta$ (in radians), is related to the wavelength of the wave λ and the diameter of the aperture d by

$$\Delta\theta(\text{rad}) \cong \lambda/d \qquad (4.1)$$

Diffraction results in the images of stars being smeared out by this angle. That means that if two stars are closer than $\Delta\theta$, their images will blend together.

EXAMPLE 4.2 Angular Resolution

Estimate the angular resolution of the eye for light of wavelength 550 nm.

Solution

We use a diameter, $d = 5$ mm, for the pupil. We use equation (4.1) to find the angular resolution in radians. We convert from radians to arc seconds to get the result in a convenient unit (1 rad $= 2.06 \times 10^{5"}$).

$$\Delta\theta = \frac{(2.06 \times 10^5)(5.5 \times 10^{-7} \text{ m})}{5.0 \times 10^{-3} \text{ m}}$$

$$= 23"$$

The eye's resolution is not quite this good, since the full diameter of the pupil is not generally used.

From equation (4.1) we see that we can improve the resolution if we use a larger aperture. A larger telescope will give us better resolution. A 10-cm-diam telescope (20 times the diameter of the pupil in the above example) will give a resolution of 1". However, diffraction is not the only phenomenon that limits resolution. The Earth's atmosphere also distorts images.

When light passes through the atmosphere from above, it is passing into increasingly dense air. As the density of air increases, the index of refraction of the air increases. Therefore, the light encounters an increasing index of refraction as it passes through the atmosphere. We can think of the atmosphere as having a large number of thin layers, each with a slightly different index of refraction. As the light passes from one layer to the one below it, its path is bent slightly toward the vertical. The star appears to be slightly higher above the horizon than it actually is.

This would be no problem if the atmosphere were stable. However, variations on time scales shorter than a second cause changes in the index of refraction in some places. The image moves around, or "twinkles." If we take a picture, we just see a blurred image. This effect is called *seeing,* and usually limits resolution to a few arc seconds. At a good observatory site, on a good night, the seeing might be as good as 1" or better. This corresponds to the diffraction limit of a 10-cm-diam telescope. Building a larger telescope does not help us get past the seeing limitation on resolution, but it still improves the light-gathering power. Hence our earlier statement that light gathering is the main purpose of large ground-based optical telescopes.

4.1.3 Image Formation in a Camera

To illustrate some basic points about the formation of images in optical systems, we look at the operation of a simple camera (Fig. 4.2). For astronomical situations, we are dealing with objects that are "at infinity," so the

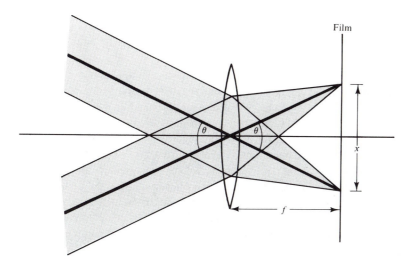

Figure 4.2   The optics of a camera. Bundles of rays from two points enter, making an angle θ with each other. The focal length of the lens is *f*.

light rays from different points are traveling parallel to one another. In the figure, we show bundles of rays coming from two different stars. The rays within each bundle arrive at the camera parallel to each other, but the bundles arrive at an angle with each other equal to the angular separation of the stars on the sky.

For a camera with a lens of focal length *f*, the rays in each bundle are brought together a distance *f* behind the lens. (The image is one focal length behind the lens, since the object is at infinity.) The images of all of the stars in a field lie in a plane, called the *focal plane*. The images of the two stars are at different points in the focal plane. We can locate the image of each star by following the *chief ray* of its bundle (the ray that passes through the center of the lens, undeflected) until it intersects the focal plane.

If the stars have an angular separation θ on the sky, then, as viewed from the lens, the two images have an angular separation θ on the focal plane. This is simply the angle between the two chief rays. The camera provides no angular magnification. As viewed from the lens, the angular separation of the stars is the same as the angular separation of the images.

We can also find the linear separation x between the two images. From the right triangle in the figure we see that

$$\tan(\theta/2) = x/2f$$

If θ is small, then tan(θ/2) is approximately θ/2, in radians. This gives us

$$\theta/2 = x/2f$$

Solving for x gives

$$x = f\theta$$

This tells us that the linear size of the image is proportional to the focal length. To get a larger image, we use a longer focal length lens. (This is what we are doing when we put a telephoto lens on a camera.)

Apart from image size, we are also concerned with the brightness of the image. We can see that the amount of light entering the camera is proportional to the area of the lens. If d is the diameter of the lens, then its area is $\pi d^2/4$. This means that the image brightness is proportional to d^2. The brightness of the image also depends on the image size. The more the image is spread out, the less light reaches any small area of the film or detector. The linear image size is proportional to f, so the image area is proportional to f^2. This means that the image brightness is proportional to $1/f^2$.

Combining these two results, we find that the image brightness is proportional to $(d/f)^2$. The quantity f/d is called the *focal ratio*, so the brightness is proportional to $(1/\text{focal ratio})^2$. We adjust the focal ratio on a camera by changing the f-stops. Since the focal length of the lens is fixed, we do it by changing the diameter of a diaphragm that controls the fraction of the total lens diameter that is actually used. Each f-stop corresponds to a change of $\sqrt{2}$ in the focal ratio, meaning that the image brightness changes by a factor of 2.

You might wonder why we use a diaphragm to block out part of the lens. It might seem more reasonable to use the whole lens and utilize the brighter images to take shorter exposures. The problem is that the images formed by lenses are not perfect. The less we use the edges of the lens, the better the images. Imperfections in the images formed by optical systems are called *aberrations*.

One type of aberration is called *spherical aberration*. It arises from the fact that spherical curves are the easiest to grind on glass surfaces. These

Figure 4.3 Lens aberrations. (a) Coma. (b) Astigmatism.

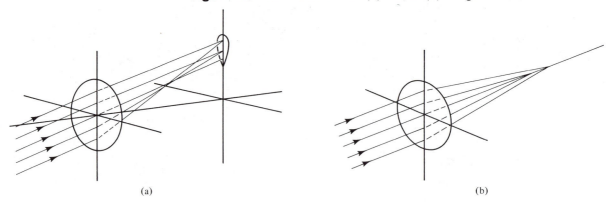

(a)   (b)

surfaces are close to the required shapes, but differ slightly from the shapes that will produce perfect images.

Another aberration, called *coma,* is shown in Fig. 4.3(a). The focal length depends on the angle φ that the incoming rays make with the axis. This results in an a conelike appearance to the images. Another type of aberration is called *astigmatism* [Fig. 4.3(b)]. It occurs when light striking the lens a certain distance above or below the center is focused a different distance from the lens than light striking to the left or right of center.

One aberration that occurs in lenses but not mirrors is called *chromatic aberration.* This is because a material's index of refraction changes with wavelength. The focal length of the lens is therefore different for different wavelengths. The images are formed in different places for different wavelengths (Fig. 4.4). We can correct, somewhat, for chromatic aberration with a two-lens system, called an *achromat.* The two lenses are made of different materials, with different indices of refraction, and different variations in the index of refraction with wavelength. An achromat only brings the images at two different wavelengths together, but the images for intermediate wavelengths are not far off.

Now that we have seen some of the basics of optical systems, we can look at astronomical telescopes. Most current astronomical research is done on reflecting telescopes. However, the basic ideas of image formation in reflecting and refracting telescopes are the same—and easier to visualize for refracting telescopes—so we consider them first.

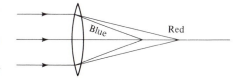

Figure 4.4 Chromatic aberration.

4.2 REFRACTING TELESCOPES

In a refracting telescope, the light first passes through a large lens, called the *objective lens.* The objective is the part that intercepts the incoming light, so it determines the light-gathering power of the telescope. The larger the objective is, the greater is the light-gathering power. The light passing through the objective is concentrated on a second lens, called the *eyepiece.* The eyepiece is used to inspect the image formed by the objective. The image formed by the eyepiece is viewed either by the eye or a camera. In practice, either the objective or the eyepiece may be a multiple lens, to correct for aberrations, but we still treat each as a single optical element.

The basic arrangement of the refracting telescope is shown in Fig. 4.5. We follow the formation of the images of two stars, just as we did with the camera. Let's assume that the focal length of the objective if f_{obj}. Since the stars are at infinity, the objective forms their images this distance behind the objective. The eyepiece has a focal length f_{eye}. We place the eyepiece a distance f_{eye} behind the image formed by the objective. (This means that the objective and eyepiece are separated by a distance equal to the sum of their focal lengths.) Since the initial images of the stars are f_{eye} from the eyepiece,

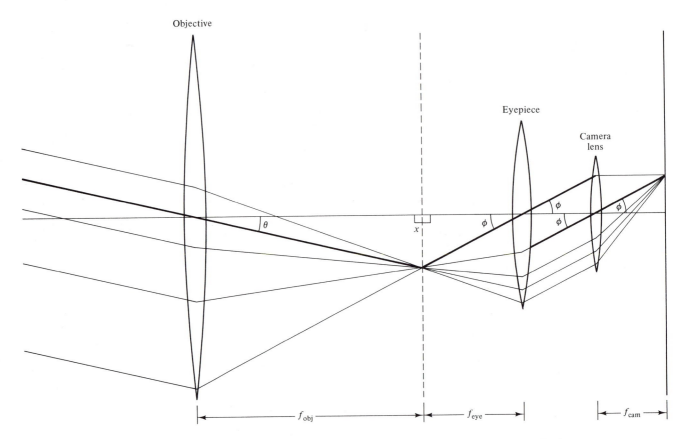

Figure 4.5 Optics of a refracting telescope. Light from a star enters, making an angle θ with the axis, and leaves the eyepiece, making a larger angle φ with the axis. The focal lengths of the objective, eyepiece, and camera lens are indicated.

the eyepiece will focus the light at infinity. This means that all of the rays in a given bundle emerge from the eyepiece parallel to one another.

If you now look through the eyepiece and focus your eye at infinity by relaxing the muscles around your eye, the rays within each bundle will be brought back together on your retina. Similarly, if you use a camera, you focus the camera at infinity, and the images of the stars fall on the film. The need to focus your eye at infinity means that the best way to look through the eyepiece is to relax both eyes and cover the unused eye rather than squinting to close the unused eye.

Let's go back to the two bundles of rays emerging from the eyepiece. Even though the rays within a given bundle are parallel to one another, the bundles make some angle with each other. If the two stars are an angle θ apart on the sky, then the two bundles will enter the objective, making this angle with each other. The bundles leave the eyepiece, making a larger angle φ with each other. We can find the angle φ by following the chief ray through the eyepiece. Note that the chief ray at the eyepiece is not the same ray that was the chief ray at the objective. However, all rays in a given bundle will emerge from the eyepiece parallel to the new chief ray.

From the two right triangles in the diagram, with a common side x, we see that

$$\tan \theta = x/f_{\text{obj}}$$

$$\tan \varphi = x/f_{\text{eye}}$$

If the angles are small, we can replace the tangent of the angle with the value of the angle in radians. If we also eliminate x in the equations, we find

$$\varphi/\theta = f_{\text{obj}}/f_{\text{eye}} \tag{4.2}$$

This means that we have an *angular magnification* equal to the ratio of the focal lengths of the two optical elements.

In general, when we want to do work with good detail in the image, we use a telescope with a long focal length objective. Of course, we can change the angular magnification of a given telescope by changing the eyepiece. There is a practical limit. You don't want to magnify the image so much that you blow up the blurring caused by atmospheric seeing.

There are some limitations in the use of a refracting telescope. One problem is the chromatic aberration in the objective. Also, the objective must be made from a piece of glass that is perfect throughout its volume, since the light must pass through it. This gets harder as you try to make larger objectives. Larger objects are also harder to support. The objective can only be supported at the edges, since light must pass through. Also, in many modern applications, we want to place instruments near the eyepiece. However, the telescope must be supported closer to the center of mass, which means far from the eyepiece. Any instrument hung at the eyepiece will exert a large torque about the mount, limiting the weight of that instrument. As a practical matter, the largest refractors such as that shown in Fig. 4.6, have objectives with diameters of, at most, 1 m.

Figure 4.6 One-meter refracting telescope at the Lick Observatory. Note the long distance over which the observer must move to keep up with the eyepiece.

4.3 REFLECTING TELESCOPES

Many of the difficulties with refracting telescopes are avoided with *reflecting telescopes*. In reflectors, the objective lens is replaced by an objective mirror. With a mirror, there is no problem of chromatic aberration, since light of all wavelengths is reflected at the same angle. Since the light doesn't pass through

(a)

(b)

Figure 4.7   Two large reflectors. (a) The 5-m-diam Hale telescope on Mt. Palomar. For over three decades, it was the largest telescope in the world. The caged part is the telescope. It has an equatorial mount. The solid piece in the foreground is part of the fork-shaped support for the telescope. To track an object as the Earth rotates, the whole fork rotates. (b) The 4-m-diam telescope at the Cerro Tololo Interamerican Observatory (along with Kitt Peak, part of our National Optical Astronomy Observatories). This telescope is identical to the 4-m-diam Mayall telescope on Kitt Peak. It is mounted in azimuth and elevation, so, it must move about two axes to track a given star. The Cassegrain focus is in a cage below the telescope. The observer doesn't stay in the cage for observing; that is done from a control room, where a television system is used to keep track of where the telescope is pointing.

the glass, the requirements are for a good surface, not a good volume, and it can be supported from behind. It is therefore possible to make reflectors larger than refractors. The largest reflector in the United States is the 5-m-diam Hale telescope on Palomar Mountain [Fig. 4.7(a)]. A 6-m-diam telescope is in operation in the Soviet Union. In addition, a 7-m-diam telescope is being planned by the University of Texas, and a 10-m-diam telescope is being built by the University of California and the California Institute of Technology.

To have good image formation, the surface of the mirror must be perfect to within approximately $\lambda/20$, where λ is the wavelength of the light being observed. For example, if you are observing light with a wavelength of 500 nm, the surface must be accurate to within 25 nm. The surface is obtained by shaping a large piece of glass, and then polishing it to the desired shape. A reflecting surface is made by depositing a thin aluminum coating on the mirror. This coating is restored every few years, and many large telescopes are equipped with aluminizing chambers for the purpose (so the mirror doesn't have to be moved very far for recoating). For most reflecting telescopes the shape of the mirror is a paraboloid. Any cross section of the mirror will be a parabola. A parabola focuses to a single point all rays coming in parallel to the axis. There are some types of telescopes (mentioned below) for which a different-shaped mirror is used.

We now look at what happens to the image formed by the objective. Replacing a lens with a mirror doesn't change any of the basic ideas of image formation. There is, however, a problem caused by the reflection of the light back along the direction from which it came. To examine the image, the eyepiece (and observer) must be placed between the stars and the mirror, blocking some of the incoming light, as shown in Fig. 4.8. If an eyepiece is put in this location, we call the focal arrangement a *prime focus*. The advantage of a prime focus is that no more mirrors are required, so light is not lost in additional reflections. However, there is some blockage of the objective. If the telescope is very large, this blockage is a small fraction of the total collecting area of the objective.

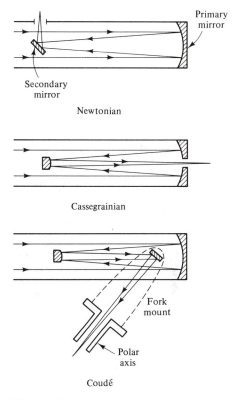

Figure 4.8 Focal arrangements in a reflecting telescope. In each case, light enters the telescope from the left.

EXAMPLE 4.3 Blockage in Prime Focus

Consider a 5.0-m-diam telescope, with a 1.0-m prime focus cage. What fraction of the incoming light is blocked by the cage?

Solution

The ratio of the areas will be the square of the ratio of the diameters. The fraction of the mirror blocked by the cage is therefore

$$\text{Fraction blocked} = (1.0 \text{ m}/5.0 \text{ m})^2$$

$$= 0.04$$

This means that only 4% of the incoming light is blocked. If we make the telescope smaller, but keep the cage the same size, the blockage gets worse. Clearly, prime focus arrangements are only suitable in large telescopes.

(a)

(b)

Figure 4.9 (a) Observer in the prime focus cage of the 5-m-diam Hale telescope (shown in Fig. 4.7a). Note that a very small fraction of the total area is blocked. (b) Observer at the Cassegrain focus of a 2.1-m-diam telescope located on Kitt Peak. The observer stays on this platform during observing, and the platform moves up and down to keep up with the eyepiece.

This problem was recognized by Newton, who devised an arrangement, called the *Newtonian focus,* in which a flat diagonal mirror is used to direct the image of the objective to the side. (This is also shown in Fig. 4.8.) The eyepiece is then mounted in the side of the tube. There is still some blockage, but it can be kept small even for small telescopes. For a larger telescope, the Newtonian arrangement is difficult to use, since the eyepiece is at the top end of the telescope. Also, the eyepiece is far from the support point of the mount, and equipment placed at the focus exerts a large torque about the support.

An alternative solution is the *Cassegrain focus,* also shown in Fig. 4.8. The prime focus cage is replaced with a mirror that directs the rays back down through a hole in the center of the primary mirror. Little light is lost by removing the center of the mirror, since it would be blocked by the prime focus cage or secondary mirror. The secondary mirror in a Cassegrain arrangement is a diverging mirror (convex), so the telescope seems to have a longer focal length than that of the objective. Since the eyepiece is just behind the primary mirror, it is a very convenient arrangement. Also, if you want to place a lot of equipment at the eyepiece position, this is not too far from the support point of the telescope. Figure 4.9 shows an observer at the prime focus and one at the Cassegrain focus.

Sometimes an astronomer will want to use equipment that cannot conveniently be mounted on a telescope. It might be too large, or it may require a room in which the temperature is kept constant. For this purpose, some large telescopes have *coudé* focal arrangements (the term comes from the French word for elbow, since the light beam is bent many times). A series of mirrors is used to direct the image into a laboratory under the telescope mount. One disadvantage of this arrangement is the large number of mirrors that must be used. No mirror is perfectly reflective, and a little light is lost at each reflection.

A general problem with any of these arrangements is that they all involve some blockage of the objective. In addition to reducing the light striking the objective, the blocking element also must be supported. Starlight passing by the supports is diffracted, creating unusual stellar images (Fig. 4.10).

Some telescopes follow the basic layout of a Cassegrain system, but have some differences in their optics. Often the goal is to provide a field of view that is relatively free of aberrations. For example, a *Schmidt camera* incorporates a glass plate shaped to provide correction for some abberations, giving a system that can take wide-field photographs. Many newer telescopes are of the *Ritchey–Cretien* design, which incorporates ellipsoidal mirrors.

Even though reflectors can be made significantly larger than refractors, it seems that a practical limit may have been reached. Until recently, it was felt that the 6-m-diam reflector would be the largest for some time. However, changes in mirror fabrication technology have allowed the planning of a mirror up to 10 m in diameter, though this will not start with a single piece of glass. A technique has been developed at the University of Arizona to

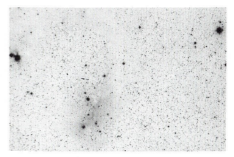

Figure 4.10 Stars act as true point sources, and their images have a diffraction pattern resulting from the supports for the secondary mirror. The pattern is evident as a cross on the brightest stars.

Figure 4.11 (a) The Multiple Mirror Telescope (MMT), operated in Arizona by the University of Arizona and the Smithsonian Astrophysical Observatory. (b) Design for a new national telescope, the National New Technology Telescope.

(a)

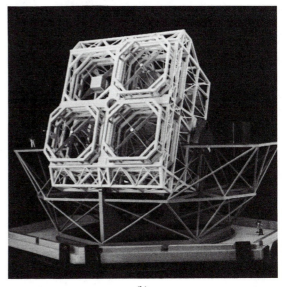

(b)

"routinely" make 8-m-diam mirrors. It involves heating glass and then spinning the glass while it cools. (The surface of a spinning liquid takes on the shape of a paraboloid.) For these mirrors, it is not necessary to use glasses with special thermal properties. This mirror is cast so that much of the glass on the backside of the mirror is missing, leaving a honeycomb pattern. When the mirror is in use, this allows the mirror to quickly reach the temperature of the outside air, reducing air currents within the telescope, a major contribution to seeing.

Astronomers still need larger telescopes. A new approach is illustrated by the Multiple Mirror Telescope (MMT) (Fig. 4.11). Instead of one large mirror, the telescope has six moderate-sized mirrors. The images from all six mirrors are brought together to produce an image that is six times as bright as the image from one mirror. Planning is now underway for a next-generation optical telescope which would have four mirrors providing the collecting area equal to that of a single mirror 15 m in diameter.

4.4   DATA HANDLING

In the previous sections we concentrated on getting as many photons to the eyepiece as possible. Now we will look at what we do with those photons once they reach the eyepiece. We will consider three different types of observations:

1. *Imaging*. This is probably the most familiar type of observation. The goal of the observations is to obtain a picture of some section of the sky.

2. *Photometry*. The name implies a measurement of light. The goal of the observations is to measure the brightness of some object. This may also include measuring brightnesses through certain filters and measuring colors.

3. *Spectroscopy*. The goal of these observations is to obtain a spectrum of some object, generally with sufficient detail to allow the study of spectral lines.

4.4.1   Detection

Whatever the type of observations, the data must be recorded in some way. In the past the most common way was to use a photographic plate. One major advantage of photography is that, in the photographic emulsion, each grain serves as a little detector of radiation. We say that the plate has a *panoramic* quality. This means that we can simultaneously record many parts of the image. There are some disadvantages to photographs. One is that a very small fraction of the photons that strike the plate are actually detected. We call the fraction of photons actually detected the *quantum efficiency* of the

(a)

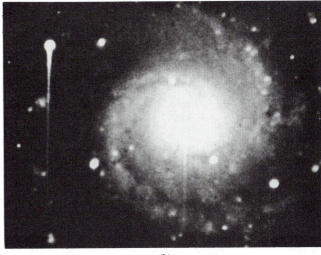

(b)

Figure 4.12 (a) A traditional photograph, and (b) a CCD image of the same galaxy, M74.

detector. For most emulsions, this efficiency is only a few percent. Another problem is *reciprocity failure*. For long exposures the film essentially "forgets" some of the photons that it has already detected. This leads to longer than expected exposures when looking at very faint objects. After many years of very slow improvements, some progress has been made in the quality of photographic plates. In some cases plates are made more sensitive by baking them in an oven filled with nitrogen.

A much higher efficiency can be obtained with *photoelectric* devices. A photon strikes a surface, causing an electron to be ejected. The electron strikes another surface, causing additional electrons to be ejected. Eventually, a sufficient number of electrons are moving for a current to be detected. In the past, such *photomultiplier* devices allowed only a single detection element in the focal plane. Efficiency was gained, but the panoramic quality was lost. A new type of detector, called a *charge-coupled device,* or *CCD,* provides a grid of detectors, all with high quantum efficiency (greater than 50%). Each element of the grid keeps an electronic record of the intensity of light striking its position. The image is later reconstructed using a computer (Fig. 4.12). Having the image in computer-readable form is actually very convenient, because many new techniques are being used to computer-enhance very faint images. This provides a large *dynamic range,* meaning that we can see faint objects in the presence of bright ones.

When photometric observations are being made one generally compares the brightness of the star under study with the brightnesses of stars whose properties have already been studied. By changing filters one can measure, for example, the U, B, and V magnitudes of a star, one after another. Some method of recording the data is still needed. One option is photographic. The brighter a star is, the larger its image will be on a photographic plate. (This is an artifact of the photographic process and atmospheric seeing.) We can measure the brightness of a star by measuring the size of its image. Photoelectric devices are well suited for photometry, since we only need to look at one star at a time. Almost all photometry is now done photoelectrically. Some of the standard colors even account for the wavelength responses of various commercially available photomultipliers.

4.4.2 Spectroscopy

In spectroscopy we need a means of bringing the image in different wavelengths to different physical locations in the focal plane. We have already seen that this can be done with a prism. Since a prism does not spread the light out very much, we say that the prism is a *low-dispersion* instrument. Dispersion is a measure of the degree to which a spectrum is spread out. Low-dispersion spectra are sometimes adequate for determining the spectral type of a star. Sometimes a thin prism is placed over the objective of a telescope, and a photograph is taken of a whole field. Instead of seeing the individual stars, the spectrum of each star appears in its place. These *objective prism spectra* (Fig. 4.13) are quite useful for classifying large numbers of stars very quickly.

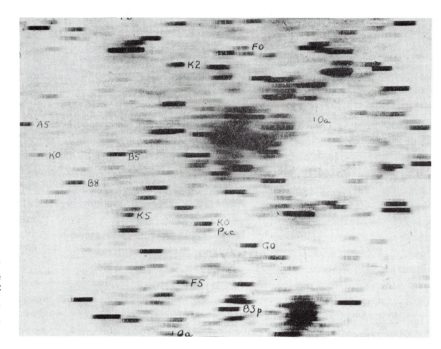

Figure 4.13 Objective prism spectra. A prism is placed over the objective lens of a refractor, turning the image of each star into a low-resolution spectrum. In each case, a few of the strongest spectral lines are visible.

When better spectral resolution is needed, we generally use a *diffraction grating*. For any wavelength λ, the grating produces a maximum at an angle θ (Fig. 4.14) given by

$$d \sin \theta = m\lambda$$

where *d* is the separation between slits or lines, and *m* is an integer, called the *order* of a maximum. The higher the order is, the more spread out is the spectrum. Suppose our grating just lets us separate (resolve) two spectral lines that are Δλ apart in wavelength. The *resolving power* of the grating is then defined as

$$R = \lambda/\Delta\lambda$$

If the grating has a total of *N* lines in it, then in order *m* the resolving power is given by

$$R = Nm$$

Some gratings have 10,000 lines per centimeter over a length of several centimeters. This means that resolving powers of 10^5 can be achieved. In general, light will go into several orders. It is possible to cut the lines of a grating so that most of the light goes into one particular order. This process is called *blazing*.

Once the spectrum is spread out, it must still be recorded. Again, the photographic plate is very useful. Once a plate is developed it can be studied directly. However, if we want a detailed graph of intensity vs. wavelength in a spectral line, a more sophisticated technique is needed. A device called a *densitometer* is used to examine the spectra on the plate (Fig. 4.15). A fine beam of light is scanned across the plate, and a photomultiplier is used to measure the fraction of the beam that passes through the plate. Less light gets through where the plate is blacker. This allows us to measure how exposed each part of the plate is.

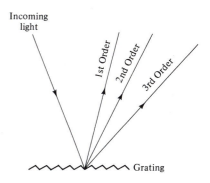

Figure 4.14 Diffraction grating. Light comes in from the upper left. The beam reflected off each step spreads out due to diffraction. However, interference effects result in maxima in the indicated directions. The angles of the steps can be adjusted (blazed) to throw most of the light into a desired order.

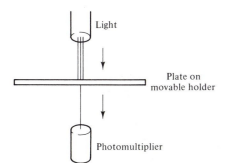

Figure 4.15 Densitometer. Light from above passes through a photographic plate (with an image already developed), and is detected by the photomultiplier. Where the plate is more exposed (dense), less light passes through.

We have only covered some of the most basic techniques for measuring spectra. Other types of devices split the light from a star into two beams and then allow the two beams to interfere. The actual spectrum can be recovered by a mathematical operation called a Fourier transform. These devices provide astronomers with a great deal of flexibility and sensitivity.

4.5 OBSERVATORIES

4.5.1 Ground-Based Observing

In the past, the convenient location of observatories was considered important. Observatories were built near universities that had astronomers, and the astronomers used whatever clear nights were available. Today, the considerable investment in large telescopes and sophisticated equipment requires a more regular utilization of the facilities. Observatories are now built only after studies to determine the quality of the site. There has also been a trend away from privately financed observatories to publicly financed observatories. (The 10-m-diam telescope is a recent exception to that trend.) Public telescopes are available to any qualified astronomer. An astronomer who has a project writes a proposal, explaining the scientific justification and the details of the observations. Generally, more time is requested than is available, and a panel of scientists decides which projects are to be done.

The selection of an observatory site depends on a number of considerations. Obviously, good weather is important. However, clear weather isn't enough. The air should be dry, since water vapor can attenuate signals, suggesting a desert or mountaintop. A mountaintop in the desert is even better. An altitude of 2 km (6600 ft) gets you above a significant amount of atmospheric water vapor. Even with a mountain in a desert, good seeing is not guaranteed. Seeing often varies with local conditions. Before an optical telescope is built, a seeing test is done, with test observations being done over the course of at least a year. An additional consideration is light pollution. Light from nearby cities is reflected up into the sky, making the sky appear to glow. The farther the site is from such lights, the better. Astronomers have found that certain types of lights are better than others. For example, low-pressure sodium vapor lights, which have a yellow appearance, give off most of their light in a narrow wavelength range, and this range can be filtered out at the telescope.

Once a good site is found, it is likely that many telescopes will be built there. A good example is Kitt Peak National Observatory (Fig. 4.16), southwest of Tucson, Arizona. This observatory has a number of different-sized telescopes, the largest being a 4-m-diam telescope. There are even telescopes on Kitt Peak operated by individual universities, not directly affiliated with the national observatory.

Figure 4.16 Kitt Peak National Observatory, southwest of Tucson, Arizona. Notice the large number of telescopes. The 4-m telescope is in the foreground.

Surprisingly, one of the best observatory sites is in the middle of the Pacific Ocean. It is on the island of Hawaii, at an elevation of 14,000 ft on an extinct volcano, Mauna Kea (Fig. 4.17). The island often has clouds, but they are generally below the altitude of the observatory, and the air above the clouds is dry. However, the lack of oxygen at this altitude makes work

Figure 4.17 Mauna Kea, on the island of Hawaii. At 14,000 feet, its summit is one of the best ground-based astronomical sites.

difficult. Many astronomers report headaches and other discomforts. Clear thinking is also difficult, and there are stories of experienced observers who left the mirror cover on, and couldn't understand why they couldn't see a bright star. Major efforts at higher sites are unlikely.

4.5.2 Observations from Space

One of the major advances in observational astronomy has been the ability to place telescopes in space. This is particularly important for observations in parts of the spectrum that don't penetrate the Earth's atmosphere. However, a telescope in space can be important even in the visible part of the spectrum. It allows us to make observations free of the blurring caused by atmospheric seeing conditions.

EXAMPLE 4.4   Diffraction-Limited Optical Telescope

What is the resolution of a 1-m-diam telescope in space for observations at a wavelength of 550 nm?

Solution

We find the diffraction limit from equation (4.1).

$$\Delta\theta = \frac{(2.06 \times 10^5)(5.5 \times 10^{-7} \text{ m})}{1 \text{ m}}$$

$$= 1.1 \times 10^{-1} \text{ arc sec}$$

A 1-m-diam telescope on the ground will never realize this resolution because of the 1″ seeing limitation. However, by putting the telescope in space, we achieve a factor of 10 improvement in the resolution over the *best* ground-based observations.

For this reason, astronomers are looking forward to the prospect of a major optical observatory in space—*Space Telescope*. Scheduled for launch in 1987, it will provide a large (2.4-m-diam) telescope continuously available in space. With this aperture, and no atmospheric seeing, the angular resolution will be 0.05″. Astronomers will be able to point the telescope with an accuracy of 0.01″. Space Telescope will be equipped with a full complement of instruments for carrying out observations involving imaging, photometry, and spectroscopy. The telescope will be an international facility, available to the general astronomical community, with time awarded on the basis of proposals. Observations will be controlled from the ground. Data will come,

Figure 4.18   Hubble Space Telescope.

via a satellite network, to the Goddard Spaceflight Center, in Greenbelt, MD, and, finally, to the Space Telescope Science Institute (STSCI), in Baltimore. It is at the STSCI where the observer will be able to view and process the data.

4.6   OBSERVING IN THE ULTRAVIOLET

The visible part of the spectrum only gives us access to a small fraction of the total radiation given off by astronomical objects. For centuries, however, this was the only information available to astronomers. We will see throughout this book that observations in other parts of the spectrum have revealed entirely new types of objects or have provided us with information crucial to understanding objects that are already observed in the visible. In this section, we discuss ultraviolet observations, because the techniques are very similar to those in optical observations.

In many ways, we can think of ultraviolet observations as being short-wavelength visible observations. The basic imaging ideas are the same (Fig. 4.19). Of course, since the wavelength is shorter, the mirror surfaces must

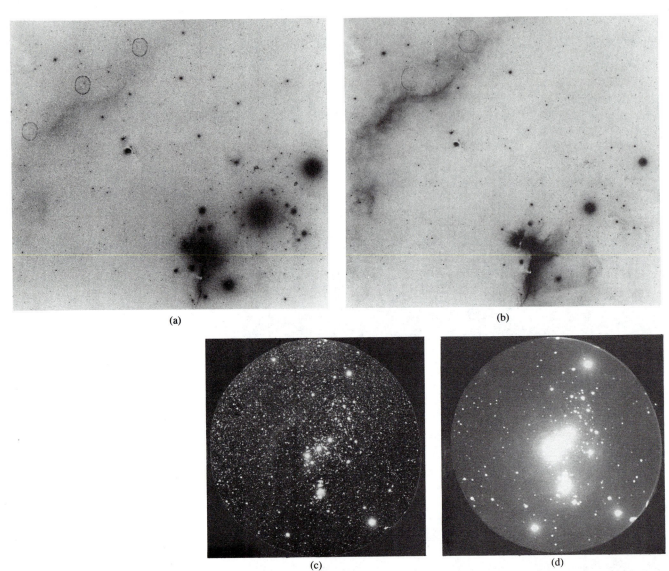

(a)

(b)

(c)

(d)

Figure 4.19 (a) and (b). A pair of negative photographs, showing a 7° × 7° part of Orion. The first is taken with blue-sensitive film, and the second is with red-sensitive film. The hot stars at the lower right show up more strongly in the blue. The glowing gas, part of Barnard's loop, at the upper left shows up more strongly in the red. (The ovals at the upper left are ghost images of the three bright stars.) (c) and (d). A visible and an ultraviolet photograph of Orion, showing an even larger region. The bright stars of the belt show up even more strongly in the violet. However, the red star, Betelgeuse, at the upper left, is hardly visible in the ultraviolet, where its position is marked by a circle. The ultraviolet image is in the wavelength range 1230–2000 Å, and was made from an Aerobie rocket.

be more accurate in the ultraviolet than in the visible. The normal coatings that we use to make mirrors reflective in the visible do not work as well in the ultraviolet, and different coatings are needed. For detection, we can use either photographic plates or photomultipliers.

The major problem is that ultraviolet radiation does not penetrate the Earth's atmosphere. If you don't go too far into the ultraviolet, some observations are possible at high altitudes. However, we have become increasingly dependent on ultraviolet satellites. A long look at the ultraviolet sky was provided by the Copernicus satellite. More recently, the International Ultraviolet Explorer (IUE) satellite has served as an ultraviolet facility available to the astronomical community.

4.7 OBSERVING IN THE INFRARED

In this section we briefly look at some of the techniques for observing in the infrared part of the spectrum. For some purposes we can simply think of infrared radiation as being long-wavelength visible radiation. In fact, much infrared astronomy is done on normal optical telescopes. The long wavelength means that surface accuracy is not a problem. A surface accurate enough for optical observations is certainly accurate enough for infrared observations. However, the longer wavelength makes diffraction a problem. For example, for a 1-m-diam telescope working at a wavelength of 10 μm, the diffraction limit is 2″, slightly worse than the seeing limit at a good site.

One problem with infrared observations not common with optical observations is radiation from the telescope itself. Parts of the telescope that are not perfectly reflecting radiate like blackbodies at temperatures close to 300 K, with a peak at 10 μm. This is not a problem for optical detectors, but is a problem for infrared detectors. (See Problem 4.24.) In an infrared telescope, the radiation paths must be carefully designed so the detector cannot "see" any hot surfaces. Some reduction in the problem can be obtained by cooling surfaces that can radiate into the detector.

Detectors used in the infrared are generally different from those in the visible. Infrared photons have less energy than do visible photons. Therefore, infrared photons are not able to expose a normal photographic emulsion. (Recently, infrared-sensitive emulsions have been developed.) Neither can they cause electrons to be ejected from a metal surface.

Originally, the most common type of infrared detector was called a *bolometer*. A bolometer is a device that heats up when radiation falls on it. We generally use a material whose electrical properties change with temperature. For example, if the resistance of the bolometer changes with temperature we can measure its temperature by measuring its resistance. By measuring the temperature increase, we can determine the total amount of energy striking the bolometer.

A bolometer does not have the panoramic quality of a photographic emulsion. We can think of the bolometer as being like one grain in the photographic emulsion. We can only get a small piece of the picture. When we point a telescope in some direction, we measure the average infrared power coming from an area of the sky determined by the resolution of the telescope. If we want to measure the radiation coming from a different position, we must point the telescope at that position and make a new observation. It can take a long time to build up a picture. Recent developments have allowed infrared astronomers to place a few bolometers close together, so that we can now see a few pieces of the picture at a time.

Spectroscopy in the infrared is different than in the visible. Prisms are not very useful. It is only recently that gratings have been employed. A crude form of spectroscopy is done by using a series of filters, each passing a

Figure 4.20 Infrared windows. In each frame, transmission as a function of wavelength is shown for an observatory at a given altitude. At low altitudes, good transmission (close to 1) is in only a few narrow wavelength regions, or windows. As one goes to higher observatory altitudes, the windows cover a wider range of wavelengths.

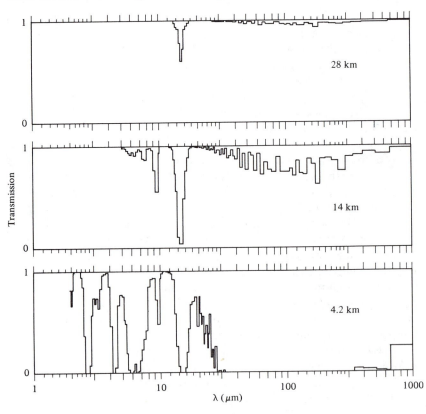

different wavelength range. We can keep changing filters while we are observing an object, and build up the spectrum one wavelength range at a time. A major recent improvement has involved devices that produce a Fourier transform of the spectrum but allow study of a full spectrum at once.

The major problem in the infrared is the Earth's atmosphere. The atmosphere is totally opaque at some infrared wavelengths, and is at best only partially transparent at all other infrared wavelengths. The opacity of the atmosphere causes two problems: (1) The atmosphere blocks infrared radiation from the sources we are studying. (2) The atmosphere emits its own infrared radiation, which can be much stronger than that received from the astronomical objects. We call the wavelengths at which some observations are possible *infrared windows* in the atmosphere. Even at these windows, observations at sea level are virtually impossible. At the very least, the 2-km altitude of many optical observatories is required. Figure 4.20 shows the major infrared windows and the altitude to which one must go to observe in them.

In general, 2 km is only sufficient for working in the near infrared, at wavelengths of a few micrometers. If we want to work farther into the infrared, higher altitudes are necessary. Some observatories are placed at higher altitudes, the highest being at about 14,000 ft (higher elevations resulting in intolerable working conditions). For example, NASA operates an Infrared Telescope Facility (IRTF) on Mauna Kea in Hawaii.

For many studies, even higher altitudes are needed. For the past ten years NASA has operated the Kuiper Airborne Observatory (KAO), shown in Fig. 4.21, a converted military transport plane. The plane carries a 0.9-m-diam infrared telescope to an altitude of 45,000 ft, for 7-hr observing sessions. The telescope is operated as a national facility, with qualified as-

Figure 4.21 The Kuiper Airborne Observatory, a 0.9-m-diam telescope mounted in a C141 aircraft. A door is opened for observing. The azimuth control of the telescope is the direction of the airplane, so flight planning must be very detailed to make sure that the telescope is pointing at the desired sources and that the plane ends up at the right base. The aircraft normally operates at 41,000 ft, but can go up 45,000 ft for part of a flight. It is operated by NASA, and is based at the NASA Ames Research Center, at Moffet Field, on the south end of San Francisco Bay.

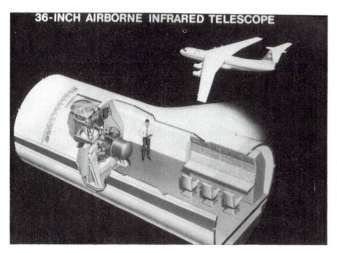

(a)

(b)

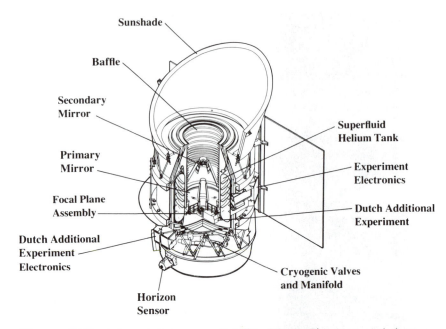

Sunshade

Baffle

Secondary
Mirror

Primary
Mirror

Focal Plane
Assembly

Dutch Additional
Experiment
Electronics

Horizon
Sensor

Superfluid
Helium Tank

Experiment
Electronics

Dutch Additional
Experiment

Cryogenic Valves
and Manifold

Figure 4.22   Infrared Astronomy Satellite (IRAS). This is an artist's conception of what the satellite looked like in orbit as it carried out its survey of the sky in the infrared.

tronomers having an opportunity to propose observations. Approximately 80 flights a year are made. Higher-altitude flights of a shorter duration are made from a smaller jet. A larger airborne observatory, with a 3-m-diam telescope placed in a Boeing 747, is being considered. For even higher altitudes, such as up to 100,000 ft, balloons are still used. (Balloons used to be the mainstay of far-infrared observers.)

Even for high-altitude observations, the atmosphere still limits what can be done. The best solution is observations from space. Infrared astronomers have recently obtained a wealth of data from the Infrared Astronomy Satellite (IRAS), a joint American–Dutch–British project (Fig. 4.22). IRAS was launched in January 1983. The 0.6-m-diam cooled telescope was primarily designed for imaging observations. It contained arrays of detectors operating in four wavelength ranges, centered roughly at 12, 25, 60, and 100 μm. From its high (560 mi) polar orbit, much of the IRAS's lifetime was devoted to a systematic survey of the sky. A certain fraction of the time was given to observations of specific objects. The large-scale survey revealed over 100,000 point sources and a network of extended infrared emission. The whole set of data is available as a resource to the general astronomical community. An astronomer interested in a particular object will be able to check the IRAS data files at the Jet Propulsion Laboratory or take data tapes back to a home institution to find out about infrared emission from that object. Some IRAS images are shown in Fig. 4.23.

Figure 4.23 IRAS image of a section of the Milky Way galaxy. Brighter areas are stronger in infrared emission.

4.8 RADIO ASTRONOMY

Radio observations provide us with very different information from optical observations and use very different techniques. The long wavelength means that the wave nature of the radiation is very apparent in the observations. The long wavelength also corresponds to low-energy photons. This means that radio observations can tell us about cool regions. For example, we will see how radio observations tell us about star formation (Part III). We will also see that there are high-energy sources that give off most of their energy at long wavelengths. Thus, radio observations also give us a way of studying high-energy phenomena.

Radio astronomy owes its origins to an accidental discovery by Karl Jansky, an engineer at the Bell Telephone Laboratories in New Jersey. In 1931, Jansky detected a mysterious source of radio interference. He noticed that this interference reached its strongest peak four minutes earlier every day. This timing suggests an object that is fixed with respect to the stars. The time of maximum interference coincided with the galactic center crossing

(a)

(b)

Figure 4.24   (a) The 43-m-diam telescope at the National Radio Astronomy Observatory, Green Bank, West Virginia. It has a surface accurate enough to allow observations at wavelengths as short as 1 cm. It is fully steerable and has an equatorial mount. The observer sits in the control room in the large building under the telescope. The telescope is controlled from there, and the astronomer can also monitor the data as it comes in. (b) A 45-m telescope in Nobeyama, Japan, with a surface accurate enough to permit observations at wavelengths of a few mm.

the local meridian. Jansky concluded that he was receiving radio waves from the galactic center. It was realized that astronomical objects can be strong sources of radio emission.

This discovery was not followed up immediately. In fact, for a long time there was really only one active radio astronomer. Grote Reber was an amateur radio astronomer in Illinois who carried out radio observations on his backyard radio telescope in the late 1930s and early 1940s. Following World War II, radio astronomy benefited from the development of radar equipment during the war. Radio observations were pursued by the British, Dutch, Australians, and a small group of Americans at Harvard. A major advancement was the ability to observe spectral lines in the radio part of the spectrum. We will discuss these lines in Chapter 16.

By the mid-1950s it was clear that a major radio observatory in this country had to be a cooperative effort, and the National Radio Astronomy Observatory (NRAO) was founded. A major role was played by Bart Bok, an optical astronomer who recognized the importance of the contributions that radio observations could make. The first telescopes of the NRAO were in Green Bank, West Virginia, far away from sources of man-made interference. Some radio telescopes are shown in Figs. 4.24 and 4.25.

We now take a look at some of the equipment. A radio telescope consists of some element that collects the radiation and a receiver to detect the radiation. Most modern radio telescopes have a large antenna, or dish, to collect the radiation and send it to a focal point. The long wavelength becomes important in this process. We have already seen that the resolution of a telescope depends on the size of the telescope relative to the wavelength (Fig. 4.26). (In the radio part of the spectrum, atmospheric seeing is not a problem.) To get good resolution we need a large surface. However, that surface doesn't have to be perfect. It can have imperfections, as long as they are smaller than approximately $\lambda/20$. For example, for a wavelength of 20 cm, 1-cm-diam holes have no effect on the performance of the telescope. The best resolution for single radio telescopes is about 30″, slightly better than the naked eye for visible viewing.

Figure 4.25 The 100-m-diam telescope of the Max Planck Institute für Radioastronomie, Effelsburg, Germany, is the largest fully steerable radio telescope. Its mount operates in azimuth and elevation. Azimuth is controlled by moving the whole structure around the circular track. As the telescope changes its elevation angle, it deforms under gravity. However, it is designed to deform from one paraboloid to another, so only its focal length changes.

Figure 4.26 Resolution for a radio telescope. On the left, the short dashed lines show what would happen if there were no diffraction. Only radiation traveling parallel to the telescope axis would reach the focus. The longer dashed lines show the effects of diffraction. Radiation coming in at a slight angle with the telescope axis can still get reflected to the focus. This means that when the telescope is pointed in one direction, it is sensitive to radiation from neighboring directions. This is shown on the right, as the telescope is sensitive to radiation coming from within a cone of angle approximately λ/D (in radians).

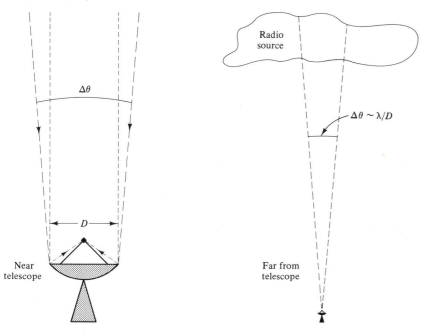

The actual detection of the radio waves takes place in a radio receiver. In general, only one receiver operates on a telescope at a given time. This is equivalent to doing optical observations with only one grain in your photographic emulsion. At any given time, the telescope is receiving radiation from a piece of sky determined by the diffraction pattern of the antenna. If we want to build up a radio image of some part of the sky, we must point the telescope at each position and take a separate observation. The receivers in radio astronomy are similar in concept to home radios. However, signals from astronomical sources are very weak by the time they reach us, so receivers for radio astronomy must be very sensitive. Sometimes they are cooled to a few degrees above absolute zero to minimize sources of background instrumental noise. Unlike bolometers, radio receivers do not simply detect all of the energy that hits them; they are also capable of preserving the spectral information.

EXAMPLE 4.5   Strength of Radio Sources

We measure the strength of radio sources in a unit called a Jansky (Jy). It is defined as 10^{-26} W/m^2/Hz reaching our telescope. For a 1-Jy source, calculate the power received by a perfect antenna with an area of 10^2 m^2, using a frequency range, or bandwidth, of 10^6 Hz.

Solution

The total power received is the power/area/Hz, multiplied by the frequency range (in Hz), and the surface area of the telescope.

$$P = (10^{-26}\ \text{W/m}^2\text{/Hz})(10^6\ \text{Hz})(10^2\ \text{m}^2)$$

$$= 10^{-18}\ \text{W}$$

This is 10^{-20} times the power used by a 100-W light bulb.

Just as with optical observations, in radio astronomy we can make continuum and spectral line observations. Continuum studies are like optical photometry. We tune our receivers to receive radiation over a wide range of frequencies, and then we measure the total amount of power received. From this information, we get the general shape of the continuous spectrum (intensity vs. frequency). In spectral line observations the radiation is detected in small frequency intervals, so the shapes of spectral lines can be determined. The spectrometers for radio observations have traditionally been electronic filters, tuned to pass various frequencies. More recently, computer-based digital systems have been used.

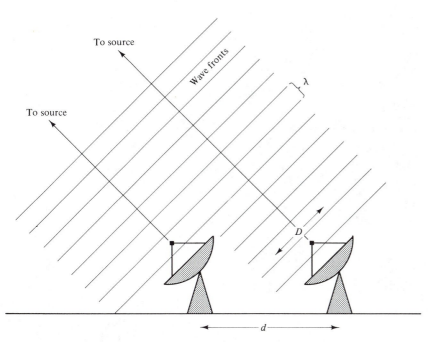

Figure 4.27   Radio interferometer. Here only two telescopes are shown, but an interferometer with any number of telescopes can be treated as a number of pairs of telescopes. The separation between the telescopes produces a phase delay, which depends on the separation *d* and the position of the source. This phase difference can be detected, providing information about source structures whose angular size is approximately λ/d (in radians). By using different telescope spacings and the Earth's rotation, information about structures of different angular sizes can be accumulated and eventually reconstructeds into a map of the source.

The problem of poor angular resolution for radio observations has been solved, in part, by using combinations of telescopes, called *inteferometers* (Fig. 4.27). Interferometers utilize the information contained in the phase difference between the signals arriving at different telescopes from the same radio source. Any pair of telescopes provides information on an angular scale approximately equal (in radians) to the wavelength, divided by the separation between the two telescopes. To get information on different angular scales, it is necessary to have pairs of telescopes with different spacings. In addition, different orientations are also needed. For this reason, interferometers generally have a number of telescopes. The Earth's rotation also helps change the orientation of any pair of telescopes, as viewed from the direction of the source.

The most useful interferometer over the past several years has been the *Very Large Array,* or *VLA,* near Socorro, New Mexico, operated by the NRAO. The VLA has 27 telescopes (each 25 m in diameter), arranged in a

Figure 4.28 Views of the Very Large Array (VLA), on the Plains of St. Augustine (at an altitude of about 7,000 ft) southwest of Socorro, New Mexico. There are 27 telescopes, each 26 m in diameter. At any instant, there are 351 pairs of telescopes. Depending on the project, the spacings can be adjusted by moving the telescopes along railroad tracks. Moving all of the telescopes takes a few days, and moves are usually done about four times a year. The VLA is operated by the National Radio Astronomy Observatory.

(a)

(b)

(c)

"Y" configuration, shown in Fig. 4.28. The telescopes are placed alongside railroad tracks, so the telescope spacings can be changed. The shortest wavelength at which the VLA operates is 1 cm. It can be used for both continuum and spectral line observations. It has proved to be a very powerful tool, providing full images of radio sources, with observing sessions ranging from a few minutes to a few hours. (The amount of data taken in this short time is so large that it takes longer for the computers to process all of the data than it does to actually observe.)

For observations requiring the best possible angular resolution, telescopes on opposite sides of the Earth are used. This is called *very long baseline interferometry,* or *VLBI*. VLBI observations have provided angular resolutions of $10^{-4''}$! To provide a dedicated group of telescopes for VLBI, the NRAO is currently building the *Very Long Baseline Array,* or *VLBA*, which extends from one end of the United States to the other (Fig. 4.29).

Figure 4.29 Arrangement of telescopes for the Very Long Baseline Array (VLBA), being built by the National Radio Astronomy Observatory. This provides baselines as long as from the east coast of the United States to Hawaii.

4.9 X-RAY ASTRONOMY

X-ray astronomy is one of the youngest fields in observational astronomy. Since x-rays do not penetrate the Earth's atmosphere, the history of x-ray astronomy is the history of high-altitude and space astronomy. Early x-ray observations were done from sounding rockets and high-altitude balloons. Of course, balloons still do not get above all of the atmosphere. The first ventures into space were on sounding rockets, providing about five minutes observing per flight. Since that time, a number of satellites have opened our eyes to the x-ray sky (Fig. 4.30). The first extensive survey was carried out by the Small Astronomy Satellite-1 (SAS-1), also known as UHURU, launched in 1970. It provided continuous x-ray monitoring and revealed hundreds of sources.

X-ray satellites were able to provide both continuous and spectral information. The spectral information came from detectors, like those used by high-energy physicists, that allowed a measurement of the energy of incoming photons. One major problem with x-ray telescopes, however, is that they were very poor imaging devices. For some materials, the index of refraction at x-ray wavelengths is less than one. This means that when striking a surface,

Figure 4.30 A map of X-ray sources, on a galactic coordinate grid. In this system, the galactic plane cuts through the center of the map from left to right.

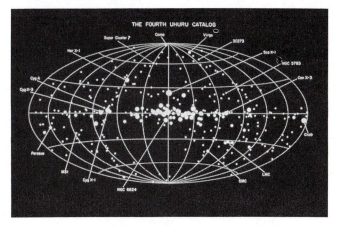

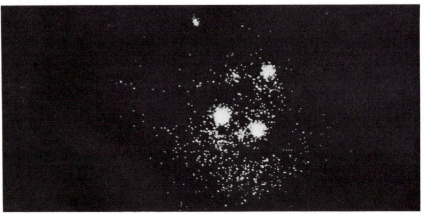

Figure 4.32   Einstein x-ray image of the cloud of ionized gas around star η Car. (An optical photograph of this region appears in Fig. 15.3)

Figure 4.31   The Einstein Observatory before launch.

after passing through a vacuum, x-rays are going from a region of higher to lower index of refraction. This means that we can have total internal reflection for grazing incidence.

The Einstein observatory had a telescope with an imaging system using the grazing incidence technique. (The x-ray beams made angles of about 1° with the surface. The optical system is outlined in Fig. 4.31, and an x-ray image is shown in Fig. 4.32. The Einstein images had a profound impact on our thinking about many types of astronomical objects. We went from just being able to probe small sections of objects, to forming whole images. In many ways, it was like having a blindfold removed.

CHAPTER SUMMARY

In this chapter we saw how different types of telescopes are used to collect data across the electromagnetic spectrum. We saw differences and similarities among the techniques used in different parts of the spectrum. Much of the progress in astronomy in the past two decades has come from our ability to make high-quality observations in parts of the spectrum other than the visible.

We looked at the important features of any telescope, the collecting area and the angular resolution. The collecting area determines how sensitive the telescope is to faint objects. The angular resolution limits how much detail we can see. In some cases the angular resolution is limited by diffraction (especially in the radio), and in other cases it is limited by atmospheric seeing (especially in the visible).

We saw the various techniques for getting information out of the radiation collected by our telescopes. Improving detector efficiency and panoramic ability has been important in all parts of the spectrum.

We saw the importance of site selection for an observatory. As an ultimate site, we saw the advantages of telescopes in space. In space, we can observe at wavelengths where the radiation doesn't penetrate the Earth's atmosphere. Even in the visible, we can achieve improved sensitivity and angular resolution.

QUESTIONS AND PROBLEMS

4.1 Describe the factors that limit the angular resolution in an optical telescope. Include estimates of the size of each effect.

4.2 What is the limiting magnitude for naked eye viewing with a 5-m-diam telescope?

4.3 Estimate the angular resolution of a 5-m-diam telescope in space.

4.4 Suppose two stars are 5″ apart on the sky. We clearly cannot resolve them with our eyes, but the angular resolution of even a modest-sized telescope is sufficient to resolve them. However, the light from that telescope must still pass through the narrow pupil of the eye. Why doesn't diffraction of the light entering the eye smear the image too much for us to resolve the two stars?

4.5 If we have two objects $\theta(″)$ apart on the sky, how far apart, x, are their images on the film of a camera with a focal length f. (Assume that we wish to express x and f in the same units.)

4.6 (a) What is the diameter of a single telescope with the same collecting area as the Multiple Mirror Telescope? (b) Astronomers have proposed a new telescope with a total collecting area equal to that of a single 25-m-diam telescope. How many 3-m-diam mirrors would be needed to make up this new telescope?

4.7 If we want to double the image size in a particular observation, by what amount would we have to change the exposure length to have a properly exposed photo?

4.8 "Faster" photographic emulsions can be made by making grains larger. Why do you think this works?

4.9 Why would you expect photographic film to be more sensitive in the blue than in the red?

4.10 Why is chromatic aberration a problem even for black-and-white photographs?

4.11 A higher-quality (more expensive) camera lens generally has smaller f-stops than an inferior lens. Why is this?

4.12 Compare the advantage and disadvantages of reflecting and refracting telescopes.

4.13 What does it mean to focus your eye or a camera "at infinity?"

4.14 If you want to photograph a planet you use a long focal length telescope; if you want to do photometry on a faint star you use a short focal length telescope. Explain.

4.15 Compare the use of a photographic plate in imaging and in photometric and spectroscopic observations.

4.16 The sodium D lines in the Sun's spectrum are at wavelengths 589.594 nm and 588.997 nm. (a) If a grating has 10^4 lines/cm, how wide a grating will be needed to resolve the two lines in first order? (b) Under this condition, what is the angular separation between the two lines? (c) How do the results in (a) and (b) change for second order?

4.17 A diffraction grating has N lines, a separation d apart. The spectrum is projected onto a screen a distance D ($>> d$) from the grating. Two lines are at λ and $\lambda + \Delta\lambda$. How far apart are they on the screen?

4.18 What are the important considerations in choosing an observatory site?

4.19 What advantages does Space Telescope have for optical observations over a ground-based telescope?

4.20 What are the similarities and differences between optical and ultraviolet observations?

4.21 Suppose some star is at the limit of naked eye visibility ($m = 6$). How much farther away can we see the same object with a telesope of diameter D? Evaluate your answer for $D = 5$ m.

4.22 (a) Using the fact that the limiting magnitude of the eye is 6, derive an expression for the limiting magnitude for direct viewing with a telescope of diameter D. (This ignores the effect of sky brightness.) (b) Use this result to derive an expression for the farthest distance at which a telescope of diameter D can be used to see an object of absolute magnitude M.

4.23 What are the similarities and differences between optical and infrared observing?

4.24 For optical observations we must still live with the fact that parts of a telescope radiate like blackbodies at about 300 K. Why isn't this a problem for optical observations?

4.25   (a) How does a bolometer work? (b) How would you use a bolometer to measure the power received in some small-wavelength range, for example, between 10 and 11 μm?

4.26   In the Earth's atmosphere the density of most constituents varies as $n(z) = n_0 \exp(-z/H)$, where n_0 is the density of that constituent at sea level, z is the altitude, and H is a constant (for each constituent) called the scale height. For oxygen, $H = 8$ km. (a) Use this expression to find the column density, $N(z)$ of a constituent above altitude z, defined as

$$N(z) = \int_z^\infty n(z')dz'$$

This quantity is important since the optical depth of the atmosphere when we look straight up from an altitude z is proportional to $N(z)$. (b) If $\tau = 3$ looking up from sea level at some wavelength, compute τ and $\exp(-\tau)$ for observatories at 2 km, 4 km, 10 km, and 20 km. (c) One of the constituents not well described by this distribution is water vapor, which is more confined to the lower altitudes. Why? (d) To the extent that water does follow a scale height distribution, we can use $H = 2$ km; redo part (b) for this value. (e) Discuss these results in terms of placement of an observatory to operate at this particular wavelength.

4.27   Suppose the optical depth of the atmosphere is τ_0 when we look straight up at some wavelength. It will be greater when we look at an angle z (called the *zenith distance*), away from the vertical. (a) Find an expression for $\tau(z)$ as a function of τ_0 and z. (b) At what values of z will τ equal $2\tau_0$ and $3\tau_0$? (c) If $\tau_0 = 0.3$, at what value of z will $\exp(-\tau) = 0.1$? (d) Discuss the significance of these results.

4.28   Why can't balloons get above all of the atmosphere?

4.29   In what ways are radio observations similar to and different from (a) optical and (b) infrared observations?

4.30   What is the angular resolution (in arc minutes) of (a) a 100-m-diam telescope operating at a 1-cm wavelength, and (b) a 25-m-diam telescope operating at a 2-mm wavelength?

4.31   We sometimes use a measure of the quality of a radio telescope as the diameter d, divided by the limiting wavelength λ_{min}. (a) Why is this quantity important? (b) If the telescope has surface errors of size Δx, give an expression for this quantity in terms of d and Δx.

4.32   How large a collecting area would you need to collect 1 W from a 1-Jy source over a bandwidth of 10^9 Hz (1 GHz)?

4.33   If an object the size of the Sun gave off a solar luminosity in radio waves, (a) what would be the power per surface area reaching us if we were (i) 1 AU away; (ii) 1 pc away? (b) If that power is uniformly spread over a frequency range of 10^{11} Hz, what is the flux density (power/surface area/Hz) in each case?

Binary Stars and Stellar Masses

If we are to understand the workings of stars, to know their masses is important. The best way to measure the mass of an object is to measure its gravitational influence on another object. For stars, we are fortunate to be able to measure the gravitational effects from pairs of stars, called *binary stars*.

5.1 BINARY STARS

Many stars we can observe appear to have companions, the two stars orbiting their common center of mass. Studies indicate that approximately half of the stars in our galaxy are binary stars. By studying the orbits of binary stars, we measure the gravitational forces that the two stars exert on each other. This allows us to determine the masses of the stars.

We classify binaries according to how the companion star manifests its presence.

1. *Optical double*. This is not really a binary star. Two stars just happen to appear along almost the same line of sight. The two stars can be at very different distances.

2. *Visual binary*. These stars are in orbit about each other and we can see both stars directly.

3. *Composite spectrum binary*. When we take a spectrum of the star, we see the lines of two different spectral type stars. From this we infer the presence of two stars.

4. *Eclipsing binary*. As we observe the light from such a system, it periodically gets brighter and fainter. We interpret the dimming as occurring

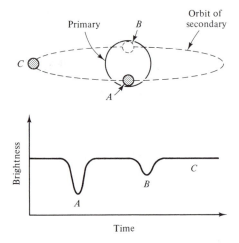

Figure 5.1 Eclipsing binary. The binary system is shown above. The light curve is shown below. The fainter secondary passes alternately in front of and behind the primary. Most of the time, as at position C, we see the light from both stars. When the secondary eclipses the primary, part of the primary light is blocked, and there is a dip in the intensity, at a point A. When the secondary passes behind the primary, its light is lost. Since the secondary is not as bright as a corresponding area of the primary, the loss of brightness is not as great, as shown at B.

Figure 5.2 Astrometric binary. Two stars orbit about a common center of mass, which, in turn, moves across the sky. The fainter star is too faint to be seen, so we only see the brighter star, moving back and forth across the path of the center of mass.

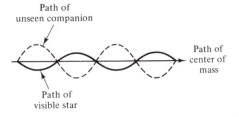

when the companion passes behind or in front of the main star. For us to see these eclipses, we must be aligned with the plane of the orbit (Fig. 5.1). A famous example of such a star is Algol (also known as β Per, indicating that it is the second brightest star in the constellation of Perseus).

5. *Astrometric binary.* Astrometry is the branch of astronomy in which the positions of objects are very accurately measured. In an astrometric binary, we can only see the brighter star. However, when we follow its path on the sky we see that, instead of following a straight line, it "wobbles" back and forth across the straight-line path. This means that the star is moving in an orbit, so we infer the presence of the companion (Fig. 5.2).

6. *Spectroscopic binary.* When we study the spectrum of a star we may see that the wavelengths of the spectral lines oscillate periodically about the average wavelength. We interpret these variations as being caused by a Doppler shift (discussed in the next section). When the star is coming toward us in its orbit, we see the lines at shorter wavelengths, and when the star is moving away from us in its orbit we see the lines at longer wavelengths.

It is possible for a given binary system to fit into more than one of these categories, depending on what we can observe.

5.2   DOPPLER SHIFT

A Doppler shift is a change in the wavelength (and frequency) of a wave, resulting from the motion of the source and/or the observer. It is most easily visualized for a sound wave or a water wave, where the waves are moving through a particular elastic medium (Fig. 5.3).

5.2.1   Moving Sources and Observers

We first look at the case of the moving observer. If the observer is moving toward the source, then the waves will be encountered more frequently than if the observer were standing still. This means that the observed frequency of the wave increases. If the frequency increases, then the wavelength must decrease. If the observer is moving away from the source, the situation is reversed. Waves will be encountered less frequently; the frequency decreases; the wavelength, therefore, increases. It should be noted that if the observer moves perpendicular to the line joining the observer and the source, no shift will be observed.

We now look at the case of a moving source. Each wavefront is now emitted in a different place. If the source is moving toward the observer, the waves will be emitted closer together than if the source were standing still. This means that the wavelength decreases. The decreased wavelength results

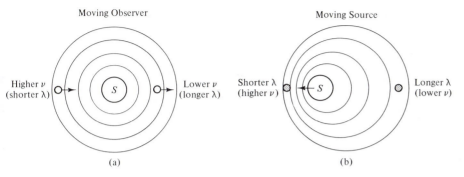

Higher ν
(shorter λ)

Lower ν
(longer λ)

(a)

Moving Source

Shorter λ
(higher ν)

Longer λ
(lower ν)

(b)

Figure 5.3 Doppler shift for waves in an elastic medium, such as sound waves. (a) Moving observer. On the left, the observer is moving toward the source, encountering wave crests more frequently than for a stationary observer. The frequency appears to increase (and the wavelength to decrease). On the right, the observer is moving away from the source, encountering waves at a lower frequency, corresponding to a longer wavelength. (b) Moving source. The motion of the source distorts the wave pattern, so the circles are no longer concentric. The observer on the left has the source approaching, producing a shorter wavelength (and a higher frequency). The observer on the right has the source receding, producing a longer wavelength (and a lower frequency).

in an increased frequency. If the source is moving away from the observer, the waves will be emitted farther apart than if the source were standing still. The wavelength increases, and the frequency decreases. Again, if the source is moving perpendicular to the line joining the source and observer, no shift results.

It is possible for both the source and observer to be moving. If their combined motion brings them closer together, the wavelength will decrease, and the frequency will increase. If their combined motion makes them grow farther apart, the wavelength will increase, and the frequency will decrease. If there is no instantaneous change in their separation, there is no shift in wavelength or frequency.

The shift depends only on the component of the relative velocity along the line joining the source and observer, since this is the only component that can change the distance r between them. We call this component the *radial velocity* (Fig. 5.4). We refer to the line joining the source and observer as the *line of sight*. From our definition of radial velocity v_r, we see that it is given by

$$v_r = dr/dt \qquad (5.1)$$

Note that if the source and observer are moving apart, r is increasing, and $v_r > 0$. If the source and observer are moving together, r is decreasing, and $v_r < 0$.

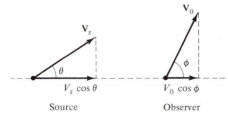

$\mathbf{V}_s$

θ

$V_s \cos\theta$

Source

$\mathbf{V}_0$

ϕ

$V_0 \cos\phi$

Observer

Figure 5.4 Radial velocity. The horizontal dashed line is the line of sight between the source and observer. The radial velocity is the difference between the line-of-sight components of the observer and source velocities.

Suppose the source is moving with a speed v_s in a direction that makes an angle θ with the line of sight, and the observer is moving with a speed v_0 in a direction making an angle φ with the line of sight. Taking the components of the two velocities along the line of sight as $v_s \cos \theta$ and $v_0 \cos \varphi$, and subtracting to get the relative radial velocity, gives

$$v_r = v_s \cos \theta - v_0 \cos \varphi \qquad (5.2)$$

In astronomy we are interested in the Doppler shift for electromagnetic waves. The underlying physics is a little different, because there is no mechanical medium for these waves to move through. They can travel even in a vacuum. (We will discuss this point further in Chapter 7.) For sound waves, the actual amount of the Doppler shift depends on whether the source or observer is moving. For electromagnetic waves, only the relative motion counts.

As long as the relative speed of the source and observer is much less than the speed of light, the results are relatively simple. If λ is the wavelength at which a signal is received, and λ_0 is the wavelength at which it was emitted (the *rest wavelength*), the wavelength shift $\Delta\lambda$ is defined by

$$\Delta\lambda = \lambda - \lambda_0 \qquad (5.3)$$

The simple result is that the wavelength shift, expressed as a fraction of the original wavelength, is equal to the radial velocity, expressed as a fraction of the speed of light. That is,

$$\Delta\lambda/\lambda_0 = v_r/c \qquad (v_r << c) \qquad (5.4)$$

If $v_r > 0$, then $\Delta\lambda > 0$. For a line in the middle of the visible part of the spectrum, a shift to longer wavelength is a shift to the red, so this is called a *red shift*. The name applies even if we are in other parts of the spectrum. A positive radial velocity always produces a red shift. If $v_r < 0$, then $\Delta\lambda < 0$, and we have a *blue shift*.

Now we look at what happens to the frequency. We remember that $\lambda = c/v$, so

$$d\lambda/dv = -c/v^2$$

$$= -\lambda/v$$

This means that $\Delta v = (-v/\lambda) \, \Delta\lambda$. Substituting this into equation (5.4) gives

$$\Delta v/v = -v_r/c \qquad (v_r << c) \qquad (5.5)$$

The shift in frequency, expressed as a fraction of the rest frequency, is the negative of the radial velocity, expressed as a fraction of the speed of light.

EXAMPLE 5.1 Doppler Shift

The Hα line has a rest wavelength, $\lambda_0 = 656.28$ nm. What is the observed wavelength for a radial velocity, where: (a) $v_r = 10$ km/s; (b) $v_r = -10$ km/s?

Solution

(a) We use equation (5.4) to find the wavelength shift:

$$\Delta\lambda = \lambda_0\,(v_r/c)$$

$$= \frac{(656.28 \text{ nm})(10.0 \text{ km/s})}{3.0 \times 10^5 \text{ km/s}}$$

$$= 0.022 \text{ nm}$$

We now add this to the original wavelength to find the observed wavelength:

$$\lambda = \lambda_0 + \Delta\lambda$$

$$= 656.28 \text{ nm} + 0.022 \text{ nm}$$

$$= 656.30 \text{ nm}$$

(b) If we take the negative of v_r, we just get the negative of $\Delta\lambda$, so

$$\Delta\lambda = -0.022 \text{ nm}$$

This gives an observed wavelength of

$$\lambda = 656.26 \text{ nm}$$

If we observe two spectral lines, their wavelength shifts will be different, since each is shifted by an amount proportional to its own rest wavelength. Thus, the spacing between the spectral lines will be shifted.

5.2.2 Circular Orbits

We now look at the Doppler shifts produced by a star in a circular orbit. The orbital speed is v, and the radius is r. The angular speed of the star in its orbit (in radians per second), is given by $\omega = v/r$. The situation is shown in Fig. 5.5. Suppose the star is moving directly away from the observer at

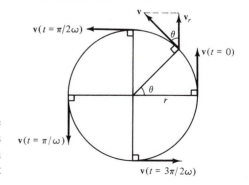

Figure 5.5 Doppler shift for a circular orbit. The speed v of the source remains constant, but the direction changes, so the radial velocity v_r changes. The angle θ keeps track of how far around the circle the source has gone. The source velocity is shown for five different values of θ.

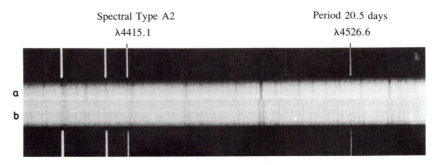

SPECTRUM OF A SPECTROSCOPIC BINARY STAR
Zeta Ursa Majoris (Mizar)

Spectral Type A2   Period 20.5 days
λ4415.1   λ4526.6

(a) June 11, 1927. Lines of the two components superimposed
(b) June 13, 1927. Lines of the two components separated by
a difference in orbital velocity of 140 km sec.

Figure 5.6   Spectra of a spectroscopic binary, taken at different points in the orbit. Note how the spectral lines shift back and forth.

time $t = 0$. At that instant the radial velocity, $v_r = v$. As the star moves, the component of its velocity along the line of sight is $v \cos \theta$. However, $\theta = \omega t$, so

$$v_r = v \cos \omega t \qquad (5.6)$$

The radial velocity changes sign every half-cycle, and repeats periodically. The period of the motion is just the circumference $2\pi r$, divided by the speed v, so $P = 2\pi/\omega$. If we substitute equation (5.6) into equation (5.4), we find that the spectral lines shift back and forth, with the shift given by

$$\Delta\lambda/\lambda_0 = (v/c)\cos \omega t$$

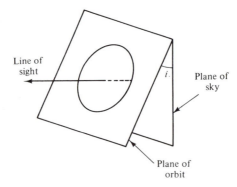

Line of
sight

i

Plane of
sky

Plane of
orbit

Figure 5.7   Inclination of an orbit. The orbit is an ellipse, which lies in a plane. This plane makes some angle i with the plane of the sky. The plane of the sky is perpendicular to the line of sight.

So far we have been considering the situation for the observer in the plane of the orbit. If the observer is not in the plane of the orbit, the Doppler shift will be reduced (Fig. 5.6). If i is the angle between the plane of the orbit and the plane of the sky, then the projection of any velocity in the plane of the orbit onto the line of sight is $v \sin i$ (Fig. 5.7). The angle i is known as the *inclination* of the orbit. This gives us a radial velocity for an orbiting star

$$v_r = v \sin i \cos \omega t \qquad (5.7)$$

5.3 **BINARY STARS AND CIRCULAR ORBITS**

In this section we will see how Newton's laws of motion and gravitation can be applied to binary stars in circular orbits. Circular orbits are not the most general case of orbital motion, but the analysis is the most straightforward, and most of the basic points are clearly illustrated. In the next section, we will go on to the general case of elliptical orbits.

We consider two stars, of masses m_1 and m_2, orbiting their common center of mass at distances, r_1 and r_2, respectively (Fig. 5.8). From the definition of the center of mass, these quantities are related by

$$m_1 r_1 = m_2 r_2 \qquad (5.8)$$

The center of mass moves through space subject only to the external forces on the binary star system. The forces between the two stars do not affect the motion of the center of mass. We will therefore ignore the actual motion of the center of mass, and view the situation as it would be seen by an observer sitting at the center of mass.

Since the center of mass must always be along the line joining the two stars, the stars must always be on opposite sides of the center of mass. This means that both stars orbit with the same orbital period P. In general, the period of an orbit is related to the radius r and the speed v by

$$P = 2\pi r/v \qquad (5.9)$$

Solving for v gives

$$v = 2\pi r/P \qquad (5.10)$$

Since the periods of the two stars must be the same, equation (5.9) tells us that

$$r_1/v_1 = r_2/v_2 \qquad (5.11)$$

Combining these with equation (5.8) gives

Figure 5.8 Binary system with circular orbits. Both stars orbit the center of mass (CM). The more massive star is closer to the center of mass. The center of mass must always be between the two stars, so the stars lie on opposite sides of it.

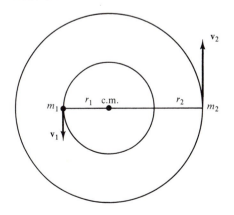

$$v_1/v_2 = r_1/r_2 = m_2/m_1 \tag{5.12}$$

(NOTE: We could have also gotten $m_1 v_1 = m_2 v_2$ directly from conservation of momentum. This is not surprising since the properties of the center of mass come from conservation of momentum.)

We now look at the gravitational forces. The distance between the stars is $r_1 + r_2$, so the force on either star is given by Newton's law of gravitation as

$$F = \frac{Gm_1 m_2}{(r_1 + r_2)^2} \tag{5.13}$$

This force must provide the acceleration associated with the change of the direction of motion in circular motion v^2/r. For definiteness, we follow star 1, so

$$F = m_1 v_1^2 / r_1 \tag{5.14}$$

Combining equations (5.13) and (5.14) gives

$$\frac{m_1 v_1^2}{r_1} = \frac{Gm_1 m_2}{(r_1 + r_2)^2} \tag{5.15}$$

Note that we can divide both sides by m_1. If we also use equation (5.10) to relate v_1 to P, we find

$$\frac{4\pi^2 r_1}{P^2} = \frac{Gm_2}{(r_1 + r_2)^2} \tag{5.16}$$

This can be simplified if we introduce the total distance R between the two stars,

$$R = r_1 + r_2$$

$$= r_1 (1 + r_2/r_1) \tag{5.17}$$

Using equation (5.12) this becomes

$$R = r_1 (1 + m_1/m_2) \tag{5.18}$$

$$= (r_1/m_2)(m_1 + m_2) \tag{5.19}$$

Substituting this into equation (5.16) gives

$$4\pi^2 R^3 / G = (m_1 + m_2)P^2 \tag{5.20}$$

Let's look at how equation (5.20) can be used to give us stellar masses. For any binary system we can determine the period directly if we watch the system for long enough. If the star is a spectroscopic binary, we can see how long it takes for the Doppler shifts to go through a full cycle. If it is an astrometric binary, we can see how long it takes to go through a full cycle of the "wobble." If it is an eclipsing binary, we can see how long it takes the light curve to go through a full cycle. If we can see both stars, we can determine R. Once we know P and R, we can use equation (5.20) to calculate the sum of the masses, $m_1 + m_2$. We can also get the ratio of the masses, m_1/m_2, either from r_1/r_2, if both stars are seen, or v_1/v_2, if both Doppler shifts are observed. Once we know the sum of the masses and the ratio of the masses, the individual masses can be determined. This situation we have outlined here is an ideal one, however. Usually, we don't have all of these pieces of information (as we will see below).

EXAMPLE 5.2 Mass of the Sun

We can consider the Sun and Earth as a binary system, so we should be able to apply equation (5.20) to find the mass of the Sun.

Solution

Since the mass of the Sun is so much greater than the mass of the Earth, we can approximate the sum of the masses as being the mass of the Sun, $M_\odot$. Equation (5.20) then becomes

$$M_\odot = \frac{4\pi^2 R^3}{GP^2}$$

$$= \frac{4\pi^2 (1.5 \times 10^{13} \text{ cm})^3}{(6.67 \times 10^{-8} \text{ dyne-cm}^2/\text{g}^2)(3.16 \times 10^7 \text{ s})^2}$$

$$= 2.0 \times 10^{33} \text{ g}$$

We call this quantity a *solar mass*. It becomes a convenient quantity for expressing the masses of other stars.

We know that for a pair of objects orbiting with a period of 1 yr, at a distance of 1 AU (defined in Section 2.6), the sum of the masses must be a solar mass. This suggests a convenient system of units for equation (5.20). If we express masses in solar masses, distances in astronomical units, and periods in years, the constants must equal one to give the above results. We can therefore rewrite equation (5.20) as

$$[R/(1 \text{ AU})]^3 = \left[\frac{m_1 + m_2}{1 \, M_\odot}\right]\left[\frac{P}{1 \, \text{yr}}\right]^2 \qquad (5.21)$$

For the solar system, we write the sum of the masses as one in these units, so the equation simply says that the cube of the radius (in astronomical units) is equal to the square of the period (in years). This is also known as *Kepler's third law of planetary motion.* This law was originally deduced observationally by Kepler, and Newton used it to show that gravity must be an inverse square law force.

For a visual binary, we don't directly measure R, the linear separation. We actually measure the angular separation on the sky, θ. If d is the distance to the binary, then R is equal to $\theta(\text{rad})d$, where $\theta(\text{rad})$ is the value of θ measured in radians. When we use this relation, R and d will come out in the same units. Therefore,

$$R(\text{AU}) = d(\text{AU})\theta(\text{rad})$$

The values of θ are so small that radians are an inconvenient quantity. We can convert to arc seconds [equation (2.16)] to give

$$R(\text{AU}) = d(\text{AU}) \, \theta('')(2.06 \times 10^5)$$

The factor of 2.06×10^5 was to convert radians to arc seconds, but it is also the factor to convert astronomical units to parsecs, so we have

$$R(\text{AU}) = d(\text{pc}) \, \theta('')$$

If we use equation (2.17) to relate the distance in parsecs to the parallax in arc seconds, this becomes

$$R(\text{AU}) = \theta('')/p('') \qquad (5.22)$$

This can be substituted directly into equation (5.21).

We will now look at the behavior of the Doppler shifts. Applying equation (5.10) to both speeds, v_1 and v_2, and remembering that the period of the orbit is the same for both stars, we have

$$r_1 + r_2 = (P/2\pi)(v_1 + v_2)$$

Using this to eliminate R in equation (5.20) gives

$$(P/2\pi G)(v_1 + v_2)^3 = m_1 + m_2 \qquad (5.23)$$

If the orbit is inclined at an angle i, then the Doppler shifts only measure the components $v_r = v \sin i$. In terms of the radial velocities, v_{1r} and v_{2r}, equation (5.23) becomes

$$\frac{(P/2\pi G)(v_{1r} + v_{2r})^3}{\sin^3 i} = m_1 + m_2 \qquad (5.24)$$

If a binary happens to be an eclipsing binary, then we know that we are close to the plane of the orbit, and i is close to $90°$. The shape of the eclipse curve can give us the value of i in that case. Otherwise, we don't know i. If a circular orbit is projected at some angle on the plane of the sky, it will appear elliptical. We will see in the next section that there are ways to determine i if we can trace the projected orbit on the sky. If i is unknown, all we can do is solve equation (5.24) with $i = 90°$. This will give us a value for $m_1 + m_2$ that is lower than the true value. (The true value would be this value divided by $\sin^3 i$, and, since $\sin^3 i$ is less than or equal to one, the value assuming $i = 90°$ is less than the true value.) We say that the calculated value is a *lower limit* to the true value. Finding lower limits is not as useful as finding actual values. However, if we study enough binary systems, we will encounter a full range of inclination angles. These statistical studies can be used to relate mass to spectral type.

EXAMPLE 5.3 Binary Star Doppler Shifts

A binary system is observed to have a period of 10 yr. The radial velocities of the two stars are determined to be $v_{1r} = 10$ km/s and $v_{2r} = 20$ km/s, respectively. Find the masses of the two stars (a) if the inclination of the orbit is $90°$ and (b) if it is $45°$.

Solution

From equation (5.24), we have

$$\frac{m_1 + m_2}{M_\odot} = \frac{(10 \text{ yr})(3.16 \times 10^7 \text{ s/yr})[(10 + 20)(1 \times 10^5 \text{ cm/s})]^3}{2 \,(6.67 \times 10^{-8} \text{ dyn-cm}^2/\text{g}^2)(2 \times 10^{33} \text{ g})(\sin^3 i)}$$

$$= 10.2/\sin^3 i$$

This means that

$$m_1 + m_2 = 10.2 \, M_\odot/\sin^3 i$$

If $i = 90°$, $\sin^3 i = 1$, so

$$m_1 + m_2 = 10.2 \, M_\odot$$

We find the ratio of the masses from the ratio of the radial velocities:

$$m_1/m_2 = v_2/v_1$$

$$= 2.0$$

This means that $m_1 = 2m_2$, so

$$2m_2 + m_2 = 10.2 \, M_\odot$$

giving $m_1 = 6.8 \, M_\odot$ and $m_2 = 3.4 \, M_\odot$. If $i = 45°$, $1/\sin^3 i = 2.8$. The ratio of the masses doesn't change, since the $\sin i$ drops out of the ratio of the radial velocities. This means that we can just multiply each mass by 2.8 to give 19.2 and 9.5 $M_\odot$, respectively.

It is often the case that only one Doppler shift can be observed. Let's assume we measure v_1, but not v_2. We must therefore eliminate v_2 from our equations. We can write v_2 as

$$v_2 = v_1 \, (m_1/m_2)$$

The sum of the velocities then becomes

$$v_1 + v_2 = v_1 \, (1 + v_2/v_1)$$

$$= v_1 \, (1 + m_1/m_2)$$

$$= (v_1/m_2)(m_1 + m_2)$$

If we now substitute into equation (5.24), we find

$$\left(\frac{P}{2\pi G}\right)(v_{1r})^3 = \frac{m_2^3 \sin^3 i}{(m_1 + m_2)^2} \tag{5.25}$$

The quantity on the right-hand side of equation (5.25) is called the *mass function*. If we can measure only one Doppler shift, we cannot determine either of the masses. We can only measure the mass function. We can, however, get information on masses of various spectral types through extensive statistical studies.

5.4 ELLIPTICAL ORBITS

5.4.1 Geometry of Ellipses

In general, orbiting bodies follow elliptical paths. A circle is just a special case of an ellipse. In this section, we generalize the results from the previous section from circular orbits to elliptical orbits. The basic underlying physical ideas remain the same.

We first review the geometry of an ellipse, as shown in Fig. 5.9. We describe the ellipse by its semimajor axis a and semiminor axis b. Each point on the ellipse satisfies the condition that the sum of the distances from any point to two fixed points, called foci (singular, focus), is constant. If r and r' are these two distances, then $r + r'$ is a constant. We can see that for a point on the major axis this sum is $2a$, so it must be $2a$ everywhere. That is

$$r + r' = 2a \qquad (5.26)$$

The *eccentricity* of an ellipse is the distance between the foci, divided by $2a$. A circle is an ellipse of eccentricity zero. The eccentricity of any ellipse must be less than one. From the right triangle in Fig. 5.9, we see that

$$\begin{aligned} b^2 &= a^2 - (ae)^2 \\ &= a^2(1 - e^2) \end{aligned} \qquad (5.27)$$

In a binary system, the center of mass of the two stars will be at one focus on the ellipse. The farthest point from that focus is called the *apastron*.

Figure 5.9 Geometry of an ellipse. The semi-major axis is a; the semi-minor axis is b. The two foci are at F and F'. The eccentricity is e, and the distances from the two foci to points on the ellipse are r and r'.

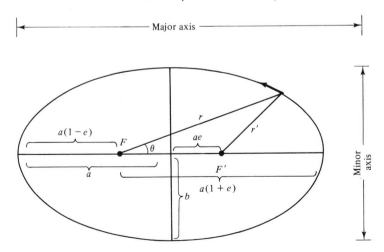

From the figure we see that the distance from the focus to this point is

$$r(\text{apastron}) = a(1 + e) \tag{5.28a}$$

The closest point to the focus is called the *periastron*. Its distance from the focus is

$$r(\text{periastron}) = a(1 - e) \tag{5.28b}$$

The average of these two values is a, the semimajor axis. This is the quantity that replaces the radius of a circular orbit in our study of binary stars.

It is useful to have an expression for the ellipse, relating the variables r and θ. From the law of cosines, we see that

$$r'^2 = r^2 + (2ae)^2 + 2r(2ae)\cos\theta \tag{5.29}$$

Using equation (5.26) gives

$$r = \frac{a(1 - e^2)}{1 + e\cos\theta} \tag{5.30}$$

5.4.2   Angular Momentum in Elliptical Orbits

The gravitational force between two objects always acts along the line joining the two objects. The center of mass also lies along this line. This means that the force on either object points directly from that object toward the center of mass. Therefore, these forces can exert no torques about the center of mass. If there are no torques about the center of mass, then the angular momentum about the center of mass is conserved, as illustrated in Fig. 5.10.

For an object of mass m, in any orbit, the angular momentum is

$$L = mvr\sin\varphi \tag{5.31}$$

where φ is the angle between the line from the center of mass to r, and the velocity vector $\mathbf{v}$ as shown in Fig. 5.10. We look at the area swept out by the line from the center of mass to r in the time interval dt. The area is the thin triangle shown in Fig. 5.10. The long side of the triangle is r, and the small side is $v\,dt$. The small right triangle shows that the height of the larger triangle is $v(dt)\sin\varphi$. The area of that triangle is then half the base × height, so

$$dA = \tfrac{1}{2}r\,(v\,dt)\sin\varphi$$

Figure 5.10 Angular momentum in an elliptical orbit. In this case, the object is a distance r from the focus F. Its velocity makes an angle φ with the line from F to the object. (If this were a circle, φ would be 90°.) The time interval over which we mark the motion is Δt.

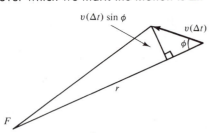

The rate at which area is swept out is

$$dA/dt = \tfrac{1}{2}rv \sin \varphi$$

However, we can use equation (5.31) to eliminate $rv \sin \varphi$, giving

$$dA/dt = \tfrac{1}{2}(L/m)$$

Since L is constant, the rate at which area is swept out is constant. Equal areas are swept out in equal times. (When applied to the planetary motions, this is known as *Kepler's second law,* as we will discuss in Section 22.1.)

The major consequence of angular momentum conservation is therefore that the objects move slower when they are farther apart and faster when they are close together.

5.4.3 Energy in Elliptical Orbits

We next look at the total energy of the binary system. Adding the kinetic energies of the two stars and their gravitational potential energy (defined to be zero when the stars are infinitely apart) gives

$$E = \tfrac{1}{2}m_1 v_1^2 + \tfrac{1}{2}m_2 v_2^2 - Gm_1 m_2/r \tag{5.32}$$

In this expression, v_1 and v_2 are the speeds relative to the center of mass. Remember, from the definition of the center of mass we found that $m_1 v_1 = m_2 v_2$. We also introduce the relative speed of the two stars

$$v = v_1 + v_2 \tag{5.33a}$$

They are added because the two stars are always moving in opposite directions, so their relative speed will be the sum of the magnitudes of their individual speeds. In terms of v, the two speeds v_1 and v_2 are

$$v_1 = \frac{m_2 v}{m_1 + m_2} \tag{5.33b}$$

$$v_2 = \frac{m_1 v}{m_1 + m_2} \tag{5.33c}$$

Substituting these into equation (5.32) gives

$$E = m_1 m_2 \left[\frac{v^2}{2(m_1 + m_2)} - \frac{G}{r} \right] \tag{5.34}$$

Since energy is conserved, we can evaluate it at any point we want. For simplicity, we choose apastron (speed v_a) and periastron (speed v_p). We can relate the speeds v_a and v_p by conservation of angular momentum. Angular momentum conservation is easy to apply at the apastron and periastron because the velocities are perpendicular to the line from the focus to the star. Since $\varphi = 90°$ at these points, the angular momentum [equation (5.31)] is then just mvr. Using equations (5.28a) and (5.28b) tells us that

$$a(1 + e)v_a = a(1 - e)v_p \tag{5.35}$$

Solving for the ratio v_p/v_a gives

$$\frac{v_p}{v_a} = \frac{1 + e}{1 - e} \tag{5.36}$$

Now that we have the ratio v_p/v_a, we need another relation between them to be able to solve for v_a and v_p individually. We can use conservation of energy to equate the energies at the apastron and periastron. Using equation (5.34) gives

$$\frac{v_a^2}{2(m_1 + m_2)} - \frac{G}{a(1 + e)} = \frac{v_p^2}{2(m_1 + m_2)} - \frac{G}{a(1 - e)} \tag{5.37}$$

Rearranging gives

$$\left[\frac{G(m_1 + m_2)}{a}\right]\left(\frac{1}{1 - e} - \frac{1}{1 + e}\right) = \tfrac{1}{2}(v_p^2 - v_a^2)$$

$$= \tfrac{1}{2}v_a^2\left[\left(\frac{v_p}{v_a}\right)^2 - 1\right]$$

We can use equation (5.36) to eliminate the ratio v_p/v_a. Solving for v_a gives

$$v_a^2 = \left[\frac{G(m_1 + m_2)}{a}\right]\left(\frac{1 - e}{1 + e}\right) \tag{5.38a}$$

We now use equation (5.36) with this to solve for v_p:

$$v_p^2 = \left[\frac{G(m_1 + m_2)}{a}\right]\left(\frac{1 + e}{1 - e}\right) \tag{5.38b}$$

If we put these into equation (5.34), the total energy simplifies to

$$E = -Gm_1m_2/2a \tag{5.39}$$

We can now use this as the left-hand side of equation (5.34). We can then solve for v at any point r.

$$v^2 = G(m_1 + m_2)(2/r - 1/a) \qquad (5.40)$$

5.4.4 Observing Elliptical Orbits

In studying the Doppler shifts of elliptical orbits as compared with circular orbits, there are three important differences:

1. The speed along an elliptical orbit is not constant.
2. In an elliptical orbit the velocity is not perpendicular to the line from the center of mass to the orbiting object.
3. Even if you are in the plane of the orbit, the radial velocity curve depends on where you are relative to the major axis of the ellipse.

These points are illustrated in Fig. 5.11. Some radial velocity curves for real binary stars are shown in Fig. 5.12.

As we said above, we must correct any Doppler shift for the inclination of the orbit i. If we take a tilted ellipse and project it onto the sky, we still have an ellipse. However, the ellipse will have a different eccentricity than the true ellipse. When we look at an elliptical orbit, how can we tell if it is tilted or not? For a tilted orbit the foci will not appear to be in the right place for the projected ellipse. Therefore, if we see two stars orbiting a point

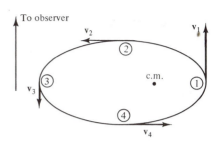

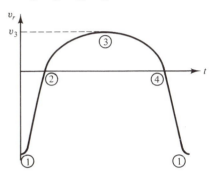

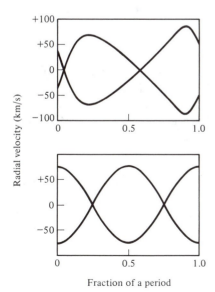

Radial velocity (km/s)

Fraction of a period

Figure 5.12 Radial velocity curves for binary systems.

Figure 5.11 Radial velocity v_r vs. t for an elliptical orbit. In contrast to the circular orbit, both the magnitude and direction of **v** change throughout the orbit. (We assume for this figure that the observer is in the plane of the orbit, or that $i = 90°$.) Four points are shown in the orbit (above) and in the radial velocity curve (below). At points 2 and 4 the motion is perpendicular to the line of sight, so $v_r = 0$. For point 1 the motion is toward the observer, producing the maximum negative v_r, and for point 3 the motion is away from the observer, producing the maximum positive v_r. The motion is also faster at 1 than at 3. In addition, going from 4 to 1 to 2 takes less time than going from 2 to 3 to 4. This accounts for the distorted shape of the radial velocity curve.

TABLE 5.1 Mass and Spectral Type (MS)

Spectral type	$M/M_\odot$
O5	40.0
B5	7.1
A5	2.2
F5	1.4
G5	0.9
K5	0.7
M5	0.2

different from the center of mass, we will know that the orbit is inclined. We can determine the inclination from the degree to which the foci appear to be displaced. (In this technique, we don't actually know where the center of mass is, so a process must be used in which we try one position for the center of mass and try to match the projected orbit, repeating the process until a good fit is achieved.)

5.5 STELLAR MASSES

As a result of studying many binary systems, astronomers have a good idea of the masses of main sequence stars. These results are summarized in Table 5.1. Just as the Sun's temperature places it in the middle of the main sequence, its mass is in the middle of the range of stellar masses. The lowest-mass main sequence stars have about 0.07 of a solar mass, and the most massive stars commonly encountered have about 60 solar masses. (Recent studies suggest that the star, η Carina, had about 200 solar masses when it was on the main sequence.) When we think of how large or small stars might have turned out to be, the observed range of stellar masses is not very large. This range is an important constraint on theories of stellar structure.

An even more stringent constraint is in the relationship between mass and temperature on the main sequence. The cooler stars are less massive, and the hotter stars are more massive. We have already said that the existence of the main sequence implies a certain relationship between size and temperature. This means that for a star on the main sequence, once the mass is specified, the radius and temperature are determined. Another way of looking at it is to say that a star's mass determines where on the main sequence it will fall.

Since the mass determines the radius and temperature of a main sequence star, it should not be surprising that the mass also determines the luminosity. The exact dependence of luminosity on mass is called the *mass–luminosity relationship*. This relationship should also be explainable from theories of stellar structure. The relationship is shown in Fig. 5.13. We can summarize it by saying that the luminosity varies approximately as some power α of the mass. It we express luminosities in terms of solar luminosities, and masses in terms of solar masses, this means that

$$L/L_\odot = (M/M_\odot)^\alpha \qquad (5.41a)$$

A single value of α does not work for the full range of masses along the main sequence. The appropriate values are

$$\alpha = 1.8 \text{ for } M < 0.3\,M_\odot \quad \text{(low mass)}$$

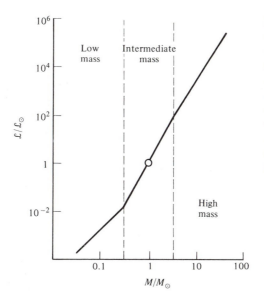

Figure 5.13 Mass–luminosity relationship. The vertical axis gives luminosities, relative to that of the Sun, and the horizontal axis gives masses, also relative to that of the Sun. Note that the axes are actually logarithmic, so a straight line means a power-law relationship. The slope of the line is different in the three mass ranges. Note, also, that the range of luminosities is much larger than the range of the masses.

$$\alpha = 4.0 \text{ for } 0.3 \, M_\odot < M < 3 \, M_\odot \quad \text{(intermediate mass)} \quad (5.41\text{b})$$

$$\alpha = 2.8 \text{ for } 3 \, M_\odot < M \quad \text{(high mass)}$$

EXAMPLE 5.4 Mass–Luminosity Relationship

Use the mass–luminosity relationship to find the relative luminosities of a $10 \, M_\odot$ and $0.1 \, M_\odot$ star, respectively.

Solution

We have to use each section of the *M–L* relationship separately.

$$\frac{L(10 \, M_\odot)}{L(0.1 \, M_\odot)} = \frac{L(10 \, M_\odot)}{L(3 \, M_\odot)} \frac{L(3 \, M_\odot)}{L(0.3 \, M_\odot)} \frac{L(0.3 \, M_\odot)}{L(0.1 \, M_\odot)}$$

$$= (10/3)^{2.8}(3/0.3)^{4.0}(0.3/0.1)^{1.8}$$

$$= 6.3 \times 10^6$$

In understanding how stars are formed (Chapter 14) we would like to know the distribution of stellar masses. That is, we would like to know the relative abundances of stars of various masses. Since we now know how to

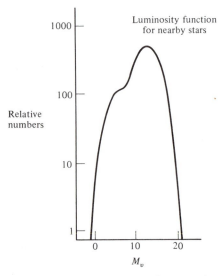

Figure 5.14 Luminosity function for nearby stars. The graph shows the relative numbers of stars with various absolute visual magnitudes, M_v.

relate mass and spectral type, we can carry out such studies by looking at the relative numbers of different spectral types or luminosities. These studies are difficult, because we can see brighter stars to greater distances. (This is called a *selection effect*.)

We find that most of the stars in our galaxy are low-mass stars. In fact, the total mass in all the low-mass stars is much more than the total mass in all of the high-mass stars, even though each low-mass star, obviously, has less mass than each high-mass star. However, the luminosity of each high-mass star is so much greater than that of each low-mass star that most of the luminosity of our galaxy comes from a relatively small number of high-mass stars.

5.6   STELLAR SIZES

We have alluded so far to stellar sizes, but we have not discussed how they are determined. In this section, we will look at various methods for measuring stellar radii.

The star whose size is easiest to measure is the Sun. This is actually quite useful. We have seen that the Sun is intermediate in its mass and temperature, so its radius is probably a fairly representative stellar radius. The angular radius of the Sun $\Delta\theta$, is 16'. The Sun is at a distance $d = 1.5 \times 10^8$ km, so its radius $R_\odot$ is given by

$$R_\odot = d \tan \Delta\theta$$

$$= 6.96 \times 10^5 \text{ km}$$

The Sun is the only star whose disk subtends an angle larger than the seeing limitations of ground-based telescopes. We therefore need other techniques for determining radii. If we know the luminosity (from its absolute magnitude) and the surface temperature (from the spectral type) of a star, we can calculate its radius, using equation (2.7). Solving for the radius gives

$$R^2 = L/4\pi\sigma T^4 \tag{5.42}$$

EXAMPLE 5.5   Luminosity Radius

Estimate the radius of an A0 star. (Use Appendix E for stellar properties.)

Solution

We can express the various quantities in solar units. Taking ratios, we can use equation (5.42) to give

$$R/R_\odot = (L/L_\odot)^{1/2}(T/T_\odot)^{-2}$$

$$= (80)^{1/2}(1.85)^{-2}$$

$$= 2.6$$

Eclipsing binaries provide us with another means of determining stellar radii (Fig. 5.15). This method derives from an analysis of the shape of the light curve and a knowledge of the orbital velocities, from Doppler shift measurements. (In an eclipsing binary, we don't have to worry about the inclination of the orbit.) Particularly important is the rate at which the light level decreases and increases as the beginning and end of the eclipses.

We can also estimate the radii of rotating stars. If there are surface irregularities, such as hot spots or cool spots, the brightness of the star will

Figure 5.15 Stellar sizes determined from eclipsing binaries. The orbit is shown above and the light curve is shown below. The lengths of the eclipses, and the steepness of the sections at the beginning and end of each eclipse (such as A–B and E–D) depend on the sizes of the stars.

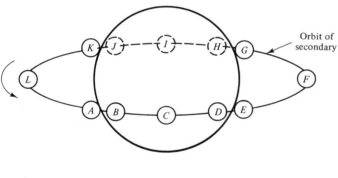

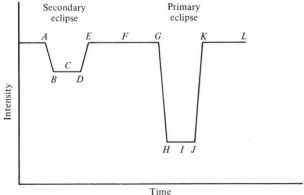

Figure 5.16 (a) Tracing of a lunar occultation of the star β Sco A. On the left we see the star before the occultation, and on the right it is occulted. The wiggles in the curve are due to diffraction effects as the star passes behind the lunar limb. The dots are the actual data, and the smooth curve is the best fit of a theoretical model to the data. (b) Theoretical calculations of what the curves in (a) would look like for stars of different angular sizes.

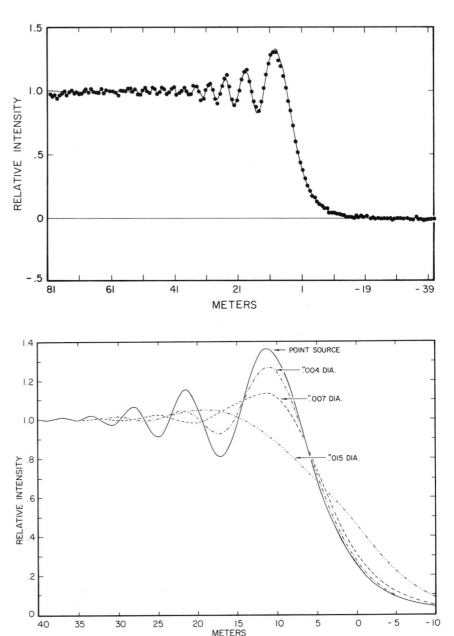

depend on whether these spots are facing us or are away from us. The brightness variations give us the rotation period P. From the broadening of spectral lines, due to the Doppler shift, we can determine the rotational speed v. This speed is equal to the circumference $2\pi R$ divided by the period. Solving for the radius gives

$$R = Pv/2\pi \qquad (5.43)$$

Sometimes the moon passes in front of a star bright enough and close enough for detailed study. An analysis of these *lunar occultations* tells us about the radius of the star. The larger the star is, the longer it takes the light to go from maximum value to zero as the lunar edge passes in front of the star. Actually, since light is a wave, there are diffraction effects as the starlight passes past the lunar limb, as shown in Fig. 5.16. The light level oscillates as the star disappears. The nature of these oscillations tells us about the radius.

There is another observational technique, called *speckle interferometry* (Fig. 5.17), that has been quite successful recently. If it were not for the seeing fluctuations in the Earth's atmosphere, we would be able to get images of stellar disks down to the diffraction limits of large telescopes. However, the atmosphere is stable for short periods of time, of the order of 0.01 s. The blurring of images comes from trying to observe for longer than this. If an image were bright enough to see in this short time, we could take a picture with diffraction-limited resolution. Unfortunately, 0.01 s is not long enough to collect enough photons from a star. However, we can collect a series of 0.01-s images, observing interference between light coming through slightly different paths in the atmosphere. The final images must be reconstructed mathematically.

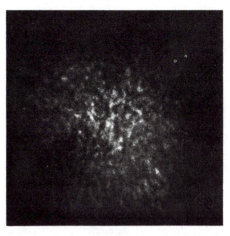

Figure 5.17 Image of a star, using speckle interferometry.

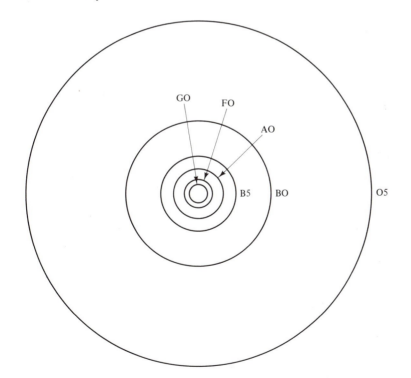

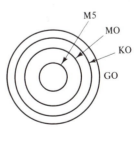

Figure 5.18 Stellar sizes for different spectral types. The largest, down to G0, are shown at the upper left. The smaller ones from G0 down, are blown up at the lower right. The circle for G0 is repeated to give the relative scale of the two parts of the figures.

The results of these various techniques are shown in Fig. 5.18. We see that, on the main sequence, stars get larger with increasing surface temperature. This is why the luminosity of stars increases with increasing surface temperature at a rate greater than T^4

CHAPTER SUMMARY

We saw in this chapter how the gravitational interactions in binary systems can be studied to determine the masses of stars.

We saw how the Doppler shift can be used to determine the radial velocities of objects. In binary systems, we can observe the wavelengths of spectral lines varying periodically as the stars orbit the center of mass. We saw what information could be obtained even if we don't know orbital inclination.

We saw how the masses of the orbiting objects are related to the orbital radii and speeds, and the period of the system, for circular orbits. We extended these ideas to elliptical orbits. For elliptical orbits, we saw how conservation of angular momentum means a change in the

speed with distance from the center of mass, and how this affects the Doppler shift we see in different parts of the cycle. We also found the total energy for elliptical orbits and saw how the kinetic energy varies as the speed varies.

We saw that the range of masses along the main sequence is much less than the range of luminosities. There is also a close relationship between mass and luminosity for main-sequence stars.

We saw how eclipsing binaries can be used to tell us something about stellar sizes. We also looked at other techniques for determining stellar sizes, including knowing the luminosity and temperature, using lunar occultations, and speckle interferometry.

QUESTIONS AND PROBLEMS

5.1 Under what conditions can we determine the masses of both stars in a binary system? Think of as many combinations of situations as you can.

5.2 For a binary that is *only* detected as an astrometric binary, are there conditions under which we can determine the masses of both stars?

5.3 (a) A source is moving away from an observer with a speed of 10 km/s, along a path that makes a 30° angle with the line of sight. At what wavelength will the Hα line be observed? (b) If the observer is also moving directly away from the source at a speed of 20 km/s, at what wavelength will the Hβ line be observed?

5.4 If the Hα line in a source is Doppler-shifted by +0.10 Å, by how much is the Hβ line shifted?

5.5 If the Hα line in a source is Doppler-shifted by −1.0 Å, what is the radial velocity of the source?

5.6 You are standing a distance d from a railroad track. A train comes past you at a constant speed v, passing you at time $t = 0$. (a) What is the radial velocity of the train as a function of time? (b) Graph your result for both positive and negative times.

5.7 A railroad train goes around a circular track with speed v, moving away from you at time $t = 0$. (a) What is the radial velocity of the train as a function of time t? (b) Graph your result. Assume that your distance from the track is much larger than the radius of the circle.

5.8 By how much can the Hα line in some object be shifted as a result of the Earth's motion around the Sun?

5.9 Why must the periods of both stars in a binary system be the same?

5.10 Show that, if we had taken the force on the second star in equation (5.13), we would still have obtained the result in equation (5.20).

5.11 Let $M = m_1 + m_2$ and $x = m_1/m_2$. (a) Find expressions for m_1 and m_2 in terms of M and x. (b) What is the significance of your result?

5.12 We observe a binary system in which the two stars are $1''$ and $2''$, respectively, from the center of mass. The system is 10 pc from us. The period is 10 yr. What are the masses of the two stars, assuming that $i = 90°$?

5.13 A star in a circular orbit has a speed of 30 km/s. The period is 10 yr. The star is $2''$ from the center of mass. How far away is this star from us?

5.14 Derive a form of equation (5.23) that relates mass in solar masses, period in years, and velocity in kilometers per second.

5.15 The Hα lines from two stars in a binary system are observed to have Doppler shifts of 0.022 and 0.044 nm, respectively. The period of the system is 20 yr. What are the masses of the two stars (a) if $i = 90°$; (b) if $i = 30°$?

5.16 Suppose that many binary systems, with randomly distributed inclinations, are observed. (a) Draw a graph showing the relative probability of finding values of $\sin^3 i$ in various small ranges from 0 to 1. (b) The average value of some function $f(x)$ over the interval 0 to L is

$$\langle f(x) \rangle = (1/L) \int_0^L f(x)dx$$

What is the average value of $\sin^3 i$ over the angle range 0 to $\pi/2$ radians? (c) What does this tell you about the relationship between the lower limits to masses and the actual masses of binary systems?

5.17 An astrometric binary is 10 pc from the Sun. The visible star orbits $2''$ from the center of mass with a period of 30 yr. What can you conclude about the mass of the unseen companion?

5.18 An eclipsing binary system has a period of 1.7 days. One star has a Doppler shift of 100 km/s. What can you conclude about the mass of the companion?

5.19 Derive equations (5.38a) and (5.38b).

5.20 A star moves in an elliptical orbit of eccentricity e. The plane of the orbit makes an angle i with the plane of the sky. The orbit is oriented so that the line joining the foci is perpendicular to the line formed by the intersection of the plane of the orbit and the plane of the sky. (a) Show that the projected orbit is an ellipse. (b) Find the eccentricity of the projected orbit.

5.21 What in Section 5.4 suggests that the semimajor axis for elliptical orbits should replace the radius for circular orbits in equation (5.20)?

5.22 What is the luminosity (in solar luminosities) of (a) a $0.5\,M_\odot$ and (b) a $5.0\,M_\odot$ star?

5.23 Show that our equation for an ellipse reduces to

$$(x/a)^2 + (y/b)^2 = 1$$

by using the transformation

$$x = r \cos \theta + a$$

$$y = r \sin \theta$$

5.24 For an elliptical orbit, calculate the angular momentum L in terms of G, m_1, m_2, and e.

CHAPTER

6

The Sun:
A Typical Star

The Sun (Fig. 6.1) is the only star we can study in any detail. It therefore serves as a guide to our pictures of other stars. Any theory of stellar structure must first be able to explain the Sun before explaining other stars. As we have seen, the Sun's spectral type places it in the midrange of main sequence stars. If we can understand the Sun, we have the hope of being able to understand a significant number of other types of stars.

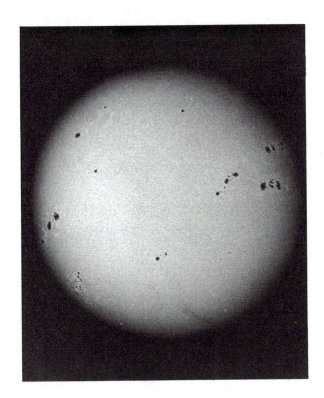

Figure 6.1 The Sun.

6.1 BASIC STRUCTURE

We have already seen that the mass of the Sun is 2×10^{33} g and that its radius is 7×10^{10} cm $(7 \times 10^5$ km). Its average density is

$$\rho = \frac{M_\odot}{(4\pi/3)R_\odot^3}$$

$$= \frac{2 \times 10^{33} \text{ g}}{(4\pi/3)(7 \times 10^{10} \text{ cm})^3}$$

$$= 1.4 \text{ g/cm}^3$$

For comparison, the density of water is 1 g/cm^3. The Sun is composed mostly of hydrogen (94.3% by number of atoms), with some helium (5.6%), and only 0.1% metals. The abundances of the individual metals are shown in Appendix G. Our best measurement of the effective temperature of the Sun is 5762 K. The solar luminosity is 3.8×10^{33} erg/s. (The effective tem-

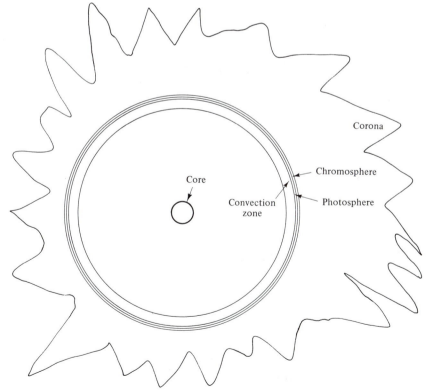

Figure 6.2 Basic structure of the Sun.

perature is the temperature that we use in the Stefan–Boltzmann law to give the luminosity.)

When we look at the Sun, we see only the outermost layers. We have to deduce the internal structure from theories of stellar structure (discussed in Chapter 9). The basic structure is shown in Fig. 6.2. In the center is the *core*. It is the source of energy generation in the Sun. Its radius is about 10% of the full solar radius. The outermost layers form the atmosphere. We divide the atmosphere into three parts. Most of the light we see comes from the *photosphere*, the bottom layer of the atmosphere. Above the photosphere is the *chromosphere*, named thus because it is the source of red light seen briefly during total eclipses of the Sun. The chromosphere is about 10^4 km thick. The outermost layer is the *corona*, which extends far into space. It is very faint, and, for most of us, can only be seen during total eclipses. Beyond the corona, we have the *solar wind*, not strictly part of the Sun, but a stream of particles from the Sun into interplanetary space.

6.2   ELEMENTS OF RADIATION TRANSFER THEORY

Radiation is being emitted and absorbed in the Sun in all layers. However, we see radiation mostly from the surface. Most radiation from below is absorbed before it reaches the surface. To understand what we are seeing when we look at the Sun, we need to understand the interaction between radiation and matter. For example much of what we know about the solar atmosphere comes from studying spectral lines, as well as the continuum. In understanding how radiation interacts with matter, known as *radiation transfer theory*, we will see how to use spectral lines to extract detailed information about the solar atmosphere.

We first look at the absorption of radiation by the atoms in matter. We can think of the atoms as acting like small spheres, each of radius r (Fig. 6.3). Each sphere absorbs any radiation that strikes it. To any beam of

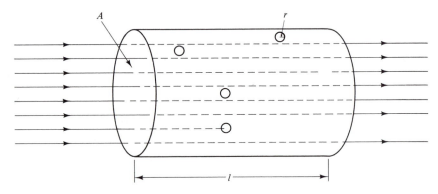

Figure 6.3 Absorption of radiation. Radiation enters from the left. Any beam striking a sphere gets absorbed.

radiation, a sphere looks like a circle of projected area πr^2. If the beam is within that circle, it will strike the sphere and be absorbed. We say that the *cross section* for striking the sphere is $\sigma = \pi r^2$. The concept of a cross section carries over into quantum mechanics. Instead of the actual size of an atom, we use the effective area over which some process (such as absorption) takes place.

We consider a cylinder of these spheres, with the radiation entering the cylinder from one end. We would like to know how much radiation gets absorbed, and how much passes through to the far end. We let n be the number of spheres per unit volume. The cylinder has length l and area A, so the volume is Al. The number of spheres in the cylinder is

$$N = nAl \qquad (6.1)$$

We define the total cross section of all the spheres as the number of spheres, multiplied by the cross section per sphere.

$$\sigma_{tot} = N\sigma \qquad (6.2)$$

$$\sigma_{tot} = nAl\sigma \qquad (6.3)$$

In making this definition we have assumed that an incoming beam can "see" all the spheres. No sphere shadows, or blocks, another. We are assured of little shadowing if the spheres occupy a small fraction of the area, as viewed from the end. That is

$$\sigma_{tot} << A \qquad (6.4)$$

Under these conditions, the fraction of the incoming radiation that will be absorbed f, is just the fraction of the total area A, that is covered by the spheres. That is

$$f = \sigma_{tot}/A \qquad (6.5)$$

Using equation (6.3), this becomes

$$f = nl\sigma \qquad (6.6)$$

We define the *optical depth* to be this quantity

$$\tau \equiv nl\sigma \qquad (6.7)$$

Our requirement in equation (6.4) reduces to $\tau << 1$. Under this restriction, the optical depth of any section of material is simply the fraction of incoming radiation that gets absorbed when the radiation passes through the material.

In general, σ will be a function of wavelength. For example, we know that at a wavelength corresponding to a spectral line, a particular atom will have a very large cross section for absorption. At a wavelength not corresponding to a spectral line, the cross section will be very small. To remind us that σ is a function of λ (or ν), we write it as σ_λ (or σ_ν). This means that the optical depth is also a function of λ (or ν), so we rewrite equation (6.7) as

$$\tau_\lambda = nl\sigma_\lambda \qquad (6.8)$$

In our discussions, the quantity nl occurs often. It is the product of a number density and a length, so its units are measured in number per unit area. It is the number of particles along the full length l of the cylinder per unit surface area. For example, if we are measuring lengths in centimeters, it is the number of particles in a column whose face surface area is 1 cm^2, and whose length is l, the full length of the large cylinder. We call this quantity the *column density*.

We can see that the optical depth depends on properties of the material—e.g., cross section and density of particles—and on the overall size of the absorbing region. It is sometimes convenient to separate these two dependences by defining the *absorption coefficient*, which is the optical depth per unit length through the material,

$$\kappa_\lambda = \tau_\lambda/l \qquad (6.9a)$$

$$= n\sigma_\lambda \qquad (6.9b)$$

If κ_λ gives the number of absorptions per unit length, then its inverse gives the mean distance between absorptions. This quantity is called the *mean free path,* and is given by

$$L_\lambda = 1/\kappa_\lambda \qquad (6.10)$$

$$= 1/n\sigma_\lambda \qquad (6.11)$$

In terms of these two quantities, the optical depth is given by

$$\tau_\lambda = l/L_\lambda \qquad (6.12a)$$

$$= \kappa_\lambda l \qquad (6.12b)$$

In the above discussion, we required that the optical depth be much less than one. Our interpretation of τ as the fraction of radiation absorbed only holds for $\tau \ll 1$. What if this is not the case? We then have to divide our cylinder into several layers. If we make the layers thin enough, we can be assured that the optical depth for each layer will be very small. We then follow the radiation through, layer by layer, looking at the fraction absorbed in each layer (Fig. 6.4).

Let's look at radiation passing through some layer with optical depth $d\tau$. The intensity of radiation striking that layer is I. Since $d\tau$ is the fraction of this radiation that is absorbed, the amount of radiation absorbed in this layer is $Id\tau$. The amount of radiation passing through to the next layer is $I(1 - d\tau)$. The change in intensity dI, while passing through the layer, is

$$dI = I(\text{out}) - I(\text{in}) \tag{6.13}$$

$$= -I \, d\tau \tag{6.14}$$

Note that dI is negative, since the intensity is decreased in passing through the layer.

Now that we know how to treat each layer, we must add up the effect of all of the layers to get the effect on the whole sample of material. We can see that we have formulated the problem so that we are following I as a function of τ. We let τ' be the opacity through which the radiation has already passed by the time it reaches a particular layer, and I' be the intensity reaching that layer. Then τ' ranges from zero at the point where the radiation enters the material, to τ, the full optical depth at the point where the radiation leaves the material. Over that range, I' varies from I_0, the incident intensity, to I, the final intensity. Using equation (6.14), we get

$$dI'/I' = -d\tau' \tag{6.15}$$

In this form, all the I' dependence is on the left, and all of the τ' dependence is on the right. To add up the effect of all the layers, we integrate equation (6.15) between the limits given above:

$$\int_{I_0}^{I} (dI'/I') = -\int_{0}^{\tau} d\tau' \tag{6.16}$$

$$\ln(I) - \ln(I_0) = -\tau$$

Using the fact that $\ln(a/b) = \ln(a) - \ln(b)$, this becomes

$$\ln(I/I_0) = -\tau \tag{6.17}$$

Raising e to the value on each side, remembering that $e^{\ln x} = x$, and multiplying both sides by I_0, gives

$$I = I_0 e^{-\tau} \tag{6.18}$$

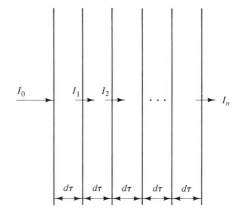

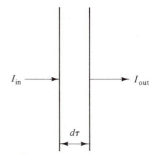

Figure 6.4 Radiation passing through several layers. Each layer has an optical depth $d\tau$. The bottom figure is for calculating the effect of each layer.

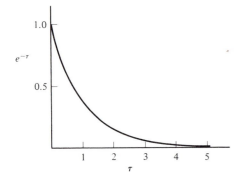

Figure 6.5 $e^{-\tau}$ vs. τ, showing the falloff in transmitted radiation as the optical depth increases. Note that the curve looks almost linear for small τ. For large τ, it asymptotically approaches zero.

We can check this result in the limit $\tau << 1$, called the *optically thin* limit, using the fact that $e^x \cong 1 + x$ for $x << 1$. In this case, equation (6.18) becomes

$$I = I_0(1 - \tau) \qquad (6.19)$$

This is the expected result for small optical depths, where τ again becomes the fraction of radiation absorbed.

As shown in Fig. 6.5, $e^{-\tau}$ falls off very quickly with τ. This means that to escape from the Sun, radiation must come from within approximately one optical depth of the surface. This explains why we only see the outermost layers. Since the absorption coefficient κ_λ is a function of wavelength, we can see to different depths at different wavelengths. At a wavelength where κ_λ is large, we don't see very far into the material. At wavelengths where κ_λ is small, it takes a lot of material to make $\tau_\lambda = 1$, so we see far into the material. We take advantage of this to study conditions at different depths below the solar surface.

So far we have only looked at the absorption of radiation passing through each layer. However, radiation can also be emitted in each layer, and the amount of emission also depends on the optical depth. In general, we must carry out complicated radiative transfer calculations to take all effects into account. To solve these problems we use large computers to make mathematical models of stellar atmospheres. In these calculations, we input the distribution of temperature, density, and composition, and predict the spectrum that we will see, including emission and absorption lines. We vary the input parameters until we find models that produce predictions that agree with observations. The results of these calculations are not unique, but they do give us a feel for what processes are important in stellar atmospheres. The more observational data we can predict with the models, the more confident we can be that the temperatures, densities, and compositions we derive are close to the actual ones.

6.3 THE PHOTOSPHERE

Most of the visible photons we receive from the Sun originate in the photosphere. One question we might ask is why we see continuum radiation at all. We have already seen how atoms can emit or absorb energy at particular wavelengths, producing spectral lines. However, we have not discussed the source of the continuum opacity. It turns out that most of the continuum opacity in the Sun at optical wavelengths is due to the presence of H^- ions. An H^- ion is an H atom to which an extra electron has been added. This extra electron is held only very weakly to the atom. Very little energy is required to remove it again. H^- ions are present because there is so much hydrogen, and there are a large number of free electrons to collide with the atoms.

If we have an H^- ion and a photon (γ) strikes it, the photon can be absorbed and the electron set free:

$$H^- + \gamma \rightarrow H + e^-$$

The H^- ion is a bound system. The final state has an H atom and a free electron. We call this process a *bound–free* process. In such a process, the wavelength of the incoming photon is not restricted, as long as the photon energy is sufficient to remove the electron. The electron in the final state can have any kinetic energy, so a continuous range of photon energies is possible. This process then provides most of the continuum opacity of the photosphere. The continuum emission results from the inverse process.

6.3.1 Appearance of the Photosphere

We have said that the Sun is the one star we can study in great detail. To do this we try to observe the photosphere with the best resolution possible. When we observe the sun, the light-gathering power of our telescope is not usually a problem. Therefore, we can try to spread the image out over as large an area as possible, making it easier to study detailed structure. We therefore want a telescope with a long focal length to give us a large image scale. The solar telescope shown in Fig. 6.6 provides this type of detailed picture.

Figure 6.6 Solar telescope at Kitt Peak observatory. The telescope has a very long focal length, so that we can produce a large image and study the detailed appearance. Since the tube is so long, it is not reasonable to move it. Instead, the large mirror at the top is moved to keep the sunlight directed down the tube.

Figure 6.7 Granulation in the Sun. Remember, the darker areas are not really dark. They are just a little cooler than the bright areas.

Figure 6.8 Granulation and convection zones. This is a side view to show what is happening below the surface. Hotter gas is being brought up from below, producing the bright regions. The cooler gas, which produces the darker regions, is carried down to replace the gas that was brought up.

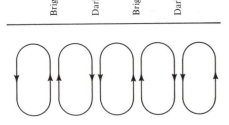

When we look on a scale of a few arc seconds, we see that the surface of the Sun does not have a smooth appearance (Fig. 6.7). We see a structure, called *granulation,* in which lighter areas are surrounded by darker areas. The darker areas are not really dark. They are just a little cooler than the lighter areas, and only appear dark in comparison to the light areas. The granules are typically about 1000 km across. The pattern of granulation also changes with time, with a new pattern appearing every 5 to 10 min.

We interpret this granulation as telling us something about the underlying structure we cannot see directly. The granulation can be explained by circulating cells of material, called *convection zones* (Fig. 6.8). (Convection is the form of energy transport in which matter actually moves from one place to another.) The bright regions are warmer gas rising up from below. The dark regions are cooler gas falling back down.

In addition to the granulation variations, there is also a variation called the *five-minute oscillation,* in which parts of the photosphere are moving up and down. We think that this oscillation results from standing acoustic waves in the upper layers of the convection zone. This type of oscillation is one of many that are studied for clues to the Sun's interior structure. This area of research is called solar seismology.

One interesting question about the photosphere concerns the sharpness of the solar limb. The Sun is a ball of gas whose density falls off continuously as one gets farther from the center. There is no sharp boundary, yet we see a definite edge to the Sun. In Fig. 6.9 we see some lines of sight through the photosphere. As the line of sight passes farther from the center of the Sun, the opacity decreases because (1) the path length through the Sun is less, and (2) it passes through less-dense regions. Since the amount of light getting through is proportional to $e^{-\tau}$, the effect of τ changing from line of sight to line of sight is enhanced by the exponential behavior. Therefore, the transition from the Sun being mostly opaque to being mostly transparent takes place over a region that is small compared to our resolution, and the edge looks sharp.

6.3.2   Temperature Distribution

Another interesting phenomenon near the solar limb can be seen in the photograph in Fig. 6.1. The Sun does not appear as bright near the limb as near the center. This *limb darkening* is also an optical depth effect, as shown in Fig. 6.10. We compare two lines of sight: (1) toward the center of the Sun from the observer, and (2) offset from the center of the Sun. In each line of sight, we can only see down to $\tau \cong 1$. Line of sight #1 is looking straight down into the atmosphere, so it gets closer to the center of the Sun before an optical depth of one is reached than does line of sight #2, which has a longer path through any layer. We see deeper into the Sun on line of sight #1 than we do on line of sight #2. (Since line of sight #1 takes us the

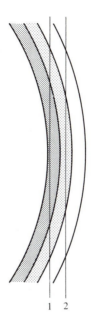

Figure 6.9 Lines of sight through the solar limb. For clarity, we think of the Sun as being composed of a series of spherical shells. The density in each shell decreases as we get farther from the center. This decreasing density is indicated by the shading. Two lines of sight are indicated. Note that most of each line of sight is in the densest layer through which that particular line passes. Even though line #2 is not shifted very far from line #1, line #1 passes through much more material.

deepest into the Sun, we define the point at which τ reaches one on this line as being the *base of the photosphere*. When we talk about the temperature of the Sun, we are talking about the temperature at the base of the photosphere.)

When we look at the Sun, it appears brighter along line of sight #1 than it does along line of sight #2. This means that the Sun is hotter at the end of #1 than at the end of #2. From this we conclude that the photosphere

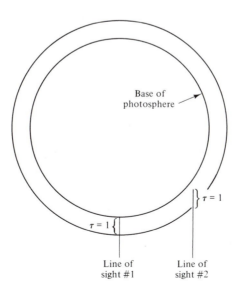

Figure 6.10 Limb darkening. Line of sight #1 is directed from the observer towards the center of the Sun, so it takes the shortest path through the atmosphere. This line allows us to see the deepest into the photosphere, and the base of the photosphere is defined to be where the optical depth τ along this line reaches 1. Line of sight #2 is closer to the edge, so it doesn't allow us to see as deep into the photosphere. If the temperature decreases with increasing height in the photosphere, then line of sight #1 allows us to see hotter material than does line of sight #2, and the edge of the Sun appears darker than does the center.

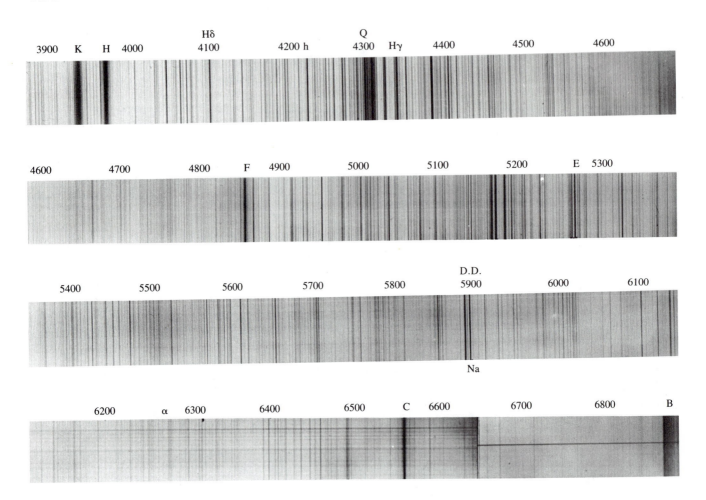

Figure 6.11   The solar spectrum.

gets cooler as one gets farther from the center of the Sun. If the photosphere got hotter as one gets farther from the center, we would see limb brightening.

We get more useful information about the photosphere by studying its spectral lines (Fig. 6.11). The spectrum shows a few strong absorption lines and a myriad of weaker ones. The stronger lines were labeled A through K by Fraunhofer in 1814. These lines have since been identified. For example, the C line is the first Balmer line (Hα); the D line is a pair of lines belonging to neutral sodium (NaI); and the H and K lines belong to singly ionized calcium (CaII). Sodium and calcium are much less abundant than hydrogen, but their absorption lines are as strong as Hα. We have already seen in Chapter 3 that this can result from the combined effects of excitation and ionization.

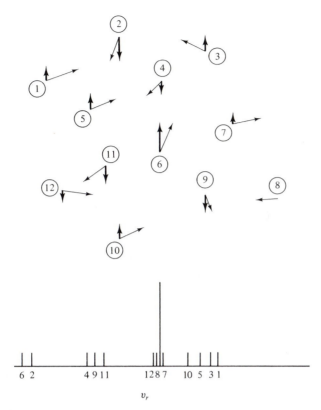

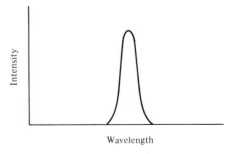

Figure 6.12   Doppler broadening. The top shows the (random) motions of a group of particles. The lighter vectors are the actual velocities; the darker vectors are the radial components, which produce the Doppler shift. For each particle, identified by a number, the radial velocities are plotted below. The line profile is the sum of all the individual Doppler-shifted signals, and with many more particles, it would have a smooth appearance.

6.3.3   Doppler Broadening of Lines

When we study the lines with good spectral resolution, we can look at the details of the line profile. The lines are broadened by Doppler shifts due to the random motions of the atoms and ions in the gas (Figs. 6.12 and 6.13). If all the atoms were at rest, all the photons from a given transition would emerge with a very small spread in wavelength. However, the atoms are moving with random speeds in random directions. We therefore see a spread of Doppler shifts, and the line is broadened. If the gas is hotter, both the spread in speeds and the broadening are greater.

We can estimate the degree of broadening as a function of temperature. If $\langle v^2 \rangle$ is the average of the square of the random velocities in a gas, and m is the mass per particle, then the average kinetic energy per particle is $\frac{1}{2}m\langle v^2 \rangle$. If we have an ideal monatomic gas, this should equal $\frac{3}{2}kT$, giving

$$\tfrac{1}{2}m\langle v^2 \rangle = \tfrac{3}{2}kT \tag{6.20}$$

Solving for $\langle v^2 \rangle$ gives

$$\langle v^2 \rangle = 3kT/m \tag{6.21}$$

Figure 6.13   Line profile. We plot intensity as a function of wavelength.

Taking the square root gives us the root mean square (rms) speed

$$v_{\text{rms}} = (3kT/m)^{1/2} \tag{6.22}$$

This gives us an estimate of the range of speeds we will encounter. (The range will be larger, since we can have atoms coming toward us or away from us, and since some atoms will be moving faster than this average, but the effective spread is reduced since only the component of motion along the line of sight contributes.) To find the actual wavelength range over which the line is spread out, we would use this speed in the Doppler shift expression; see equation (5.4).

EXAMPLE 6.1   Doppler Broadening

Estimate the wavelength broadening in the Hα line in a gas composed of hydrogen atoms at $T = 5500$ K.

Solution

Using equation (6.22) gives

$$v_{\text{rms}} = \left[\frac{(3)(1.38 \times 10^{-16} \text{ erg/K})(5.5 \times 10^3 \text{ K})}{1.7 \times 10^{-24} \text{ g}} \right]^{1/2}$$

$$= 1.2 \times 10^6 \text{ cm/s}$$

From equation (5.4) we estimate the linewidth as

$$\Delta\lambda = \lambda v_{\text{rms}}/c$$

$$= \frac{(6562.8 \text{ Å})(1.2 \times 10^6 \text{ cm/s})}{3.0 \times 10^{10} \text{ cm/s}}$$

$$= 0.3 \text{ Å}$$

In any spectral line, smaller Doppler shifts are more likely than larger ones, so the optical depth is greatest at the center of the line, and falls off to either side. At different Doppler shifts away from the line center, we will see different distances into the Sun. The farther we are from the line center, the deeper we see. By studying line profiles in detail we learn about physical conditions at different depths. Also, each spectral line has a different optical depth in the line center, so different lines allow us to see down to different depths in the photosphere. When we perform model atmosphere calculations we try to predict as many of the features of the spectrum as possible, including the relative strengths of various lines and the details of certain line profiles.

From observations of various spectral lines, the temperature profile for the photosphere has been derived. It is shown in Fig. 6.14. Note that the temperature falls as one goes up from the base of the photosphere. This is what we might expect, since we are getting farther from the heating source. However, an interesting phenomenon is observed. The temperature reaches a minimum 500 km above the base of the photosphere and then begins to rise with altitude. We will see, below, that this temperature rise continues into the higher layers.

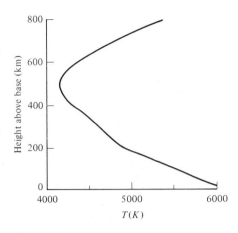

Figure 6.14 Temperature vs. height in the photosphere.

6.4 THE CHROMOSPHERE

At most wavelengths the chromosphere is optically thin, so we see right through it to the photosphere. Under normal conditions the continuum radiation from the photosphere overwhelms that from the chromosphere. However, during a total eclipse of the sun, just before and after totality, the moon blocks the light from the photosphere, but not from the chromosphere. For that brief moment we see the red glow of the chromosphere. The red glow comes from Hα emission. The optical depth of the Hα line is sufficiently large that we can study the chromosphere by studying that line. At the center of the Hα line we see down only to 1500 km above the base of the photosphere.

We can also study the large-scale structure of the chromosphere by taking photographs through filters that only pass light from the Hα line. One such photograph is shown in Fig. 6.15. In this picture we see granulation on an even larger scale than in the photosphere. This is called *supergranulation*. The super granules are some 30,000 km across. As with the granules in the photosphere, the matter in the center of the supergranules is moving up and the matter at the edges is moving down. These motions can be determined from Doppler shifts of spectral lines. We also see smaller-scale irregularities in the chromosphere, called *spicules*. These are protrusions from the surface some 700 km across and 7000 km high.

When the chromosphere is visible just before and after totality, we see only a thin sliver of light. The effect is the same as if the light has been passed through a curved slit. If we then use a prism or grating to spread out the different wavelengths, we will get a line spectrum, though each line will appear curved (Fig. 6.16). This spectrum is called a *flash spectrum* because it is only visible for the brief instant that the Moon is covering all of the photosphere but not the chromosphere. Note that the spectrum shows emission lines. This is because there is no strong continuum to be absorbed.

When we study the spectra of the chromosphere we find that it is hotter than the photosphere. The chromospheric temperature is about 15,000 K.

Figure 6.15 Supergranulation is seen in this image showing the Doppler shift in the FeI line at 5103 Å.

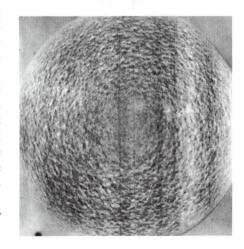

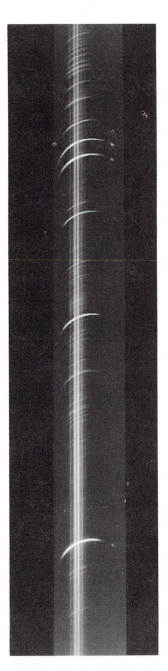

Figure 6.16 Flash spectrum. Just before or after totality in a solar eclipse, the thin edge of the chromosphere can produce a spectrum in a spectrograph, with no slit needed.

(The Sun doesn't appear to be this hot because the chromosphere is optically thin and doesn't contribute much to the total radiation that we see.) We are faced with the question of the temperature rising as we get farther from the center of the Sun. We will discuss this point further in Section 6.5 when we talk about the corona.

6.5   THE CORONA

6.5.1   Parts of the Corona

The corona is most apparent during total solar eclipses (Fig. 6.17), when the much brighter light from the photosphere and chromosphere is blocked out. The corona is simply too faint to be seen when any photospheric light is present. You might think that we can simulate the effect of an eclipse by holding a disk over the sun. If you try this, light that would come directly from the photosphere to your eye will be blocked out. However, some photospheric light that is originally headed in a direction other than directly at you will scatter off the atoms and molecules in the Earth's atmosphere, and reach you anyway (Fig. 6.18). This scattered light is only a small fraction of the total photospheric light, but is still enough to overwhelm the faint corona. This is not a problem during solar eclipses because the Moon is outside the atmosphere, so there is nothing to scatter light around it.

Therefore, solar eclipses still provide us with unique opportunities to study the corona. For this purpose, we are fortunate that the Moon subtends almost the same angle as the Sun, as viewed from the Earth. The Moon can exactly block the photosphere and chromosphere, but not the corona. Unfortunately, we do not get an eclipse of the Sun every month. We would if the Moon's orbit were in the same plane as the Earth's orbit around the Sun. However, the Moon's orbit is inclined by 5°, so total eclipses of the Sun are rare events. The average time between total solar eclipses is about $1\frac{1}{2}$ years. Even when one occurs, the total eclipse is observable from a band not more than 300 km wide, and totality lasts only a few minutes.

Solar astronomers take advantage of eclipses whenever possible, but also look for other ways to study the corona. It turns out that some ground-based, noneclipse observations are possible. Telescopes, called *coronagraphs*, have disks to block out the photospheric light, and cut down as much scattered light as possible. They are placed at high-altitude sites, with very clear skies. For example, the Haleakala Crater on the island of Maui (operated by the University of Hawaii) is at an altitude of 3,000 m (10,000 ft). Other important coronagraphs are at Sacramento Peak in New Mexico and the Pic du Midi, in the French Pyrenees.

One way to get around the scattering in the Earth's atmosphere is to put a coronagraph in space. This is not quite as good as a solar eclipse, since

Figure 6.17 Total solar eclipse, showing the corona.

space probes are not totally free of escaping gases. Some photospheric light is scattered by these gases. However, the results are much better than for a ground-based telescope. They also have an advantage over eclipse studies in that they allow for continuous study of the corona. For example, the orbiting Orbiting Solar Observatory 7 (OSO-7) provided observations of the corona over the period 1971–1974. In 1973 and 1974, observations of the corona were made by the astronauts in Skylab.

There is one additional ground-based technique for studying the corona. It involves studying radio waves. Radio waves pass through the Earth's atmosphere and are not appreciably scattered. We, therefore, don't have to worry about the radio waves behaving like the visible light in Fig. 6.18.

In discussing what we have learned so far about the corona, we divide the corona into three parts:

1. The *E-corona* is a source of emission lines directly from material in the corona. These lines come from highly ionized species, such as FeXIV (thirteen times ionized iron). If we look at the Saha equation, discussed in Chapter 3, we see that highly ionized states are favored under conditions of high temperature and low density. The high temperature provides the energy necessary for the ionization. The low density means that collisions leading to recombinations will be rare.

2. The *K-corona* (from the German *Kontinuierlich*, for continuous) is the result of photospheric light scattered from electrons in the corona.

3. The *F-corona* (for Fraunhofer) is not really part of the sun. It comes from photospheric light scattered by interplanetary dust. Since we are just seeing the reflected photospheric light, this light still has the Fraunhofer spectrum. Both the F- and K-coronas appear at approximately the same an-

Figure 6.18 Effect of scattered light on corona studies. We can use an occulting disk, but sunlight can scatter off particles in the Earth's atmosphere and reach our telescope. Since the corona is very faint, and the Earth's atmosphere is very efficient at scattering, especially for blue light, this scattered light can overwhelm the direct light we get from the corona.

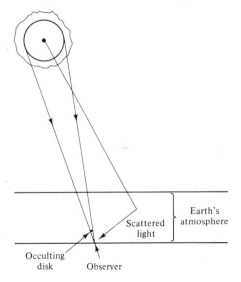

Earth's atmosphere

Scattered light

Occulting disk

Observer

gular distance from the center of the Sun, but there are experimental ways of separating their contributions to the light that we see.

6.5.2 Temperature of the Corona

When we analyze the abundance of highly ionized states, the Doppler broadening of lines, and the strength of the radio emission, we find that the corona is very hot, about 2×10^6 K. As we have stated, the density is very low, approximately 10^{-9} times the density in the Earth's atmosphere.

Again, we must explain why a part of the atmosphere farther from the center of the Sun is hotter than a part closer to the center. We should note that it is not necessarily hard to keep something hot if it can't lose heat efficiently. For example, in a well-insulated oven, once the required temperature has been reached the heat source can be turned off and the temperature of the oven still stays high. An ionized gas (plasma) like the corona can lose energy only through collisions between the particles. For example, an electron and an ion could collide, with some of the kinetic energy going to excite the ion into a higher state. The ion can then emit a photon and return to its lower-energy state. The emitted photon escapes and its energy is lost to the gas. If the lost energy is not replaced the gas will cool.

Since collisions play an important role in the above process, the rate at which the gas can lose energy will depend on the rate at which collisions occur. The rate of collisions should be proportional to the product of the densities of the two colliding species. However, the density of each species is roughly proportional to the total density ρ of the gas (each being some fraction of the total). This means that the collision rate is proportional to ρ^2. Of course, the amount of gas that has to be cooled is proportional to ρ. We can define a *cooling time,* which is the ratio of the gas volume to be cooled $\alpha\rho$ to the gas collision rate $\alpha\rho^2$. The cooling time is then proportional to $1/\rho$. Therefore, low-density gases take longer to get rid of their stored heat.

Further, if the density is very low, a high temperature doesn't necessarily mean a lot of energy stored. The energy is $\frac{3}{2}kT$ per particle. Even if the quantity kT is very large, if the total number of particles is very small, the energy stored is not as large as if the density were much higher.

EXAMPLE 6.2   Energy Density in the Corona and the Earth's Atmosphere

Compare the energy density (energy per unit volume) in the corona with that in the Earth's atmosphere.

Solution

For each, the energy density is proportional to the density of particles n and to the temperature T. (All other constants will drop out when we take a ratio.) Therefore

Energy density corona/energy density Earth

$$= [n(\text{corona})/n(\text{Earth})][T(\text{corona})/T(\text{Earth})]$$

$$= (1 \times 10^{-9}) [(2 \times 10^{6})/(3 \times 10^{2})]$$

$$= 7 \times 10^{-6}$$

Even though the corona is so hot, its very low density gives it a lower energy density.

We must still come up with some explanation for getting energy into the corona. Some have suggested mechanisms in which oscillations near the surface of the Sun send supersonic sound waves (shock waves) into the Sun's upper atmosphere. In addition, there are mechanisms for heating that involve the Sun's magnetic field. These theories are still under study, and we still do not have a definitive picture of the energy balance in the corona.

When we study photographs of the corona, we find its structure is very irregular. We often see long streamers (Fig. 6.19) whose appearance varies with time. We think these phenomena are related to the Sun's magnetic field, as are other aspects of solar activity (to be discussed in the next section).

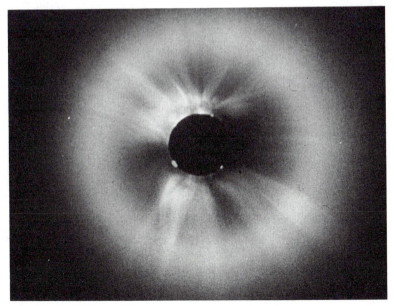

Figure 6.19   Coronal streamers.

6.6 SOLAR ACTIVITY

6.6.1 Sunspots

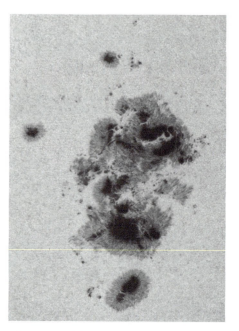

Figure 6.20 *Closeup of a sunspot group. The darker inner area of each spot is the umbra, and the lighter outer area of each spot is the penumbra. The spots appear dark because the surrounding areas are brighter.*

When we look at photographs of the Sun (Figs. 6.1 and 6.20) we note a pattern of darker areas. As with the granulation, the dark areas are not really dark. They are just not as bright as the surrounding areas. The gas in these darker areas is probably at a temperature of about 3800 K. A closeup of the *sunspots* shows that they have a darker inner region (the *umbra*) surrounded by a lighter region (the *penumbra*).

The number of sunspots on the Sun is not even approximately constant. It varies in a regular way, as shown in Fig. 6.21. It was realized in the mid-19th century that sunspot numbers follow an eleven-year cycle. The number of sunspots in a peak year is not the same as in another peak year. However, the peaks are easily noticeable.

We see more regularity to the pattern when we plot the height of each sunspot above or below the Sun's equator as a function of when the sunspot appeared. This was done in 1904 by E. Walter Maunder. An example of such a diagram is shown in Fig. 6.22. Early in an eleven-year cycle, sunspots appear far from the equator. Later in the cycle they appear closer to the equator. This results in a butterflylike pattern, and, in fact, these diagrams are sometimes called butterfly diagrams.

When Maunder investigated records of past sunspot activity, he found that there was an extended period when no sunspots were observed. This period, from 1645 to 1715, is known as the *Maunder minimum*. This minimum has recently been reinvestigated by John A. Eddy, who found records to indicate that no aurorae were observed for many years during this period. Also, during this time, weak coronae were reported during total solar eclipses.

Figure 6.21 Numbers of sunspots per year as a function of year. The 11-yr pattern is evident. Note that the number in a peak year is not the same from cycle to cycle.

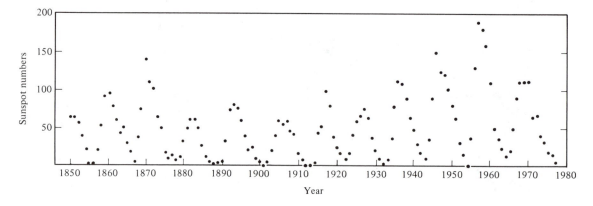

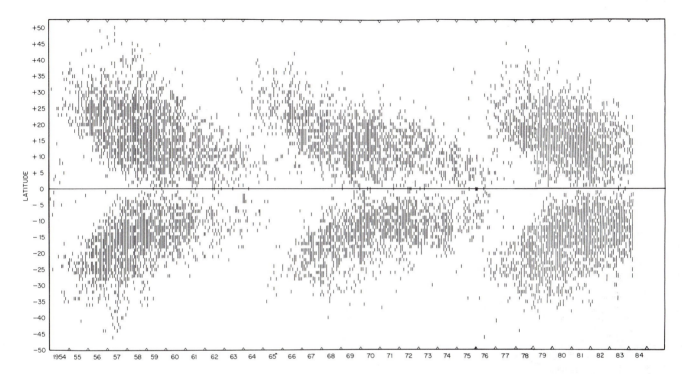

Figure 6.22 Butterfly diagram. This shows the distribution of sunspots in solar latitude as a function of time in the sunspot cycle. Early in a cycle, spots appear at higher latitudes; late in a cycle, they appear close to the equator. Note that a new cycle starts before the old cycle completely ends.

An unusual correlation was also found in the growth rings in trees, suggesting that the altered solar activity had some effect on growth on Earth. The mechanism by which this takes place is poorly understood, but we can use the growth rings as an indicator of solar activity farther in the past than our records. A study of growth rings in old trees indicates that the Maunder minimum was not unique. There may have been several periods in the past with extended reduced solar activity.

Sunspots appear to be regions where the magnetic field is higher than on the rest of the Sun. In discussing sunspots, we should briefly review properties of magnetic fields and matter. Magnetic fields arise from moving (including spinning) charges. Magnetic field lines form closed loops (Fig. 6.23). This is equivalent to saying that there are no point sources of magnetic charge, magnetic monopoles. (We will discuss the implication of the possible existence of magnetic monopoles in Chapter 21.)

The magnetic field strength **B** is defined such that the magnetic force on a charge q, moving with velocity **v,** is (in cqs units)

$$\mathbf{F} = q\mathbf{v} \times \mathbf{B}/c \qquad (6.23)$$

where the × indicates a vector cross product. There is no magnetic force on a charge at rest or on a charge moving parallel to the magnetic field lines.

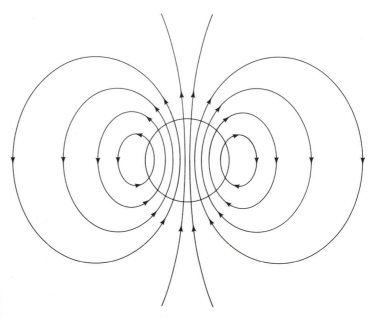

Figure 6.23 Field lines for a dipole magnetic field.

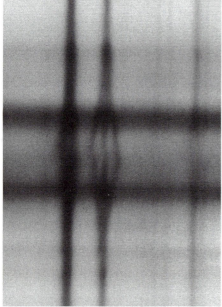

Figure 6.24 Zeeman effect in sunspots. The spectrometer slot was placed across a sunspot. In the spectrum, on the left and right, away from the spot, the spectral lines are unsplit. In the center, near the spot, some of the spectral lines are split into three. The stronger the magnetic field, the greater the splitting.

The force is maximum when the velocity is perpendicular to the magnetic field. The force is perpendicular to both the direction of motion and to the magnetic field. This means that the component of motion parallel to the field line is not altered, but the component perpendicular to the field line is. Charged particles will move so that they spiral around magnetic field lines.

In addition to exerting a force on a moving charge, a magnetic field will exert a torque on a current loop. We can think of a current loop as having a magnetic dipole moment (Fig. 6.23), and the torque will cause the dipole to line up with the field lines. In this way, the dipole moment of a compass needle lines up with the Earth's magnetic field.

How do we measure the Sun's magnetic field? For certain atoms placed in a magnetic field, energy levels will shift. This is known as the *Zeemann effect* (Fig. 6.24). Different energy levels shift by different amounts. Some transitions that normally appear as one line split into a group of lines. The amount of splitting is proportional to the strength of the magnetic field. (We can check the amount of splitting in different atoms in various magnetic fields in the laboratory.)

Measurements by George Ellery Hale in 1908 first showed that the magnetic fields are stronger in sunspots. He also found that sunspots occur in pairs, with one corresponding to the north pole of a magnet and the other to the south pole. In each sunspot pair, we can identify the one that ''leads'' as the Sun rotates. In a given solar hemisphere, the polarity of the leading spot of all pairs is the same. That polarity is different in the two hemispheres.

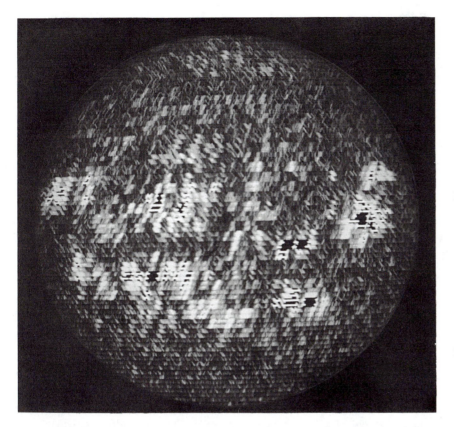

Figure 6.25   Solar magnetic field. This shows the result of observations of the Zeeman effect, brighter areas correspond to stronger fields.

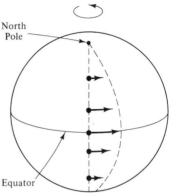

This polarity reverses in successive eleven-year cycles. During one cycle all the leading sunspots in the northern hemisphere will be magnetic north, while those in the south will be magnetic south. In the next cycle all the leading sunspots in the northern hemisphere will be magnetic south. The Sun's magnetic field reverses every eleven years! This means that the sunspot cycle actually is a 22-year cycle in the Sun's magnetic field.

If the Sun's magnetic field arose in the core, as the Earth's does, we would expect it to be quite stable. (There is geological evidence that the Earth's magnetic field reverses periodically, but on geological time scales, not every eleven years.) We now think that the Sun's magnetic field arises · below the surface. We also know that the Sun does not rotate as a rigid body. By following sunspot groups (Fig. 6.26), we see that the material at the equator takes less time to go around than material at higher latitudes. For example, it takes 25 days for material at the equator to make one circuit, while it takes 28 days at 40° latitude.

As the Sun rotates differentially, the magnetic field gets distorted. This

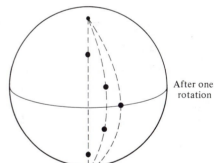

Figure 6.26   Differential rotation of the Sun, as traced out by sunspots. Since the Sun rotates faster at the equator than at higher latitudes, sunspots at the equator take less time to make one rotation than do those at higher latitudes. In this schematic diagram, a group of sunspots starts on the same meridian, but after one rotation of the Sun, they are on different meridians.

Figure 6.27   This shows how the solar magnetic field gets twisted and kinked by the differential rotation of the Sun.

is because the charged particles in the matter cannot move across field lines, so the field lines are carried along with the material. (We say that the magnetic field is frozen into the material.) The development of the magnetic field is shown in Fig. 6.27. As the field lines wind up, the field gets very strong in places. Kinks in the field lines develop and break through the surface. The sunspots apparently arise through some, as yet poorly understood, dynamo motion, involving convective motion and the magnetic fields.

6.6.2   Other Activity

Sunspots are just one manifestation of solar activity related to the sun's magnetic field. Another form of activity is the *solar flare,* shown in Fig. 6.28. A flare involves a large ejection of particles. Flares develop very quickly and last tens of minutes to a few hours. Temperatures in flares are high, up to 5×10^6 K. They also give off strong Hα emission, and flares are seen well when the Sun is photographed through an Hα filter. Flares have been detected to give off energy in all parts of the electromagnetic spectrum. The cause of flares is not well understood, but they appear to be related to strong magnetic fields and the flow of particles along field lines.

Solar activity is also manifested in *plages* (from the French word for "beach"). Plages are bright regions around sunspots. They show up in Hα photos of the Sun. They remain after sunspots disappear. Plages are apparently chromospheric brightening caused by the strong magnetic fields.

Filaments are dark bands near sunspot regions. They can be up to 10^5 km long. Filaments appear to be boundaries between regions of opposite magnetic polarity. When filaments are seen projected into space at the limb of the Sun, they appear as *prominences* (Fig. 6.29). Some prominences vary on short time scales, and other evolve more slowly.

The *solar wind* is a stream of particles from the Sun into interplanetary space. We can see the effects of the solar wind when we look at a comet that is passing near the Sun. The tail of the comet always points away from

Figure 6.28   In this photograph, taken through a Hα filter, we are looking down on a large solar flare. This flare appears brighter than its surroundings in the Hα line.

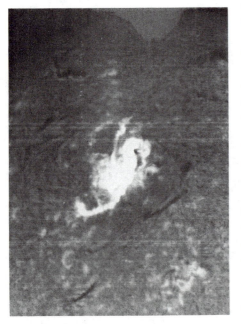

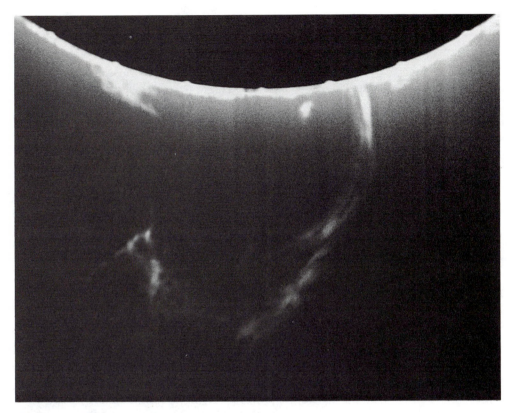

Figure 6.29 This large prominence was photographed during a solar eclipse.

the Sun. This is because the material in the tail was driven out of the head of the comet by the solar wind. The rate at which the Sun is losing mass is $10^{-14}\ M_\odot/\mathrm{yr}$. The wind is still accelerating in its 5- to 10-day trip from the Sun to the Earth. At the Earth's orbit, its speed is about 400–450 km/s, and the density of particles is 5–10 cm^3. The particles in the solar wind are all positive ions and electrons. It is thought that the solar wind originates in lower-density areas of the corona, called *coronal holes*.

The solar wind can have an effect on the Earth. Most of the solar wind particles directed at the Earth never reach the surface of the Earth. The Earth's magnetic field serves as an effective shield, since the charged particles cannot travel across the field lines. Some of the particles, however, travel along the field lines and come closest to the surface of the Earth near the magnetic poles. These charged particles are responsible for the aurora displays. When solar activity is increased, the aurorae become more widespread. The increased abundance of charged particles in the atmosphere also creates radio interference.

CHAPTER SUMMARY

We looked in this chapter at the one star we can study in detail, the Sun.

Most of what we see in the Sun is a relatively thin layer, the photosphere. In studying radiative transfer, we saw how we can see to different depths in the photosphere by looking at different wavelengths. The distance we see corresponds to approximately one optical depth.

The photosphere doesn't have a smooth appearance. Instead, it has a granular appearance, with the granular pattern changing on a time scale of several minutes. This suggests convection currents below the surface. Supergranulation suggests even deeper convection currents.

The chromosphere is difficult to study. In the chromosphere the temperature begins to increase as one gets farther from the center of the Sun. This trend continues dramatically into the corona. The best studies of the conrona have come during total solar eclipses and from space.

We saw how sunspots appear to be places of intensified magnetic fields, as evidenced by the Zeemann effect. The number of spots goes through a 22-year cycle, which represents a complete reversal of the Sun's magnetic field. The structure of the magnetic field is also related to other manifestations of solar activity, such as prominences.

QUESTIONS AND PROBLEMS

6.1   Appendix G gives the composition of the Sun, measured by the fraction of the number of nuclei in the form of each element. Express the entries in this table as the fraction of the mass that is in each element. (Do this for the ten most abundant elements.)

6.2   Calculate the effective temperature of the Sun from the given solar luminosity and radius, and compare your answer with the value given in the chapter.

6.3   Assume that for some process the cross section for absorption of a certain wavelength photon by H is 10^{-16} cm^2, and the density of H is 1 g/cm^3. (a) Suppose we have a cylinder that is 1 m long and has an end area of 1 cm^2. What is the total absorption cross section for this? How does it compare with the area of the end? (b) What is the absorption coefficient (per unit length)? (c) What is the mean free path? (d) How long a sample of material is needed to produce an optical depth of one?

6.4   Suppose we have a uniform sphere of 1 $M_\odot$ of hydrogen. What is the column density through the center of the sphere? (Assume $R = R_\odot$.)

6.5   How large must the optical depth through a material be for the material to absorb: (a) 1% of the incident photons; (b) 10% of the incident photons; (c) 50% of the incident photons; (d) 99% of the incident photons?

6.6   If we have a material that emits uniformly over its volume, what fraction of the photons that we see come from within one optical depth of the surface?

6.7   In studying radiative transfer effects we let τ be a measure of where we are in a given sample, rather than a position x. Explain why we can do this, and how τ and x are related.

6.8   Suppose we divide a material into N layers, each with optical depth $d\tau = \tau/N$, where τ is the total optical depth through the material and $d\tau << 1$. (a) Show that if radiation I_0 is incident on a layer and I emerges from that layer, then

$$I = I_0(1 - d\tau)^N$$

(b) Show that this reduces to $I = I_0 e^{-\tau}$ [equation (6.19)] in the limit of large N. (*Hint:* You may want to look at various representations for the function e^τ.)

6.9   For what value of x does the error in the approximation $e^x \cong 1 + x$ reach 1%?

6.10   Suppose we have a uniform density sphere of radius R and absorption coefficient κ. We look along various paths, passing different distances p from the center of the sphere at their point of closest approach to the center. (a) Find an expression for the optical depth τ as a function of p. (b) Calculate $d\tau/dp$, the rate of change of τ with p. (c) Use your results to discuss the sharpness of the solar limb.

6.11 (a) Explain how we can use two different optical depth spectral lines to see different distance into the Sun. (b) Why is the Hα line particularly useful in studying the chromosphere?

6.12 (a) Consider a charge Q near a neutral object. If the object is a conductor, charge can flow within it. The presence of the charge Q induces a dipole moment in the conductor, and there is a net force between the dipole and the charge. Show that this force is attractive. (b) How does this apply to the possibility of the existence of the H$^-$ ion?

6.13 Explain how absorption and emission by the H$^-$ ion can produce a continuum, rather than spectral lines.

6.14 What is the thermal Doppler broadening of the Hα line in a star whose temperature is 20,000 K?

6.15 Explain why we see a range of Doppler shifts over a spectral line.

6.16 If the corona has $T = 2 \times 10^6$ K, why don't we see the Sun as a blackbody at this temperature?

6.17 Why can't you see the corona when you cover the Sun with your hand?

6.18 What advantages would a coronagraph on the Moon have over one on the Earth?

6.19 (a) Why does the F-corona still show the Fraunhofer spectrum? (b) Would you expect the light from the Moon to show the Fraunhofer spectrum?

6.20 Why is the low density in the corona favorable for high levels of ionization?

6.21 (a) Why are collisions important in the cooling of a gas? (b) Why does the cooling rate depend on the square of the density? (c) How would you expect the heating rate to depend on the density?

6.22 Compare the *total* thermal energy stored in the corona and photosphere.

6.23 (a) At what wavelength does the continuous spectrum from sunspots peak? (b) What is the ratio of intensities at 5500 Å in a sunspot and in the normal photosphere? (c) What is the ratio of energy per second per surface area given off in a sunspot and in the photosphere?

6.24 Explain why charged particles drift parallel to magnetic field lines.

6.25 How long does it take before material at the solar equator makes one more revolution than that at 40° latitude?

6.26 How are various forms of solar activity related to the Sun's magnetic field?

6.27 Calculate the energy per second given off in the solar wind.

6.28 To describe completely the radiative transfer problem, we must take emission into account as well as absorption. The *source function S* is defined so that $S\,d\tau$ is the increase in intensity due to emission in passing through a region of optical depth $d\tau$. This means that the radiative transfer equation should be written

$$dI/d\tau = -I + S$$

(a) If S is a constant, solve for I vs. τ, assuming an intensity I_0 enters the material. (b) Discuss your result in the limit $\tau \ll 1$ and $\tau \gg 1$.

PART II

STELLAR EVOLUTION

In this part we look at the stellar life cycle. We see how stars are born out of the interstellar medium, how they live on the main sequence, and how they die. In the process of dying, they return some or most of their mass to the interstellar medium, to be incorporated in the next generation of stars. There is no real beginning. It is like a moving merry-go-round. We can pick up the cycle at any point.

We start with a mass of interstellar material that has begun to collapse to form a star. (We will see in Part III how the material got together.) We then study the stellar prime of life and stellar death. In studying stellar death we encounter a variety of unusual objects which provide some of the most extreme conditions known to physicists.

In discussing various aspects of stellar evolution, we deal with conditions so extreme we have to include the special and general theories of relativity in our considerations. For that reason, we begin this section with a chapter presenting those theories.

Relativity

7

Einstein's theory of relativity has caused us to rethink the meaning of both space and time, concepts that had been taken for granted for centuries. The foundation of this revolution is the *special theory of relativity,* which Einstein published in 1905. The *general theory of relativity,* published in 1916, is really a theory of gravitation set in the foundations of the special theory; it also allows us to analyze the properties of frames of reference that are accelerating.

7.1 FOUNDATIONS OF SPECIAL RELATIVITY

7.1.1 Problems with Electromagnetic Radiation

The problems that lead to special relativity start with Maxwell's theory of electromagnetic radiation. Maxwell's equations, presented in 1873, allow for the existence of waves of oscillating electric and magnetic fields. All waves known before electromagnetic waves required a medium in which to travel. For example, sound waves can travel through air, but not through a vacuum. There is no obvious medium necessary for the propagation of electromagnetic waves. Physicists postulated a medium that is difficult to detect, called the *luminiferous ether,* or simply the *ether.* The ether supposedly fills all of space. Once we have a medium, then we have a reference frame for the motion of the waves. For example, the speed of sound is measured with respect to the air through which it is moving. An observer moving through the medium will detect a different speed for the waves than an observer at rest in the medium.

Einstein's questions about Maxwell's equations involve the appearance of electromagnetic waves to different observers, who are moving at different speeds. Einstein started with the postulate that *the laws of physics, properly stated, should be independent of the velocity of the observer.* It may be that the values of certain quantities change with the motion of the observer, but the relationships among physical quantities do not change.

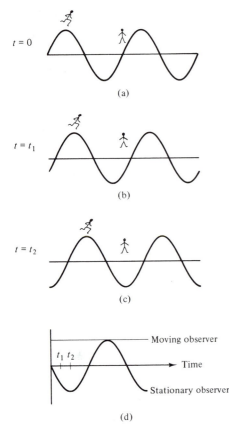

Figure 7.1 Observers of electromagnetic waves. One observer is stationary and the other is moving with the wave. (a) The moving observer is at a crest, and the stationary observer is at a null. (b) The moving observer is still at a crest, and the stationary observer sees a negative value. (c) The moving observer is still at the crest, and the stationary observer is at a dip. (d) We plot what each sees as a function of time. The moving observer sees a constant value, while the stationary observer sees a sinusoidally varying value.

Einstein examined Maxwell's equations to see if they obeyed this simple rule. His reasoning is illustrated in Fig. 7.1. The solution of Maxwell's equations gives us waves that vary sinusoidally in both space and time. That is, the waves vary with position, repeating each wavelength and, with time, repeating each cycle. How would an electromagnetic wave appear to an observer moving along with the wave at the speed of light? The wave would appear sinusoidal in space, but would appear constant in time, since the observer is moving along with the wave. However, there is no mathematical solution to Maxwell's equations that is constant in time, but which varies sinusoidally in space. This seemed to be a contradiction.

Two possibilities were left: (1) Maxwell's equations were correct in only one reference frame, the ether; or (2) there was something wrong with the basic concepts of space and time. Einstein chose the second, saying that, for some reason, it must be impossible to move with an arbitrary speed relative to an electromagnetic wave. He concluded that *the speed of light in a vacuum is the same for all observers, independent of their motion.* This suggests that electromagnetic waves must be different from the familiar mechanical waves. There must also be something wrong with the concept of the ether.

When Einstein was working on this problem, experiments had already been done which cast the existence of the ether into doubt. An experiment originally done by A. A. Michelson in 1881, and in an improved fashion by Michelson and Morley in 1887, is shown in Fig. 7.2. The experiment was designed to measure the motion of the Earth through the ether by measuring the speed of light in two directions perpendicular to each other. No change in the speed of light with direction was found. This meant that the Earth could not be moving through the ether. If the ether existed, it must be dragged along with the Earth.

There is another observation, which rules out this dragging of the ether. This involves the *aberration of starlight*, depicted in Fig. 7.3. Aberration is a slight change in the angle at which light appears to be arriving. For stars, the shift is always in the direction of the motion of the Earth, so it changes throughout the year. The positions of some stars shift by as much as 20 arc seconds from their true positions. Properly interpreted, this has actually been used to determine the speed of light. In the ether theory there is no way for aberration to be observed if the Earth is dragging the ether. Aberration should only exist if the Earth is moving through the ether. The combination of the Michelson–Morley experiment and the observation of aberration conclusively rules out the existence of the ether.

The fact that the speed of light is independent of the velocity of the observer contradicts our everyday experience, in which relative velocities are additive. Einstein began to look at the underlying cause for the speed appearing to be constant. In measuring a speed, we measure a distance and a time interval. Einstein suspected that the problem lay in our traditional concepts of space and time. Physicists such as Newton simply assumed that

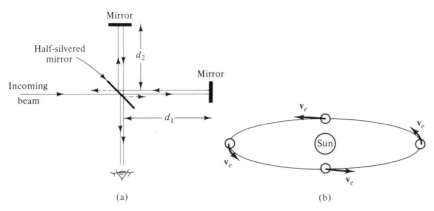

(a)   (b)

Figure 7.2 The Michelson–Morley experiment. (a) The basic apparatus. Light enters from the left, striking a mirror which transmits some light and reflects the rest. The reflected beam goes directly to the observer. The transmitted beam continues on to a regular mirror, and is reflected back. (For clarity, we have displaced the reflected beam from the incoming beam.) The reflected beam again strikes the partially transparent mirror. We don't care about the transmitted beam. The reflected beam goes up to another mirror, and is reflected back. This beam is again partially reflected and partially transmitted. We don't care about the reflected part, but the transmitted part goes to the observer, where it can interfere with the first beam that went to the observer. If the whole apparatus is traveling through the ether, then speed of light is not the same in all directions. Therefore, the speed of light traveling the leg d_1 is different from that traveling the leg d_2. If we then rotate the apparatus, the relative speeds on the two legs changes, causing a shift in the interference pattern seen by the observer. When this was done, no shift was seen, meaning that the Earth is not moving through the ether. (b) It could have been that the experiment was done at the exact time when the Earth's orbital motion exactly cancelled the Sun's motion through the ether, giving a null result for the experiment. For this reason, the experiment was repeated at different times of the year, when the Earth's motion is in a different direction. The result was still no shift in the fringes.

space and time were given. Einstein suggested that they might not be absolute, but might depend on the motion of the observer. Einstein examined the idea of an absolute time and looked at whether time might actually be a quantity that depends on the motion of the observer.

7.1.2 Problems with Simultaneity

Einstein realized that an absolute time was tied to the concept of *absolute simultaneity*. By absolute simultaneity we mean that if two events appear simultaneous to one observer, they appear simultaneous to all observers. This

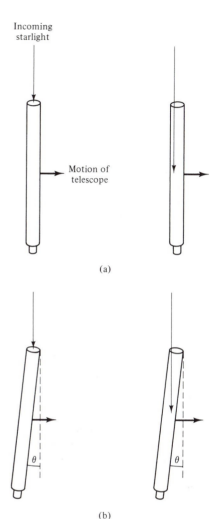

(a)

(b)

Figure 7.3 Aberration of starlight. (a) Assume that the telescope is moving to the right as the beam of light enters, with the telescope tube lined up with the beam of light. Since the speed of light c is finite, the telescope moves as the light passes through, and the light strikes the side. (b) To observe the light, we must tilt the telescope slightly. Thus, as the telescope moves over, the beam is always centered in the tube. We must tip the telescope in the direction in which it is moving.

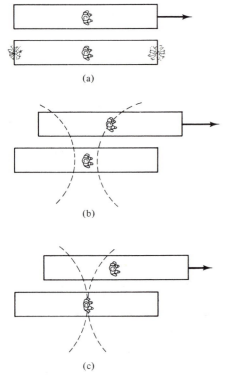

Figure 7.4 Flashes in railroad cars and simultaneity. The top car is moving past the bottom car. (a) When observers in the centers of each car are closest to each other, the flashes go off in the bottom car. (b) The motion of the top car means that the observer in that car sees the right flash first. (c) The observer in the bottom car sees both flashes at the same time.

is important because telling time is actually noting the simultaneity of two events. For example, if we say that the train left the station at 7:00, we are saying that two events are simultaneous. The first event is the train leaving the station, and the second event is the clock showing 7:00. If those two events are simultaneous for one observer, but not for all observers, then the concept of absolute time has no meaning.

The experiment depicted in Fig. 7.4 shows that two events can be simultaneous for one observer, but not another. The two observers are at the center of identical railroad cars. One car is at rest with respect to the station. The other is moving past at some speed. When the two observers are opposite each other, two flashes go off at the ends of one car. The flashes are judged to be simultaneous by the observer at rest. How are they seen by the other observer? The figure shows that the flash that the observer is moving toward is seen first. The flashes are not simultaneous for the moving observer. Simultaneity is not absolute. Therefore, time is not absolute.

With this as a starting point, we now go on to investigate how different types of situations appear to observers with different velocities. In special relativity, we deal only with reference frames in which there are no accelerations or gravitational fields. Such a reference frame is called an *inertial reference frame*. An inertial frame might be provided by a space station in free-fall. Einstein's postulate about the laws of physics being the same in different reference frames only applies to inertial frames. (We know that accelerating frames must be different, because they have pseudoforces, such as centrifugal force.) Another way of stating Einstein's postulate is that *there is no experiment we can perform to tell us which inertial frame is moving and which is at rest. There is no "preferred" inertial frame.* All we can talk about is the relative motion of two inertial frames.

7.2   TIME DILATION

Now that we know time is relative, we can see how a clock appears to two different observers. One observer is at rest with respect to the clock, and the other observer sees the clock moving. Time viewed in the frame in which the clock is at rest is called the *proper time* for that clock. The word "proper" does not denote anything superior about this frame. It just happens to be the frame in which the clock is at rest. We can think of proper time as the time interval between two events at the same place.

A simple clock is shown in Fig. 7.5. In this clock a light beam bounces back and forth between two mirrors, separated by a distance L. The time for the light pulse to make a round trip is

$$t_0 = 2L/c \tag{7.1}$$

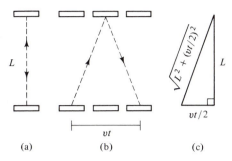

Figure 7.5 Light clock. (a) In the rest frame of the clock, the light bounces back and forth. (b) In the laboratory frame, with the clock moving, the light beam travels a longer path. (c) Calculating the extra distance traveled.

In the frame in which the clock is moving, the light beam takes a longer path. Since the speed of light is the same in both frames, the beam must take longer to make the roundtrip. From the figure we see that the distance traveled is $2[L^2 + (vt/2)^2]^{1/2}$, so the time is

$$t = (2/c)[L^2 + (vt/2)^2]^{1/2}$$

Squaring this gives

$$t^2 = (4L^2/c^2)[1 + (vt/2L)^2]$$

We use equation (7.1) to eliminate L, giving

$$t^2 = t_0^2 + (v^2/c^2)t^2$$

We want to solve for t in terms of t_0:

$$t^2[1 - (v/c)^2] = t_0^2$$

Taking the square root of both sides and solving for t, we have

$$t = \frac{t_0}{[1 - (v/c)^2]^{1/2}} \qquad (7.2)$$

The significance of this result is that the time interval measured in the frame in which the clock is moving is greater than that measured in the frame in which the clock is at rest. Suppose we have two identical clocks. If we keep one at rest and let the other move, the moving clock appears to run slow. It is important to realize that the situation is perfectly symmetric. If there is an observer traveling with each clock, each observer sees the other clock as running slow. This effect is called *time dilation*.

From equation (7.2) we can see that time dilation depends on the quantity $1/[1 - (v/c)^2]^{1/2}$. This quantity is generally designated γ, and is plotted as a function of v/c in Fig. 7.6. Note that this quantity is close to one for

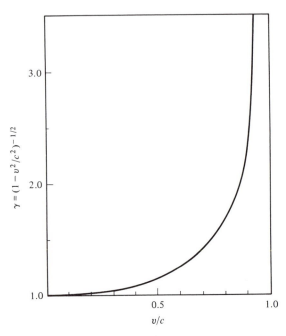

Figure 7.6   The quantity γ vs. v/c. For v/c small, γ is close to 1. As v/c approaches 1, γ becomes infinite.

small velocities, and only gets large when v is very close to c. This confirms our intuition that the results of special relativity should reduce to the familiar everyday results when the speeds are much less than c.

Time dilation is not an artifact of the light clock we have depicted in Fig. 7.5. It applies to all clocks. For example, it applies to the decay of unstable elementary particles. We note that particles moving close to the speed of light should appear to live much longer than the same types of particles at rest. This is tested almost daily at particle accelerators around the world.

EXAMPLE 7.1   Time Dilation

How fast must a particle be traveling to live ten times as long as a particle at rest?

Solution

We simply set

$$10 = \frac{1}{[1 - (v/c)]^{1/2}}$$

Squaring gives

$$100 = \frac{1}{[1 - (v/c)^2]^{1/2}}$$

Solving for $(v/c)^2$ gives

$$(v/c)^2 = 0.99$$

Taking the square root gives $v/c = 0.995$. The particle must be within a half-percent of the speed of light.

Time dilation applies to biological clocks. A person traveling at a high speed will not age as fast as a person at rest. Of course, the situation must be symmetric. Each person sees the other age slower. This leads to a puzzle known as the *twin paradox*. For example, two twins start on Earth. One is an astronaut who goes on a trip at a speed close to c. The other stays on Earth. From the point of view of the one who stayed on Earth, the astronaut is moving and will not age as much as the one on Earth. The astronaut will appear younger upon returning. However, the astronaut sees the one on Earth as moving away. Therefore, the one on Earth should appear younger. It is alright for two moving observers to see each other age slower. However, we have a problem if we bring the two astronauts together—both at rest, we can see which one is really younger and decide which was really moving. This would seem to violate Einstein's postulate. However, if the two twins start and end together at rest, then one twin must accelerate to get to high speed. That acceleration produces pseudoforces which can be felt by only one twin. This breaks the symmetry of the problem without any logical contradiction. (Remember, a pseudoforce is really an inertial response to an acceleration of the reference frame.)

7.3 LENGTH CONTRACTION

Once the concept of time becomes suspect, the concept of length must also be reinvestigated. Think how we measure the length of an object. We measure the positions of the two ends and take the difference between the two positions. For this procedure to have any meaning, the measurements must be carried out simultaneously. (If I measure the position of the front of a car today and the back of the car tomorrow, after the car has traveled 1000 km, I cannot conclude that the car is 1000 km long.) Unfortunately, we have seen that observers in different inertial frames cannot agree on the simultaneity of events separated in space.

Figure 7.7 Length contraction. (a) Perpendicular to the direction of motion? Assume that objects shrink perpendicular to the direction of motion. A and B are the same height when both are at rest. Both hold swords parallel to the ground, and B moves past. If B shrinks, A's sword will miss B, but B's sword will cut A. However, from B's point of view, as shown in the right figure, A is moving, and should shrink, resulting in injury to B but not to A. (b) To measure the length of a stick, we must first measure its speed. We do this by timing the time for one point of the stick to go a known distance between two stationary clocks. In the upper frame, the front of the stick starts at the right clock. In the lower frame, it reaches the left clock. The time difference is noted, and the speed is calculated. (c) Knowing the speed of the stick, we measure its length by seeing how long it takes to pass a single clock. In the top half of the frame, the front of the stick is at the clock, and the measurement starts. In the bottom half, the back of the stick reaches the clock, and the measurement ends.

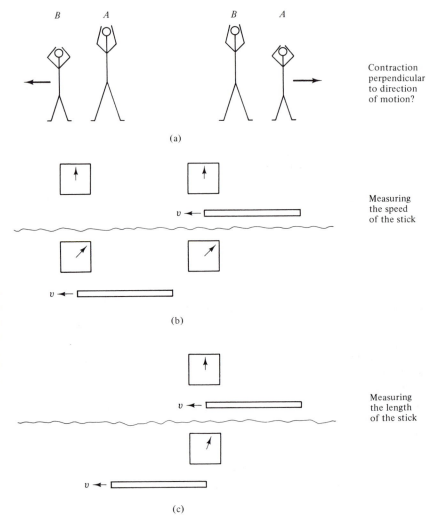

It is therefore not surprising that lengths will appear different to observers in different inertial frames. In fact, physicists had been playing with this idea before Einstein's 1905 paper. H. Lorentz had proposed it as a way around the results of the Michelson–Morley experiment. He said that the ether could be preserved if the lengths of objects depended in a particular way on their state of motion.

In considering changes in length, we look separately at lengths perpendicular to and parallel to the direction of motion. We can first show that there can be no change of length perpendicular to the direction of motion. Let's assume that there was such a change and that moving objects shrink. We now consider an experiment. Two people of identical height are standing as in Fig. 7.7a. Each has a sword held out at the level of the top of the head.

Now person B is carried past person A at a high speed. According to A, B is moving, and B gets shorter. B's sword cuts A's head off, while A's sword passes safely over B. According to B, A is moving and the situation is reversed. We have a true contradiction. Each person is dead in their own rest frame but survives in the other. The only way out of this is to say that there can be no change in length perpendicular to the direction of motion.

We can think of no such examples to rule out length changes parallel to the direction of motion. Here, there is actually a change of length. Moving objects appear to shrink. We call this effect the *Lorentz contraction*. To see this, we use Figure 7.7b to show how we might measure the length of a moving object. The length of an object, measured in the frame in which the object is at rest, is called the *proper length L_0*. This can be measured in the usual way, since the ends aren't going anywhere. We now measure its length in a frame in which it is moving. We can tell its speed v by having two markers at rest in our frame, and measuring the time for the front of the object to travel from one marker to the other. We can then measure its length by timing the passage of the object past one marker. This time interval Δt, as measured in the frame of the object, is

$$\Delta t = L_0/v$$

In our frame it takes

$$\Delta t' = \Delta t/\gamma$$

To get the relation between Δt and $\Delta t'$, we have used the time dilation result.

We now say that the length of the object is

$$L = v \, \Delta t'$$

$$= v \, \Delta t/\gamma$$

Finally, substituting $\Delta t = L_0/v$ gives

$$L = L_0/\gamma \qquad (7.3)$$

Not surprisingly, the length contraction has a similar dependence on v/c as has time dilation.

As with time dilation, length contraction is symmetric. If A and B are carrying meter sticks parallel to the direction of their relative motion, A will see B's stick shrink, and B will see A's stick shrink. There is no contradiction here, since we cannot compare the ends of the sticks simultaneously for both observers. This symmetry has provided some interesting puzzles that start with seeming contradictions, but end with logical resolutions. (See Problem 7.12.)

7.4 THE DOPPLER SHIFT

With lengths and times appearing different to different observers, it is also necessary to take a closer look at the Doppler shift, since wavelengths obviously involve length, and frequencies obviously involve time. Since there is no ether, the Doppler shift for electromagnetic waves can only depend on the relative motions of the source and observer. This is different than for the case of sound waves, for which the shift is slightly different, depending on which is moving. We can show that the result for electromagnetic waves doesn't depend on which is moving by considering the source moving toward the receiver at some speed v, and then considering the receiver moving toward the source at v. In both cases we denote quantities as measured in the rest frame of the source as unprimed and quantities as measured in the rest frame of the receiver as primed.

7.4.1 Moving Source

The first case is the source moving toward the receiver. The source emits N waves in time $\Delta t'$, as measured by the receiver. In this time, the first wave travels a distance $c\Delta t'$, and the source travels a distance $v\Delta t'$. The wavelength will then be the distance between the source and the first wave and the source, divided by the number of waves. That is

$$\lambda' = \frac{c\Delta t' - v\Delta t'}{N}$$

The frequency is then given by

$$\nu' = c/\lambda'$$

$$= \frac{c}{c - v}\frac{N}{\Delta t'} \tag{7.4}$$

We would like to relate this to the frequency in source frame ν. It is given as the number of waves N, divided by the time interval Δt, as measured in the source frame:

$$\nu = N/\Delta t$$

We can use the time dilation formula, $\Delta t' = \gamma \Delta t$, to make this

$$\nu = N\gamma/\Delta t'$$

This can now be used to eliminate $\Delta t'$ in equation (7.4), giving

$$\nu' = \frac{1}{1 - v/c} \frac{\nu}{\gamma}$$

$$= \nu \frac{[1 - (v/c)^2]^{1/2}}{1 - v/c}$$

$$= \nu \left[\frac{(1 - v/c)^{1/2}(1 + v/c)^{1/2}}{1 - v/c} \right]$$

If we multiply numerator and denominator by $(1 + v/c)^{1/2}$, this simplifies to

$$\nu' = \frac{\nu(1 + v/c)}{[1 - (v/c)^2]^{1/2}}$$

$$= \nu \left(1 + \frac{v}{c} \right) \gamma \qquad (7.5)$$

This is like the classical Doppler shift formula, except for the extra factor of γ, which comes from the time dilation.

7.4.2 Moving Observer

We now consider the receiver moving toward the source. In the source's frame, in time Δt, the receiver will receive all waves in a length $(c + v)\Delta t$. This number is the length divided by the wavelength:

$$N = \frac{(c + v)\Delta t}{\lambda}$$

$$= \frac{(c + v)\Delta t}{c/\nu}$$

According to the receiver, this takes place in time $\Delta t' = \Delta t/\gamma$, so the frequency is observed to be

$$\nu' = \frac{N}{\Delta t'}$$

$$= \frac{(1 + v/c)\Delta t}{\Delta t'}$$

$$= \frac{(1 + v/c)v}{[1 - (v/c)^2]^{1/2}}$$

$$= v\left(1 + \frac{v}{c}\right)\gamma$$

This is identical to equation (7.5), proving that the Doppler shift is independent of whether the source or the receiver is moving, or whether both are moving.

7.4.3 General Result

We now generalize the result. If we had considered the source and observer moving apart, we would have gotten $1 - v/c$ in the numerator of equation (7.5). The v in the $1 \pm v/c$ is just the radial velocity, the component of the velocity along the line of sight. We should therefore replace that v in equation (7.5) with v_r. However, the other v in equation (7.5) comes from the time dilation, which is independent of the direction of motion. It must remain as the total relative speed of the object. This means that it is possible to have a Doppler shift, even when the motion is perpendicular to the line of sight. This is simply a result of time dilation, and is not important until v is close to c. With these generalizations, equation (7.5) becomes

$$v' = v(1 - v_r/c)\gamma \tag{7.6}$$

We can derive a corresponding expression for wavelength, remembering that $\lambda' = c/v'$,

$$\lambda' = \frac{\lambda}{[\gamma(1 - v_r/c)]} \tag{7.7}$$

In Chapters 19–21, we will encounter distant objects moving away from us at speed close to c. For these, the radial velocity is very close to the total speed v. This allows us to make a simplification:

$$\lambda' = \frac{\lambda[1 - (v/c)^2]^{1/2}}{1 - v/c}$$

$$= \frac{\lambda[(1 - v/c)(1 + v/c)]^{1/2}}{1 - v/c}$$

which simplifies to

$$\lambda' = \lambda\left(\frac{1 + v/c}{1 - v/c}\right)^{1/2} \tag{7.8}$$

EXAMPLE 7.2 Relativistic Doppler Shift

Find the wavelength at which we will observe the Hα line if it is emitted by an object moving away with $v/c = 0.3$.

Solution

From equation (7.8), we find that

$$\frac{\lambda'}{\lambda} = \left(\frac{1 + 0.3}{1 - 0.3}\right)^{1/2}$$

$$= 1.36$$

This means that

$$\lambda' = (1.36)(6562.8 \text{ Å})$$

$$= 8925.4 \text{ Å}$$

The line is shifted from the visible into the near infrared!

7.5 SPACE-TIME

7.5.1 Four-Vectors and Lorentz Transformation

Phenomena such as time dilation and length contraction are not simply illusions. They are real effects. Our failure to appreciate this previously comes from a failure to appreciate the true nature of space and time. Classical physicists assumed that space and time were there, just like a blank piece of graph paper, and that the laws of physics were laid down on top of them. Einstein realized that the laws of physics were intimately entwined with space and time.

This required the abandonment of time as an entity independent of space. We must now treat it as another coordinate. Instead of our familiar three-dimensional space, we have a four-dimensional *space-time*. To remind us that time is just another way of measuring distance, we sometimes write the time coordinate as *ct*, so that it has the same dimensions as the other coordinate. In this way we could measure time in meters. What is a time of one meter? It is the time that it takes light to travel one meter. (Note that we have previously used time as a measure of distance when we introduced the light-year.)

An interesting sidelight to this has come from the organizations that set international standards such as the meter and second. It used to be that such units were defined independently, and c was just a measured quantity. The speed of light is now taken as a defined quantity, where all decimal places beyond the most accurate measured value are taken to be zero. It gives the conversion from meters to seconds. This means that we now only need a standard for the second or the meter, but not both.

In space-time, we speak of *four-vectors* to distinguish them from the ordinary three-dimensional vectors. Any event is characterized by the four coordinates t, x, y, z. Observers in different inertial frames will note different coordinates for events, but these coordinates are related. If one inertial system is moving with respect to the other at speed v in the x-direction, the coordinates in the two frames are found by assuming that they are linear in the coordinates, and must give the correct results for length contraction and time dilation. The result is

$$x = \gamma(x' + vt') \qquad y = y'$$
$$t = \gamma(t' + vx'/c^2) \qquad z = z' \tag{7.9}$$

The reverse transformation is given by

$$x' = \gamma(x - vt) \qquad y' = y$$
$$t' = \gamma(t - vx/c^2) \qquad z' = z \tag{7.10}$$

These relationships together are called the *Lorentz transformation*.

We interpret the Lorentz transformation as telling us that the rules of geometry are different for space-time than they are for ordinary space. To illustrate this point, we use a *space-time diagram* like that shown in Fig. 7.8. For simplicity, we plot only one space coordinate x as well as the time coordinate. We can keep track of events in such a diagram by plotting the coordinates of the event. By convention, we have time running vertically. The effect of the Lorentz transformation is to rotate the axes through an angle whose tangent is v/c. The unusual feature is that the x-axis and t-axis rotate in opposite directions, so that the axes are no longer perpendicular. Note that $v = c$ puts both axes in the same place. It should not surprise us that something funny happens when $v = c$, because this is where the quantity γ becomes infinite.

We know that in ordinary three-dimensional space, a rotation changes the coordinates of an object, but the lengths of things are changed. That is, if we have two objects, as shown in Fig. 7.9, whose separations are given in one coordinate system by $\Delta x, \Delta y, \Delta z$ and in another by $\Delta x', \Delta y', \Delta z'$, the distance between the two, which is the square root of the sum of the squares of the coordinate differences, doesn't change. That is,

$$(\Delta x)^2 + (\Delta y)^2 + (\Delta z)^2 = (\Delta x')^2 + (\Delta y')^2 + (\Delta z')^2 \tag{7.11}$$

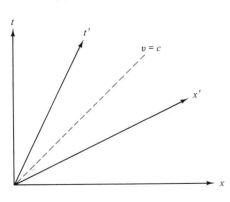

Figure 7.8 Lorentz transformation on a space-time diagram. The transformation looks like a rotation of the axes, except that the time and space axes rotate in opposite directions.

We say that *a length is invariant under rotation*.

Since the Lorentz transformation has properties of a rotation, is there a corresponding concept in space-time? The answer is "yes," but the quantity is slightly different, because the time axis rotates in the opposite direction to the space axes. We define the *space-time interval* as

$$(\Delta s)^2 = (c\,\Delta t)^2 - (\Delta x)^2 - (\Delta y)^2 - (\Delta z)^2 \qquad (7.12)$$

This is the quantity that is invariant under a Lorentz transformation. (Note that the Lorentz transformations can be derived by assuming that this quantity is invariant and that the transformation is linear in the coordinates. When this is done, time dilation and length contraction can then be derived from the transformation, rather than the other way around.)

To get a feeling for the physical meaning of Δs, consider an observer moving from one place to another in time Δt, as measured in that observer's rest frame. This means that Δt is the proper time interval. In that observer's rest frame, there is no change of position, so $\Delta x = \Delta y = \Delta z = 0$. This means that $\Delta s = c\,\Delta t$. Therefore, Δs is just the proper time interval (in units of length).

We can define three types of space-time intervals (Fig. 7.10), depending on whether $(\Delta s)^2$ is zero, positive, or negative. Suppose our two events are the emission and absorption of a photon. We know exactly how long it will take a photon to travel a given distance, and it will always work out that $(\Delta s)^2$ is zero. We call such intervals *lightlike*.

Intervals for which $(\Delta s)^2$ is positive are called *timelike*. The positions are close enough in space that a photon would have more than enough time to get from the first event to the second. This means that the first event might have caused the second. In the opposite case, when $(\Delta s)^2$ is negative, we call the interval *spacelike*. A photon cannot traverse the distance in the time given. Unless a signal can be sent faster than the speed of light, there is no way that one event could have caused the other.

If we extend our space-time diagram into more dimensions, we call the surface defined by $(\Delta s)^2 = 0$ the *light cone*. Events that could have caused the event at the origin of the cone are inside the past light cone. Those that might be caused by the event at the origin are inside the future light cone. Events that are outside the light cone can have no causal connection with the event at the origin.

7.5.2 Energy and Momentum

The space-time coordinates of an event are not the only quantities that transform according to the Lorentz transformation. For example, another important four-vector involves energy and momentum. Since positions and times appear different from different reference frames, velocities and accelerations should also appear different. This means that Newton's second law, expressed as $\mathbf{F} = m\mathbf{a}$, is not in the right form. Instead, we must use the form $\mathbf{F} =$

$$(\Delta x)^2 + (\Delta y)^2 = (\Delta x')^2 + (\Delta y')^2$$

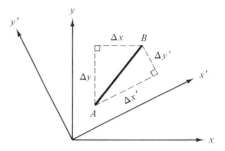

Figure 7.9 Invariance of lengths under rotation. The dark line represents the distance between points A and B. The components of this length with respect to the x and y axes are Δx and Δy, respectively, and with respect to the x' and y' axes are $\Delta x'$ and $\Delta y'$, respectively. Independent of the components used, the length of the dark line is the same.

Figure 7.10 Space-time intervals.

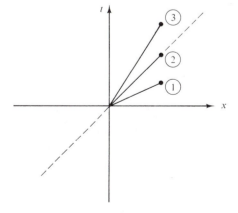

$d\mathbf{p}/dt$. We will not prove it here, but, in order to have momentum conservation still hold when we apply the velocity transformations that come from the Lorentz transformation on position and time, the relativistic momentum must be defined as

$$\mathbf{p} = \gamma m\mathbf{v}$$

Note that for $v \ll c$, γ is approximately one, and the new definition is the same as our old definition, mv.

If we start with an object at rest and accelerate it to some speed, the work we do is still equal to the kinetic energy E_k of the object. We find that this is

$$E_k = mc^2(\gamma - 1)$$

Again, we can see that this definition is consistent with our classical definitions. When $v \ll c$, we can write

$$\gamma = (1 - v^2/c^2)^{-1/2}$$

$$\cong 1 + v^2/2c^2$$

where we have used the fact that for $x \ll 1$, $(1 + x)^n \cong 1 + nx$. The kinetic energy for $v \ll c$ then becomes

$$E_k = mc^2(1 + v^2/2c^2 - 1)$$

$$= \tfrac{1}{2}mv^2$$

which is our classical definition.

When we apply the Lorentz transformation, we find that the quantity that is conserved in collisions is $E_k + mc^2$, where E_k is defined as above. We identify this quantity as the total energy of the object. That is,

$$E = \gamma mc^2 \tag{7.13}$$

Note that our energy doesn't go to zero for an object at rest. The *rest energy* of an object is therefore

$$E(\text{rest}) = mc^2$$

When we apply the Lorentz transformations to velocity, we find that momentum p transforms like the position x, and the energy E transforms like the time t. We therefore refer to an energy-momentum four-vector, whose transformation properties are identical to those of the space-time coordinate four-vector. There is also an invariant length associated with the energy-momentum four-vector. It is

$$E^2 = [(cp_x)^2 + (cp_y)^2 + (cp_z)^2]^2$$

In a rest frame moving with a particular object, the momentum of that object is zero and $\gamma = 1$, so the energy is mc^2. The length of the four-vector is then simply mc^2, the rest energy. (Since the length is invariant it must be mc^2 for any observer. We just chose to work it out in an easy frame.) Therefore, we see that rest energy behaves like proper time (which is the length of the space-time four-vector).

Note that as v/c approaches one, γ approaches infinity. This means that the energy approaches infinity. It takes an infinite amount of energy to accelerate an object with nonzero rest mass to the speed of light. This means that the speed of light is a limiting speed.

Some physicists have speculated on particles that can travel faster than light. These particles have been given the name *tachyons*. The trick is that these particles, if they exist, can never go slower than the speed of light. The speed of light would still be a barrier. If tachyons exist, they can interact with photons, and they could make their presence known. All experiments to look for tachyons have indicated that they do not exist.

7.6 GENERAL RELATIVITY

General relativity is Einstein's theory of gravitation that builds on the geometric concepts of space-time introduced in special relativity. Einstein was looking for a more fundamental explanation of gravity than the empirical laws of Newton. Besides coming up with a different way of thinking about gravity (in terms of geometry), general relativity makes a series of specific predictions of observable deviations from Newtonian gravitation, especially under strong gravitational fields. These predictions provide a stringent test of Einstein's theory.

7.6.1 Curved Space-Time

According to Einstein the presence of a gravitational field alters the rules of geometry in space-time. The effect is to make it seem as if space-time is "curved." To see what we mean by geometry in a curved space, we look at the example of geometry on the surface of a sphere. The surface is two-dimensional. (We need only two coordinates, say latitude and longitude, to locate any point on the surface.) However, it is curved into a three-dimensional world, and that curvature can be detected.

To discuss geometry of a sphere, we must first extend our concept of a

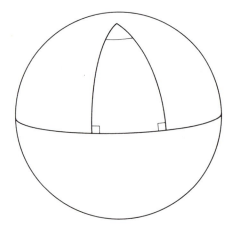

Figure 7.11 Geometry on the surface of a sphere. The shortest distances beteen two points are along great circles. In this figure we look at the triangle bounded by the equator and two meridians. The meridians cross the equator at right angles, so the sum of the angles in the triangle is greater than 180°.

straight line. In a plane the shortest distance between two points is a straight line. On the surface of a sphere, it is a great circle. Examples of great circles on the Earth are the equator and meridians. (A great circle is the intersection of the surface of the sphere with a plane passing through the center of the sphere.)

People on the surface of the Earth can tell that it is curved, and can even measure the radius, without leaving the surface. For example, two observers can measure the different position of the Sun as viewed from two different places at the same time (Fig. 7.11). Surveying the surface will also tell you that the rules of geometry are different.

For example, let's consider a triangle. In a plane, a triangle has three sides, each made up of a straight line. The sum of the angles is 180°. On the surface of a sphere we must replace straight lines, the shortest distance between two points in flat space, by great circles, the shortest distance between two points on a sphere. A triangle would therefore have its sides made up of great circles. Figure 7.11 shows one triangle composed of sections of two meridians plus a section of the equator. Each meridian intersects the equator at a right angle, so the sum of those two angles is 180°. When we add the third angle, between the two meridians, the sum of the angles exceeds 180°. The results of Euclidean (flat space) geometry no longer apply. The greater the curvature of the sphere, the more non-Euclidean the geometry appears. On the other hand, if we stick to regions of the surface that are much smaller than the radius of the sphere, the geometry will be very close to Euclidean.

We now look at what we mean when we say that gravity curves the geometry of space-time. In the absence of gravity all objects would move in straight lines at constant speeds. In the presence of gravity, trajectories curve. (These curved trajectories are still geodesics.)

This can be illustrated by the simple situation of throwing a ball up in the air. We can follow the progress of the ball on a space-time diagram, as in Fig. 7.12. The first diagram shows what happens if there is no gravity. The world line for the ball is straight. If we now turn on gravity, the world line looks like a parabola. We can say that it follows this path because the space-time surface on which it must stay is curved. The path is still the

Figure 7.12 Space-time diagram for ball thrown up from the ground. (a) With no gravity, the space-time trajectory is a straight line. (b) With constant gravity, the trajectory is a parabola.

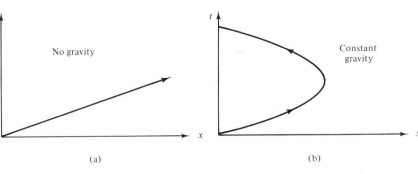

(a)   (b)

shortest distance through the curved space-time. Note that time marches onward in essentially the same way in both diagrams. The curvature shows up in the fact that the curve goes through each value of x twice. This is because the curvature of the space part of space-time is greater than that of the time part by a factor of c.

In this geometric interpretation of gravitation, we need two ingredients to a theory. The first is a way to calculate the curvature of space-time caused by the presence of any particular arrangement of masses. The second is a way to calculate the trajectories of particles in a given curved space-time. Einstein's theory of general relativity provides both. However, the mathematical complexity of the theory goes well beyond the level of this book. (Supposedly, even Einstein was upset when he realized the area of formal mathematics into which his theory had taken him.) However, we can still appreciate the underlying physical ideas, and we can even carry out some simple calculations that get us close to the right answers.

7.6.2 Principle of Equivalence

The starting point of general relativity is a statement called the *principle of equivalence,* which states that *a uniform gravitational field in some direction is indistinguishable from a uniform acceleration in the opposite direction.* Remember, an accelerating reference frame introduces pseudoforces in the direction opposite to the true acceleration of the reference frame. For example, if you are driving in a car and step on the brakes, the car has a backward acceleration (Fig. 7.13). Inside the car, you have a forward acceleration relative to the car.

We can illustrate the principle of equivalence by looking at the forces on a person standing on a scale in the elevator, as illustrated in Fig. 7.14. In the first case, we have the elevator being supported so there is no acceleration, but there is gravity. We take the acceleration of gravity to be $-g$. (Upward forces and accelerations are positive; downward forces and accelerations are negative, and we have taken g to be a positive number.) We now want to add up all of the forces on the person and equate them to ma, where m is the person's mass and a is the person's acceleration. The forces are the person's weight $-mg$ and the upward force of the scale on the person's feet F_S. The acceleration is zero, so

$$mg + F_S = 0$$

Solving for F_S gives us

$$F_S = mg$$

By Newton's third law, the force the scale exerts on the person has the same magnitude as the force the person exerts on the scale. Therefore, F_S also

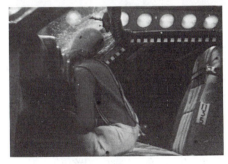

Figure 7.13 Pseudo-force in an accelerating car. In this case, the car is coming to a sudden stop. The dummies are not wearing seat belts, so there is no force to slow them down with the car. They continue forward at a constant speed until they hit the window and dashboard. From the point of view of an observer in the car, the dummies appear to accelerate forward. This apparent acceleration is opposite to the actual acceleration of the car.

Real forces   Gravity   Acceleration

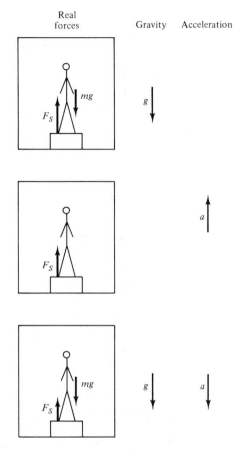

Figure 7.14 Person in an accelerating elevator. When gravity is present, it is indicated by a downward arrow, marked g. When the elevator is accelerating, it is indicated by an arrow marked a.

gives us the reading of the scale. In this case it is simply the weight of the person—the expected result.

We now look at the case of no gravity, but with an upward acceleration a. The only force on the person is F_S. Applying $F = ma$ gives

$$F_S = ma$$

If we arrange the acceleration so its value is equal to g, we have

$$F_S = mg$$

This is the same result we had in the first case. As far as the person in the elevator is concerned, there is no way to tell the difference between a gravitational field with an acceleration g downward and an upward acceleration of the reference frame.

To illustrate the point further we look at the third case, in which there is gravity, but the elevator is in free-fall. The force on the person is F_S upward and mg downward, and the acceleration is mg downward. This gives us

$$F_S - mg = -mg$$

This tells us that F_S is zero. The person is "weightless." The acceleration of the elevator has exactly canceled the gravitational field. For the person inside the elevator, there is no way to distinguish this situation from that of a nonaccelerating elevator with no gravitational field. This is the same weightlessness that is felt by astronauts in orbiting space vehicles. Orbiting objects are also in free-fall, but the horizontal component of their velocity is so great that they never reach the ground; they just follow the curvature of the Earth.

If you look carefully at the above discussion, you will see that we have really used the concept of mass in two different ways. In one case, we said that a body of mass m, subjected to force F, will have an acceleration F/m. In this sense, mass is a measure of the ability of an object to resist the effects of an applied force. When we use mass in this sense, we refer to it as *inertial mass*. However, we have also used mass as a measure of the ability of an object to exert and feel gravitational forces. In this context, we speak of *gravitational mass*. In the same sense, we use electric charge as a measure of the ability of an object to exert and feel electrical forces.

The principle of equivalence is really a statement that the ratio of inertial and gravitational masses is the same for all objects. If m_g is the gravitational mass and m_i is the inertial mass, then the weight of an object is $m_g g$. The acceleration is therefore $m_g g / m_i$. (If the ratio is constant, we can always find a system of units in which the ratio is one.) If the two masses are equal, then all masses have the same acceleration g in a gravitational field. This fundamental point was supposedly first realized by Galileo. It is not obvious on

Figure 7.15 Astronaut on the Moon, where in the absence of air resistance, a hammer and feather fall with the same acceleration.

the surface on the Earth, since air resistance affects how objects fall. However, as shown in Fig. 7.15, a hammer and feather fall with the same acceleration on the surface of the Moon, where there is no air resistance. It is the fact that all objects fall with the same acceleration in a uniform gravitational field that allows us to replace a gravitational field with an acceleration in the opposite direction.

It is important to remember that, just because we call both quantities "mass," there is no obvious reason for gravitational and inertial mass to have the same numerical value. In the same way, we expect no equality between the electric charge of an object and its mass. If inertial and gravitational mass are the same, this tells us that gravity must in some way be special. As we will see in the next section, considerable effort has gone into verifying the principle of equivalence.

7.7 TESTS OF GENERAL RELATIVITY

Over the years since Einstein's publication of general relativity, a number of exacting tests have been carried out to test observational predictions of the theory. Some of the tests are really only tests of the principle of equivalence, while others are true tests of general relativity.

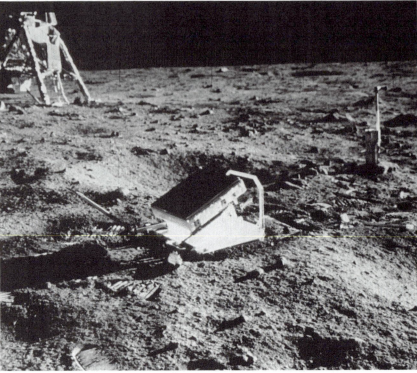

Figure 7.16 The 107-in. telescope of the McDonald Observatory, Texas, has been used to fire a laser beam at a reflector on the Moon, and then detect the weak return. By timing the round trip, the distance to the moon is very accurately determined.

A direct test of the principle of equivalence involves the measurement of the attraction of two different objects by some third body. A class of such experiments are called Eotvos experiments, after the person who devised the original experiment around the turn of the century. The most accurate recent versions of the experiment were carried out by a group at Princeton in the 1960s and by a group at Moscow University in the 1970s. Their findings indicate that the principle of equivalence is good to one part in 10^{11}.

The equivalence principle we have discussed applies strictly to objects that are so small we can ignore differences from one side to the other in the gravitational field they feel. We can treat them as point objects. However, there is a stronger form of the equivalence principle that says that it also applies to objects with a substantial gravitational binding energy, such as planets and stars. This has been tested by closely measuring the motion of the Moon (Fig. 7.16). A series of mirrors have been left on the Moon by the Apollo astronauts. Laser signals can be sent from Earth, bounced off these small mirrors, and then detected as very weak return signals. By timing the roundtrip we can measure the distance to the Moon very accurately, to within a few centimeters! These studies have indicated that the Earth and Moon fall toward the Sun with the same acceleration to within 7 parts in 10^{12}.

7.7.1   Orbiting Bodies

One series of tests of general relativity involves the behavior of orbiting bodies. The paths are slightly different than predicted by Newtonian gravitation. The most important feature involves elliptical orbits. In an elliptical orbit, the distance of the orbiting body from the body exerting the force on it is changing. The orbiting body is therefore passing through regions of different curvature. Figure 7.17 has a diagram which may help in visualizing this. The effect of the changing curvature is to cause the orbit not to close. After each orbit the position of perihelion (closest approach) has moved around slightly.

The effect will be greatest for orbits of the highest eccentricity, since the widest range of curvatures will be covered. Also, the smaller the semimajor axis is, the greater is the effect. This is because the gravitational field changes faster with radius when you are closer to the object exerting the force (see Problem 7.30). In the solar system, both of these points make the effect most pronounced for Mercury. It is closest to the Sun and, except for Pluto, has the most eccentric orbit.

The perihelion of Mercury's orbit advances by some 5600 arc seconds per century. However, of this, all but 43 arc seconds per century can be accounted for by Newtonian effects and the perturbations due to the motions of the other planets. The Newtonian effects could be calculated accurately and subtracted off. Einstein was able to explain the 43 arc seconds per century exactly in his relativistic calculations. This was considered to be an interesting result of relativity but not a crucial test, since Einstein explained something after it had been observed. A crucial test for a theory involves predicting things that haven't been observed yet.

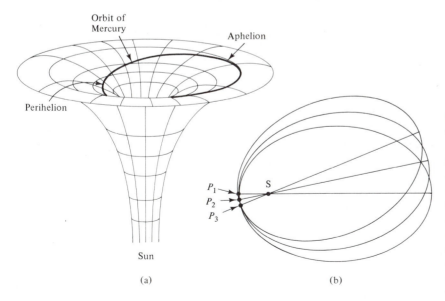

(a)   (b)

Figure 7.17 (a) Curved space-time for Mercury's orbit around the Sun. The closer to the Sun you get, the greater the curvature of space-time. Since Mercury's orbit is elliptical, its distance from the Sun changes. It therefore passes through regions of different curvature. (b) This causes the orbit to precess. We can keep track of the precession by noting the movement of the perihelion, designated P_1, P_2, and P_3 for three successive orbits.

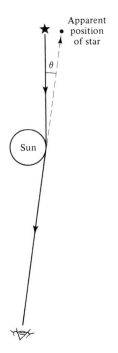

Figure 7.18 Bending of starlight passing by the Sun. The actual position of the star is marked by an asterisk. The observer thinks that the star is straight back along the received ray.

In recent years, a controversy has grown out of this test of general relativity. A group at Princeton in the 1960s measured the shape of the Sun and found a significant flattening. A nonspherical Sun would have a certain effect on the orbit of Mercury, reducing the general relativistic effect by enough to say that Einstein's calculation was wrong. Further measurements have indicated that the original experiment on the Sun was probably in error, but a new experiment has suggested that some flattening is still present. However, astronomers are still waiting for more evidence.

7.7.2   Bending of Electromagnetic Radiation

Einstein's chance to predict an effect that had not been seen came in the bending of light passing by the edge of the Sun. He said that the warping of space-time alters the path of light as it passes near the source of a strong gravitational field. The light will then appear to be coming from a slightly different direction. If the light is coming from a star, the position of the star will appear to be slightly different than if the bending had not taken place, as indicated in Fig. 7.18.

According to Einstein, the angle θ (in radians) through which light passing a distance b from an object of mass M is given by

$$\theta = 4GM/bc^2 \qquad (7.14)$$

If we set b equal to the radius of the Sun (7×10^{15} cm) we get an angle of 8.47×10^{-6} rad, which is equal to 1.74 arc seconds. This is a very small angle and hard to measure.

The measurement is made even more difficult by the fact that we cannot see stars close to the Sun in the sky. Therefore, the test must be made during a total eclipse of the Sun, when the sky is photographed, and then the same field is rephotographed approximately six months later. The positions of the stars on the two photographs are then compared. The first attempt to carry this out was by a German team trying to get to a Russian viewing site for a 1914 eclipse. They were thwarted by the state of war between the two countries. The next try was in 1919, in an effort headed by Sir Arthur Eddington. In the intervening years, Einstein had found a mistake in his calculations, so it is probably just as well that the observations weren't done until the theoretical prediction was finalized. The result was a confirmation of Einstein's prediction. The recognition of the magnitude of Einstein's contribution was immediate, both among physicists and the general public.

The solar eclipse experiment is a hard one, and the original one had a 10% uncertainty associated with it. More recent tries have reduced that uncertainty to about 5%. Different types of experiments are needed for greater accuracy. A major improvement can be made by using radio waves. The bending applies equally to electromagnetic radiation of all wavelengths. The advantage of radio waves is that the Earth's atmosphere doesn't scatter them.

We can observe any radio source as the Sun passes in front of it and watch the position of the source change. These tests have confirmed Einstein's predictions to a greater accuracy than the eclipse experiments.

There is another effect related to the bending of light. The curvature of space-time around the sun causes a delay in the time for a signal to pass by the Sun. Two types of observations have been done to test this. One involves the reflection of radio waves from Mercury and Venus as they pass behind the Sun. We know the positions of the planets very accurately, so we know how long it should take for a signal to make a roundtrip. The other type of measurement involves spacecraft that we have sent to various parts of the solar system, especially Mariners 6, 7, and 9, and the Viking orbiters and landers on Mars. We simply follow the radio signals from the spacecraft. Since we know where the spacecraft are, we can determine the time delay as the spacecraft pass behind the Sun. Using this technique, Einstein's predictions have been confirmed to an accuracy of 0.1%.

There is another interesting result related to the bending of the paths of electromagnetic waves. A massive object can bend the rays so well that it can act as a gravitational lens. Physicists have speculated on this possibility for some time. Recent observations of quasars, discussed in Chapter 19, have revealed a few sources in which double images are seen as a result of this gravitational lens effect.

7.7.3 Gravitational Red Shift

The wavelengths of photons change as they pass through a gravitational field. This effect is called the *gravitational red shift* (Fig. 7.19). It is really a consequence of the principle of equivalence.

To see how the energy of the photon changes, we consider the following experiment. Suppose a mass m is initially at rest, a distance r_2 from a mass M. We allow the object to fall to a distance r_1. The energy of the object is initially

$$E_2 = mc^2$$

When the object reaches r_1, gravity has done work on it, increasing its energy by the change in the gravitational potential energy, so

$$E_1 = mc^2 + GMm(1/r_1 - 1/r_2) \tag{7.15}$$

When the object reaches r_1, we convert all of the energy into a single photon, moving outward. The wavelength of the photon λ_1 is given by

$$hc/\lambda_1 = E_1$$

If the photon is not red-shifted, it will arrive back at r_2 with more energy

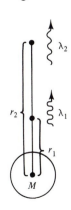

Figure 7.19 Gravitational red shift. As the photon moves farther from the mass, its wavelength increases.

than the original object had when it left r_2. Therefore, the photon must be red-shifted so that its energy equals E_2. That is

$$hc/\lambda_2 = mc^2$$

Taking a ratio of the two wavelengths gives

$$\frac{hc/\lambda_1}{hc/\lambda_2} = \frac{E_1}{mc^2}$$

or

$$\frac{\lambda_2}{\lambda_1} = 1 + \left(\frac{GM}{c^2}\right)\left(\frac{1}{r_1} - \frac{1}{r_2}\right) \tag{7.16}$$

This derivation should not be considered rigorous—it is more of a dimensional analysis. However, it gives a result that agrees with the full general relativistic calculation for shifts that are not too large. The actual result is

$$\frac{\lambda_2}{\lambda_1} = \left(\frac{1 - 2GM/r_2c^2}{1 - 2GM/r_1c^2}\right)^{1/2} \tag{7.17}$$

If we use the fact that $(1 - x)^{1/2}$ is approximately equal to $1 - x/2$ for $x \ll 1$, equation (7.17) gives the same result as equation (7.16).

If we use equation (7.16) and take $r_2 = \infty$, we get

$$\frac{\lambda_2}{\lambda_1} = 1 + \frac{GM}{r_1 c^2} \tag{7.18}$$

If we compute the wavelength shift $\Delta\lambda$, we find

$$\frac{\Delta\lambda}{\lambda} = -\frac{GM}{rc^2} \tag{7.19}$$

EXAMPLE 7.3 Gravitational Red Shift

Find the gravitational red shift for radiation emitted from the surface of the Sun and for radiation emitted from the surface of a $1\,M_\odot$ neutron star.

Solution

For $M = 2\times10^{33}$ g, we use equation (7.18) to get

$$\frac{\lambda_2}{\lambda_1} = \frac{1}{1 - (1.5\times10^5 \text{ cm}/r)}$$

For $r = 7 \times 10^{10}$ cm, $\lambda_2/\lambda_1 = 1 + 2.14 \times 10^{-6}$, and for $r = 1 \times 10^6$ cm, $\lambda_2/\lambda_1 = 1.18$.

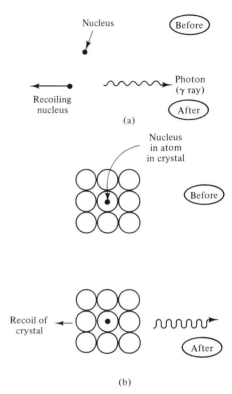

The shift for spectral lines in the Sun is very small. The shift for neutron stars is quite large, but we don't observe conventional optical spectral lines in neutron stars. However, some x-ray and gamma-ray lines are observed to be shifted by about 20% in neutron stars. The effect is measurable for white dwarfs, and red shifts have been measured. The two best cases are for Sirius B (3×10^{-4}) and 40 Eriadni B (6×10^{-5}). These results are accurate to about 15%.

There is an interesting way to measure the gravitational red shift on earth. It utilizes a phenomenon known as the *Mossbauer effect* (Fig. 7.20). This involves the emission of a gamma ray by a nucleus held firmly in place by a solid crystal. In a free nucleus, the gamma ray would lose a little energy due to the recoil of the nucleus. (The recoil is to conserve momentum.) When the nucleus is in a crystal, the whole crystal takes up the recoil. It moves very little, because of its large mass, and the energy lost by the gamma ray is small. This means that the gamma ray energy is well defined.

If the gamma ray is emitted by a nucleus in one crystal, it can be absorbed by a nucleus in an identical crystal, as long as there was no wavelength shift while the photon is in motion. A group of physicists tried an arrangement in which the gamma rays were emitted in the basement of a building and they were absorbed on the roof. The small gravitational redshift was enough for the gamma rays to arrive at the roof at the wrong wavelength to be absorbed. The gamma rays could be red-shifted back to the right wavelength by moving the crystal on the roof toward that in the basement, producing a small Doppler shift. By seeing what Doppler shift is necessary to offset the gravitational red shift, the size of the gravitational red shift can be measured. The result agrees with the theoretical prediction.

There is another manifestation of the gravitational red shift. It is *gravitational time dilation*. All oscillators or clocks run slower in a strong gravitational field than they do in a weaker field. If we have two clocks at r_1 and r_2, the time they keep will be related by the same expression as the gravitational red shift. That is

$$\frac{T_2}{T_1} = \left(\frac{1 - 2GM/r_2c^2}{1 - 2GM/r_1c^2}\right)^{1/2} \tag{7.20}$$

From this we see that $T_2 > T_1$. This effect has been tested by taking identical clocks, leaving one on the ground and placing the other in an airplane. (Of course, you must correct for the special relativistic effect due to the motion of the airplane). The airplane experiments have yielded results that agree with theory. Even more recently, tests on rockets have yielded even more accurate results.

Figure 7.20 The Mossbauer effect. (a) Emission of a gamma ray by a free nucleus. To conserve momentum, the nucleus recoils. The recoiling nucleus carries away some energy. Therefore, the energy of the gamma ray is less than the energy difference between the two levels involved in the particular nuclear transition. (b) If the nucleus is part of an atom, which is, in turn, bound into a solid crystal, the whole crystal must recoil. Since the crystal is much more massive than the nucleus, its recoil is negligible. This means that the energy of the recoil is very small, and the energy of the gamma ray is always equal to the difference between the two levels involved in the nuclear transition.

Figure 7.21 To detect gravity waves, physicists are hoping to measure very small shape changes in very massive objects.

7.7.4 Gravitational Radiation

Just as the classical theory of electricity and magnetism predicts that accelerating charges will give off electromagnetic radiation, general relativity predicts that certain types of systems should give off *gravitational radiation*. Gravitational radiation is more complicated than electromagnetic radiation. When a gravity wave passes by, the geometry of space-time briefly distorts. The types of systems that might produce gravitational radiation are orbiting bodies or collapsing objects. Some groups have attempted to directly detect gravitational radiation from astronomical sources, hoping to detect small changes in the size of large detectors, such as that in Fig. 7.21. None of these experiments has been successful yet. (In Chapter 12, we will discuss a system which has provided strong indirect evidence for the existence of gravitational radiation.)

7.7.5 Competing Theories

One of the reasons there has been so much interest in testing general relativity as accurately as possible is that there are some competing theories to Einstein's. These theories generally have the same starting point, but differ in their details. The result of these competing theories is that the experimental foundations of general relativity are now much stronger than they were when the theory was initially studied.

7.8   BLACK HOLES

One of the exciting aspects of astronomy is the possibility of studying a variety of fascinating objects. By our earthly standards, even a normal star contains extreme conditions. However, there are other objects that make the conditions on stars seem commonplace. Among these objects are *black holes*, regions of space from which no light can escape.

7.8.1 Schwarzschild Radius

Shortly after Einstein published his general theory of relativity, Karl Schwarzschild worked out the solution for the curvature of space-time around a point mass. He found that there is a critical radius at which a *singularity* occurs. A singularity is a place where some quantity becomes infinite. This critical radius is called the *Schwarzschild radius*. For a mass M this radius R_S is given by

$$R_S = 2GM/c^2 \qquad (7.21)$$

Real objects are not pointlike, but have some finite extent. The interpretation of Schwarzschild's result is that if an object of mass M is completely contained within its Schwarzschild radius, the singularity will occur.

We can understand the significance of this critical radius by recalling the discussion of the gravitational red shift in Section 7.7. We saw that if a photon is emitted at wavelength λ_1 at a distance r_1 from mass M, and is detected at r_2, its wavelength λ_2 is

$$\frac{\lambda_2}{\lambda_1} = \left(\frac{1 - 2GM/r_2 c^2}{1 - 2GM/r_1 c^2}\right)^{1/2} \tag{7.22}$$

If we set $r_1 = 2GM/c^2$ (the Schwarzschild radius), we find that λ_2 is infinite, even if r_2 is only slightly greater than r_1. This means that no electromagnetic energy can escape from the Schwarzschild radius. We call an object that is contained within its Schwarzschild radius a *black hole*.

EXAMPLE 7.4 Schwarzschild Radius

Find the Schwarzschild radius for an object of one solar mass.

Solution

From equation (7.21) we have

$$R_S = \frac{(2)(6.67 \times 10^{-8} \text{ dyn-cm}^2/\text{s}^2)(2 \times 10^{33} \text{ g})}{(3 \times 10^{10} \text{ cm/s})^2}$$

$$= 3.0 \times 10^5 \text{ cm}$$

$$= 3.0 \text{ km}$$

Since the Schwarzschild radius varies linearly with mass and has a value of 3 km for a 1-$M_\odot$ object, we can write an expression for R_S for any mass object. It is

$$R_S = (3 \text{ km})(M/M_\odot) \tag{7.23}$$

Remember, every object has its Schwarzschild radius. However, it can only be a black hole if it is contained within this radius. For example, the Sun is much larger than 3 km, so it is not a black hole.

The density of a 1-$M_\odot$ black hole would be quite high, almost 10^{17} g/cm^3. This is higher than the density of a neutron star. However, as we consider more massive black holes, the density goes down. This is because the radius is proportional to the mass, but the volume is proportional to the radius cubed (and therefore to the mass cubed). This means that the density is proportional to $1/M^2$. Since we know the density for a 1 $M_\odot$ black hole, we can write the density for any other mass black hole as

$$\rho = (1 \times 10^{17} \text{ g/cm}^3)(M/M_\odot)^{-2} \tag{7.24}$$

By the time the mass reaches $10^8\ M_\odot$, the density is only a few g/cm^3, just a few times the density of water.

We would expect the region just outside black holes to be characterized by strong tidal forces. That is, the gravitational force should fall off very quickly with small changes in distance from the surface. If we write the acceleration of gravity as a function of radius,

$$g(r) = GM/r^2 \tag{7.25}$$

Differentiating with respect to r gives

$$dg(r)/dr = -GM/r^3 \tag{7.26}$$

Though the gravitational force falls off as $1/r^2$, the tidal effects fall off as $1/r^3$, meaning that they are most important for small values of r.

EXAMPLE 7.5   Black Hole Tidal Forces

Find the difference between the acceleration of gravity at the feet and head of an astronaut just outside a 1-$M_\odot$ black hole.

Solution

The change in g, Δg, is given by

$$\Delta g = (dg/dr)\Delta r$$

where Δr is the distance over which the change is found. In this case Δr is the height of the astronaut, which we will take to be 2 m. We find dg/dr from equation (7.26) to be

$$dg/dr = -(6.67 \times 10^{-8} \text{ dyn-cm}^2/\text{s}^2)(2 \times 10^{33} \text{ g})(3 \times 10^5 \text{ cm})^3$$

$$= -5 \times 10^9 \text{ (cm/s}^2)/\text{cm}$$

(Note that the minus sign means that gravity is stronger at the feet than at the head.) For $\Delta r = 2 \times 10^2$ cm, we have

$$\Delta g = -1 \times 10^{12} \text{ cm/s}^2$$

This is 10^9 times the acceleration of gravity at the surface of the Earth. The astronaut will be pulled apart with a force equal to 10^9 times the astronaut's weight!

The tidal force dg/dr is proportional to M/r^3, just as is the density. Therefore, the tidal force will be less for more massive black holes, falling to more tolerable values for very massive black holes (see Problem 7.34).

7.8.2 Approaching a Black Hole

What is it like to fall into a black hole? We consider two astronauts. One approaches the black hole, and the other stays a safe distance away. The various steps are indicated in Fig. 7.22.

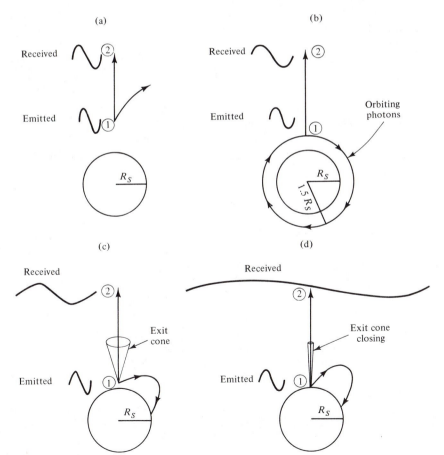

Figure 7.22 Approaching a black hole. Astronaut #1 approaches the black hole while astronaut #2 stays behind. In each frame, the emitted and received waves correspond to a beam sent from #1 to #2.

We assume the astronaut approaching the black hole can send out signals in various directions, including back to the other astronaut. As the first astronaut approaches the black hole, the first thing the distant astronaut would notice is a red shift in signals received. The magnitude of the red shift increases as the first astronaut gets closer to the Schwarzschild radius.

Before the Schwarzschild radius is reached, another effect becomes noticeable. The paths of photons sent out by the first astronaut are not straight lines. They bend. The only direction in which the astronaut can aim a beam and not have it bend is straight up. If the beam is not aimed sufficiently close to the vertical, the bending will be so great that the light will not escape. Only light aimed into a cone about the vertical, called the *exit cone,* will escape. As the first astronaut gets closer to the Schwarzschild radius the exit cone gets smaller. At a distance equal to $\frac{3}{2}R_S$, photons aimed horizontally go into orbit about the black hole. The sphere of orbiting photons is called the *photon sphere*.

The second astronaut never actually sees the first astronaut reach the Schwarzschild radius. The gravitational time dilation is so great as R_S is approached, that the second astronaut thinks it takes an infinite amount of time for the first astronaut to reach R_S. Another way of looking at the situation is to say the photon that would record the instant of arrival at the black hole gets red-shifted to zero energy, so it is never detected by the second astronaut. The time dilation makes the first astronaut appear to slow down as R_S is approached.

From the point of view of the first astronaut, there is no such respite. The Schwarzschild radius is reached very quickly. If the black hole is of sufficiently small mass, the tidal forces would tear the first astronaut apart. However, if the black hole is massive enough, the tidal forces might be survived and the astronaut crosses R_S. When this happens we say the astronaut has crossed the *event horizon.* If the black hole is massive enough, the astronaut might not notice anything unusual, except for the fact that escape is impossible!

Once inside the black hole, the inevitable journey to the center continues. The gravitational time dilation is so great that time passes slowly. However, the headlong rush through space continues. Outside the black hole, it is time that rushes on while distance is covered slowly. It is as if crossing the event horizon has interchanged the roles of space and time.

The second astronaut can tell nothing about what is going on inside the black hole. In fact, the only properties of a black hole that can be deduced are its mass, radius, electric charge, and angular momentum. (In this section, we have assumed zero angular momentum. Rotating black holes will be discussed in the next section.) The external simplicity of black holes is summarized in a theorem that states that *black holes have no hair*.

So far, we have been discussing nonrotating black holes. The structure of a rotating black hole is somewhat more complicated than that of a non-

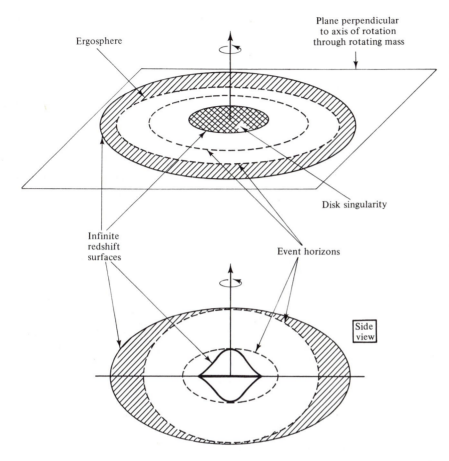

Plane perpendicular
to axis of rotation
through rotating mass

Ergosphere

Disk singularity

Infinite
redshift
surfaces

Event horizons

Side
view

Figure 7.23 Rotating black hole. The structure of the surfaces is complicated, so we show two cuts. In the upper figure we see the intersection of various surfaces with the plane perpendicular to the axis of rotation. At the center is a disk singularity. (This is just a disk, and doesn't extend above or below this plane.) There are two infinite red-shift surfaces and two event horizons. Between the event horizons, the roles of space and time are reversed. The region between the outer infinite red-shift surface and the outer event horizon is called the ergosphere. The lower figure shows a side view.

rotating black hole, and is depicted schematically in Fig. 7.23. The situation shown is for the case in which the angular momentum per unit mass J/M is less than GM/c. For the case shown, there are two infinite red shift surfaces, instead of the single event horizon. Between the two surfaces, the roles of space and time are interchanged, just as inside the event horizon in the nonrotating case. The region between the outer infinite red shift surface and the event horizon is called the *ergosphere*. The name results from the fact that there is a way to extract energy from the black hole by moving particles through the ergosphere in the correct trajectory.

7.8.3 Stellar Black Holes

In Chapter 11 we will see that some types of stars evolve to a point where nothing can support them. Such a star will collapse right through the Schwarzschild radius for its mass, and will become a black hole. Black holes would be a normal state for the evolution of some stars. How would we detect a stellar black hole? We obviously could not see it directly. We could not even see it in silhouette against a bright source, since the area blocked would be only a few kilometers across. We have to detect stellar black holes indirectly. We hope to see their gravitational effects on their surrounding environment. This is not a hopeless task, since we might expect to find a reasonable number in binary systems. We will discuss the probable detection of black holes in binary systems in Chapter 12.

Figure 7.24 Pair production. (a) The process in free space. An electron and positron are created out of nothing, but quickly come back together and annihilate. (b) Near a black hole, one of the particles can be captured before they can annihilate, and the other escapes, carrying energy away from the black hole.

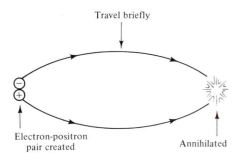

Travel briefly

Electron-positron
pair created

Annihilated

(a)

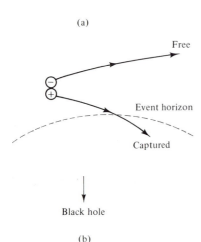

Free

Event horizon

Captured

Black hole

(b)

7.8.4 Nonstellar Black Holes

Black holes that have masses much less than a solar mass are called *mini black holes*, and those that are much more massive are called *maxi black holes*.

We think that mini black holes might have formed when the density of the universe was much higher than it is now. (The conditions in the early universe will be discussed in Chapter 21.) These may still exist. The British physicist Stephen Hawking has found that there is a mechanism by which mini black holes can actually evaporate. Hawking is studying the relationship between gravity and quantum mechanics, and the process he has proposed is a quantum-mechanical one.

This mechanism involves a different concept of a vacuum than we are accustomed to seeing. In classical physics, a vacuum is simply nothing. In quantum mechanics, it is possible to make something out of nothing, if you don't do it for long. It amounts to borrowing energy for a brief time interval. The more energy you borrow, the less time you can borrow it for. It is related to the uncertainty principle. We have talked about the uncertainty principle as it relates to momentum and position. However, it also relates to the energy and lifetime of a state. It says that if a state has a lifetime Δt, then the energy of the state is uncertain by an amount ΔE, given by

$$\Delta t \, \Delta E \geq \hbar \qquad (7.27)$$

The longer-lived a state, the more accurately its energy can be determined. Since the energy of a state is uncertain by ΔE, it is possible for us to have this extra amount of energy and not be able to detect it.

As a result of the uncertainty principle, a quantum-mechanical vacuum is a very busy place. At any place it is possible to spontaneously create a particle–antiparticle pair (Fig. 7.24). This requires an energy equal to $2mc^2$, where m is the mass of the particle (or antiparticle). This pair can only exist for at most a time $\hbar/2mc^2$. Before that time is up they must find each other and annihilate. Since electrons have masses that are much less than the proton mass, an electron–positron pair will live longer than a proton–antiproton pair.

We can therefore think of the vacuum as being made up of continuously appearing and disappearing electron–positron pairs (with a small contribution from heavier particle–antiparticle pairs). This phenomenon is called *vacuum polarization*.

When an electron–positron pair is created just outside a black hole, it is possible for one of the particles to be pulled into the black hole before the two recombine. The other particle will continue moving away from the black hole. The two cannot recombine. The particles then exist for much longer than the time limit for violating conservation of energy. We must therefore make up the energy from somewhere. The process actually reduces the mass of the black hole. The black hole shrinks slightly. For mini black holes this energy loss can be a significant fraction of the mass of the black hole. Eventually, the black hole shrinks to the point where it disappears in a small burst of gamma radiation. The more massive a black hole is when it starts out, the longer it will live. In the lifetime of the universe, black holes smaller than some given mass should already have disappeared. Those at that mass should just be dying now. Some physicists have suggested that we should be able to see the burst of gamma radiation when this happens.

Maxi black holes are probably the result of large amounts of matter gathering together in one spot. In Chapter 19 we will see the evidence for 10^8–10^9 $M_\odot$ black holes being present in the centers of some galaxies.

CHAPTER SUMMARY

In this chapter we saw how the special and general theories of relativity have changed our thinking about the nature of space and time.

We saw how the requirement that the laws of physics be the same in all inertial reference frames leads to the idea that the speed of light is the same for all observers. This, in turn, leads us to the phenomena of time dilation and length contraction. These phenomena are only large when the speeds involved are close to c.

We saw that we can no longer think of space and time as being separate entities, but must consider a four-dimensional coordinate system, called space-time. We defined a space-time interval, which is invariant under Lorentz transformation (and is equal to the proper time).

We saw that energy and momentum must be treated like space and time. This leads to the relativistic energy, $E = \gamma mc^2$, and the idea of a rest energy mc^2.

We then saw how the ideas of space-time carry over to a theory of gravitation—general relativity. The interpretation of gravitational fields is that they alter the geometry of space-time, causing it to behave like that on a curved surface. The starting point for general relativity is the principle of equivalence, which tells us that inertial mass and gravitational mass are the same

We saw that there are several effects of general relativity that can be tested. These include the advancement of the perihelion of orbiting bodies, the bending of electromagnetic radiation, the gravitational red shift and time dilation (which are really consequences of the principle of equivalence), and gravitational radiation.

We also saw how the gravitational red shift leads to the concept of black holes, objects from which nothing can escape.

QUESTIONS AND PROBLEMS

7.1 What are the differences between sound waves and electromagnetic waves?

7.2 How does the speed of light being independent of the velocity of the observer eliminate the problem that Einstein found with Maxwell's equations?

7.3 The angular displacement of an image (in radians) due to aberration is approximately v/c, as long as $v \ll c$. Use the fact that the Earth orbits the Sun once per year at a distance of 1.5×10^8 km to find the maximum displacement of a star's image resulting from the motion of the Earth. Express your answer in arc seconds.

7.4 What is the relationship between simultaneity and absolute time?

7.5 What do we mean by the terms "proper length" and "proper time"?

7.6 You and a friend carry identical clocks. Your friend passes by on a rapidly moving train. As your clock ticks off 1.00 s, you see your friend's clock tick off 0.50 s. How much time would your friend see your clock tick off in the time it takes your friend's clock to tick off 1.00 s?

7.7 How fast must a clock be moving to appear to run at half the rate of an identical clock at rest?

7.8 Some radioactive particles are traveling at $0.999\,c$. If their lifetime is 10^{-20} s when they are at rest, what is their lifetime at this speed? How far do they travel in that time?

7.9 You and a friend carry identical meter sticks and identical clocks. Your friend goes by on a fast-moving train, holding the meter stick parallel to the direction of motion of the train. If in the time it takes your clock to tick off 1.00 s you see 0.5 s tick off on your friend's clock, how long does your friend's stick appear to you? How long does your stick appear to your friend, assuming the sticks are parallel to each other?

7.10 How fast does an object have to be going so that it appears to be 10% of its original length?

7.11 A painter's assistant is carrying a 10-m ladder parallel to the ground. The assistant is moving at $0.99\,c$. The painter is up on a high ladder and drops a cloth with a 5-m hole in it, parallel to the ground. From the point of view of an observer on the cloth, the ladder shrinks to less than 5 m in length and fits through the hole. From the point of view of the assistant, the hole shrinks, so it is even smaller than the ladder. Yet, we know that the ladder must get through in all reference frames if it gets through in one. How does it get through as viewed by the assistant?

7.12 A source of radiation is moving away from you at 10% of the speed of light. At what wavelength is the Hα line seen?

7.13 How fast must an object be moving for all spectral lines to be seen at twice their rest wavelength?

7.14 We define the *red shift* z as the shift in wavelength $\Delta\lambda$, divided by the rest wavelength λ_0. On the assumption that only radial motions are involved, find an expression for z as a function of v/c.

7.15 Show that equation (7.8) reduces to the classical result when $v \ll c$.

7.16 Think about how the length of an object is determined, and show that the Lorentz transformations [equation (7.9)] give Lorentz contraction.

7.17 How is the geometry of space-time in special relativity different from that of the geometry of three-dimensional space?

7.18 Suppose we have two events that take place at (t_1, x_1, y_1, z_1) and (t_2, x_2, y_2, z_2) in one reference frame, and at (t_1', x_1', y_1', z_1') and (t_2', x_2', y_2', z_2') in the other reference frame. The coordinates in the two frames are related by the Lorentz transformations, equation 7.9. Show that the space-time interval between the two events is the same in both reference frames.

7.19 Show that in the limit $v \ll c$, the expression for the energy of a particle, equation 7.13, reduces to the sum of the rest energy plus the familiar kinetic energy, $\frac{1}{2}mv^2$.

7.20 Show that if tachyons exist, their rest mass must

be an imaginary number if their energy is to be real for $v > c$. (An imaginary number is the square root of a negative number.)

7.21   What do we mean when we say the gravity alters the geometry of space-time?

7.22   Would the principle of equivalence still hold if inertial mass were twice the gravitational mass? (*Hint:* Work through the accelerating elevator example.)

7.23   Consider an object with the same density as the Sun. Find an expression for the bending of starlight past the edge of this object as a function of the size of the object.

7.24   For a neutron star with 1 $M_\odot$ and a radius of 10 km, through what angle is light bent as it passes the edge?

7.25   Why don't you need a solar eclipse to measure the bending of radio waves past the edge of the Sun?

7.26   List the tests of general relativity discussed in this chapter.

7.27   For a sphere of a given mass M show that there is a radius such that, if a photon is emitted from that radius, it will be red-shifted to an infinite wavelength (zero energy).

7.28   For a white dwarf with 1 $M_\odot$ and a radius of 5 $\times$ 10^3 km, find the wavelength to which the Hα line will be red-shifted by the time it is seen by a distant observer.

7.29   (a) For the test of the gravitational red shift involving the Mossbauer effect, calculate the shift in going from the Earth's surface to 50 m above the Earth's surface. (b) How fast would the receiver have to move toward the source to compensate for the red shift? (c) Compare your answer in (b) with the speed that an object falling from the roof would acquire just before striking the ground.

7.30   Show that the Schwarzschild radius can also be found by taking the escape velocity from an object of mass M and radius R, and setting it equal to the speed of light.

7.31   (a) Compute your Schwarzschild radius. (b) What would the density be for a black hole of your mass?

7.32   For what mass black hole does the density equal 1 g/cm^3?

7.33   For what mass black hole does the difference between the acceleration of gravity at an astronaut's feet and head equal the acceleration of gravity on the Earth (1000 cm/s^2)?

7.34   Since we cannot run a ruler from the center of a black hole to the Schwarzschild radius, how would you ''measure'' the radius of a black hole? (*Hint:* Think in terms of a measurement that doesn't involve crossing the event horizon.)

7.35   Find an expression for dg/dr at the surface of a black hole, as a function of the mass of the black hole. Your expression should be usable in the way that equation (7.23) is usable for finding the mass.

7.36   How does the rate of a clock 1.5 R_S from a 3-$M_\odot$ black hole compare with the rate of a clock far from the black hole?

7.37   (a) What is the exit cone? (b) When the exit cone closes, what happens to photons aimed straight up?

7.38   Is there some place near a black hole where you could look straight ahead and see the back of your head? Explain.

7.39   If an electron–positron pair forms outside of the vacuum, how long can they live before they must annihilate?

8

Protostars

8.1 STAR FORMATION

The space between the stars is not completely empty. It is filled with diffuse clouds of gas and dust, called the *interstellar medium*. (We will discuss the interstellar medium and star formation in more detail in Chapters 14 and 15. In this chapter, we will just assume that a sufficient amount of material has been brought together to form a star.) The dust is easier to see because it efficiently blocks the light from distant stars. However, the gas (mostly hydrogen) provides 99% of the mass in the interstellar medium. The interstellar gas and dust are concentrated in clouds of varying sizes, temperatures, and densities. The clouds have even denser condensations within them. These condensations are the potential sites of star formation. (As we will see in Chapter 14, even the densest observed interstellar clouds have densities that would constitute a good vacuum in the laboratory. However, the clouds are so large that they are quite massive.)

The clouds are kept together by the gravitational attraction among all of the particles in the cloud. If a cloud is to be *gravitationally bound,* the gravitational forces holding it together must be greater than the other forces driving it apart. We can think of the random thermal motions in the gas as resisting the collapse. In order for the cloud to be bound, we want the thermal speeds to be less than the escape velocity.

We can also put this argument in terms of energies. For the cloud to be bound, we want the total energy (gravitational potential plus thermal) to be negative. To apply this, we must be able to calculate the gravitational potential energy and thermal energy of a cloud. We do this for a simple geometry, a spherical cloud.

8.1.1   Gravitational Potential and Kinetic Energy of a Sphere

We are generally dealing with systems of large numbers of particles. Therefore, rather than thinking in terms of individual particles of mass m, we can think in terms of a fluid of average density ρ. (The density is simply the mass per particle, multiplied by the number of particles per unit volume.) In this section, we will evaluate the kinetic and potential energies for a uniform (constant-density) sphere. Even though real systems might not be exactly uniform or spherical, the results will generally only change by numerical factors of order unity.

We begin by calculating the gravitational potential energy of a uniform sphere of mass M, radius R, and density ρ. These three quantities are related by

$$M = (4\pi/3)R^3\rho \tag{8.1}$$

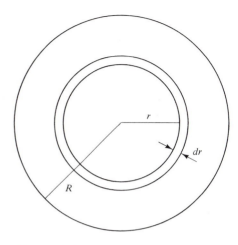

Figure 8.1   We model stars by studying spherical shells.

The gravitational potential energy is the work required to bring all of the material in from far away (infinity) to the final configuration. The final result should not depend on the order in which the various parts of the sphere are assembled, so we do the calculation in the easiest way that we can envision. We can think of the sphere as being made up of shells (Fig. 8.1). We can assemble the sphere one shell at a time, starting with the smallest.

Let's assume we have already assembled shells through radius r. We now want to calculate the work to bring in the next shell. The thickness of the shell is dr. The volume of the shell is its surface area multiplied by its thickness:

$$dV = 4\pi r^2 \, dr \tag{8.2}$$

The mass contained in the shell is the volume multiplied by the density:

$$dM = 4\pi r^2\rho \, dr \tag{8.3}$$

The total mass of material already assembled interior to radius r is

$$M(r) = (4\pi/3)r^2\rho \tag{8.4}$$

The quantity $M(r)$ is important since the shell that ends up at radius r will only feel a net force from material inside it. Even after we bring in more material outside this shell, the net force exerted on any particle in the shell by matter outside radius r is zero. Also, for the mass in the shell, the mass $M(r)$ exerts a force equal to that which would be exerted by the same mass all located at the center of the sphere.

For any two point masses, remember that the gravitational potential energy (relative to infinity) is given by

$$U = -Gm_1m_2/r \qquad (8.5)$$

We let m_1 equal $M(r)$ and m_2 equal the mass of the shell dM. The work to bring in this shell is then

$$
\begin{aligned}
dU(r) &= -GM(r)dM/r \\
&= -G(4\pi/3)(4\pi)\rho^2 r^4 \, dr
\end{aligned}
\qquad (8.6)
$$

To find the effect of all the shells, we integrate the quantity dU from $r = 0$ to $r = R$:

$$
\begin{aligned}
U &= \int_0^R dU(r) \\
&= -G(4\pi/3)(4\pi)\rho^2 \int_0^R r^4 \, dr \\
&= -G(4\pi/3)(4\pi)\rho^2 \tfrac{1}{5} R^5 \qquad (8.7) \\
&= -\tfrac{3}{5}G[(4\pi/3)\rho R^3]^2/R \\
&= -\tfrac{3}{5}GM^2/R
\end{aligned}
$$

For other shaped objects, the gravitational potential energy is generally proportional to $-GM^2/R$, where R is some average length. The constant of proportionality is generally close to one.

The thermal energy is $\tfrac{3}{2}kT$ per particle. Total thermal energy is then

$$K = \tfrac{3}{2}NkT \qquad (8.8)$$

where N is the total number of particles in the cloud. If the mass per particle is m, and the total mass of the cloud is M, then

$$N = M/m \qquad (8.9)$$

so

$$K = \tfrac{3}{2}(M/m)kT \qquad (8.10)$$

8.1.2 Collapsing Clouds

The condition for gravitational binding (total energy negative) is then

$$\tfrac{3}{5}GM^2/R \geq \tfrac{3}{2}kT(M/m)$$

Dividing both sides by M and multiplying by $\frac{5}{3}$ gives

$$(M/R) \geq \tfrac{5}{2}(kT/Gm) \tag{8.11}$$

The mass and radius of a cloud are not independent, since they are related by the density $\rho = M/(4\pi/3)R^3$. We might therefore like to use equation (8.11) to estimate the smallest-size cloud of a given ρ, m, T for which the cloud is gravitationally bound. This minimum radius is called the *Jeans length, R_J*. Jeans obtained essentially the same result with a more sophisticated analysis. We therefore eliminate M in equation (8.11), and change the inequality to an equality, since we are looking for the value of R that is on the boundary between bound and unbound. This gives

$$(4\pi/3)R_J^3\rho/R_J = 5kT/2Gm$$

Solving for R_J,

$$R_J = (15kT/8\pi GmR)^{1/2} \tag{8.12}$$

Note that $(15/8\pi)^{1/2} = 0.77$, which is close to one. As the geometry of the cloud changes, the exact value of the constant will change, but it will still be close to one, so we drop these constants in our estimate and replace them by one. We then write

$$R_J \cong (kT/Gm\rho)^{1/2} \tag{8.13}$$

We can rewrite this in terms of n, the number of particles per unit volume ($n = \rho/m$),

$$R_J \cong (kT/Gm^2n)^{1/2} \tag{8.14}$$

We can also use equation (8.11) to give us the minimum mass for which a cloud of given ρ, m, T will be bound. This minimum mass is called the *Jeans mass*. It is the mass of an object whose radius is R_J, so

$$M_J = (4\pi/3)R_J^3\rho$$

$$= (4\pi/3)(kT/Gm\rho)^{3/2}\rho$$

$$= 4(kT/Gm)^{3/2}\rho^{-1/2}$$

$$= 4(kT/Gm)^{3/2}(nm)^{-1/2} \tag{8.15}$$

EXAMPLE 8.1 Jeans Length and Mass

Find the Jeans length and mass in a cloud with 10^5 H atoms per cm^3 and a temperature of 50 K.

Solution

We use equation (8.14) to find R_J:

$$R_J = \frac{(1.38 \times 10^{-16} \text{ erg/K})(50 \text{ K})}{(6.67 \times 10^{-8} \text{ dyn-cm}^2/\text{g}^2)(1.67 \times 10^{-24} \text{ g})^2(10^5 \text{ cm}^{-3})}$$

$$= 6.1 \times 10^{17} \text{ cm}$$

$$= 0.2 \text{ pc}$$

We find the mass by multiplying the density by the volume.

$$M_J = \frac{(1.67 \times 10^{-24} \text{ g})(10^5 \text{ cm}^{-3})}{(4\pi/3)(6.1 \times 10^{17} \text{ cm})^3}$$

$$= 1.5 \times 10^{35} \text{ g}$$

$$= 76 \, M_\odot$$

We could have obtained the same mass directly from equation (8.15). As we will see below, not all this mass will end up in the star.

Once a cloud becomes gravitationally bound, it will begin to collapse. We would like to be able to estimate the time for the collapse to take place. We begin by considering a particle a distance r from the center of the cloud. It will accelerate toward the center under the influence of the mass closer to the center than r. The acceleration is given by

$$a(r) = GM(r)/r^2$$

$$= G(4\pi/3)r^3\rho/r^2 \qquad (8.16)$$

$$= (4\pi/3)G r \rho$$

If the acceleration of this particle stayed constant with time, then the time for it to free-fall a distance r would be

$$t_{ff} = \left[\frac{2r}{a(r)} \right]^{1/2}$$

$$= \left[\frac{2r}{(4\pi/3)(G r \rho)} \right]^{1/2} \qquad (8.17)$$

Note that the constant $(\frac{3}{2}\pi)^{1/2} = 0.7$, which we can approximate as one, since we are just making an estimate of the time. This gives

$$t_{ff} \cong 1/(G\rho)^{1/2} \tag{8.18}$$

The free-fall time is independent of the starting radius. Therefore, all matter in a constant-density cloud has approximately the same free-fall time. However, as the cloud collapses, the density increases. The collapse proceeds faster. The free-fall time for the original cloud is therefore an upper limit to the actual collapse time. However, the result is not very different, since most of the time will still be taken up in the early stages of the collapse, when the acceleration is not appreciably different from the one we have calculated. Therefore, we use the free-fall time as a reasonable estimate of the time it will take a cloud to collapse.

There is one important difference between our idealized cloud and a real cloud. A real cloud will probably have a higher density in the center. We can see this as follows. If the cloud is initially of uniform density, all points will have the same initial inward acceleration. This means that all particles will cover the same inward distance dr (where $dr < 0$), in some time interval dt. We can see how this changes the density for different volume spheres. If the initial density is ρ_0, then the density of a constant-mass collapsing sphere that shrinks from r_0 to r is

$$\rho = \rho_0(r_0/r)^3$$

The change in the density $d\rho$ is found by differentiating to give

$$d\rho = -3\rho_0(r_0^3/r^4)dr$$

The fractional change in density $d\rho/\rho$ is

$$d\rho/\rho = -3 \; dr/r \tag{8.19}$$

This means that the smaller the initial sphere we consider, the faster its density will grow.

With a higher density at the center, the free-fall time for material near the center will be less than for material near the edge. The material from the edge will lag behind the material closer in. This will enhance the density concentration in the center. The net result is that we end up with a strong concentration in the center. The concentration will eventually become the star, but material from the outer parts of the original cloud will continue to fall in on the star for quite some time.

EXAMPLE 8.2 Free-Fall Time

Calculate the free-fall time for the cloud in the above example.

Solution

Using equation (8.18) gives

$$t_{ff} = [(6.67 \times 10^{-8} \text{ dyn-cm}^2/\text{g}^2)(1.67 \times 10^{-24} \text{ g})(10^5 \text{ cm}^{-3})]^{1/2}$$

$$= 9.5 \times 10^{12} \text{ s}$$

$$= 3 \times 10^5 \text{ yr}$$

If the cloud is rotating, then the collapse will be affected by the fact that the cloud's angular momentum must remain constant. The angular momentum L is the product of the moment of inertia I and the angular speed ω,

$$L = I\omega \tag{8.20}$$

For a sphere, the moment of inertia is

$$I = \tfrac{2}{5}Mr^2 \tag{8.21}$$

If I_0 and ω_0 are the original moment of inertia and angular speed, and I and ω are their values at some later time, conservation of angular momentum tells us that

$$I_0\omega_0 = I\omega \tag{8.22}$$

Using equation (8.21) to eliminate I and I_0, we have

$$\omega/\omega_0 = (r_0/r)^2 \tag{8.23}$$

To see what effect this has on collapse, we again look at a particle a distance r from the center of a collapsing cloud. The force on that particle is still $GM(r)/r^2$. However, the radial acceleration now has two parts: (1) $a(r)$ is associated with the change of the magnitude of the radius; and (2) the acceleration is associated with the change of direction $r\omega^2$. Therefore,

$$GM(r)/r^2 = a(r) + r\omega^2 \tag{8.24a}$$

Solving for $a(r)$ gives

$$a(r) = GM(r)/r^2 - r\omega^2 \qquad (8.24b)$$

By comparing this with equation (8.16) we see that the acceleration $a(r)$ is less for the rotating cloud than for the nonrotating cloud. The effect of the rotation is to slow down the collapse perpendicular to the axis of rotation.

The effects of rotation will be the most significant when the second term on the right-hand side of equation (8.24) is much greater than the first term, in which case

$$GM(r)/r^2 = r\omega^2$$

Multiplying both sides by r^2 gives

$$GM(r) = r^3\omega^2$$

$$= r^3\omega^2(r_0/r)^4 \qquad (8.25)$$

$$= (\omega_0 r_0)^2(r_0/r)r_0$$

Noting that $v_0 = \omega_0 r_0$, where v_0 is the speed of a particle a distance r_0 from the center,

$$GM(r) = v_0^2 r_0(r_0/r)$$

We now solve for r/r_0, the amount by which the cloud collapses before the rotation dominates:

$$r/r_0 = v_0^2 r_0/GM(r) \qquad (8.26)$$

For the cloud given in the two previous examples, with an initial rotation speed $v_0 = 1$ km/s, $r/r_0 = 0.6$. This means that by the time the cloud reaches half its initial size the rotation can completely stop the collapse perpendicular to the axis of rotation. Collapse parallel to the axis of rotation can proceed unimpeded, and the cloud will flatten. However, the rotational angular momentum is sufficient to keep much of the material from becoming a star.

More of the material can end up in stars if the cloud breaks up into a multiple star system. The angular momentum can then be taken up in the orbital motion, but individual clumps can continue contracting. This fragmentation process is probably responsible for the high incidence of binary systems. If a cloud shrinks to half its original size, the average density will go up by a factor of 8. (The density is proportional to $1/$volume, and the volume is proportional to r^3.) From equation (8.15), we see that the Jeans mass in the denser cloud will be approximately one-third of the original Jeans mass. This means that it is possible for the smaller clumps to be bound and to continue their collapse. This fragmentation process (Fig. 8.2) may be repeated until stellar mass objects are reached.

8.2 THE VIRIAL THEOREM

In a gravitationally bound system, such as a collapsing cloud, in addition to the total energy being negative, the kinetic and potential energies are related in a very specific way. This relationship is known as the *virial theorem*. In this section we derive the virial theorem.

We begin with a collection of N particles. (We can think of each star in a cluster or each atom in a gas cloud as being represented by a particle.) To simplify the calculation we will assume that all particles have the same mass m. The final result would be the same if we allowed for different masses. (See Problem 8.10.) We let the position of the ith particle, relative to some origin, be $\mathbf{r}_i$. If we have two particles, i and j, the vector giving their separation is $\mathbf{r}_j - \mathbf{r}_i$, as shown in Fig. 8.3. We let $\mathbf{F}_i$ be the net force on the ith particle. We can therefore write the equation of motion for this particle as

$$\mathbf{F}_i = m \, d^2\mathbf{r}_i/dt^2 \tag{8.27}$$

We are looking for a relationship between various types of energy. The vector dot product between a force and distance will give us an energylike quantity. We therefore take the dot product of $\mathbf{r}_i$ with both sides of equation (8.27), and then sum the resulting quantities over all of the particles to give

$$\sum_{i=1}^{N} \mathbf{r}_i \cdot \mathbf{F}_i = m \sum_{i=1}^{N} \mathbf{F}_i \cdot (d^2\mathbf{r}_i/dt^2) \tag{8.28}$$

We can rearrange this, using the following:

$$\frac{d^2}{dt^2}(\mathbf{r}_i^2) = \frac{d}{dt}\left(2\mathbf{r}_i \cdot \frac{d\mathbf{r}_i}{dt}\right) = 2\left(\frac{d\mathbf{r}_i^2}{dt}\right) + 2\mathbf{r}_i \cdot \left(\frac{d^2\mathbf{r}_i}{dt^2}\right) \tag{8.29}$$

where we have used the fact that for any vector x

$$\mathbf{x}^2 = \mathbf{x} \cdot \mathbf{x} = x^2$$

This gives

$$\sum_{i=1}^{N} \mathbf{r}_i \cdot \mathbf{F}_i = \frac{1}{2}\frac{d^2}{dt^2}\left(\sum_{i=1}^{N} m\mathbf{r}_i^2\right) - \sum_{i=1}^{N} m\left(\frac{d\mathbf{r}_i}{dt}\right)^2 \tag{8.30}$$

We now introduce the velocity of the ith particle $\mathbf{v}_i$, defined as

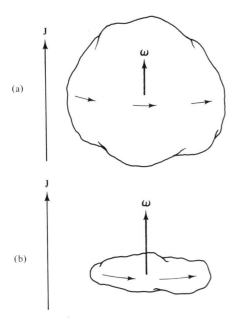

(a)

(b)

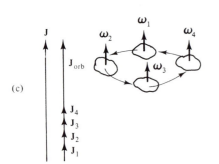

(c)

Figure 8.2 Fragmentation of a collapsing interstellar cloud. (a) The cloud is initially rotating as shown. As it collapses, the angular momentum **J** is conserved. (b) As the cloud gets smaller, its angular speed ω must increase to keep the angular momentum fixed. The rotation inhibits collapse perpendicular to the axis of rotation, and the cloud flattens. (c) Unable to collapse any further, the cloud breaks up, with the total angular momentum being divided among the spin and orbital angular moments of the individual fragments.

$$\mathbf{v}_i = d\mathbf{r}_i/dt \qquad (8.31)$$

We also introduce the quantity I, defined as

$$I = \sum_{i=1}^{N} m\mathbf{r}_i^2 \qquad (8.32)$$

If $\mathbf{r}_i$ were the distance of particle i from some axis, then I would be the moment of inertia about that axis. However, an axis is a line, and $\mathbf{r}_i$ is measured from a point. Therefore, I is a quantity that is similar to, but not the same as, the familiar moment of inertia. We also set the total kinetic energy K equal to the sum of the kinetic energies of the individual particles,

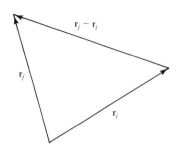

Figure 8.3   Position vectors $\mathbf{r}_i$ from the origin to the ith particle, and $\mathbf{r}_j$ from the origin to the jth particle.

$$K = \tfrac{1}{2} \sum_{i=1}^{N} m v_i^2 \qquad (8.33)$$

In terms of these quantities, equation (8.30) becomes

$$\sum_{i=1}^{N} \mathbf{r}_i \cdot \mathbf{F}_i = \frac{1}{2}\frac{d^2 I}{dt^2} - \sum_{i=1}^{N} m v_i^2 \qquad (8.34)$$

$$= \frac{1}{2}\frac{d^2 I}{dt^2} - 2K \qquad (8.35)$$

We evaluate the $\Sigma \, \mathbf{r}_i \cdot \mathbf{F}_i$ for the $\mathbf{F}_i$ being the gravitational forces between particles. We let

$$\mathbf{r}_{ij} = \mathbf{r}_j - \mathbf{r}_i \qquad (8.36)$$

be the vector distance between two particles. The net force on a given particle is the sum of the forces exerted on it by all of the other particles. Therefore

$$\mathbf{F}_i = \sum_{\substack{j=1 \\ i \neq j}}^{N} \frac{m^2 G \mathbf{r}_{ij}}{|\mathbf{r}_{ij}|^3} \qquad (8.37)$$

which gives us

$$\sum_{i=1}^{N} \mathbf{r}_i \cdot \mathbf{F}_i = \sum_{i=1}^{N} \sum_{\substack{j=1 \\ i \neq j}}^{N} G m^2 \left(\frac{\mathbf{r}_i \cdot \mathbf{r}_{ij}}{|\mathbf{r}_{ij}|^3} \right) \qquad (8.38)$$

Since i and j both go over the full range, 1 to N, we could interchange the i and j on the right-hand side of equation (8.38) without really changing anything. If we rewrite the right-hand side, interchanging i and j, and then adding it to the right-hand side, we will have a quantity equal to twice the original right-hand side. This gives us

$$\sum_{i=1}^{N} \mathbf{r}_i \cdot \mathbf{F}_i = \frac{1}{2} \sum_{\substack{i=1 \\ i \neq j}}^{N} \sum_{j=1}^{N} Gm^2 \left(\frac{\mathbf{r}_i \cdot \mathbf{r}_{ij}}{|\mathbf{r}_{ij}|^3} + \frac{\mathbf{r}_j \cdot \mathbf{r}_{ji}}{|\mathbf{r}_{ji}|^3} \right) \tag{8.39}$$

$$= \frac{1}{2} \sum_{\substack{i=1 \\ i \neq j}}^{N} \sum_{j=1}^{N} Gm^2 \left(\frac{\mathbf{r}_i \cdot (\mathbf{r}_j - \mathbf{r}_i) + \mathbf{r}_j \cdot (\mathbf{r}_i - \mathbf{r}_j)}{|\mathbf{r}_{ij}|^3} \right) \tag{8.40}$$

where we have also used the fact that $|\mathbf{r}_{ij}| = |\mathbf{r}_{ji}|$. This procedure actually allows for a simplification. To see this, we first multiply out the numerator on the right-hand side to get

$$\sum_{i=1}^{N} \mathbf{r}_i \cdot \mathbf{F}_i = \frac{1}{2} \sum_{i=1}^{N} \sum_{\substack{j=1 \\ i \neq j}}^{N} Gm^2 \left(\frac{\mathbf{r}_i \cdot \mathbf{r}_j - \mathbf{r}_i^2 - \mathbf{r}_j^2 + \mathbf{r}_j \cdot \mathbf{r}_i}{|\mathbf{r}_{ij}|^3} \right) \tag{8.41}$$

We also note that

$$|\mathbf{r}_{ij}|^2 = (\mathbf{r}_i - \mathbf{r}_j) \cdot (\mathbf{r}_i - \mathbf{r}_j)$$
$$= r_i^2 + r_j^2 - \mathbf{r}_i \cdot \mathbf{r}_j - \mathbf{r}_j \cdot \mathbf{r}_i \tag{8.42}$$

Substituting into equation (8.41) gives

$$\sum_{i=1}^{N} \mathbf{r}_i \cdot \mathbf{F}_i = \frac{1}{2} \sum_{\substack{i=1 \\ i \neq j}}^{N} \sum_{j=1}^{N} Gm^2 \left(-\frac{|\mathbf{r}_{ij}|^2}{|\mathbf{r}_{ij}|^3} \right) \tag{8.43}$$

which simplifies to

$$\sum_{i=1}^{N} \mathbf{r}_i \cdot \mathbf{F}_i = -\frac{1}{2} \sum_{\substack{i=1 \\ i \neq j}}^{N} \sum_{j=1}^{N} \left(\frac{Gm^2}{|\mathbf{r}_{ij}|} \right) \tag{8.44}$$

The term on the right-hand side is the sum of the gravitational potential energies of each pair of particles. Note that each pair appears twice in the sum, since the energy of the pair is independent of which particle of the pair we count first. For example, for particles 1 and 2, both the quantities $Gm^2/|\mathbf{r}_{12}|$ and $Gm^2/|\mathbf{r}_{21}|$ appear. This means that the double sum on the right-hand side of equation (8.44) gives us twice the gravitational potential energy, but there is a factor of one-half in front, so the right-hand side is equal to the gravitational potential energy U. We can therefore rewrite equation (8.35) as

$$\tfrac{1}{2}d^2I/dt^2 = 2K + U \tag{8.45}$$

If we take a time average of these quantities over a sufficiently long time, the left-hand side approaches zero. This leaves

$$\cdot \ 0 = 2\langle K \rangle + \langle U \rangle \tag{8.46}$$

where the $\langle \ \rangle$ represents the time average of the enclosed quantity. Equation (8.46) is the simplest form of the virial theorem.

The virial theorem applies to any gravitationally bound system that has had sufficient time to come to equilibrium. Even simple systems, like binary stars, obey the virial theorem (see Problem 8.13). If the orbits are circular, then $K = -U/2$ at all points. For elliptical orbits, r and v are changing, so K and U are changing, while their sum E is fixed. This means that we have to average over a whole orbit to get $\langle K \rangle = -\langle U \rangle/2$.

Remember, for any system, the total energy is

$$E = K + U \tag{8.47}$$

So for a system to which the virial theorem, in the form of equation (8.46) applies, we set $\langle K \rangle = -\langle U \rangle/2$, to give

$$E = \langle U \rangle/2 \tag{8.48}$$

(We don't have to take the time average of E, since E is constant.) Remember, the gravitational potential energy, defined so that it is zero when the particles are infinitely far apart, is negative. Therefore, the total energy of a bound system is negative. This means that we have to put energy in to break up the system.

8.3   LUMINOSITY OF COLLAPSING CLOUDS

As a cloud collapses the gravitational potential energy decreases. This is because particles within the cloud are moving closer to the center. The decrease in potential energy must be offset by energy radiated away or by an increase in kinetic energy. This increased kinetic energy can show up in two forms: (1) it can go into the faster in-fall of the particles in the collapsing cloud; or (2) it can go into heating the cloud.

To see what happens to the energy in a collapsing cloud from equations (8.46) and (8.48), we see that

$$E = -\langle K \rangle \tag{8.49}$$

This tells us that as the cloud collapses its internal kinetic energy K will increase. However, only half the potential energy loss shows up as increased kinetic energy. We can see from equations (8.46) and (8.47) that the total energy of the collapsing cloud is decreasing. This means that the cloud must be radiating energy away. The virial theorem tells us that half the lost potential energy shows up as kinetic energy, and half the energy is radiated away.

We can relate the luminosity of a contracting cloud to its total energy. Using equation (8.48), the total energy is

$$E = -\tfrac{3}{10}GM^2/R \tag{8.50}$$

The energy lost in radiation must be balanced by a corresponding decrease in E. The luminosity L must therefore be equal to dE/dt. Differentiating equation (8.50) gives

$$dE/dt = \tfrac{3}{10}(GM^2/R^2)(dR/dt) \tag{8.51}$$

We can also solve for dR/dt to find the collapse rate for a given luminosity.

$$dR/dt = \tfrac{10}{3}(R^2/GM^2)(dE/dt) \tag{8.52}$$

(Remember, for a collapsing cloud, both dR/dt and dE/dt are negative numbers.) If we solve equation (8.50) for R, and substitute that solution for one of the R's in equation (8.51) or (8.52), we find that

$$(1/E)(dE/dt) = (1/R)(dR/dt) \tag{8.53}$$

This tells us that, in any time interval dt, the fractional change in the energy dE/E is equal to the fractional change in the radius dR/R. These results tell us that the rate of collapse can be limited by the rate at which energy can be radiated.

We now look at the luminosity in various stages of the collapse. As the collapsing cloud heats, it is still well below normal stellar temperatures, so most of the radiation is given off in the infrared. Therefore, the opacity of the cloud in the infrared plays an important role in determining the nature of the collapse.

When the collapse begins, the material is mostly atomic and molecular hydrogen and atomic helium. As the collapse continues, half the liberated energy goes into the internal energy of the gas. However, this doesn't increase the temperature. Instead, the energy goes into the ionization of these neutral species. Following this, the liberated energy goes into heating the gas, and the pressure can eventually slow the collapse. For a 1-$M_\odot$ protostar the free-fall phase ends when the radius is about 500 $R_\odot$. (This radius varies approximately linearly with mass.) During the free-fall stage, the luminosity increases since $|dR/dt|$ increases.

EXAMPLE 8.3 Luminosity of a Collapsing Cloud

For a 1-$M_\odot$ protostar that has collapsed to 500 $R_\odot$, (a) calculate the energy that has been liberated to this point; (b) use this to calculate the average luminosity if most of the energy is liberated in the last 100 years of the collapse.

Solution

From the virial theorem, the energy radiated will be $\frac{1}{2}$ times the current gravitational potential energy.

$$E = (-\tfrac{1}{2})(-\tfrac{3}{5})\, GM^2/R$$

$$= \tfrac{3}{10}(6.67 \times 10^{-8}\ \text{dyn-cm}^2/\text{g}^2)(2 \times 10^{33}\ \text{g})^2$$

$$= (500)(7 \times 10^{10}\ \text{cm})$$

$$= 2 \times 10^{45}\ \text{erg}$$

The average luminosity is this energy divided by the time over which it is radiated.

$$L(\text{avg}) = \frac{2 \times 10^{45}\ \text{erg}}{(100)(3 \times 10^7\ \text{s})}$$

$$= 7 \times 10^{35}\ \text{erg/s}$$

$$= 170\, L_\odot$$

This is the average luminosity over the 100-yr period, but the actual luminosity at the end of that period is higher, since $|dR/dt|$ is greatest then.

Once a cloud is producing stellar luminosities by gravitational collapse, it is "officially" a protostar. Once the cloud becomes opaque the radiation can only escape from near the surface. (When the opacity is low a photon can escape from anywhere within the volume.) Since energy escapes slowly, the temperature rises quickly. Also, a large temperature difference can exist between the center and the edge. Under these conditions, the most efficient form of energy transport from the center to the outside is by convection. This point was first realized in 1961 by the Japanese astrophysicist Chushiro Hayashi. During this stage the surface temperature stays roughly constant at about 2500 K. Since the radius is decreasing, and the temperature is approximately constant, the luminosity decreases.

During this stage the central temperature is still rising. When it is high enough, nuclear reactions start. The contraction goes on for some time in the outer parts, as the pressure builds up in the core. Eventually the pressure in the core is sufficient to halt the collapse, and the star is ready to settle into its main sequence existence.

For a protostar, the continuous spectrum peaks in the near-infrared part of the spectrum. The dust in the collapsing cloud surrounding the protostar will absorb some of the radiation. The dust will be heated, but will not be at the same temperature as the star. (We'll discuss the properties of the dust more in Chapter 14.) The emission from the dust will be in the far infrared. From this we see that protostars are best observed in the infrared part of the spectrum.

8.4 EVOLUTIONARY TRACKS FOR PROTOSTARS

When we plot an H–R diagram with stars we see now, we are plotting the distribution of L, T as they are now. However, as a star evolves, its luminosity and temperature change. Therefore, its location on an H–R diagram changes. If $L(t)$ is the luminosity of the star as a function of time and $T(t)$ is the temperature as a function of time, we can plot a series of points $L(t)$, $T(t)$ and connect them to follow the evolution of the star. Such a series of points is called an *evolutionary track*. Stars evolve so slowly compared to human lifetimes that we cannot deduce the evolutionary track for a star by observing that one star. However, by observing many stars, each at a different stage, we can infer the evolutionary tracks.

We can also predict evolutionary tracks from theoretical models of protostars and stars. We use basic physics to calculate the physical conditions and see how the star's radius and temperature change with time. Since the luminosity, $L = (4\pi R^2)(\sigma T^4)$, we can relate changes in R and T to changes in L and T. When we calculate model tracks, we find that the evolutionary track of a protostar depends on its mass. This is not surprising, since we have already seen that the mass determines where a star will appear on the main sequence.

Some evolutionary tracks for protostars are shown in Fig. 8.4. Note that the protostars appear above the main sequence. This means that for a given T, protostars are more luminous than main sequence stars of the same tem-

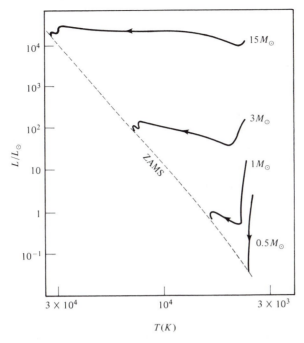

Figure 8.4 Evolutionary tracks for protostars on an *H–R* diagram. Tracks are marked by the mass used in the model. The dashed line represents the zero age main sequence (ZAMS).

perature. Protostars are larger than main sequence stars of the same temperature. This is not surprising, since protostars are still collapsing. We also call protostars *pre-main-sequence* stars.

Let's look more closely at the track for a 1-$M_\odot$ protostar (Fig. 8.5). At first the cloud is cool, and it contracts and heats. As discussed above, the T^4 increase is greater than the R^2 decrease, and the luminosity of the protostar

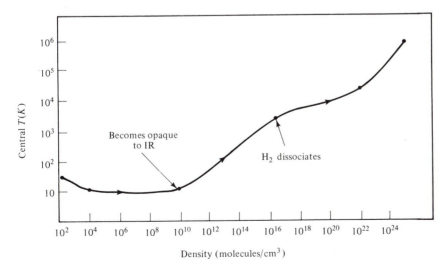

Figure 8.5 Model for the collapse of an intrastellar cloud into a protostar.

increases. The peak luminosity is reached when the temperature reaches 600 K. As the protostar gets denser, its opacity increases. Eventually, it is harder for radiation from the center to escape, and the luminosity begins to decrease. During this stage, energy transport in the star is mostly by convection. The part of the evolutionary track in which the luminosity is decreasing quickly while the temperature increases slightly is called the *Hayashi track*. After this collapse slows, the star begins to approach the main sequence. Eventually, it reaches the luminosity of a main sequence star, though it might vary somewhat before settling down.

8.5   T TAURI STARS AND RELATED OBJECTS

We would like to know if we observe any objects with the properties of the pre-main-sequence stars described above. There is a class of objects, called *T Tauri stars* (Fig. 8.6), which we think fall into this category. T Tauri is a variable star in Taurus, and T Tauri stars are variable stars with properties similar to T Tauri. Light curves are shown in Fig. 8.7, and a spectrum is shown in Fig. 8.8. These stars are of spectral class K, and appear above the main sequence on the *H–R* diagram. They show an irregular variability. Their spectra are also characterized by the presence of emission lines.

There are three possible sources of the variability that we see.

1.   The variability could arise in the photosphere. One model for this involves star spots. These are dark areas like sunspots, only larger. As the star rotates, a different fraction of the observable surface is covered with spots, and the brightness changes.

2.   The variations may arise in the chromosphere.

3.   The variations may actually result from changes in the opacity of the dust shell surrounding the star.

The emission line profiles show Doppler-shifted absorption wings, like those shown in Fig. 8.9. This suggests material both falling into the star and material coming off the star. The in-fall may be close to the star as a final stage of collapse, while the outflow is a wind (like the solar wind, but stronger) further away from the star. Alternatively, the in-fall may be in the form of a disk around the star's equator, while the outflow is along the polar axes.

From studies of the spectral lines, we think that the winds may have speeds of about 200 km/s. The mass loss in the wind, dM/dt, is about 10^{-7} $M_{\odot}$/yr. The total luminosity in the wind is the rate at which kinetic energy is carried away by the wind,

$$L_w = \tfrac{1}{2}(dM/dt)v^2 \tag{8.54}$$

Using the numbers given, we find a wind luminosity of about 1 $L_{\odot}$. That is,

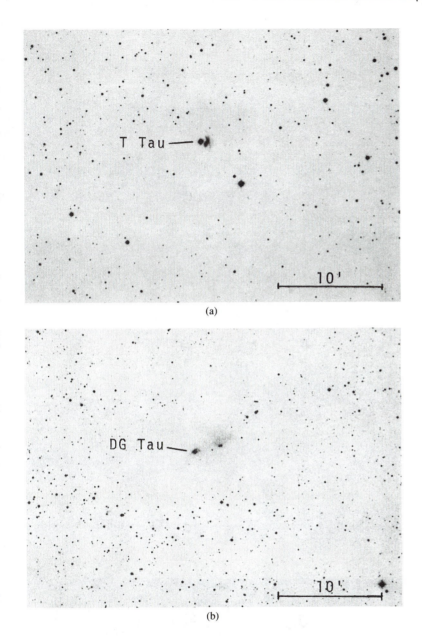

Figure 8.6   Photographs (negative) of two T Tauri stars. (a) T Tauri. The star is unremarkable, but there is a small nebula to the right. The star is in a dust cloud. The presence of the dust can be deduced from the fact that it blocks the light from the background stars. There are therefore fewer background stars in the center of the photograph than around the edges. (b) DG Tauri. Again, there is some nebulosity near the star, and the star is in a dust cloud.

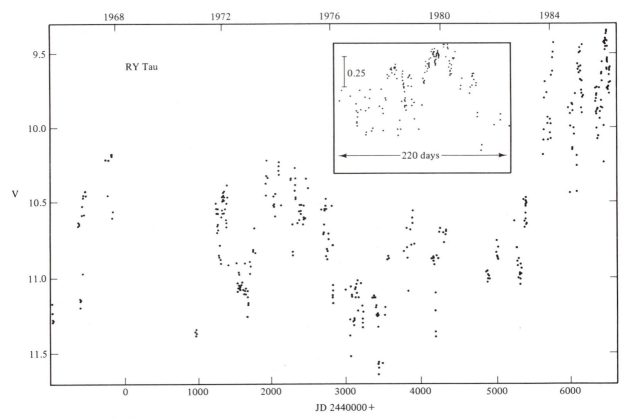

Figure 8.7 Light curve for the T Tauri stars, RY Tau. The main figure shows the long-term variability and the insert shows the short-term variability.

Figure 8.8 Spectrum of some T Tauri stars (positive) image). The bright bands are emission lines. The unlabeled spectrum at top is a laboratory comparison.

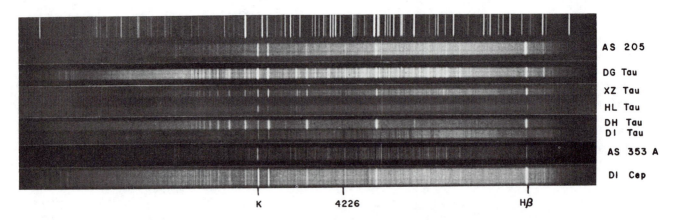

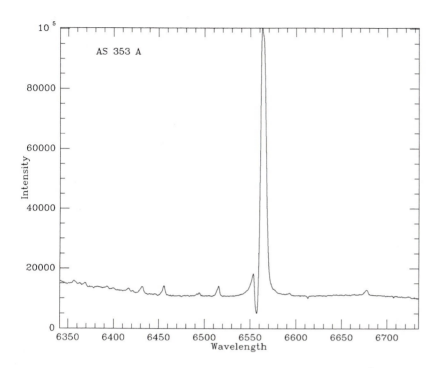

AS 353 A

Figure 8.9   Profile of the Hα line in the T Tauri star A5353A. The line is in emission, but has some absorption on the long wavelength side of the line.

the star gives off as much energy per second in its wind as the sun gives off at all wavelengths. However, the wind phase is a short-lived one. The wind does sweep out some of the dust that has collected around the star. We think that a similar wind from the sun was important in clearing debris out of the early solar system.

Another interesting class of objects that we think are associated with pre-main-sequence stars are the Herbig–Haro objects, or HH objects, shown in Fig. 8.10. They were identified independently by George Herbig of the Lick Observatory and Guillermo Haro of the Mexican National Observatory. HH objects appear as bright nebulosity on optical photographs. Their spectra sometimes resemble those of stars, and usually show emission lines, but no star is present in the nebulosity. We now think that the wind from a pre-main-sequence star clears a path through a cloud. The part of the cloud where the wind runs into the cloud is heated, and glows. We also see starlight reflected from the dust, explaining the stellar spectrum. The exciting star is deep within the cloud, and is not seen directly.

Another type of star that suggests pre-main-sequence behavior is FU Orionis (Fig. 8.11). The star brightened by six magnitudes in 1936. At that time it changed both its size and spectral type. Since then it has been undergoing irregular variations in its light. We think that it might represent the last stages of a pre-main-sequence star reaching the main sequence.

The IRAS satellite has provided as with some images, like Fig. 8.12, which may be the result of infrared emission from dust shells around recently formed stars.

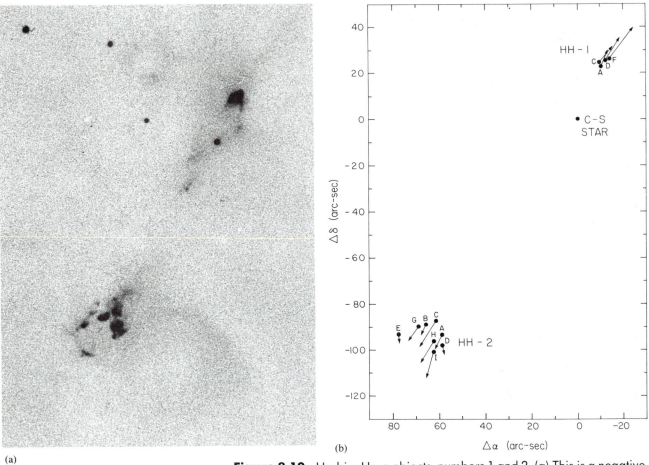

(a)

(b)

Figure 8.10   Herbig–Haro objects, numbers 1 and 2. (a) This is a negative photograph. The HH objects appear as a bright nebulosity in a dark cloud. (b) The proper motions of these features are indicated by the arrows. The arrows show how far each knot of nebulosity moves in 100 years. These objects appear to be moving away from some common center. It is thought that material has been ejected from the location of the T Tauri star, marked "C-S Star."

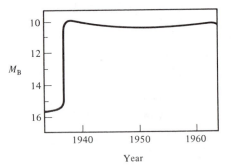

Figure 8.11   Light curve for FU Orionis, showing the sudden brightening by 5 magnitudes (a factor of 100).

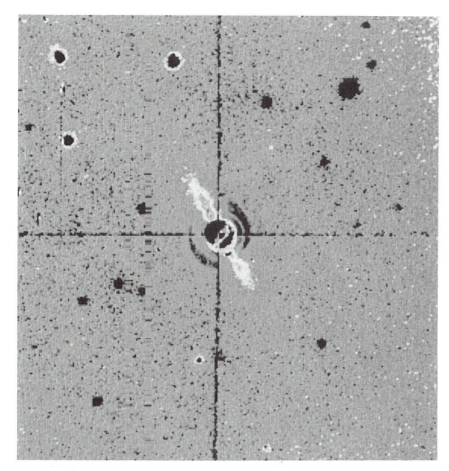

Figure 8.12 IRAS image of a dust disk around the star β Pictoris. It is thought that this type of disk is a common residue of star formation, and may be the material out of which planets can form.

CHAPTER SUMMARY

In this chapter we investigated the physical conditions in clouds that are collapsing to become stars.

We first looked at the gravitational effects in a collapsing cloud. The condition for the cloud to be collapsing is that the total energy (kinetic plus potential) be negative. We introduced the Jeans length as the minimum size for a cloud of a given density and temperature to collapse.

We looked at the effects of rotation on the collapse. In a rotating cloud, collapse is inhibited perpendicular to the axis of rotation. This might result in fragmentation of the cloud.

Once the collapse has started, the potential and kinetic energies are related by the virial theorem. This relationship limits what can happen in the future collapse. As the cloud collapses, half the change in potential energy goes into heating the cloud, and the other half must be radiated away. Most of this radiated energy is in the infrared.

We saw how to trace the behavior of a protostar (or any other star) on an *H–R* diagram, using an evolutionary track. On the *H–R* diagram, protostars appear above the main sequence.

We looked at examples of objects that are probably representative of various stages of protostellar evolution. These include T Tauri stars and Herbig–Haro objects. All these objects are characterized by strong winds or outflows.

QUESTIONS AND PROBLEMS

8.1 Explain why cool dense regions are the most likely sites of star formation.

8.2 Find the Jeans length and mass in a cloud with 10^3 H atoms/cm^3 and a temperature of 10 K.

8.3 What density material (H atoms/cm^3) will produce a Jeans mass of 1 $M_\odot$ at $T = 100$ K?

8.4 As a cloud collapses, what happens to the Jeans mass of new individual fragments?

8.5 If a cloud has a higher density in the center than at the edge when its collapse starts, explain what happens to the density contrast between the center and outer part as the collapse continues. (*Hint:* Think of the free-fall time.)

8.6 As a cloud collapses, the acceleration of the particles increases. Therefore, the free-fall time will be less than that which we calculate on the basis of constant acceleration. However, we still use the initial free-fall time as an estimate of the total free-fall time. Why does this work? (*Hint:* Think in terms of the time for the radius to halve, and halve again, and so on.)

8.7 Explain why a rotating cloud flattens as it collapses.

8.8 As a cloud collapses, is it likely that it will rotate as a rigid body (ω independent of *r*)?

8.9 Verify that a binary star system obeys the virial theorem. (a) Show that $K = -U/2$ at all times for circular orbits. (b) Use the equations for potential, kinetic, and total energies over an elliptical orbit to show that it holds even though *K* is not equal to $-U/2$ at all points on the orbit.

8.10 Rederive the virial theorem, allowing the masses of individual particles to be different.

8.11 Calculate the quantity *I* in the virial theorem for a uniform-density sphere of mass *M* and radius *R*, and compare it with the moment of inertia.

8.12 What is an evolutionary track?

8.13 What features of T Tauri stars lead us to believe they are still in the formative process?

8.14 (a) For the typical T Tauri wind described in this chapter, what is the momentum per second carried away by the wind? (b) If the wind drives away dust and slows by conservation of momentum, and the wind is effective at driving dust away until it slows to 5 km/s, what is the rate at which dust can be driven away?

8.15 Calculate the energy used up in ionizing 1 $M_\odot$ of atomic H. To what radius must a 1-$M_\odot$ protostar collapse for this much energy to be released in the change in gravitational potential energy?

The Main Sequence

In this chapter we look at the inner workings of a star once it has settled into a main sequence existence. We start by looking at the sources of stellar energy, and then we look at the physical processes that govern stellar structure.

9.1 STELLAR ENERGY SOURCES

In the last chapter we saw that protostars convert the energy of their gravitational collapse into radiation. We might wonder how long this process can go on. For example, could the Sun be powered in this way even now? To see, we estimate the gravitational energy lifetime t_g. This lifetime is the stored energy divided by the rate at which the energy is being lost. That is

$$\begin{aligned} t_g &= \frac{E}{dE/dt} \\ &= \frac{E}{L} \end{aligned} \qquad (9.1)$$

where L is the luminosity.

We now have to estimate the energy stored E. We know the Sun is no longer contracting (since we can see its size remaining constant). Whatever energy the Sun has stored from gravitational contraction must have been stored as the Sun collapsed from a large cloud to its current size. This energy is just half (from the virial theorem) of the negative of the Sun's current gravitational potential energy.

$$E = \frac{(\frac{1}{2})(\frac{3}{5})GM^2}{R}$$

$$= \frac{(0.3)(6.67 \times 10^{-8} \text{ dyn-cm}^2/\text{g}^2)(2 \times 10^{33} \text{ g})^2}{7 \times 10^{10} \text{ cm}} \qquad (9.2)$$

$$= 1.1 \times 10^{48} \text{ erg}$$

This gives a lifetime, called the *Kelvin time*,

$$t_g = \frac{1.1 \times 10^{48} \text{ erg}}{4 \times 10^{33} \text{ erg/s}}$$

$$= 3 \times 10^{15} \text{ s}$$

$$= 1 \times 10^7 \text{ yr}$$

This is a lifetime of 10 million years. However, we know from geological evidence that the Earth has been around for over four billion years. Therefore, the Sun (and presumably other stars) cannot exist in a stable configuration on stored gravitational energy.

An alternative source of energy is chemical reactions. After all, we use chemical reactions to make our automobiles go on Earth. We can estimate the amount of energy stored in the Sun capable of being released in chemical reactions. Typical energies for chemical reactions should be equivalent to some fraction of the binding energy of molecules. We can estimate that we might get approximately 1 eV per atom in the Sun. The total energy available is then 1 eV multiplied by the total number of atoms, M/m. The energy available is then

$$E = \frac{(1 \text{ eV})(1.6 \times 10^{-12} \text{ erg/eV})(2 \times 10^{33} \text{ g})}{1.67 \times 10^{-24} \text{ g}}$$

$$= 2 \times 10^{45} \text{ erg}$$

This is less energy than is stored in gravitational potential energy, so chemical reactions clearly cannot provide a long-term energy source for the Sun.

The answer to our problem is nuclear reactions. The typical energies available in nuclear reactions are about 1 MeV per atom, instead of 1 eV. This is an improvement of a factor of 10^6. To see how nuclear reactions provide energy for the Sun, we have to look at some of the elements of nuclear physics.

9.2 NUCLEAR PHYSICS

9.2.1 Nuclear Building Blocks

We have already seen that the positive charge in atoms is confined to the small nucleus (10^{-15} m across). The nucleus is composed of protons and neutrons. Because they are the building blocks of the nucleus, we call them *nucleons*. The proton has a charge $+e$ (where $-e$ is the charge on the electron). Since atoms are neutral, the number of electrons orbiting the nucleus equals the number of protons in the nucleus. The chemical properties of an atom depend on the number of orbiting electrons. Therefore, the number of protons in the nucleus ultimately determines the identity of an element. We designate the number of protons with the symbol Z, called the *atomic number*. The charge on the nucleus is $+Ze$. The highest naturally occurring value of Z is 92 (uranium), with approximately a dozen "man-made" elements with higher values of Z.

Neutrons are electrically neutral. The mass of the neutron is slightly greater than that of the proton. Nuclei with the same numbers of protons can have different numbers of neutrons N. The total number of nucleons in the nucleus is called the *mass number*, $A = Z + N$. The mass of the nucleus is approximately Am_p. Two nuclei with the same number of protons but different numbers of neutrons are called *isotopes* of the same element. We generally designate an element with a letter symbol (e.g., H for hydrogen), preceded by a superscript giving A, followed by a subscript giving Z. (This letter subscript is redundant since the symbol for the element tells us Z, and is sometimes left out.) For example, the three known isotopes of hydrogen are written as $^1H_1$, $^2H_1$, $^3H_1$, or, simply, ^1H, ^2H, ^3H.

For any given number of protons, we do not have an arbitrary number of neutrons. In general, we find that stable elements have approximately equal numbers of protons and neutrons. For larger values of Z the stable nuclei have a slightly larger number of neutrons than protons.

We now look at the force that holds nuclei together. We generally refer to four forces in nature. In order of decreasing strength, they are *strong nuclear, electromagnetic, weak nuclear,* and *gravity*. The nuclear forces are short-range, as opposed to the electromagnetic and gravitational forces which can be felt at long ranges. For example, if we look at the force between two protons, when they are far apart, the electric force dominates, and the protons repel each other. When we get the protons very close together, the attractive nuclear force becomes stronger. The strong nuclear force between two protons is the same as that between two neutrons and is the same as that between a proton and a neutron (Fig. 9.1).

This gives us an idea of the role of a neutron in a nucleus. If we just have protons the electric repulsion will be too strong, and the nucleus will not be stable. If we add neutrons, we get additional binding from the nuclear

Figure 9.1 Properties of the nuclear and electrical forces between neutrons and protons. (a) The nuclear force F_N is the same for a proton and a proton, a neutron and a neutron, and a proton and a neutron. The electric force only acts between the proton and the proton. The magnitude of this force is the same as that between a proton and an electron. (b) How the forces vary with distance. As the two protons are brought closer together, the electric repulsion gets stronger, but the nuclear attraction gets stronger faster.

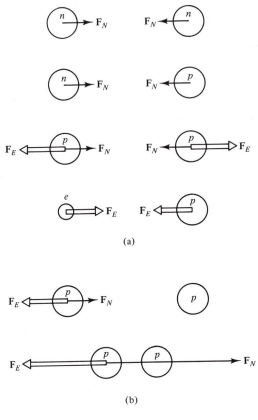

(a)

(b)

force, but no additional electrical repulsion. (The electric repulsion is actually reduced, since the neutrons keep the protons farther apart.) This explains why a nucleus needs approximately at least as many neutrons as protons.

9.2.2 Binding Energy

Since the nuclear force is attractive, we must do work to get two nucleons apart. The work required to disassemble a nucleus is the *binding energy* of the nucleus (Fig. 9.2). The binding energy is analogous to the ionization energy of an atom—the energy required to separate an electron from the rest of the atom, doing work against the electrical attraction. Since the nuclear force is so strong, nuclear binding energies are much greater than atomic binding energies by a factor of about 10^6. We measure nuclear binding energies in MeV, rather than eV. The greater the binding energy of a nucleus, the more work we must do to get the nucleus apart. Therefore, a larger binding energy means a more stable nucleus.

The binding energies of nuclei are so large we can measure them directly by comparing the mass of a nucleus with the masses of its components. We

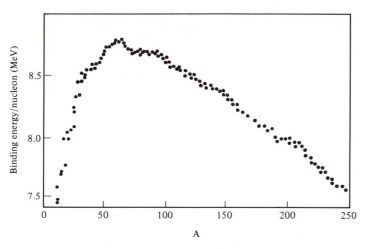

Figure 9.2 Nuclear binding energies. The horizontal axis is the mass number A, and the vertical axis is the total binding energy of the nucleus, divided by the number of nucleons in the nucleus. Nuclei with a higher binding energy per nucleon are the most stable.

will always find the nucleus to have measurably less mass than its initial components. This follows from Einstein's realization that mass and energy are equivalent. This equivalence is described quantitatively by the expression

$$E = mc^2 \qquad (9.3)$$

This means that the mass of a nucleus is less than the masses of its components by BE/c^2, where BE is the binding energy. This is also true for atoms. However, the binding energy for atoms is so small that the mass difference is negligible.

EXAMPLE 9.1   Rest Energy

Compute the energy of a proton at rest, and express the result in MeV.

Solution

The energy is

$$E = (1.67 \times 10^{-24} \text{ g})(3 \times 10^{10} \text{ cm/s})^2$$

$$= 1.5 \times 10^{-3} \text{ erg}$$

$$= \frac{1.5 \times 10^{-3} \text{ erg}}{1.6 \times 10^{-12} \text{ erg/eV}}$$

$$= 9.4 \times 10^{8} \text{ eV}$$

$$= 940 \text{ MeV}$$

EXAMPLE 9.2   Binding Energy

Calculate the binding energy of the deuteron, an isotope of hydrogen containing one proton and one neutron, given the following data:

$$m_p = 1.6726 \times 10^{-24} \text{ g}$$

$$m_n = 1.6749 \times 10^{-24} \text{ g}$$

$$m_d = 3.3436 \times 10^{-24} \text{ g}$$

Solution

The binding energy is given by

$$E = (m_p + m_n - m_d)c^2$$

$$= (3.9 \times 10^{-27} \text{ g})(2.9979 \times 10^{10} \text{ cm/s})^2$$

$$= 3.5 \times 10^{-6} \text{ erg}$$

$$= 2.2 \text{ MeV}$$

By comparing the result of this example with the previous example, we see that the nuclear binding energy is a few tenths of a percent of the total rest energy. This mass difference is relatively easy to measure.

9.2.3 Nuclear Reactions

We have seen that there are large amounts of energy involved in holding a nucleus together. However, we must still have a way of liberating that energy. If we have a nuclear reaction in which the products have a greater binding energy than the reactants, the difference in energy will be available to heat the surroundings. Before we discuss the particular nuclear reactions that work in stars, we look at the types of nuclear reactions that can take place.

The first type of reaction is a *decay*. In a decay, a nucleus emits a particle. Depending on the type of particle that is emitted, the results is a different nucleus or the same nucleus in a lower energy state. We identify three different types of decays, depending on the type of particle that is emitted. The three types are called alpha (α), beta (β), and gamma (γ).

An *alpha* particle is a nucleus of the most common form of helium, ^{4}He. This means that the alpha particle has two protons and two neutrons. This particular combination is very stable and can be emitted as a single group. The final nucleus has two fewer protons and two fewer neutrons than the original nucleus. This means that the element changes as the atomic number Z decreases by two. The mass number A decreases by four. The process can only take place if the resulting nucleus is more stable than the original nucleus. This means that the binding energy of the final nucleus is greater than that of the original nucleus. Once an alpha particle gets far enough from the nucleus, it no longer feels the nuclear attraction. However, since it has a charge of $+2e$, it feels the Coulomb repulsion of the remaining nucleus. The alpha particle accelerates away from the nucleus. The energy liberated in the reaction can be carried away in the form of the kinetic energy of the alpha particle.

A *gamma*-ray is simply a high-energy photon. Since it carries away neither mass nor charge, A and Z do not change. The identity of the nucleus does not change in gamma decay. The photon does carry away energy. The protons and neutrons in the nucleus must also be treated quantum-mechanically as being waves. This means that there are allowed states for the motions of nucleons within the nucleus. A gamma-ray is emitted when a transition is made from a higher energy state to a lower energy state. The energy of the gamma-ray is equal to the energy difference between the two states. We can learn a lot about the internal structure of nuclei by studying the energies of gamma-rays that are emitted, just as the wavelengths of visible photons tell us about the structure of atoms.

A *beta* particle is an electron (or a positron, which is the antiparticle to the electron). In a beta decay, an electron or positron is emitted. The charge of the nucleus increases by $+e$ (or $-e$ if a positron was emitted), but the mass number A remains unchanged. The net result is to change a neutron into a proton. (The reverse, changing a proton into a neutron, is also possible, but requires a source of energy, since the neutron is more massive than the proton.) When a neutron changes to a proton, Z increases by one, but N decreases by one, leaving A unchanged.

In beta decays an additional particle, called the *neutrino*, is emitted. Neutrinos were originally postulated because an analysis of certain beta decays indicated that some energy and angular momentum were being lost in the process. It was therefore assumed that a massless neutral particle was carrying away energy and angular momentum. If neutrinos are truly massless (and we will have more to say on this point Part V of this book) then they travel at the speed of light. The existence of the neutrino was verified experimentally in the 1950s. Neutrinos do not interact with other matter via the strong nuclear or the electromagnetic forces. This means that a reaction in which a neutrino is involved must proceed by the weak nuclear force. The weak force is so weak that neutrinos rarely interact with matter. Weak decays also take much more time than do strong decays.

The basic beta-decay reaction is

$$ n \rightarrow p + e^- + \bar{\nu} \tag{9.4} $$

The bar over the ν actually indicates an antineutrino. (Every particle has a corresponding antiparticle. A particle and antiparticle have identical masses, but all other properties, such as charge, are the negative of each other. When a particle and antiparticle come together, they can convert all of their mass into energy—in other words, annihilate each other—without violating any conservation laws.) In free space the average time for this reaction to occur is 11 min. In nuclei this reaction can take place only if the resulting proton ends up in a lower-energy state than the original neutron. This places an upper limit on the number of neutrons that a nucleus with a given Z can have and still remain stable. If we have too many neutrons, some will beta decay and become protons in lower-energy states.

Another type of nuclear reaction is *fission*, in which a nucleus breaks up into smaller parts. The controlled fission of uranium provides the energy in current nuclear power plants. Generally, in fission, a very heavy nucleus breaks up to form more stable, middle-mass nuclei. From Fig. 9.2, we see that the most stable nuclei are those with intermediate masses.

The final type of nuclear reaction we will consider is *fusion*, in which lighter nuclei come together to build heavy nuclei. It is fusion that is important in stellar energy generation. One problem with fusion is that it is hard to get started. When two nuclei are far apart the nuclear force between them has no effect. They only feel the electrical repulsion. To get the nuclei close enough together for the nuclear force to take over, we must do work against the electrical repulsion. This requires accelerating the particles to high energies and then letting them collide. Since they have high energies, they will be slowed by the repulsion, but not stopped altogether. On the Earth, we can accomplish this acceleration for small quantities of matter in particle accelerators, but this does not work for large quantities of material. This is the major problem that must be overcome before we can realistically think about using fusion as a source of energy on the Earth.

9.2.4   Overcoming the Fusion Barrier

One way to increase the energies of particles is to raise their temperature. If the temperature is high enough, particles will be moving fast enough to overcome the electrical repulsion, and fusion reactions can take place. On the Earth this poses problems since we have trouble containing a gas at the required temperature, tens of millions of degrees Kelvin. However, in stars, the material is confined by gravity. If sufficiently high temperatures are reached, the nuclear reactions can take place. If the temperatures are not high enough to support nuclear reactions, then we don't have a star.

We can estimate the temperature required for nuclear fusion to take place. Let's consider the case of two protons. Their electric potential energy when they are a distance r apart is

$$U = e^2/r \tag{9.5}$$

Suppose we start two protons far apart, so we can take their potential energy as zero. We let K be the kinetic energy of the particles at this point. We would like the particles to come to rest (zero kinetic energy) a distance r_p, the radius of the proton, apart. From conservation of energy, the kinetic energy when the protons are far apart must be

$$K = e^2/r_p$$

$$= \frac{(4.8 \times 10^{-10} \text{ esu})^2}{1 \times 10^{-13} \text{ cm}}$$

$$= 2 \times 10^{-6} \text{ erg}$$

If we divide this energy by the Boltzmann constant k, we get an estimate of the temperature at which the average kinetic energy of the particles in the gas is equal to this energy. This gives a temperature of 2×10^{10} K, a very high temperature. If we had considered the case of nuclei with charges $+Z_1 e$ and $+Z_2 e$, the potential energy becomes $Z_1 Z_2 e^2/r$. This means that even higher temperatures are needed for higher-charge nuclei.

Actually, the temperature doesn't have to be as high as our calculation suggests. If the temperature is as high as we calculate, then the energy of the *average* particle will be high enough for it to participate in nuclear fusion. We have already seen that not all the particles in a gas have the average energy. Even at lower temperatures there will be some particles with a high enough energy to undergo fusion. For particles with a Maxwell–Boltzmann velocity distribution, the probability of finding a particle with energy between E and $E + dE$ is proportional to a term like the $\exp(-E/kT)$ in the Boltzmann equation. That is,

$$P(E)\, dE \propto E^{1/2} \exp(-E/kT)\, dE \qquad (9.6)$$

There is another effect which allows nuclear reactions to take place at a lower temperature than we calculated. Figure 9.3 shows the potential energy of two protons as a function of the distance between them. For the most part the force is repulsive, but if we could get the protons close enough the force would be attractive. Suppose E_0 is the minimum energy for a particle to be able to overcome the Coulomb repulsion. We can see that this energy E_0 is just the "height" of the potential energy barrier. Particles with energy greater than E_0 can get through the barrier, and particles with energy less than E_0 will reach a point where their energy equals the height of the barrier at some value r. At this point all of the energy is in the form of potential energy, so the kinetic energy is zero. The particle has stopped and is about to head back in the other direction. The point at which a particle turns back is called the *turning point*.

We have been looking at the situation from the point of view of classical

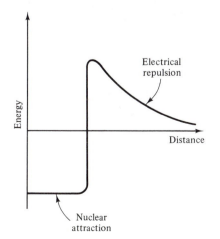

Figure 9.3 The potential energy for two protons as a function of distance. This includes the electrical force and a model for the nuclear force.

mechanics. Quantum-mechanically, we should talk about the probability of finding a particle in various places. Let's again look at the particle with energy less than E_0. It has some probability for being farther from the proton than the classical turning point. We would then say that the probability of it being closer than the turning point is zero. However, the probability is related to a wave phenomenon, and wave phenomena cannot suddenly go from some finite value to zero. They must fall off gradually. That means that there is actually some probability of finding the incoming particle closer to the proton than the classical turning point. This phenomenon is called *barrier penetration* or *tunneling*.

In general, a particle can penetrate to a distance approximately equal to its wavelength h/mv. More precisely, the probability of penetrating a distance x is related to the wavelength by

$$P(x) \propto \exp(-ax/\lambda)$$
$$= \exp(-axmv/h)$$

(9.7)

where a is some constant. Suppose we have to tunnel a distance x, equal to the classical turning point r_0, defined by initial kinetic energy being equal to the electrostatic potential energy at r_0, or

$$\tfrac{1}{2}mv^2 = Z_1Z_2e^2/r_0$$

Substituting this into equation (9.7), we have the probability of penetrating r_0

$$P(r_0) \propto \exp(-aZ_1Z_2e^2/2hv)$$

$$\propto \exp(-bE^{1/2})$$

where b is a constant. The effective area of the nucleus for a reaction is approximately λ^2 and is proportional to $1/E$. When we combine this with the Maxwell–Boltzmann velocity distribution, we find the probability of a reaction by nuclei of energy E is proportional to

$$\exp(-E/kT - b/E^{1/2})$$

This means that at a given T there will be a most likely E for reactions to take place. This is known as the *Gamow peak* shown in Fig. 9.4. It turns out that most reactions involve particles that are on the low-energy side of the velocity distribution. These lower-energy particles have a longer wavelength, and can therefore penetrate deeper into the barrier. The exponential behavior of the reaction rates also makes them very sensitive to temperature.

Figure 9.4   The probability of a nuclear fusion, as a function of the particle energy E, at a given gas temperature. This shows the combined effects of the location of the classical turning point and quantum-mechanical tunneling.

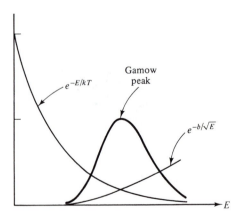

9.3 NUCLEAR ENERGY FOR STARS

When a star is on the main sequence, its basic source of energy is the conversion of hydrogen into helium. We start with four protons and end up with one ^{4}He nucleus. However, it is unlikely that four protons will get close enough to directly form a ^{4}He nucleus in a single reaction. There are different series of reactions that achieve this net result, and they will be discussed below.

We can calculate the energy released by converting four protons to one ^{4}He by comparing their masses. We find that

$$4m_p - m(^4\text{He}) = 0.007(4m_p) \tag{9.8}$$

This means that 0.007 of the mass of each proton is converted into energy.

EXAMPLE 9.3   Lifetime of the Sun

Estimate the lifetime of the Sun for producing energy at its current rate from nuclear fusion.

Solution

If 0.007 of the mass of each proton in the Sun is converted into energy, and if we assume that most of the mass of the Sun is in the form of protons, then 0.007 of the Sun's total mass is available for conversion into energy. The total energy available is therefore

$$E = 0.007 M_\odot c^2$$

$$= (0.007)(2 \times 10^{33} \text{ g})(3 \times 10^{10} \text{ cm/s})^2$$

$$= 1.3 \times 10^{52} \text{ erg}$$

The lifetime is then this energy divided by the luminosity.

$$t_n = E/L$$

$$= \frac{1.3 \times 10^{52} \text{ erg}}{4 \times 10^{33} \text{ erg/s}}$$

$$= 3.2 \times 10^{18} \text{ s}$$

$$= 1 \times 10^{11} \text{ yr}$$

However, we think that only 10% of the mass of the Sun is in a region hot enough for nuclear reactions, so we must lower our estimate by a factor of ten. This leaves us with a lifetime of ten billion years. We think the Sun has already lived half of this time.

We now look at actual nuclear reactions, the net result of which is to convert four protons into a ^{4}He nucleus, plus energy. The basic series of reactions in stars like the Sun is called the *proton–proton* chain, because it starts with the direct combination of two protons:

$$p + p \rightarrow d + (e^+) + \nu \qquad (9.9)$$

which we can also write as

$$^1\text{H} + {}^1\text{H} \rightarrow {}^2\text{H} + (e^+) + \nu$$

In this reaction the (e^+) is a positron, the antiparticle to the electron. We can see by the presence of the neutrino that this process is a weak interaction, and therefore goes very slowly. This process requires temperatures of about 10^7 K. Once the deuteron has been created, it can quickly react with another proton:

$$d + p \rightarrow {}^3\text{He} + \gamma \qquad (9.10)$$

which we can also write as

$$^2\text{H} + {}^1\text{H} \rightarrow {}^3\text{He} + \gamma$$

If each of these reactions takes place twice, then we have started with six protons and now have two ^{3}He nuclei. (^{3}He is the isotope of helium with two protons and one neutron. It is one of the few stable nuclei with fewer neutrons than protons.) The ^{3}He nuclei can combine to give

$$^3\text{He} + {}^3\text{He} \rightarrow {}^4\text{He} + {}^1\text{H} + {}^1\text{H} \qquad (9.11)$$

Note that we have gotten two of the protons back. The net result is that we have converted four protons into a ^{4}He nucleus, along with two positrons, two gamma rays, and two neutrinos. The energy given off by this chain is carried away by the positrons, gamma rays, and neutrinos. The positrons will immediately scatter off (or annihilate) other particles in the gas, resulting in a heating of the gas. The gamma ray will travel a small distance before being absorbed, also heating the gas. However, the hot gas will emit new photons. The neutrino interacts so weakly that it will escape from the star completely.

A process that is important in stars more massive than the Sun is responsible for the buildup of elements heavier than helium. It is called the *triple-alpha* process, because the net result is to convert three alpha particles into a ^{12}C nucleus. The repulsion between two alpha particles is four times that between two protons, so higher temperatures, about 10^8 K, are required for this process. The first step in the chain is

$$^4\text{He} + {}^4\text{He} \rightarrow {}^8\text{Be} + \gamma \qquad (9.12a)$$

We should note that the binding energy of two ^{4}He nuclei exceeds that of the ^{8}Be nucleus, because the ^{4}He is such a stable nucleus. This tells us that the ^{8}Be should be unstable and break up. However, there will be so many alpha-particles around that some ^{8}Be nuclei will capture an alpha-particle before breaking up. If this did not happen the buildup of heavier elements would be blocked. The combination of the ^{4}He and ^{8}Be gives

$$^4\text{He} + {}^8\text{Be} \rightarrow {}^{12}\text{C} + \gamma \qquad (9.12b)$$

In massive stars, where the temperature is higher than in the Sun, there is another cycle that is important in converting four protons into one ^{4}He nucleus. The process is called the *CNO cycle*. The cycle is indicated schematically in Fig. 9.5, and the steps are:

$$^{12}\text{C} + {}^1\text{H} \rightarrow {}^{13}\text{N} + \gamma \qquad (9.13)$$

$$^{13}\text{N} \rightarrow {}^{13}\text{C} + (e^+) + \nu \qquad \text{(beta decay)} \qquad (9.14)$$

$$^{13}\text{C} + {}^1\text{H} \rightarrow {}^{14}\text{N} + \gamma \qquad (9.15)$$

$$^{14}\text{N} + {}^1\text{H} \rightarrow {}^{15}\text{O} + \gamma \qquad (9.16)$$

$$^{15}\text{O} \rightarrow {}^{15}\text{N} + (e^+) + \nu \qquad \text{(beta decay)} \qquad (9.17)$$

$$^{15}\text{N} + {}^1\text{H} \rightarrow {}^{12}\text{C} + {}^4\text{He} \qquad (9.18)$$

We see that the net result is the conversion of four protons into one ^{4}He nucleus plus two positrons, two neutrinos, and three photons. All nuclei created as intermediate products are used up in the next step. In addition, the last step returns the ^{12}C we needed to start the cycle, so the cycle can go again. In this sense, we can think as the ^{12}C as a catalyst for the reaction. (A catalyst is something which helps something happen, but which is not itself changed in the process.) There are more complicated versions of this cycle involving even heavier elements, but the basic ideas are the same.

As we try to build heavier and heavier elements through fusion, the electrical repulsion becomes stronger. This becomes an effective barrier to the formation of heavier elements. However, no matter how high the Z a

Figure 9.5 The CNO cycle. Solid arrows represent reactions. Symbols over these arrows indicate emitted particles. Dashed lines indicate when a created particle participates in another reaction.

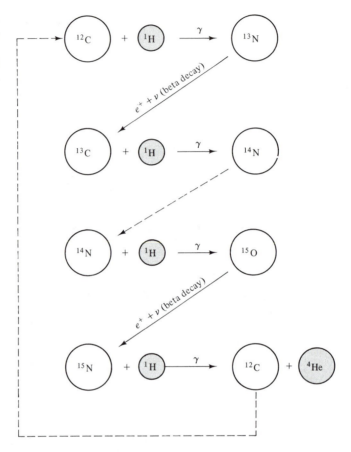

nucleus has, we can always get a neutron near it with no electrical repulsion. If the neutron is moving slowly, it can be captured by the nucleus. This can be important in stars, because some reactions produce free neutrons, so there are generally some free neutrons available. We can schematically represent what happens when a nucleus (Z, A) captures a neutron:

$$(Z, A) + n \rightarrow (Z, A + 1) \qquad (9.19)$$

What happens next depends on whether the rate of neutron capture is slow or rapid, compared to the rate of beta decay. If neutron capture is slow, we call the sequence of reactions an *s-process* (Fig. 9.6). In this situation, the new nucleus $(Z, A + 1)$ will beta decay before it can capture another neutron:

$$(Z, A + 1) \rightarrow (Z + 1, A) + (e^-) + \bar{\nu} \qquad (9.20)$$

If neutron capture is rapid we call the sequence of reactions an *r-process*

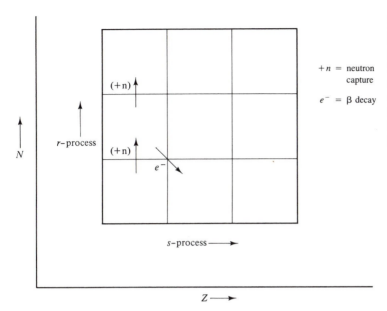

Figure 9.6 The r and s processes. The horizontal axis indicates increasing proton number Z, and the vertical axis is increasing neutron number N. In an r-process, a neutron is captured, and in an s-process, the capture is followed by a beta decay. The +n next to an arrow indicates neutron capture. The e⁻ next to an arrow indicates beta decay.

(Fig. 9.6). The nucleus $(Z, A + 1)$ will capture another neutron before it beta decays:

$$(Z, A + 1) + n \rightarrow (Z, A + 2) \tag{9.21}$$

In either case, the resulting nucleus can either beta decay or capture a neutron, depending on the relative rates. When we have a string of nuclei for which neutron capture is favorable, the r-process allows the buildup of neutron-rich nuclei. This will go on until so many neutrons are added that a beta decay breaks the chain. The r- and s-processes can explain the abundances of many of the heavier nuclei. (It should be noted that these are not equilibrium processes.)

The various nuclear processes that we have discussed are responsible for the presence of the heavy elements around us. We will see in later chapters how this material is spread out into interstellar space. The net result is to produce the abundances shown in Fig. 9.7. The nuclear physics determines which elements are most abundant.

When we study stellar structure in the next section, we will treat the nuclear physics as something that is known. We assume that once we know the composition of some region of a star and the temperature, we can specify which nuclear reactions will be important. Moreover, we assume that we know how the rates of the reactions depend on temperature. This is a very important point. We have already seen, in this section, that tunneling and the fact that the higher than average energy nuclei are the ones most likely to react, places a strong temperature dependence on the reaction rates. For

Figure 9.7 Cosmic abundances of the elements as a function of atomic number Z.

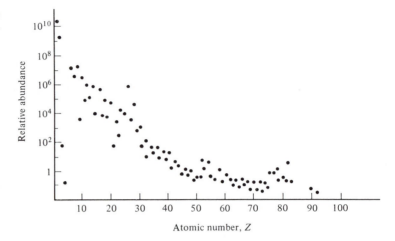

example, the rates of some important reactions depend on temperature T, as T^7.

We say that a collapsing object makes the transition from protostar to star when its primary source of energy generation is nuclear reactions, rather than gravitational collapse. This changeover is not a sudden one, as the material closer to the center heats up fastest. Eventually, enough energy is generated internally for the collapse to halt. When this happens, the star reaches a stable condition. It is on the main sequence.

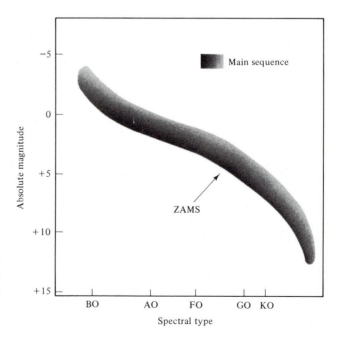

Figure 9.8 Zero age main sequence. The main sequence on this *H–R* diagram appears as a band, since stars on the main sequence get slightly brighter as their composition changes. The lower edge of this band represents the points where stars first appear on the main sequence, the ZAMS.

As nuclear reactions take place in the star, the composition of the star is actually changing. This change could affect the spectral type and luminosity of the star while the basic structure does not change very much. These changes result in a main sequence that is a band in the *H–R* diagram, rather than a thin line. However, there is a line that we can identify as connecting the points on the *H–R* diagram where stars of each spectral type first appear on the main sequence. We call this line the *zero-age main sequence*, or *ZAMS*. This line is shown relative to the main sequence band in Fig. 9.8.

9.4 STELLAR STRUCTURE

The basic philosophy of stellar structure studies is that stars obey the laws of physics, so we should be able to predict and explain their structure by applying those laws. To do this we must identify the basic physical processes that are important in stars, such as nuclear physics for energy generation. We must also be able to perform large numbers of intricate calculations. This latter facility is provided by modern computers.

Once we carry out stellar structure calculations, we can compare the predictions of the theories with observations. For example, if we put $1 \, M_{\odot}$ of material into model calculations, we should come out with a star whose radius, temperature, and luminosity match these of the Sun. If we put in different amounts of material, our calculations should reproduce the main sequence. We should also find stars in the same mass range as those on the main sequence. Stellar models should allow us to predict stellar evolution. They should also be able to tell us how changes in composition lead to changes in structure.

Stars are easier to analyze than some other astronomical objects because they have very simple shapes. They are spheres. We also assume that their structure is spherically symmetric. That is, if we draw any two radii for the sphere and look at the conditions—temperature, density, composition—as a function of distance from the center, the results will be the same for both radii. We can rotate the star through any single about any axis and not change the result. (This is not strictly true if the star is rotating.) To study a star we divide it into spherical shells, each of thickness dr, as shown in Fig. 9.9. If the density a distance r from the center is $\rho(r)$, then the mass contained in a shell of radius r is

$$dM = \rho(r) \, dV \qquad (9.22)$$

where dV ($= 4\pi r^2 \, dr$) is the volume of the shell. This gives

$$dM = 4\pi r^2 \rho(r) \, dr$$

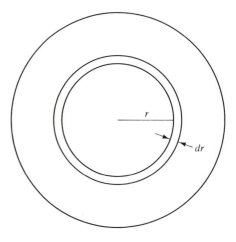

Figure 9.9 Shell in a star.

Dividing by dr gives us the rate at which we add mass as we go farther out from the center of the star:

$$dM/dr = 4\pi r^2 \rho(r) \qquad (9.23)$$

This condition is called *mass continuity*, and simply tells how the rate of change of $M(r)$, mass interior to r, is related to the density at r.

9.4.1 Hydrostatic Equilibrium

The material in a shell of radius r is pulled toward the center of the star by the gravitational attraction of all of the mass interior to that shell. Something must support the matter in the shell or else the star will collapse. That something is the pressure difference between the bottom of the shell and the top of the shell. This condition is called *hydrostatic equilibrium* (Fig. 9.10). Hydrostatic equilibrium applies in the Earth's atmosphere as well as in the oceans and in a glass of water. The weight of each layer of the fluid is supported by the pressure difference between the bottom and the top.

We can see how much of a pressure difference is needed by considering

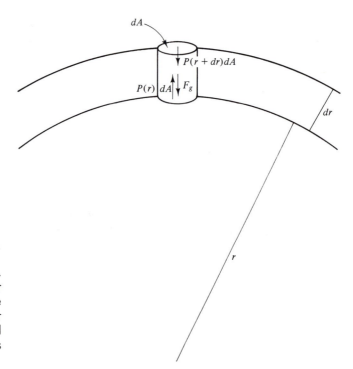

Figure 9.10 Hydrostatic equilibrium. The distance from the center of the star is r, and the shell thickness is dr. The density in the shell is $\rho(r)$. We consider the forces on a cylinder of height dr and the end area dA. The pressure at r is $P(r)$.

a small cylinder, of height dr and area dA, as shown in Fig. 9.10. The mass of the cylindrical element is

$$dm = \rho(r)\, dr\, dA \qquad (9.24)$$

The gravitational force depends on $M(r)$. In any particular model, we can find $M(r)$ by integrating equation (9.23):

$$M(r) = 4\pi \int_0^r \rho(r')r'^2\, dr \qquad (9.25)$$

The gravitational force is given by

$$F_G = -\,GM(r)\, dm/r^2 \qquad (9.26)$$

We use the minus sign $(-)$ to indicate that the force is directed downward. Taking dm from equation (9.24) gives

$$F_G = -[GM(r)/r^2]\rho(r)\, dr\, dA \qquad (9.27)$$

We now look at the force exerted on the top and bottom of the cylinder by the pressure of the fluid. The difference between the upward force on the bottom and the downward force on the top is called the *buoyant force* F_B. If $P(r)$ is the pressure at the bottom of the cylinder and $P(r + dr)$ is the pressure at the top, then

$$F_B = P(r)\, dA - P(r + dr)\, dA$$
$$= [P(r) - P(r + dr)]dA \qquad (9.28)$$

We have used the fact that the force is pressure $\times$ area. We define the pressure difference between the top and bottom dP as

$$dP = P(r + dr) - P(r) \qquad (9.29)$$

Note that the pressure will decrease as r increases (since the pressure at the bottom of the shell must be greater than the pressure at the top), so dP is a negative number. Equation (9.28) then becomes

$$F_B = \{P(r) - [P(r) + dP]\}\, dA$$
$$= -dP\, dA \qquad (9.30)$$

The condition for hydrostatic equilibrium is

$$F_G + F_B = 0 \qquad (9.31)$$

Substituting from equations (9.27) and (9.30) gives

$$-[GM(r)/r^2]\rho(r)\ dr\ dA\ -\ dP\ dA\ =\ 0 \qquad (9.32)$$

We are interested in obtaining the rate at which the pressure changes with radius dP/dr, so we divide both sides by dA, add dP to both sides, and then divide both sides by dr to give

$$dP/dr\ =\ -[GM(r)/r^2]\rho(r) \qquad (9.33)$$

This is sometimes called the equation of hydrostatic equilibrium. We can rewrite it more simply by noting that the quantity $GM(r)/r^2$ is equal to the local acceleration of gravity $g(r)$, so

$$dP/dr\ =\ -\rho(r)g(r) \qquad (9.34)$$

The equation of hydrostatic equilibrium tells us that the denser the fluid is, the more rapidly P_C changes with r. This is because a denser fluid means a higher-mass shell, and a stronger gravitational force pulling it in. This requires a larger pressure difference between the top and the bottom to support it. Also, the greater $g(r)$ is, the greater the gravitational force is pulling the shell in. This means that a larger $g(r)$ also requires a faster rate of change in the pressure.

EXAMPLE 9.4   Central Pressure of the Sun

Use the equation of hydrostatic equilibrium to estimate the central pressure of the Sun by considering the whole Sun as one shell.

Solution

If we consider a whole star to be one shell, then $\Delta R = R$ is the radius of the star, and $\Delta P = P_C$ is the central pressure (taking the pressure at the surface to be zero). The equation of hydrostatic equilibrium then gives

$$P_C\ =\ (GM/R^2)\rho R$$

where ρ is the average density, and is approximately M/R^3, so

$$P_C\ =\ (GM/R)(M/R^3)$$

$$=\ GM^2/R^4$$

Substituting values for the Sun gives

$$P_C = \frac{(6.67 \times 10^{-8} \text{ dyn-cm}^2/\text{g}^2)(2 \times 10^{33} \text{ g})^2}{(7 \times 10^{10} \text{ cm})^4}$$

$$= 1 \times 10^{16} \text{ dyn/cm}^2$$

The actual value, obtained from stellar models, is about 20 times this value.

Another equation that we use in modeling stars is the *equation of state*. The state of the gas is described by the pressure, density, and temperature. The equation of state is an equation that relates those three quantities. We can write it in the general form

$$P = f(\rho, T) \tag{9.35}$$

The actual form of the function f depends on the nature of the gas. For an ideal gas, the equation of state has the simple form

$$P = (\rho/m)kT \tag{9.36}$$

where m is the mass per particle. For gases with relativistic particles, the equation of state is different.

EXAMPLE 9.5 Central Temperature of Sun

Use the result of the previous example and the equation of state for an ideal gas to estimate the central temperature of the Sun.

Solution

To use the ideal-gas law we need an estimate for the density. We simply use the average density, which is the mass, divided by the volume. Since the hydrogen is completely ionized, we take the mean mass per particle m to be $\frac{1}{2}m_p$.

$$T = mP/\rho k$$

$$= mP(4\pi/3)(R_\odot)^3/M_\odot k$$

$$= \frac{(0.5)(1.67 \times 10^{-24} \text{ g})(1 \times 10^{16} \text{ dyn/cm}^2)(4\pi/3)(7 \times 10^{10} \text{ cm})^3}{(2 \times 10^{33} \text{ g})(1.38 \times 10^{-16} \text{ erg/K})}$$

$$= 4.4 \times 10^7 \text{ K}$$

9.4.2 Energy Transport

In making a stellar model we must also consider how energy gets from the inside of the star to the outside. In general, energy can be transported by conduction, convection, and radiation. In stellar interiors conduction is not generally important. The energy transport must be such that the temperature $T(r)$ does not change with time. For this to happen the energy entering any shell per second must equal the energy leaving that shell per second.

If radiation transport dominates, we can calculate the required temperature distribution $T(r)$. We let $f(r)$ be the flux of radiation through a surface at radius r. If the surface emits like a blackbody, then

$$f(r) = \sigma T(r)^4 \tag{9.37}$$

We can find the rate at which $f(r)$ changes with $T(r)$,

$$df/dT = 4\sigma T^3 \tag{9.38}$$

If we interpret df as the small change in f due to a small change in T, dT, then

$$df = 4\sigma T^3 \, dT \tag{9.39}$$

However, the change in flux passing through a given layer of the star must depend on the ability of the layer to absorb radiation. In Section 6.2, we saw that this is given by the absorption coefficient κ. In stellar structure it is more convenient to deal with the absorption coefficient per density of material. We therefore let $\kappa'(r)$ be the opacity per unit mass at r. This means that $\kappa'(r)\rho(r)$ gives the fraction of radiation absorbed per centimeter. Using these definitions

$$df = -\kappa'(r)\rho(r)f(r) \, dr \tag{9.40}$$

We define the luminosity of a given layer as the flux f multiplied by the surface area of the layer,

$$L(r) = 4\pi r^2 f(r) \tag{9.41}$$

If we use equations (9.39) and (9.40) to eliminate df and equation (9.41) to eliminate $f(r)$, we have

$$4\sigma T(r)^3 \, dT = -\frac{\kappa'(r)\rho(r)L(r) \, dr}{4\pi r^2} \tag{9.42}$$

Solving for the luminosity gives

$$L(r) = -\left[\frac{16\pi r^2 \sigma T(r)^3}{\kappa'(r)\rho(r)}\right]\frac{dT}{dr} \tag{9.43}$$

This tells us how the rate of energy flow depends on the rate at which the temperature changes with r. In general, dT/dr is a negative number (the temperature drops off with distance from the center), so the luminosity is positive. A more exact calculation gives essentially the same result as equation (9.43) but with the 16 replaced by $64/3$.

From equation (9.43) we can see that the opacity per unit mass $\kappa'(r)$ is very important in determining the energy transfer, and therefore the structure of the star. This opacity depends on the composition of the star. Accurate stellar structure calculations require good knowledge of the opacity as a function of composition and temperature.

In addition to energy transport we must consider energy generation. if energy is generated in a particular shell, then the energy leaving that shell will exceed the energy entering the shell by the amount of energy generated. We let $\epsilon(r)$ be the energy generation rate per unit mass within the shell at radius r. The increase in luminosity in that layer, due to energy generation, is then

$$dL = 4\pi r^2\rho(r)\epsilon(r)\, dr \tag{9.44}$$

The rate at which the luminosity changes with radius is then

$$dL/dr = 4\pi r^2\rho(r)\epsilon(r) \tag{9.45}$$

To carry out model calculations we must be able to specify $\epsilon(r)$ as a function of composition and temperature. This is where the input from nuclear physics is particularly important.

One might wonder why stars don't explode with all the energy they produce. The answer is that their stability comes from their negative heat capacity. (Heat capacity is the energy required to raise the temperature by a given amount.) To see this we look at the total energy $E = U + K$ (where K is now the kinetic energy of the thermal motions of particles in the gas), and we recall the virial theorem, which tells us that $E = -K$. Suppose we now add energy. This increases E, meaning that it makes E less negative. This makes K less positive, so the total thermal kinetic energy decreases. The gas cools. Therefore, if the star produces too much energy for its equilibrium configuration, it can expand and cool to adjust. The above argument doesn't apply to degenerate gases, since their internal energy is essentially independent of the temperature. Therefore, as we encounter degenerate stars, or parts of stars, we will see that explosions are possible.

9.5   STELLAR MODELS

In the previous section we saw a group of equations that describe the physics that governs stellar structure: (1) mass continuity; (2) hydrostatic equilibrium; (3) equation of state; (4) energy transport; and (5) energy generation.

The inputs to the stellar models are the mass of the star and the composition. We must also specify the nuclear physics, which gives the energy generation as a function of these conditions. We must also put in the information about the opacities. For the purposes of calculation, we break the star into spherical shells. We must solve for the distribution of density $\rho(r)$ and temperature $T(r)$ which satisfy the conditions imposed by the equations of stellar structure. This is such a complicated process that realistic stellar models are all calculated on computers.

Stellar model calculations can also be used to predict stellar evolution. As nuclear reactions take place in the hot central core of the star, the composition changes. This changes the energy generation rate and the opacity, meaning that the structure will change. We do a model calculation for the initial composition. We then determine the rate at which various nuclei are produced and destroyed. We then know what the composition will be some time, say 50,000 years, later. We now calculate a model with the altered composition, and repeat the process. We follow the evolution of the star in these time steps. We choose the time steps so that the composition changes somewhat, but not too much, during that time step.

To see the results of a model calculation, we look at a model for the Sun. The distribution of temperature and density is shown in Fig. 9.11.

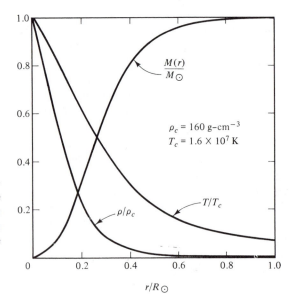

Figure 9.11 Temperature and density as a function of distance from the center of the Sun, as calculated from the solar model which best agrees with the global properties (radius, surface temperature) of the Sun.

9.6 SOLAR NEUTRINOS

Though our model for the Sun gives the correct radius and temperature, there are certain aspects we cannot check directly. Almost all of the direct information we get from the Sun comes from photons emitted in the solar atmosphere. We cannot directly observe the photons that are emitted in the nuclear reactions in the center. These photons are quickly absorbed and their energy takes about 10^7 years to reach the surface. We have no direct observations of the solar core now.

There is one opportunity to make a direct observation of the solar core. Neutrinos created in nuclear reactions in the core escape at the speed of light, virtually unattenuated. They reach us 8.5 min after they are created. We could use our solar models to predict the rate at which neutrinos are emitted, and then measure the flux of neutrinos at the Earth. This would be a direct test of the solar model. The problem is that neutrinos are very hard to detect. If the whole Sun cannot absorb many neutrinos, a detector on Earth will absorb even fewer. The neutrino produced in the first reaction of the proton–proton chain ($p + p \rightarrow d + e^+ + v$) has a relatively low energy, and this makes it particularly hard to detect.

There is, however, a source of higher-energy neutrinos. Once ^{3}He is formed, most of it reacts to form ^{4}He, as discussed in Section 9.2. However, in a small fraction of the cases, the following reaction can take place:

$$^3\text{He} + {}^4\text{He} \rightarrow {}^7\text{Be} + \gamma \qquad (9.46)$$

The ^{7}Be then captures a proton:

$$^7\text{Be} + p \rightarrow {}^8\text{B} + \gamma \qquad (9.47)$$

The boron then beta decays, emitting a neutrino in the process:

$$^8\text{B} \rightarrow {}^8\text{Be} + (e+) + v \qquad (9.48)$$

The ^{8}Be then breaks apart into two alpha particles:

$$^8\text{Be} \rightarrow {}^4\text{He} + {}^4\text{He} \qquad (9.49)$$

The neutrino emitted in the beta decay of boron-8 has enough energy to provide us with some chance of detecting it.

We can detect this neutrino using an isotope of chlorine, ^{37}Cl. About 25% of all naturally occurring chlorine is in the form of this isotope. When struck by a sufficiently high-energy neutrino, it can absorb the neutrino:

$$^{37}\text{Cl} + v \rightarrow {}^{37}\text{Ar} + e^- \qquad (9.50)$$

This particular isotope of argon is radioactive, and its decay can be detected by normal particle detectors in the laboratory. If we start with a tank of chlorine (and no radioactive argon), and we end up with a small amount of ^{37}Ar, we can determine the rate at which neutrinos are hitting the tank.

This is the basic idea behind an experiment being conducted by R. Davis of the Brookhaven National Laboratory (Fig. 9.12). Their liquid source of chlorine is perchloroethylene (cleaning fluid). Since neutrinos interact infrequently, a large quantity of this "detector" is used—about 10^5 gallons. The argon that is produced is an inert gas, so it will form bubbles in the fluid, allowing it to be removed. The experiment is run for some period of time, typically a month. The gas is then collected, and is measured for ^{37}Ar activity.

Even if some radioactive argon is found, we won't know that it results from solar neutrinos. There are also high-energy particles in the Earth's atmosphere, coming from space, called cosmic rays, which can produce a similar result. The Earth can shield the tank from cosmic rays but not from neutrinos. The tank is therefore placed 1.5 km underground in the Homestake gold mine, near Lead, South Dakota. Another source of possible contamination is natural radioactivity from the rocks in the mine. For this reason the tank is surrounded by a larger tank containing water. The water blocks the high-energy particles from radioactive decays in the mine, but doesn't block the neutrinos. You can see from this very brief description that the experiment is a very difficult one.

The results of the experiment (Fig. 9.13) from over ten years of observing, have been astounding. It is convenient to express the rate of detecting solar neutrinos in *solar neutrino units*, or *SNU*s. The standard model of the Sun predicts that we should be measuring 6 SNUs. The experiment reveals

Figure 9.12   Solar neutrino experiment. This tank, in the Homestake gold mine in Lead, South Dakota, contains 400,000 liters of cleaning fluid. The chlorine in the fluid acts as the neutrino detector.

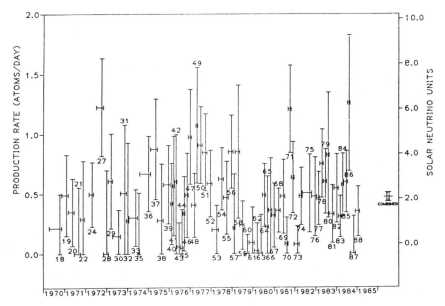

Figure 9.13 Results of the solar neutrino experiment. The horizontal bars indicate the time over which a given measurement was made. The vertical bars indicate the uncertainty in the result. The rightmost data point with the small error bars is the average of all the results.

a much lower number, only 1 or 2 SNUs. This leads to what we call the *solar neutrino problem*. Astrophysicists have so much faith in our understanding of stellar interiors that such a large discrepancy is an indication of a severe problem.

A number of solutions have been proposed. One possibility is that there is something wrong with the experiment itself. Possibly, the detector is not as sensitive to neutrinos as originally thought. However, various aspects of the experiment have been checked and refined over the years, and there is a general feeling that the experiment is correct.

If the experiment is correct, then the solar model must be examined. It is possible that some of the inputs are not correct. For example, the nuclear reaction rates have a very strong temperature dependence. If we have that dependence slightly wrong then significant changes can occur in the solar model. It is also possible that the opacities (as a function of composition and temperature) are slightly off. The solar neutrino problem has stimulated work in those areas. We now have improved nuclear physics data and opacity data. A better solar model has been calculated, but the discrepancy remains.

It has been suggested that the Sun may go through a cycle in its energy generation, and that right now it is generating less than the average amount of energy. In this cycle, at some point the core cools, reducing the rate of nuclear reactions. The pressure decreases, and the core contracts. As the core contracts, it converts gravitational potential energy into kinetic energy, and begins to heat. As the core gets hotter, the rate of nuclear reactions increases. The pressure increases, and the core expands. The cycle then starts again. If this is the answer, then the neutrino observations are really giving us a good view of what is happening in the Sun now. In this picture, it is possible for the cycle to take place with very little variation in the solar luminosity. Since photons are scattered many times before they can get from the center to the photosphere, the light from the photosphere reflects the average energy production over the Kelvin time. It should be pointed out that no mechanism for such oscillations has been found.

It has also been suggested that we don't know as much as we thought we did about neutrino physics. We now think that there are three different neutrino types, each with its own antineutrino. (We'll discuss particle physics more in Chapter 21.) The type of neutrino that we have been discussing is called an electron neutrino, because it always appears in reactions with electrons (or positrons). This is the type of neutrino produced in the Sun, and the type that will interact with ^{37}Cl. However, some theories have suggested that neutrinos can change their identities. If this is the case, then a neutrino can be created as an electron neutrino in the Sun, but change its identity by the time it reaches the Earth, thereby eluding our detectors. According to this idea, only about one-third of the neutrinos produced in the Sun are capable of being absorbed by ^{37}Cl. These identity changes, or *neutrino oscillations*, are also related to the suggestions that neutrinos have a very small (but not zero) mass. The experimental evidence for this is still being studied, and new experiments are underway.

Other explanations of the solar neutrino problem are even more speculative and involve very nonstandard solar models. For example, it has been suggested that there is a small black hole at the center of the Sun. Astrophysicists are waiting to exhaust the more conventional possibilities before pursuing the nonstandard explanations too vigorously.

In the meantime, a new solar neutrino experiment is being planned. Since the ^{8}B neutrino comes from a relatively unimportant branch in the Sun's nuclear reaction chain, it may be that there is some small error in our calculations which just happens to be magnified for this minor branch. There is another reaction that produces a detectable neutrino. The reaction is not as important as the main proton–proton chain, but it is more significant than the branch involving the ^{8}B. This reaction is called the *p–e–p* reaction:

$$p + e^- + p \rightarrow d + \nu \qquad (9.51)$$

It occurs once for every 400 direct *p–p* reactions. It is rare because it is much harder to bring three particles together at the same time than it is to bring two particles together.

The neutrino produced by the *p–e–p* reaction is not as high in energy as the ^{8}B beta-decay neutrino. Therefore, it cannot be absorbed by ^{37}Cl.

However, it has a higher energy than the p–p neutrino. The p–e–p neutrino can be absorbed by gallium. As with the chlorine, large quantities of gallium will be needed for the experiments. For now, we must wait for the new results.

CHAPTER SUMMARY

In this chapter we looked at the processes responsible for the structure of main sequence stars.

We started by looking at energy sources. Nuclear reactions are the only source capable of giving stars their inferred lifetimes. We saw that the temperatures required for nuclear reactions to take place are in excess of 10^7 K, even with barrier tunneling to help get the nuclei together.

The basic source of energy on the main sequence is the conversion of hydrogen to helium. In low-mass stars this takes place primarily via the proton–proton chain. In more massive stars, with higher central temperatures, other cycles, such as the CNO cycle, are important.

We also looked at the basic processes that govern stellar structure. We saw that normal stars are in hydro-

static equilibrium, supported by the pressure difference between lower layers and higher layers. We also saw that the temperature distribution is determined by the requirement that the temperature of each layer be constant.

Once the basic laws of stellar structure are outlined, stellar models can be computed, generally using computers. In a model, we start with a certain mass and composition, and calculate an equilibrium configuration.

We saw how much of stellar structure seems to be understood, but encountered the puzzle of the solar neutrino experiment. The neutrinos allow us to see what the core of the Sun is doing now, and it does not appear to be doing as much as the models predict.

QUESTIONS AND PROBLEMS

9.1 Estimate the lifetime for a 10 $M_{\odot}$ star on the main sequence to give off energy stored from gravitational collapse.

9.2 (a) Why is 1 eV/atom a reasonable estimate for the energy available in chemical reactions? (b) Is the estimate likely to be wrong by more than a factor of 10 in either direction? Explain. (c) If the estimate is wrong by a factor of 10 in either direction, will it change the conclusion that the Sun cannot exist on chemical reactions?

9.3 It has been said that is we did not know that $E = mc^2$, then we would not know about nuclear energy. Explain why this is not true.

9.4 Calculate the mass corresponding to the binding energy of an H atom. What fraction of the mass of the atom is this?

9.5 What is the rate at which the Sun is converting mass to energy?

9.6 Estimate the nuclear reaction lifetime of a star as a function of its mass. Assume that 10% of the mass is in the core and available for nuclear reactions. (*Hint*: Use the mass–luminosity relationship.) (b) Plot your results for the range of masses encountered along the main sequence.

9.7 Explain the factors that place upper and lower limits on the number of neutrons that can go into a nucleus with some specific number of protons.

9.8 In some cases in this chapter we have ignored the difference between m_p and m_n, and in some cases the difference is important. Explain both situations.

9.9   What is the difference in mass between the neutron and proton, expressed in MeV? (b) How does this relate to the energy available in the beta decay of a neutron?

9.10   Calculate the binding energy of a ^{4}He nucleus.

9.11   How much energy per proton is given off in the p–p chain? (Express your answer in MeV.)

9.12   Suppose we have Z protons and have to distribute them into two nuclei, one with Z_1 protons and the other with Z_2 protons ($Z = Z_1 + Z_2$). (a) What arrangements give the maximum and minimum coulomb repulsion? (b) What does this tell you about the types of fusion that are most likely to take place in stars?

9.13   What are the similarities between gamma emission by nuclei and visible light emission by atoms?

9.14   Why is there no Coulomb barrier to fission if there is one to fusion?

9.15   Why are the rates of certain fusion reactions very sensitive to temperature?

9.16   (a) How close can two protons get if one is at rest and the other has a kinetic energy equal to the average energy at $T = 10^7$ K? (b) What is the wavelength of the moving proton, and how does it compare to the minimum separation between the two protons? (c) Repeat the calculations for a proton with ten times the average energy at this temperature.

9.17   Why are the r- and s-processes important?

9.18   (a) What are the parameters that we put into a stellar model? (b) What calculations do we perform? (c) How do we test the results?

9.19   (a) What do we mean by spherical symmetry? (b) Why will a rotating star not be spherically symmetric?

9.20   Suppose the density in a star is given by

$$\rho(r) = \begin{cases} \rho_0 & r < r_0 \\ \rho_0(r_0/r)^2 & r_0 < r < R \\ 0 & R < r \end{cases}$$

(a) Find an expression for $M(r)$. (b) If the mass of the star is $1\,M_\odot$ and $R = R_\odot$, and $r_0 = 0.1\,R$, what is the value of ρ_0?

9.21   For the density distribution in Problem 9.20, find $P(r)$.

9.22   In equation (9.26) we have used the form for the gravitational force between two point particles. These particles have masses $M(r)$ and dm and are a distance r apart. However, $M(r)$ represents an extended mass and dm represents a whole shell, so that neither represents a point. How can we use the simple formula? [*Hint*: Treat the justification for $M(r)$ and dm separately.]

9.23   What effect would a slight increase in opacity at all layers have on the structure of a star?

9.24   Explain how we simulate stellar evolution.

9.25   Why is the solar neutrino experiment so important?

9.26   When we discussed explanations of the solar neutrino experiment, we said that the Sun may be generating some energy now through gravitational collapse. However, earlier in the chapter we ruled out gravitational collapse as a stellar energy source. Why isn't this a contradiction?

9.27   The equilibrium structure of a star is ultimately governed by its mass and composition. Show that the structure of the star determines the rate of energy generation and not the other way around.

Stellar Old Age

We have already seen that the mass of a star is the most important property in determining the star's structure. For a main sequence star the mass determines the size and temperature. The lifetime of a star on the main sequence depends on the available fuel and the rate at which the fuel is being consumed—the luminosity. Both of these quantities depend on the star's mass, so the lifetime on the main sequence also depends on the mass. When that star uses up its basic supply of fuel, its ultimate fate also depends on its mass.

10.1 EVOLUTION OFF THE MAIN SEQUENCE

10.1.1 Low-Mass Stars

We first look at stars whose mass is less than about 5 $M_\odot$. Eventually a star will reach the point where all the hydrogen in the core has been converted to helium. For a low-mass star the central temperature will not be high enough for the helium to fuse into heavier elements. There is still a lot of hydrogen outside the core, but the temperature is not high enough for nuclear reactions to take place. The core begins to contract, converting gravitational potential energy into kinetic energy, resulting in a heating of the core. The hydrogen just outside the original core is heated to the point where it can fuse to form helium, and this takes place in a shell at the outer edge of the core. We refer to this as a *hydrogen-burning shell*, where the word "burning" refers to nuclear reactions, rather than chemical burning (Fig. 10.1). As the core contracts, the rate of energy generation in the shell increases. This shell can actually give off energy at a greater rate than the core did during the normal lifetime.

While all of this is happening in the interior, the outer layers of the star are changing. Energy transport from the core is radiative, and is limited by the rate at which photons can diffuse through the star. The outer layers of the star get hotter and expand. As the gas expands, it cools. The star's radius

Figure 10.1   Star with an H-burning shell. (a) The temperature in this star is not hot enough to fuse the helium in the center, but is hot enough to keep the H in the shell burning. (b) In this star, the temperature is hot enough to keep both burning. (Remember, by "burning" we are talking about nuclear reactions.)

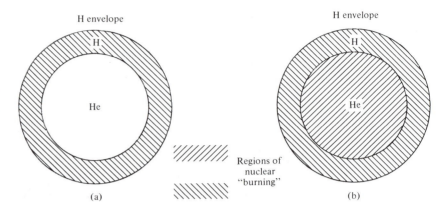

has increased, but its temperature has decreased, so the luminosity only increases slightly. The behavior of the star's track on the *H–R* diagram is shown in Fig. 10.2. The track moves to the right, and the star appears as a subgiant.

There is a mechanism that keeps the surface temperature from getting too low. The rate of photon diffusion increases as the absolute value of dT/dr increases. Remember, dT/dr is negative, so we are saying that the greater the temperature difference is between some point on the inside and the sur-

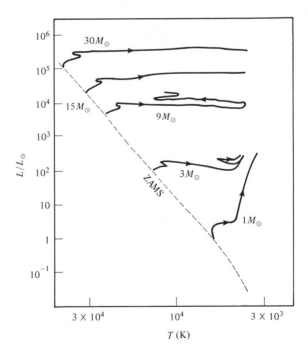

Figure 10.2   Evolutionary tracks away from main sequence on *H–R* diagram. Each track is marked by the mass for the model. The dashed line is the zero age main sequence (ZAMS).

face, the greater is the energy flow between those two points. (In the winter, a larger temperature difference between the inside and outside of your house results in a faster heat loss.) If the surface temperature starts to fall too much, the photon diffusion is faster, delivering more energy to the surface, and raising the surface temperature. Therefore, as the radius continues to increase, the surface temperature remains approximately constant. The luminosity increases, and the evolutionary track moves vertically. The star is a *red giant*.

By the time the star becomes a red giant the energy transport in the envelope is convective. This is because of the large value of $-dT/dr$. The analogous situation on the Earth involves the heating of the atmosphere. Sunlight heats the ground, and then infrared radiation from the ground heats the air. (This explains why the air is cooler at high altitudes; it is farther from the direct heat source, the ground.) In this situation, we say that the energy transport is radiative. However, if $-dT/dr$ becomes larger, then $-dP/dr$, the rate at which the pressure falls, also becomes large. The air that is heated near the ground expands slightly and becomes very buoyant, being driven upward by the pressure difference between the bottom and top of any parcel of air. The hot air rising, being replaced by cool air falling, known as convection, becomes the dominant mode of energy transfer.

We now look at the evolution of the core while the star is becoming a red giant. The temperature of the core climbs to 10^8 K. This is hot enough for the triple-alpha process to take place [equations (9.11) and (9.12)], fusing the helium into carbon. The density is so high that the helium no longer behaves like an ideal gas. It is called a *degenerate* gas. We will discuss degenerate gasses in Section 10.4, but for now we note that the equation of state is very different for a degenerate gas. In an ideal gas, when the triple-alpha process starts, the extra energy generated causes an increase in pressure, which causes the gas to expand, slowing the reaction rate. This keeps the reactions going slowly. In a degenerate gas (discussed in Section 10.4) the pressure doesn't depend on temperature and no such safety valve exists. The conversion of helium to carbon takes place very quickly. We call this sudden release of energy the *helium flash*. The energy released causes a brief increase in the stellar luminosity.

Following the helium flash the energy production decreases. The core is no longer degenerate, and steady fusion of helium to carbon takes place. This region is surrounded by a shell in which hydrogen is still being converted into helium. At this point the star reaches the horizontal branch on the *H–R* diagram. The outer layers of the star are weakly held to the star since they are so far from the center. The star begins to undergo mass loss. The subsequent evolution depends on the amount of mass that is lost.

Eventually all the helium in the core is converted into carbon and oxygen. The temperature is not high enough for further fusion, and the core again begins to contract. A helium-burning shell develops, and the rate of

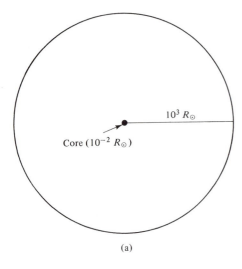

(a)

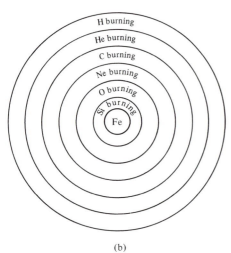

(b)

Figure 10.3   Shells in the core of a high-mass star as it evolves away from the main sequence. (a) The core is only a small fraction of the total radius. (b) In the core, there is a succession of shells of different composition. Each shell has exhausted the fuels that are still burning in shells further out.

energy production again increases. The envelope of the star again expands. On the *H–R* diagram the evolutionary track ascends the giant branch again, reaching what is called the *asymptotic giant branch*. Stars on the asymptotic giant branch are more luminous than red giants. The star can briefly become large enough to be a red supergiant at this stage. The star can also undergo oscillations in the rate of nuclear energy generation.

10.1.2   High-Mass Stars

More massive stars live a shorter lifetime on the main sequence than lower-mass stars. As with the lower-mass stars, the main sequence lifetime for higher-mass stars ends when the hydrogen in the core is used up. The core begins to contract, and the temperature for helium fusion to heavier elements is quickly reached. The helium fusion takes place before the core can become degenerate. Therefore, in contrast to the helium flash in lower-mass stars, the helium burning in more massive stars takes place steadily. At this point the star has a helium-burning core with a hydrogen-burning shell around it (Fig. 10.3).

When the helium in the core is exhausted, the temperature is high enough for the carbon and oxygen to fuse into even heavier elements. At this time, we have a carbon- and oxygen-burning core, surrounded by a helium-burning shell, which is in turn surrounded by a hydrogen-burning shell. As heavier elements are built up, the core develops more layers.

As the luminosity of the core increases, the outer layers of the star expand. The atmosphere cools with the expansion, but the size increases sufficiently for the luminosity to increase. Eventually, the radius of the star reaches about $10^3 R_\odot$. At this point we call the star a *red supergiant*.

10.2   CEPHEID VARIABLES

10.2.1   Variable Stars

If we monitor the brightnesses of certain stars we find that many change with time. These are known as *variable stars*. The periods of variability range from seconds to years. We have already seen that eclipsing binaries appear as variables. However, many variables have luninosity variations associated with physical changes.

Since we will be using specific stars as examples, we will briefly mention systems for naming normal and variable stars. The bright stars are named, in order of brightness within their own constellation, by a Greek letter followed by the Latin genative form of the constellation name. An example is α-Orionis (abbreviated as α-Ori), the brightest star in the constellation of Orion. Some bright stars are also still known by their ancient names. Variable stars are listed in order of discovery within a given constellation. The first is designated R (e.g., R-Ori), the next S, and so on to Z. After that, two letters are used, starting with RR, RS, to RZ, then SS to SZ, and so on, until ZZ is reached. Then comes AA through AZ, BB through BZ, and so on to QZ. (The letter J is never used because of possible confusion with I.) This gives a total of 334 stars per constellation. Beyond that, numbers, starting with 335, preceded by a V (for variable), are used (e.g. V335-Ori, V336-Ori, etc.).

For any particular star, we are interested in getting a light curve, a graph of its magnitude as a function of time. Studies of variable stars often require very long term monitoring. In some cases it is possible to recover information on a star's variability from plate archives. When photographic plates are taken at an observatory, the astronomer who took them is often required to return them to the observatory when that astronomer's work has been completed. The astronomer may be interested in only one star on the plate, but it contains a record of many stars. This record can be used as a basis for comparison with future observations. Observations of many-variable stars can be so time-consuming that it has become an area of astronomy where amateur observers have been able to make significant contributions, generally coordinated by the American Association of Variable Star Observers (AAVSO). For any particular star, we are interested in getting a light curve—a graph of its magnitude as a function of time. (In these studies, we measure time in Julian days, the number of days since noon on 1 January 4713 B.C., or modified Julian days, the number of days since the beginning of the Besselian year, 1 January 1950. See Appendix F for a further discussion of timekeeping systems.)

We distinguish different types of variables by such things as their period and the magnitude range. A particular class of variable is generally named after the prototype of the class, either the first or most prominent star with the distinguishing properties of the class. In this section, we look at a few examples of the most important types of variables. Different types of variables appear in different parts of the *H–R* diagram, as shown in Fig. 10.4.

Mira variables are named after the prototype (a star also known as Ceti). These stars have periods of about three months to two years, or even longer, and are called *long-period variables*. Any individual Mira variable may show fluctuations in its period. These stars change their magnitude by about 6 mag (over a factor of 100 in brightness). For example, the apparent magnitude of Mira ranges from 9 mag to 3 mag. These changes in brightness are accompanied by changes in spectral type. Mira changes from M9 to M5. This means that a temperature change is accompanying the luminosity change.

Cepheid variables are named after the prototype δ Cephei. Its period is 5.4 days and its apparent magnitude varies from 3.6 to 4.3 mag. In general, Cepheids have periods from 1 to 100 days. We know of only about 700 in our galaxy. Another familiar Cepheid is Polaris, which changes by only 0.1 mag, from 2.5 to 2.6. (For a long time, astronomers did not know that Polaris

Figure 10.4 *H–R diagram, showing the location of various types of variable stars.*

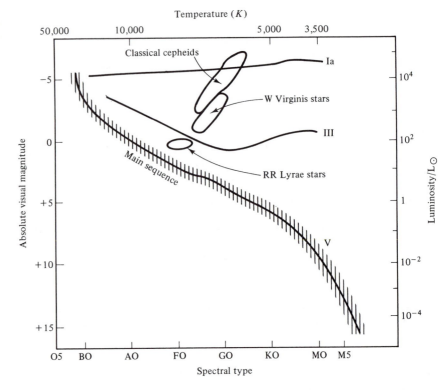

(the North Star) is a variable, and it was used as a reference star for measuring the magnitudes of other stars.) Cepheids have masses of approximately 6 $M_\odot$, and radii of about 25 $R_\odot$.

10.2.2 Cepheid Mechanism

When we study the spectral lines of Cepheids, we can detect Doppler shifts that vary throughout the light cycle (Fig. 10.5). The Doppler shifts go through a cycle in the same period as the light. This means that the surface of the star is moving. The size of the star changes as the luminosity changes. The spectral type also changes throughout the cycle. The luminosity change is therefore associated with changes in radius and surface temperature.

A star may become a Cepheid variable when it reaches the stage described at the end of the last section. To see how a Cepheid oscillates, let's consider the oscillations of a normal star. These oscillations are radial. They involve only the inward and outward motion of the outer layers of the star. Suppose we are able to perturb a star by decreasing its radius R. The density then increases, and the pressure increases. This excess pressure will make the layers expand back. However, just as a swing overshoots its lowest point as it returns from its maximum height, the star can overshoot its equilibrium radius R_0. Now the star is larger than its equilibrium radius, and the pressure decreases, allowing material to fall back. This process can then repeat itself.

In the above analysis, we ignored the effects of opacity. In a normal star the opacity decreases when the temperature increases. We again start with a perturbation reducing R below R_0. This causes P and T to increase. The increase in T decreases the opacity. The reduction of opacity allows some of the excess pressure to be relieved by allowing heat to flow out of the denser regions as radiation. This reduces the tendency of the star to overshoot. If we had started with a perturbation in which $R > R_0$, P and T would have decreased. The opacity would have increased, and the tendency to fall back too fast would be reduced. The result of the opacity is to quench the oscillation.

For a narrow range of conditions the opacity increases as the temperature increases. The source of the opacity is the ionization of He^+ to form He^{2+}. If we start with a perturbation in which $R < R_0$, the pressure and temperature increase. Now the opacity also increases, so the excess pressure does not get relieved, except by driving the star back. The tendency to overshoot is enhanced. Similarly, with $R > R_0$, the pressure and temperature decrease. The opacity also decreases, reducing the pressure even further. The material falls back quickly and overshoots. The oscillation can continue indefinitely, rather than being quenched. These are the conditions that produce a Cepheid variable.

10.2.3 Period–Luminosity Relation

An important feature of Cepheids is that they provide us with a method of measuring distances. The method involves a *period–luminosity relation* (Fig. 10.6). This relation was discovered by Henrietta Leavitt, who was studying Cepheids in the Large and Small Magellenic Clouds, two small galaxies near the Milky Way. The advantage of studying the Cepheids in either of these objects is that the Cepheids are all at almost the same distance. For the Small Magellenic Cloud it was found that there is a relationship between the period

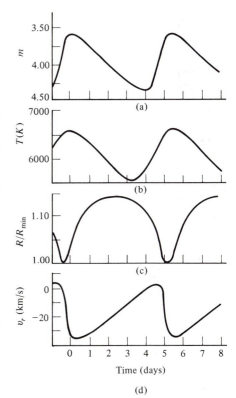

Figure 10.5 Radial velocity and light curves for δ Cep, the prototypical Cepheid. (a) Apparent magnitude as a function of time within the period. (b) Temperature as a function of time. (c) Radius, relative to the minimum radius, as a function of time. (d) Radial velocity of the surface, as a function of time. Note that the radial velocity is 90° out of phase with the radius.

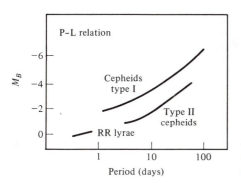

Figure 10.6 Period–luminosity relationship for Cepheids.

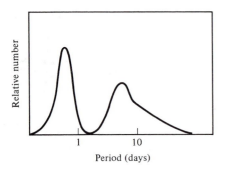

Figure 10.7 Distribution of periods for Cepheids. Note that there are two distinct groupings.

of the Cepheid and its mean apparent magnitude. Since all of the stars are at the same distance, this means that there is a relationship between the period and the mean absolute magnitude.

If we know the exact relationship between period and absolute magnitude, then when we observe a Cepheid we can measure its period and convert it into an absolute magnitude. We can always measure the apparent magnitude. The difference $m - M$ is the distance modulus, and gives us the distance. This technique is important because Cepheids are bright enough to be seen in other galaxies, providing us with distances to those galaxies.

However, before we can use the $P–L$ relationship, it must be calibrated. We need independent methods of measuring the distances to some Cepheids. This is difficult, since there are none nearby. Statistical studies have been done to achieve this calibration.

If we plot a histogram indicating how many Cepheids have various periods (Fig. 10.7), we find an interesting result: The distribution has two peaks in it. This suggests that there are actually two different types of Cepheids. The group with the shorter periods are typical of those studied in the Magellanic Clouds, and are called *classical Cepheids*. The group with the longer periods are called *type II Cepheids* or *WW Virginis stars* (named after their prototype). Type II Cepheids are found in globular clusters in our galaxy. In general, a type II Cepheid is 1.5 mag fainter than a classical Cepheid of the same period. Also, the $P–L$ relation is slightly different for the two types.

The original calibration of the Cepheid distance scale was done for type II Cepheids, since we can study them in our galaxy. However, when we look at a distant galaxy, we can more easily study the brighter classical Cepheids. Therefore, the Cepheids studied in other galaxies were 1.5 mag brighter than assumed. This means that other galaxies are farther away than originally assumed.

EXAMPLE 10.1   Cepheid Distance Scale

By how much does the calculated distance of a galaxy change when we realize we are looking at classical, rather than type II, Cepheids?

Solution

We have seen already that the Cepheids are 1.5 mag brighter than originally assumed. This increases the distance modulus, $m - M$, by 1.5 mag. By equation (2.17), this increases the distance by a factor of $10^{(15/5)} = 2$. Thus, galaxies are twice as far away as originally thought. The difference between the two types of Cepheids was realized in the 1950s, and people talked about the size of the universe doubling.

Another type of variable star that is useful in distance determinations is the *RR Lyrae* variable. They are found in globular clusters, and are sometimes called *cluster variables*. They have very short periods, generally less than one day. The absolute magnitudes of all RR Lyrae stars are close to 0. Actually, they fall between 0 and 1, and obey a weak *P–L* relation of their own. The absolute magnitudes were established by using clusters whose distances were known from other techniques. Once the absolute magnitudes are calibrated, we can use RR Lyrae stars as distance indicators.

It should not be surprising that stars with pulsations have a *P–L* relation. For radial oscillations, we expect the period to be roughly equal to $(G\rho)^{-1/2}$, where ρ is the average density of the star. We can understand this qualitatively by noting that a star pulsating under its own gravity is like a large pendulum. The period of a pendulum is $2\pi(L/g)^{1/2}$. For a star, $L = R$ and $g = GM^2/R^2$, so the period is approximately $(GM/R^3)^{-1/2}$, and M/R^3 is approximately the density. Therefore, since the period is related to ρ (ρ is approximately M/R^3) and the luminosity is related to the radius, the period should be related to the luminosity (see Problem 10.19).

10.3 PLANETARY NEBULAE

We have already said that the outer layers of a red giant are held very weakly. Remember, the gravitational force on a mass m in the outer layer is GmM/R^2, where M is the mass of the star and R is its radius. As the star expands, M stays constant, so the pull on the outer layer falls off as $1/R^2$. Since the outer layer is weakly held, it is subject to being driven away. The actual mechanism for driving material away is still not fully understood. It may involve pressure waves moving radially outward. It may also involve radiation pressure. Photons carry energy and momentum. When photons from the inside of the star strike the gas in the outer layers and are absorbed, their momentum is also absorbed. By conservation of momentum, the shell must move slightly outward.

We do observe shells that are ejected. They have a fuzzy appearance in small telescopes, just like planets, and when originally observed, were called *planetary nebulae* (Fig. 10.8). (Their name has nothing to do with their properties.) From the photograph in Fig. 10.8, we see that some planetary nebulae have a ringlike appearance. However, these are spherical shells. We see them as rings because our line of sight through the edge of the shell passes through more material than a line of sight through the center (Fig. 10.9). Thus, the center appears to be quite faint. When we look at spectral lines in planetary nebulae, we see two Doppler shifts. One line is red-shifted and the other is blue-shifted. The blue-shifted line comes from the part of the shell that is moving toward us, and the red-shifted line comes from the

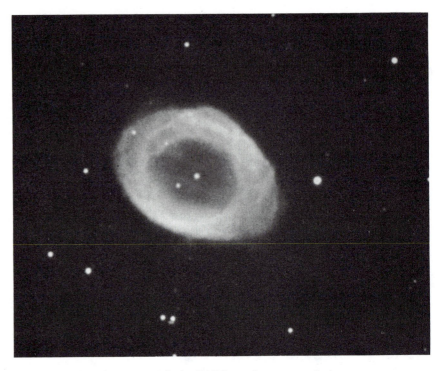

Figure 10.8   The Ring Nebula (M57), a planetary nebula.

Figure 10.9   Lines of sight through a planetary nebula. (a) Appearance. The shaded regions represent the places where the lines of sight pass through the shell. The line of sight near the edge passes through more material than that through the center. This is responsible for the ringlike appearance. (b) Doppler shifts. Material on the near side is moving toward the observer, producing a blue shift, and material on the far side is moving away, producing a red shift.

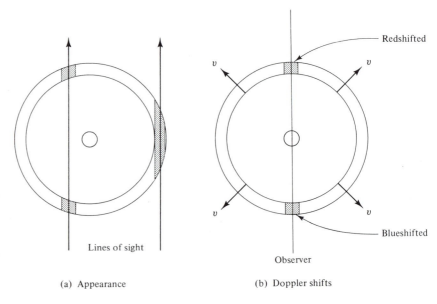

(a) Appearance

(b) Doppler shifts

NGC 6720

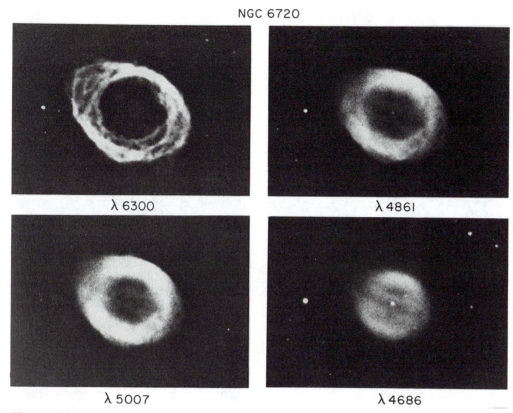

λ 6300

λ 4861

λ 5007

λ 4686

Figure 10.10 The ring nebula, shown in Fig. 10.8, is shown here in four different colors of light, highlighting gas with different physical conditions. The wavelengths are indicated below each frame.

part of the shell that is moving away from us. From these Doppler shifts we find shells expanding as velocities of a few tens of kilometers per second.

The physical conditions in planetary nebulae are determined from observations of various spectral lines. Different lines are sensitive to different temperature and density ranges. For example, Fig. 10.10 shows photographs of a planetary nebula taken at the wavelengths of certain emission lines. Different lines reveal different aspects of the nebula structure. Information is also obtained from studies of radio waves emitted by the nebulae.

From these studies we find that masses of planetary nebulae are of the order of 0.1 $M_\odot$. The temperatures are about 10^4 K. The mass tells us that up to ten percent of the stellar mass is returned to the interstellar medium in the ejection of the nebula. This material will be included in the next generation of stars to form out of the interstellar medium. (This is in addition to mass lost through winds.)

10.4 WHITE DWARFS

10.4.1 Electron Degeneracy

The material left behind after the planetary nebula is ejected is the remnant of the core of the star. It is mostly carbon or oxygen, and its temperature is not high enough for further nuclear fusion to take place. The gas pressure is not high enough to support the star against gravitational collapse. This collapse would continue forever if not for an additional source of pressure when a high enough density is reached. This pressure arises from *electron degeneracy*.

Electron degeneracy arises from the *Pauli exclusion principle*, which says that no two electrons can be in exactly the same state. For two electrons to be in the same state, all of the quantum numbers describing that state must be the same. This must take account of the fact that the electron has angular momentum, called "spin." That spin can have two opposite orientations. For convenience we call them "up" and "down." An up electron and a down electron in the same energy level are considered to be in different states. However, two is the limit. We can only put two electrons in each energy level.

We can see how this affects the properties of atoms with many electrons. Suppose we build the atom by adding electrons one at a time. The first electron goes into the lowest energy level. The second also goes into the lowest energy level, but with the opposite spin orientation. The first level is now full. The third electron must go into the next level. After we have added all of the electrons, we can add up the excitation energies of all of the electrons. We will find that the average excitation energy of the electrons in the atom is much greater than kT. This means that electrons are in higher energy levels than we would guess by just considering the thermal energy available. The problem is that the electrons cannot jump into the filled lower states.

We can apply the same idea to a solid, which we can think of as a structure with many energy levels. The electrons fill the lowest energy levels first, but as they fill, the electrons end up in higher and higher levels, as shown in Fig. 10.11. The average energy of the electrons is, again, much greater than kT. In fact the distribution of electrons in a solid at room temperature is negligibly different from that in a solid at absolute zero. We call an electron gas in which all of the electrons are in the lowest-energy states allowed by the exclusion principle a *degenerate gas*.

In a degenerate gas, most of the electrons will have energies much greater than they would have in an ordinary gas. These high-energy electrons also have high momenta. They can therefore exert a pressure considerably in excess of the pressure exerted by an ideal gas at the same temperature. This higher pressure is called *degeneracy pressure*. We have everyday ex-

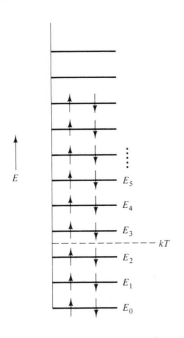

Figure 10.11 Energy levels in degenerate gas. The energies of the levels are indicated on the right. In each level, an upward arrow represents an electron with one spin sense, and a downward arrow represents an electron with opposite spin sense. The dashed line indicates the average thermal energy per particle. The total energy is the sum of the energies of the individual electrons.

$$E = 2\{E_0 + E_1 + E_2 + E_3 + E_4 + \cdots\} \gg NkT$$

amples of this pressure. For example, it is responsible for the hardness of metals.

We can also describe this pressure in terms of the uncertainty principle. In Chapter 3, we saw that we must think of electrons as having wave properties. We can only talk about the probability of finding an electron in a given place, or moving with a given speed. As a result of this wave property, we cannot simultaneously make an accurate determination of the position and momentum of an electron. If we can determine the momentum with an uncertainty Δp, and the position with an uncertainty Δx, the uncertainty principle tells us that

$$\Delta p \, \Delta x \geq \hbar \tag{10.1}$$

where $\hbar$ is Planck's constant, divided by 2π. For a given Δx, the uncertainty in momentum is

$$\Delta p \geq \hbar / \Delta x \tag{10.2}$$

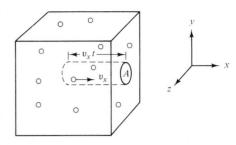

Figure 10.12   Pressure in a degenerate gas. We consider the force on the section of area A of the right-hand wall of the box, due to the x-component of the motions of the particles.

When the density gets very high, we are trying to force the electrons close together. This means that we are trying to confine them to a small Δx. Therefore, the uncertainty in the momentum must be large. Therefore, very large momenta are possible. These high-momentum electrons are responsible for the increased pressure. Figure 10.12 shows a container with density n and particles moving with speed v_x in the x-direction, the number of particles hitting a wall per second per unit surface area is nv_x. The momentum per second per unit surface area delivered to the wall is then $nv_x p_x$, where p_x is the x-component of the momentum. The momentum per second per unit surface area is just the pressure exerted by the gas on the wall,

$$P = nv_x p_x \tag{10.3}$$

If we have n_e electrons per unit volume, then there is one electron per box with volume $1/n_e$. The side of such a box is $1/n_e^{1/3}$, so the average spacing between electrons is

$$\Delta x = 1/n_e^{1/3} \tag{10.4}$$

If we say that the average momentum is of the order of the momentum uncertainty, then

$$P_x \cong \hbar / \Delta x$$
$$= hn_e^{1/3} \tag{10.5}$$

The speed of each electron is its momentum, divided by its mass, so

$$v_x = p_x/m_e \tag{10.6}$$

This gives a pressure, using equation (10.3),

$$P \cong \hbar^2 n_e^{5/3}/m_e \tag{10.7}$$

This is just an estimate of the pressure. A more detailed calculation yields a pressure that is about a factor of 2 higher than that given by equation (10.7).

Equation (10.7) gives the pressure in terms of the electron density. We would like to express it in terms of the total mass density ρ. If the density of positive ions with charge Ze is n_z, then in a neutral gas the density of electrons must be

$$n_e = Zn_Z \tag{10.8}$$

Each positive ion has a mass of Am_p, if we ignore the difference between the proton and neutron masses. The total density of the gas then is

$$\begin{aligned} \rho &= Am_p n_Z + m_e n_e \\ &\cong Am_p n_Z \end{aligned} \tag{10.9}$$

In going to the second line we have ignored the mass of the electrons relative to the mass of the nucleons. Using equations (10.8) and (10.9), the electron density is related to the total density by

$$n_e = (Z/A)(\rho/m_p) \tag{10.10}$$

Substituting this into equation (10.7) and adding the factor of 2 to account for the difference between our estimate and the detailed calculation, we have

$$P = (2)(\hbar^2/m_e)(Z/A)^{5/3}(\rho/m_p)^{5/3} \tag{10.11}$$

Note that in a degenerate gas, the pressure depends on the density, but not on the temperature. This is because the average energy is much greater than kT, so the thermal pressure is small compared to the degeneracy pressure. We have already seen that this point is important in deciding whether the triple-alpha process will take place in a controlled way (normal gas) or in a flash (degenerate gas).

10.4.2   Properties of White Dwarfs

A star supported by electron degeneracy pressure will be quite small, since it must collapse to a high density before the degeneracy pressure is large enough to stop the collapse. These objects are still quite hot, being the remnants of the core of a star. These objects are the stars that appear on the H–R diagram as *white dwarfs*.

EXAMPLE 10.2   White Dwarf Density and Luminosity

Estimate the density of a white dwarf if it has a solar mass packed into a sphere the size of the Earth as we found in Section 3.5.

Solution

We find the density by dividing the mass by the volume

$$\rho = \frac{2 \times 10^{33} \text{ g}}{(4\pi/3)(6.4 \times 10^8 \text{ cm})^3}$$

$$= 2 \times 10^6 \text{ g/cm}^3$$

(Remember, the density of water is only 1 g/cm³, so a white dwarf is very dense.)

EXAMPLE 10.3   White Dwarf Degeneracy Pressure

For a white dwarf of density 1.0×10^6 g/cm³, and $Z/A = 0.5$, estimate the degeneracy pressure and compare it with the thermal pressure of a gas at a temperature of 1.0×10^7 K.

Solution

We find the pressure from equation (10.11):

$$P = (2)\left[\frac{(1.05 \times 10^{-27} \text{ erg-s})^2}{9.11 \times 10^{-28} \text{ g}}\right]\left[\frac{(0.5)(1.0 \times 10^6 \text{ g/cm}^3)}{1.67 \times 10^{-24} \text{ g}}\right]^{5/3} \quad (10.12)$$

$$= 3.2 \times 10^{22} \text{ dyn/cm}^2$$

For an ideal gas the pressure is given by

$$P = (n_e + n_Z)kT$$

where $n_e + n_Z$ represents the total density. However, each atom of atomic number Z contributes Z electrons, so

$$n_e = Zn_Z$$

We therefore have

$$P = (Z + 1)n_e kT$$

$$\cong Zn_e kT$$

We now relate this to the density ρ. If A is the mass number of the nuclei, then (ignoring the difference between the proton and neutron masses)

$$\rho = An_Z m_p$$

This gives

$$P = (Z/A)(R/m_p)kT$$

$$= (0.5)\left(\frac{1.0 \times 10^6 \text{ g/cm}^3}{1.67 \times 10^{-24}}\right)(1.38 \times 10^{-16} \text{ erg/K})$$

$$\times (1.0 \times 10^7 \text{ K})$$

$$= 4.1 \times 10^{20} \text{ dyn/cm}^2$$

The degeneracy pressure is a factor of about 100 higher than the normal thermal pressure, even at that very high temperature.

We can also use the expression for degeneracy pressure [equation (10.11)] to relate the mass and radius of a white dwarf. We saw in Example 9.4 that we could use hydrostatic equilibrium to approximate the central pressure by

$$P = GM^2/R^4$$

If we put this in equation (10.11), and substitute $M/4R^3$ for the density, we find

$$GM^2/R^4 = (2)(\hbar^2 m_e)(Z/Am_p)^{5/3}(M/4R^3)^{5/3} \tag{10.13}$$

Rearranging gives

$$GM^{1/3}R = (2)(\hbar^2/m_e)(Z/4m_pA)^{5/3} \tag{10.14}$$

The right-hand side is all constants, so for a given R, a mass can be calculated (Problem 10.9).

A degenerate gas has a very low opacity to radiation. For a photon to

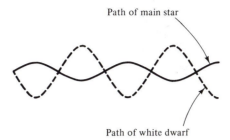

Path of main star

Path of white dwarf

Figure 10.13 Even if we cannot see a white dwarf in a binary system, we can detect its presence. The visible star will appear to wobble as it and the white dwarf orbit their center of mass.

be absorbed, an electron would have to jump to a higher-energy state. However, such a transition would have to be to an already empty state, and may require much more energy than that carried by an optical photon. In addition, degenerate gases are good heat conductors, which explains why metals are good heat conductors. The low opacity and high thermal conductivity mean that a degenerate gas cannot support a large temperature difference. The internal temperature of a white dwarf is approximately constant across the star, at about 10^7 K. The outermost 1% of the star is not degenerate, and it is in that thin layer that the temperature falls from 10^7 K to the roughly 10^4 K indicated by its spectral type. These conditions make a white dwarf very different from a normal star. In addition, Zeeman splitting measurements suggest very strong magnetic fields, about 10^7 gauss in some cases.

What happens to a star after it becomes a white dwarf? As it radiates, it must get cooler. This is because it is giving off energy, but has no source of new energy. The degeneracy pressure doses not depend on the temperature, so the star will maintain its size even as it cools. Eventually, it will be too cool to see. It will take tens of billions of years for a star that is now visible as a white dwarf to become that cool. At that stage, we would call the star a *black dwarf* (not to be confused with a black hole). We can understand the long lifetime when we realize that a white dwarf radiates like a 10^4 K blackbody, but has a thermal reservoir at 10^7 K.

Even at their current temperatures, their small size makes white dwarfs difficult to see. We sometimes detect their presence in binary systems by measuring their gravitational influence on a star that we can see. We deduce the mass of the white dwarf from its influence on the visible companion's orbit (Fig. 10.13).

10.4.3 Relativistic Effects

The treatment of degeneracy pressure must be modified by considerations introduced in the special theory of relativity. This was first realized by S. Chandrasekhar (who shared the 1983 Nobel Prize in physics for his work on stellar structure). Chandrasekhar found that these corrections reduce the degeneracy pressure. This provides an upper limit to the mass that can be supported by electron degeneracy pressure.

The modification arises from the fact that electrons cannot travel faster than the speed of light. In using equation (10.3), we can still say that p_x is $\hbar n_e^{1/3}$. However, we can no longer say that $v_x = p_x/m_e$. We have to use the correct relativistic expression. To find the maximum degeneracy pressure, we can take $v_x = c$. This gives

$$P_{\max} = \hbar c n_e^{4/3} \tag{10.15}$$

A more detailed calculation gives approximately 0.8 times this. Using this, the expression analogous to equation (10.11) is

$$P = (0.8)(\hbar c)(Z/A)^{4/3}(\rho/m_p)^{4/3} \tag{10.16}$$

If we use the same assumptions that we used to get the mass–radius relation [equation (10.14)], we find that the radius drops out, and we simply have an expression for the mass (Problem 10.18)

$$M = (0.13)(\hbar c/G)^{3/2}(Z/Am_p)^2 \tag{10.17}$$

This mass corresponds to the maximum pressure, so it is the maximum mass that can be supported by electron degeneracy pressure. A more accurate calculation, which takes into account the variations in pressure and density with distance from the center of the star, gives a mass that is a factor of about 25 higher. The resulting mass, called the *Chandrasekhar limit*, is 1.44 $M_\odot$. A star whose nuclear processes have stopped and whose mass is greater than 1.44 $M_\odot$ will continue to collapse beyond the white dwarf phase. The fate of such stars will be discussed in Chapter 11.

CHAPTER SUMMARY

In this chapter we saw what happens to stars that use up their supply of hydrogen fuel in their cores. Low-mass stars evolve into red giants. Higher-mass stars evolve into red supergiants.

We saw how some stars at this stage are unstable to pulsations and become Cepheid variables. The Cepheids are particularly important because of a period–luminosity relation. This relation serves as an important distance indicator.

In the red giant or supergiant phase, the outer layers are loosely bound to the star. Mass loss takes place. One form of mass loss is the ejection of a planetary nebula.

We saw how the remnant of a low-mass star is a white dwarf, a star supported by electron degeneracy pressure. There is a limit, the Chandrasekhar limit, on the mass that can be supported by electron degeneracy pressure. Best estimates of this limit are 1.44 $M_\odot$.

QUESTIONS AND PROBLEMS

10.1   In which aspects of post-main sequence evolution does the mass enter into consideration? How are low-mass stars really different from high-mass stars?

10.2   Why is there hydrogen left in a shell around the core after it is used up in the core?

10.3   The momentum of a photon is E/c. (a) Calculate the momentum per second delivered to the outer layers of a $10^2 \, L_\odot$ star if all of the photons are absorbed in this layer. (b) How does the force on the layer compare with the gravitational force on the layer if the layer has a radius $R = 100 \, R_\odot$ and mass $M = 0.1 \, M_\odot$, and the rest of the star has a mass of $1 \, M_\odot$?

10.4   What is the energy in a 0.1-$M_\odot$ planetary nebula expanding at 10^3 km/s?

10.5   Suppose a planetary nebula is a spherical shell whose thickness is 10% of its radius. Compare the length of the longest and shortest lines of sight through the shell, and relate your answer to the appearance of planetary nebulae.

10.6   How does electron degeneracy pressure support a white dwarf?

10.7   (a) If we try to confine an electron to within an atom, $\Delta x \cong 1$ Å, what is Δp, the uncertainty in its momentum? (b) What velocity would the electron need to have this momentum?

10.8   (a) At white dwarf densities, what is the average separation between electrons? (b) What is their momentum uncertainty Δp? (c) What is the velocity corresponding to that momentum?

10.9   (a) Use the mass–radius relationship for white dwarfs to calculate the radius of a 1-$M_\odot$ white dwarf. (b) Use this result to rewrite equation (10.14) in a form that gives R in km when M is expressed in solar masses.

10.10   (a) What is the thermal energy stored in a 10^7 K

white dwarf of $1 \, M_\odot$? (b) Taking its luminosity, and assuming it lives like a 10^4K blackbody, estimate its lifetime as a luminous object.

10.11   Derive equation (10.15).

10.12   Suppose a planetary nebula expands at a speed v, determined from Doppler shift observations. The angular size θ is observed to be increasing at a rate $d\theta/dt$. In terms of these quantities, find an expression for the distance to the planetary nebula.

10.13   Calculate the escape speed from a 1-$M_\odot$ white dwarf. Compare it with that for a main sequence star of the same mass.

10.14   Suppose we find a classical Cepheid with a period of 10 days and an apparent magnitude $+6$. How far away is it?

10.15   Suppose we find a classical Cepheid with a period of 8 days and $m = +7$. What is the period of a type II Cepheid at the same distance with the same apparent magnitude?

10.16   Suppose a Cepheid has period $P = 10$ days. If the surface oscillates sinusoidally with a maximum amplitude, $\Delta R_{max} = 10^3$ km. (a) Write an expression for ΔR vs. t, assuming that $\Delta R = 0$ at $t = 0$, and that the star is expanding at $t = 0$. (b) Find the speed of the surface $v(t)$. (c) Find the wavelength observed for the Hα line vs. t.

10.17   List the distance measurement techniques that we have encountered so far in this book, and estimate the distance range over which each is useful.

10.18   Derive equations (10.16) and (10.17).

10.19   Use the fact that $P \propto \rho^{-1/2}$, $\rho \propto M/R^3$ and $L = 4\pi R^2 \sigma T^4$, to derive a relationship between period and luminosity.

11 The Fate of High-Mass Stars

In Chapter 10 we saw how stars evolve to the red giant or supergiant stages, and how low-mass stars (less than 5 $M_\odot$) lose enough mass to leave behind a white dwarf as the final stellar remnant. We also saw that electron degeneracy pressure can only support a 1.44-$M_\odot$ remnant. In this chapter we will see what happens to higher-mass stars.

It is important to remember that stars lose mass as they evolve. This mass loss can be through winds or the ejection of planetary nebulae. (In the next chapter, we will see that stars in binary systems can transfer mass to a companion.) Though we still only have estimates for the total amount of mass loss, it seems likely that the more massive stars can lose more than half of their mass by the time they have passed through the red supergiant phase. A star's evolution will depend on how much mass it starts with, and how much mass it loses along the way.

11.1 SUPERNOVAE

11.1.1 Core Evolution of High-Mass Stars

In the core of a high-mass star the buildup of heavier elements continues. If we look at the nuclear binding energies (Fig. 9.2) we see that the isotope of iron ^{56}Fe is the most stable nucleus. This means that any reaction involving ^{56}Fe, be it fission or fusion, requires an input of energy. When all of the mass of the core of the star is converted to ^{56}Fe (and other stable elements, such as nickel), nuclear reactions in the core will stop.

At this stage, the core will start to cool and the thermal pressure will not be sufficient to support the core. As long as the mass of the iron core is less than the Chandrasekhar limit, the core can be supported by electron degeneracy pressure. However, once the core goes beyond that limit, there is nothing to support it, and it collapses. In the collapse some energy, previously in the form of gravitational potential energy, is liberated. Since the energy is available, the ^{56}Fe can react by using up the energy. This means that the core does not get any hotter. It continues to collapse. A runaway situation develops in which the iron and nickel consume any liberated energy. As the iron is destroyed, protons are liberated from the nuclei. The electrons in the star then combine with these protons to form neutrons and neutrinos. The reaction can be written:

$$e^- + p \rightarrow n + \nu \qquad (11.1)$$

The core is driven to a very dense state in a short time, about 1 s. What happens next is not completely understood, but the collapse results in an explosion in which most of the mass in the star is blown away. The neutrons created in reaction (11.1) probably play a role in this. They also obey the exclusion principle, and exert a degeneracy pressure (the details of which we will discuss in the next section). This pressure can stop the collapse and cause the material to bounce back. In addition, so many neutrinos are created and the material is so dense that a sufficient number of neutrinos interact with the matter, forcing material outward.

Such an exploding star is called a *supernova*. This type of supernova is actually called a *type II* supernova. Another type of supernova, type I, seems to be associated with older objects in our galaxy. (The mechanism for type I supernovae probably involves white dwarfs in close binary systems, and is discussed in Chapter 12.) During the explosion, nuclear reactions take place very rapidly, and elements much heavier than iron are created. This material is then spread out into interstellar space, along with the results of the normal nucleosynthesis during the main sequence life of the star.

The light from a supernova explosion can exceed that of an entire galaxy (Fig. 11.1). The energy output in a type II supernova is about 10^{53} erg. Of

Figure 11.1 A supernova in another galaxy can be almost as bright as the whole galaxy. These images show NGC7331 before and after a supernova outburst.

Figure 11.2 Light curves for type I and type II supernovae.

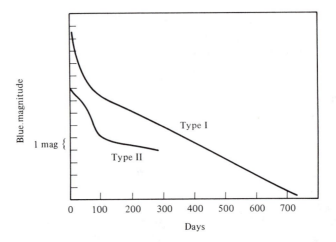

this, about 10^{51} erg shows up as kinetic energy of the shell, and 10^{50} erg as light. (Most of the energy is in the escaping neutrinos.) After a rapid increase in brightness, the supernova fades gradually, on a time scale of several months (Fig. 11.2).

11.1.2 Supernova Remnants

The material thrown out in a supernova explosion is called a *supernova remnant*. It contains most of the material that was once the star. In young supernova remnants we can actually see the expansion of the ejected material. These remnants are important because they spread the products of nucleosynthesis in stars throughout the interstellar medium. There, this material, enriched in "metals," will be incorporated into the next generation of stars. This explains why stars that formed relatively recently in the history of our galaxy have a higher metal abundance than older stars. In the later stages of a supernova remnant's expansion, we still see a glowing shell, like that in Fig. 11.3. These shells also serve to stir up the interstellar medium. This point will be discussed further in Chapter 14.

Supernova remnants generally have magnetic fields that are strong by interstellar standards and a supply of high-energy electrons. In Chapter 6, we discussed the motion of charged particles in a magnetic field. The component of an electron's velocity along the field lines is not changed, since the force must be perpendicular to the field direction and to the electron's velocity. The electron traces out a helix, as it circles around the field lines, and drifts along in the direction of the field, as shown in Fig. 11.4. Since the velocity of the electron is changing (in direction), the electron must be accelerating. Classical electromagnetic theory tells us that accelerating charges must give off radiation. This radiation is called *synchrotron radiation*.

Synchrotron radiation has a number of distinguishing characteristics.

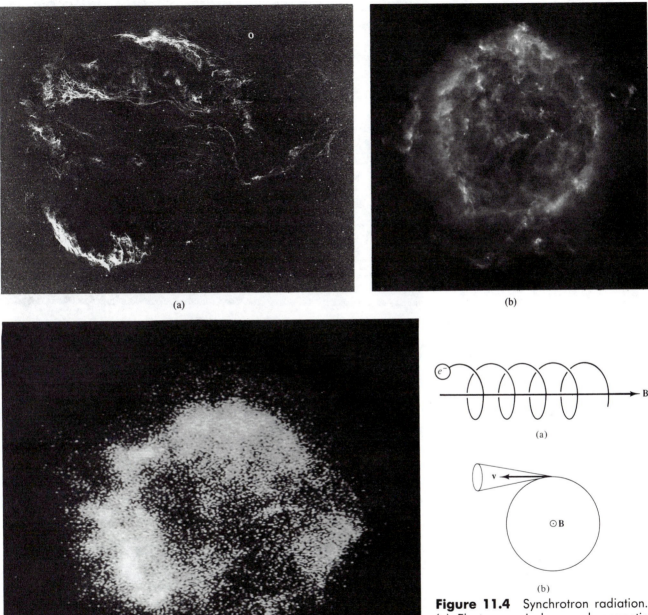

(a)

(b)

(a)

(b)

Figure 11.3 Three views of supernova remnants. (a) An optical photograph of the Veil nebula in Cygnus. (b) A radio image of the source known as Cas A (the brightest radio source in Cassiopia), taken with the VLA. (c) An X-ray image of Cas A, made with the Einstein observatory.

Figure 11.4 Synchrotron radiation. (a) Electrons spiral around magnetic field lines. (b) In their circular motion, they are constantly accelerating, and therefore give off radiation. At any instant, most of the radiation is emitted in a very small cone centered on the velocity of the electron. This beaming of the radiation is a relativistic effect.

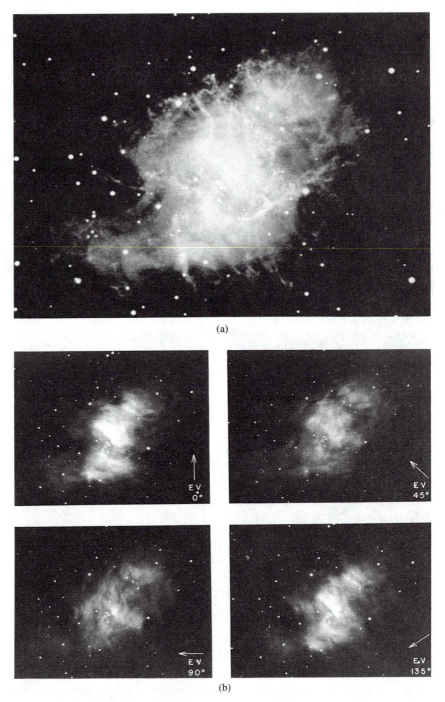

(a)

(b)

Figure 11.5 The Crab Nebula, a young supernova remnant (a) in normal light and (b) through a polarizing filter.

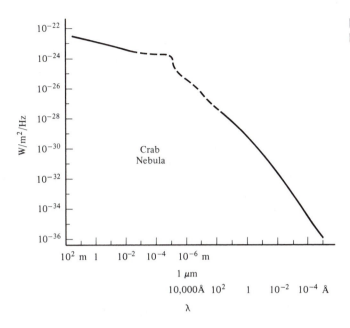

Figure 11.6 Spectrum of the Crab Nebula, from radio to gamma rays.

One is that the radiation is polarized. If we have a radio telescope receiver that detects only one direction of linear polarization, and we change the orientation of the detector, we see that the intensity of the radiation varies with the angle of the detector. This means that the electric field vector of the incoming radiation is preferentially aligned along some direction. Figure 11.5 shows that even visible light from the Crab Nebula is polarized.

The spectrum of synchrotron radiation is also quite distinctive. Most of the radiation is given off at long wavelengths. The intensity of radiation falls off as the wavelength raised to some power. This is called a *power law spectrum*. Figure 11.6 shows the spectrum of a prominent supernova remnant, the Crab Nebula, in the constellation of Taurus. The intensity of radiation is much greater in the radio part of the spectrum than in other parts of the spectrum, but the source is so strong that we are able to detect synchrotron radiation in all parts of the spectrum, even at gamma-ray energies. So much energy is given off in synchrotron radiation that the high-energy electrons should be losing their energy in a very short time. However, they are still radiating strongly some 1000 years after the supernova explosion. Until recently, a problem in astronomy has been explaining where the energy comes from to maintain the high energy of the electrons in the Crab Nebula.

11.2 NEUTRON STARS

In this section we look at what happens to the core that is left behind in the supernova explosion. The core is compressed so that normal gas pressure cannot support it. We have already seen that if its mass is more than 1.4

$M_\odot$, electron degeneracy pressure cannot support it. The collapse of the core continues beyond even the high densities associated with white dwarfs. As the density increases, electrons are forced together with protons to make neutrons. The resulting object is called a *neutron star*.

11.2.1 Neutron Degeneracy Pressure

Neutrons have spin properties similar to those of electrons. They therefore also obey the Pauli exclusion principle. Neutrons are therefore capable of exerting a degeneracy pressure if the density gets high enough. We can estimate the neutron degeneracy pressure as we did the electron degeneracy pressure in the last chapter. The result corresponding to equation (10.7), including the factor of 2, is

$$P \cong (2)\hbar^2 n_n^{5/3}/m_n \tag{11.2a}$$

Since the star is all neutrons, $\rho = n_n m_n$, so

$$P = (2)\hbar^2 \rho^{5/3}/m_n^{8/3} \tag{11.2b}$$

By comparing this with equation (10.7) we see that at a given density electron degeneracy pressure will be greater by a factor of m_n/m_e, or about 2000. However, in a neutron star there are no free electrons, so neutron degeneracy pressure is all we have to support the star. Because the density of a neutron star is much higher than that of a white dwarf, the neutron degeneracy pressure in a neutron star is greater than the electron degeneracy pressure in a white dwarf. The neutron degeneracy pressure will halt the collapse of the star.

Let's consider some of the properties of neutron stars. We can estimate the radius from a mass–radius relationship, like that for white dwarfs (Problem 11.6). We find a radius of about 15 km for a mass of 1 $M_\odot$. This means that a neutron star concentrates more than a solar mass in a sphere smaller than the island of Manhattan.

EXAMPLE 11.1   Density of a Neutron Star

Estimate the density of a neutron star and compare it with that of a neutron. Take the mass of the star to be 1.4 $M_\odot$.

Solution

The density is the mass, divided by the volume:

$$\rho = \frac{(1.4)(2 \times 10^{33} \text{ g})}{(4\pi/3)(1.5 \times 10^6 \text{ cm})^3}$$

$$= 2 \times 10^{14} \text{ g/cm}^3$$

The density of the neutron is

$$\rho_n = \frac{1.7 \times 10^{-24} \text{ g}}{(4\pi/3)(1.0 \times 10^{-13} \text{ cm})^3}$$

$$= 4 \times 10^{14} \text{ g/cm}^3$$

So we see that the density of a neutron star is very close to that of a neutron. This means that the neutrons in a neutron star must be packed close together, with very little empty space.

EXAMPLE 11.2 Neutron Degeneracy Pressure

Compare the neutron degeneracy pressure in a neutron star with the electron degeneracy pressure in a white dwarf.

Solution

Using equations (11.2b) and (10.7)

$$P_{ns}/P_{wd} = (\rho_{ns}/\rho_{wd})^{5/3}(m_e/m_n)$$

$$= [(2 \times 10^{14})/(2 \times 10^6)]^{5/3}(1/2000)$$

$$= 1 \times 10^{10}$$

EXAMPLE 11.3 Acceleration of Gravity on a Neutron Star

For the star in the above example find the acceleration of gravity at the surface.

Solution

The acceleration of gravity is

$$g = GM/R^2$$

$$= \frac{(6.67 \times 10^{-8} \text{ dyn-cm}^2/\text{g}^2)(1.4)(2.0 \times 10^{33} \text{ g})}{(1.5 \times 10^6 \text{ cm})^2}$$

$$= 8.3 \times 10^{13} \text{ cm/s}^2$$

This is 8.5×10^{10} times the acceleration of gravity at the surface of the Earth! (With such strong gravitational fields, we should really be using general relativity to calculate particle motions.) You can calculate your weight on the surface of a neutron star.

Another interesting effect comes from the fact that g changes very quickly with radius R. Differentiating the expression for g gives

$$dg/dR = -2GM/R^3$$

If we use the numbers in the above example, we find $dg/dR = -1.1 \times 10^8$ (cm/s^2)/cm. This is equal to a change of 10^5 times the acceleration of gravity on the Earth per centimeter. If you were floating near the surface, your feet would be pulled to the center with a much greater acceleration than your head. Your body would be pulled apart by these tidal forces. By tidal forces we mean effects that depend on the difference between forces at opposite sides of an object. Some astrophysicists have jokingly noted that an astronaut should visit a neutron star in a prone position to minimize the tidal effects.

The large acceleration of gravity also has another interesting effect. The equation of hydrostatic equilibrium [equation (9.34)] tells us that the rate at which the pressure in the atmosphere of the neutron star falls off, dP/dR, is proportional to g. The atmospheric pressure thus falls off very quickly. This leads to an atmosphere that is only about 1 cm thick. (The thin atmosphere is another reason for an astronaut to stay in a prone position.)

11.2.2 Rotation of Neutron Stars

In the process of the collapse of a core to become a neutron star, any original rotation of the core will be amplified. If the angular momentum of the core is conserved, the core must rotate faster as it gets smaller. Since the core shrinks by a large amount, the rotation speed is increased by a large amount. The angular momentum J is given by

$$J = I\omega \tag{11.3}$$

where I is the moment of inertia, and ω is the angular speed. If we put in the moment of inertia for a sphere, we find

$$J = \tfrac{2}{5}MR^2\omega \tag{11.4}$$

To get a feel for how fast the neutron star can rotate, let's assume that the angular momentum of the neutron star is equal to that of the Sun. (This is probably a conservative estimate since the Sun is not rotating very rapidly.) Using equation (11.4) we have

$$\omega_{\odot} R_{\odot}^2 = \omega R^2 \qquad (11.5)$$

Solving for $\omega/\omega_{\odot}$, we have

$$\omega/\omega_{\odot} = (R_{\odot}/R)^2$$

$$= [(7 \times 10^{10})/(1.5 \times 10^6)]^2 \qquad (11.6)$$

$$= 2 \times 10^9$$

The rotation period of the Sun is about 30 days. The period of a neutron star would therefore be

$$P_{ns} = \frac{2.6 \times 10^6 \text{ s}}{2 \times 10^9}$$

$$= 1.3 \times 10^{-3} \text{ s}$$

In addition to all of its other extreme properties a neutron star might be rotating 1000 times per second!

11.2.3 Magnetic Fields in Neutron Stars

We might also expect a neutron star to have a strong magnetic field. This is a consequence of Faraday's law, which can be written as

$$\oint \mathbf{E} \cdot d\mathbf{l} = -d\Phi_B/dt$$

In this expression, Φ_B is the magnetic flux through a surface. (For a small surface element the flux is the dot product of the magnetic field and a vector whose magnitude is the surface area and whose direction is perpendicular to the surface.) The integral on the left is performed around a closed path that forms that boundary of the surface. **E** is the electric field induced by the change in the flux. The integral on the left is the induced electromotive force (emf) around the closed path. If there is an induced emf around the path, and the material has some conductivity, currents will flow to oppose the change in flux. When the flux through a surface stays constant, we say that the flux is frozen in the material.

The flux is proportional to the strength of the magnetic field and to the surface area. This means that the quantity

$$BR^2 = \text{constant} \qquad (11.7)$$

The magnetic field should therefore be proportional to $1/R^2$, just as is the rotation rate. If the neutron star started off with a solar magnetic field, it would end up with a magnetic field of about 2×10^9 times that of the Sun. It should be noted that, even though flux conservation arguments give us what might be the right order or magnitude for **B**, there is some disagreement over the actual mechanism for the buildup of large fields in neutron stars. Some theories involve dynamo mechanisms, in which the rotation plays an important role.

Most of the interior of a neutron star is thought to be a fluid. Because of certain quantum-mechanical properties, this fluid can flow with no frictional resistance or viscosity. A fluid with no viscosity is called a *superfluid*. (This is analogous to a superconductor which allows the flow of electricity with no resistance.) Actually, the existence of vortices (like whirlpools) in the rapidly rotating superfluid leads to some viscosity. This is what couples the fluid to the outside layers. Outside the fluid is a solid crust made up of heavy elements, such as iron. The crust is probably less than 1 km thick. We have already seen that the atmosphere is even thinner.

The possibility of the existence of neutron stars was first realized in the 1930s. However, they were objects that existed in theory only. It was felt that their small size would make them too hard to see. In 1967 an accidental discovery in radio astronomy revived interest in neutron stars.

11.3   PULSARS

11.3.1   Discovery

In 1967, Antony Hewish and his graduate student, Jocelyn Bell Burnell, were looking for the twinkling of radio sources. This twinkling, called *scintillation*, is analogous to the twinkling of stars. However, it is not caused by the Earth's atmosphere. It is caused by charged particles in interplanetary space. The significance of these observations is that they required observations with short time resolution. For most radio observations the sources are so weak we have to observe for a long time to detect a signal. There is generally no call to see how fast things are changing. However, Hewish and Bell were looking for changes on time scales that were much shorter than for typical radio observations. This good time resolution allowed them to make an unusual accidental discovery.

They found a rapidly varying source. When they looked at it in more detail they noted that the signal was a string of pulses, like that in Fig. 11.7, with a very regular separation. By counting a large number of pulses, they determined the period to be 1.3373011 s. A few other *pulsars* were found, suggesting a general phenomenon. Since then, hundreds of pulsars have been discovered. Most of them are seen close to the plane of our galaxy, as shown

Figure 11.7   Structure of a neutron star.

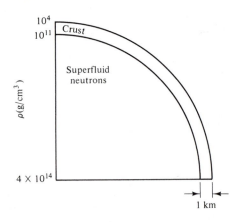

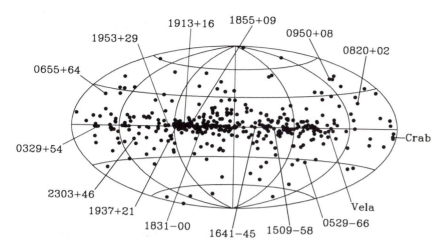

Figure 11.8 Distribution of pulsars on a galactic coordinate system. Their tendency to concentrate near the galactic plane indicates that they are within our own galaxy.

in Fig. 11.8. This suggests that they are in the galaxy. If they were extragalactic, their distribution would not correlate with anything within our galaxy.

We see some of the properties of pulsar signals in Fig. 11.9. Note that each pulse is not exactly the same as the previous one. The period may even vary slightly from one pulse to the next. However, if we take the average of 100 pulses and compare it with the average of the next 100 pulses we find

Figure 11.9 Sequence of pulses from a pulsar and the average of those pulses. Note that successive pulses may have a different appearance.

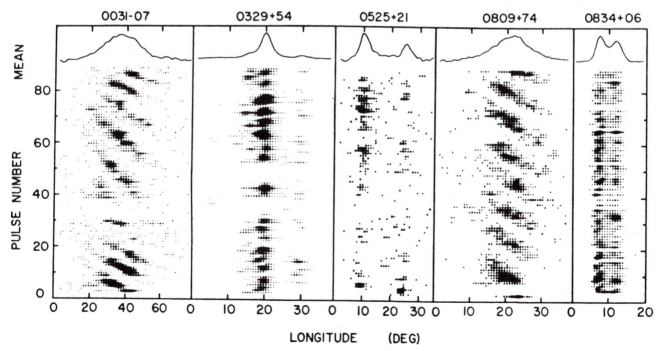

that the two averages agree quite well. We can see that the pulsar is actually "off" most of the time. The pulse is only on for a small fraction of the period. We define the *duty cycle* as the fraction of the pulse period in which the pulse is actually on. For most pulsars the duty cycle is about 5%. This means that the peak brightness is about 20 times the average brightness. Notice that there may also be a smaller pulse somewhere in the middle of the cycle. This is called an *interpulse*, and fewer than 1% of pulsars have one.

11.3.2   What Are Pulsars?

When the discovery of pulsars was announced, astronomers immediately attacked the problem of what they are. Initial guesses even included the possibilities that they were signals sent by intelligent civilizations (called LGM theories (for *little green men*). However, as more pulsars were discovered in different parts of the galaxy it seemed that a more natural explanation was needed. Efforts were concentrated on trying to find the clock mechanism. Three basic mechanisms were tried: (1) pulsation; (2) orbital motion; and (3) rotation. Three different types of objects were considered: (1) normal stars; (2) white dwarfs; and (3) neutron stars.

Pulsation of stars is reasonably well understood. In Chapter 10, we saw that for radial oscillations the period is roughly proportional to $(G\rho)^{-1/2}$, where ρ is the average density of the star. The pulsations of normal stars have periods of hours to days, and would not explain pulsars. For a period of 0.1 s, a density of about 10^9 g/cm^3 is required. This is about 10^3 times the density of a white dwarf, but only 10^{-5} of the density of a neutron star. No stellar object has a density even close to the right range. Therefore, radial pulsations in stars are ruled out.

We next consider orbital motion. The period and radius of an orbit are related by [equation (5.20)]

$$4\pi^2 R^3/G = (m_1 + m_2)P^2 \tag{11.8}$$

Solving for the radius gives

$$R = [(G/4\pi^2)(m_1 + m_2)P^2]^{1/3} \tag{11.9}$$

For a period of 1 s, the radius is 2000 km; for a period of 0.1 s, the radius is 400 km; for a period of 0.01 s, the radius is 100 km. For the range of pulsar periods, the orbital radii would have to be smaller than a normal star, or even a white dwarf. If we are to have orbiting objects, we would need two neutron stars.

There is a problem with orbiting systems. The energy of the orbit is [by equation (5.39)]

$$E = Gm_1m_2/2R$$

As the pulsar gives off radiation, to conserve energy, the orbital energy must decrease. This means that E gets more negative, or the absolute value of E gets larger. For this to happen, R must get smaller. As R gets smaller, the period must decrease, since $P \propto R^{3/2}$ [equation (11.7)]. However, we observe pulsar periods to increase rather than decrease. There is an additional problem with orbiting neutron stars. We saw in Chapter 7 that the general theory of relativity predicts that such a system would lose energy very quickly by giving off gravitational radiation. We would be able to detect this as a significant decrease in the orbital period.

This leaves rotation as the mechanism producing the regularity in the pulses. For any rotating star the rotation must not be so fast that objects on the surface lose contact with the surface. The gravitational force must be greater than the force required to keep a point moving in a circular path. If not, an object at the surface will go into orbit just above the surface. This gives the same constraint on the size of the object as equation (11.9), with the sum of the masses replaced by the mass of the single rotating star. This again rules out a rotating normal star and a rotating white dwarf. (White dwarfs might actually work for some of the slower pulsars, but not for the fastest.) This leaves rotating neutron stars as the best candidates for pulsars.

In describing the emission from pulsars, an analogy has been drawn with a lighthouse. The light in the lighthouse is always on, but you can only see it when the beam points in your direction. If you stay in one place, you will see the light appear to flash on briefly once per cycle. In this model of neutron stars, there is some emission point or "hot spot" on the surface of the star, producing a beam of radiation, like a lighthouse beam. We can only see the radiation when the beam is pointed in our direction.

The emission mechanism may be related to the strong magnetic field that we think many neutron stars must have. The magnetic axis of the star is probably not aligned with the rotation axis. (This is not unusual. It also happens on the Earth.) If the beam of radiation is somehow collimated along the magnetic axis, we only see it when the beam points in our direction. We may even see two pulses, as shown in Fig. 11.10.

The details of the mechanism are not clear. One possibility is that the rotation and strong magnetic field result in an observer near the surface seeing a rapidly changing magnetic field. Farday's law tells us that a rapidly changing magnetic field produces an electric field. This electric field can accelerate charged particles to large energies. These particles than spiral around the magnetic field lines, producing synchrotron radiation. This is only a very general outline of what may be taking place. A considerable effort is still going on to try to understand the emission of radiation under the extreme conditions that exist near the surface of a neutron star.

The lighthouse nature of the pulsar emission mechanism has another consequence. The beam only traces out a cone on the sky. If the observer is

Figure 11.10 Role of a magnetic field in a pulsar emission mechanism. If the magnetic axis is not aligned with the rotation axis, then the magnetic axis can act like a searchlight beam.

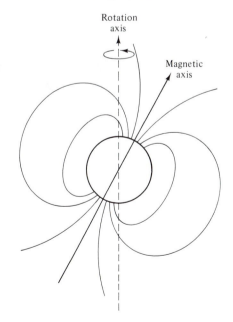

Rotation axis

Magnetic axis

Figure 11.11 Schematic diagram of an instrument for looking at optical pulses. The rotating disk is synchronized with the radio pulses. It can be synchronized with any part of the cycle.

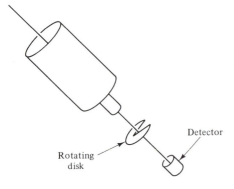

Detector

Rotating
disk

not on that cone, the pulsar will never be visible. This is also true for a lighthouse. If you are directly above the lighthouse you will never see the light. Any given pulsar will only be seen by about 20% of the potential observers. This means that our galaxy contains many more pulsars than the few hundred that we actually observe. This is especially true, since we haven't surveyed the whole galaxy for pulsars. Current estimates place the number of pulsars in the galaxy at about 200,000.

Probably the most extensively studied pulsar is one in the center of the Crab Nebula, the remnant of the supernova explosion observed in 1054. We have already seen that the emission from the Crab Nebula is polarized, suggesting synchrotron radiation, and that it gives off most of its energy in the

Figure 11.12   The results of using equipment like that depicted in Fig. 11.11. This shows the Crab pulsar (NP 0532) in visible light. Each frame shows an image of the field at the point in the cycle indicated by the light curve below. When the pulsar is on, it is the brightest object in the field. When it is off, we cannot see it.

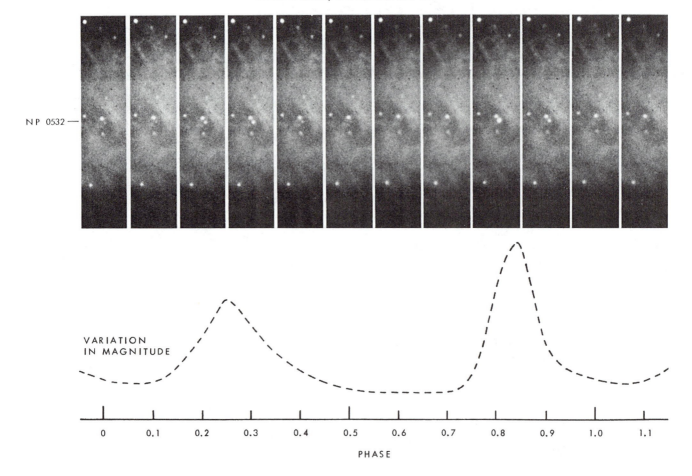

NP 0532

VARIATION
IN MAGNITUDE

0   0.1   0.2   0.3   0.4   0.5   0.6   0.7   0.8   0.9   1.0   1.1

PHASE

radio part of the spectrum. We have also seen that the energy loss via synchrotron radiation is so great that the nebula should have faded considerably in its optical and x-ray luminosity. Something is replenishing the energy that is radiated away. The total energy loss rate of the nebula is about 3×10^{38} erg, or $10^5 L_\odot$. It has been known that there is a star with an unusual spectrum at the center of the nebula. This star is known as Baade's star, after Walter Baade who first noted its peculiarity.

A very rapid pulsar was discovered at the position of this star. Its period is 0.033 s. This pulsar was very important in ruling out rotating white dwarfs as the source of pulsars. After the radio pulsar was discovered, it was suggested that it might be possible to observe optical pulses from the star. However, to catch the star in different parts of the cycle would require exposures of about 10^{-3} s, much too short to see anything. An interesting technique was used to get around this problem. The star was observed for many cycles, but the image was recorded only during a small part of each cycle. The part of the cycle was the same cycle after cycle. For example, we expose for the same 10^{-3} s every period until we get a good image. We then shift our exposure by 10^{-3} s and repeat the process. We use the radio signals to synchronize our observing with the pulsing of the star.

The experimental setup is outlined in Fig. 11.11, and the results are shown in Fig. 11.12. The optical pulsations are clearly visible. When the star is ''on'' it is the brightest star in the field. When it is ''off'' we cannot see it at all. If we just take a normal photograph of the nebula, we just see the average brightness of the star. If the duty cycle is 5%, then the average brightness is about 5% of the peak brightness.

11.3.3 Period Changes

Just as astronomers got used to the idea of pulsars as dependable clocks in the sky, it was discovered that all pulsars are slowing down. Some sample data are shown in Fig. 11.13. It was also found that the fastest pulsars were slowing down with the greatest rate of change of the period. This suggests that the fastest ones must be younger. The increase in the pulse period means that the star is not rotating as fast as it once was. This means that the kinetic energy of the star is decreasing. This is not surprising, since the star is giving off energy in the pulses.

The energy of a rotating sphere is given by

$$E = \tfrac{1}{2}I\omega^2 \qquad (11.10)$$

The rate of change of energy is found by differentiating both sides with respect to time:

$$dE/dt = I\omega(d\omega/dt) \qquad (11.11)$$

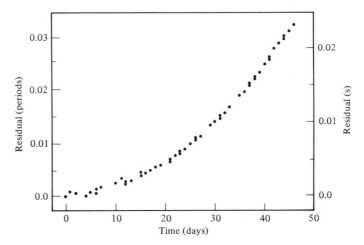

Figure 11.13 The slowing of the pulsar PSR0329 + 54 is shown by the increasing delay in pulse arrival times.

We can relate this to the fractional change in energy, using equation (11.10):

$$(1/E)(dE/dt) = (2/I\omega^2)(I\omega)(d\omega/dt) \qquad (11.12)$$
$$= 2(1/\omega)(d\omega/dt)$$

The fractional change in energy is therefore equal to twice the fractional change in the frequency. We can relate the change in frequency to the change in period:

$$\omega = 2\pi/P \qquad (11.13)$$

Differentiating, we obtain

$$(d\omega/dt) = (-2\pi/P^2)(dP/dt)$$

Using equation (11.13) to eliminate one of the factors of P on the right-hand side gives

$$(1/\omega)(d\omega/dt) = -(1/P)(dP/dt) \qquad (11.14)$$

The fractional rate of energy change then becomes

$$(1/E)(dE/dt) = -(1/P)(dP/dt) \qquad (11.15)$$

Thus, when we observe the rate at which a pulsar is slowing down, we can directly relate it to the energy loss.

EXAMPLE 11.4 Pulsar Energy Loss

Consider a rotating neutron star with a mass $M = 2M_\odot$, a radius $R = 15$ km, a period $P = 0.1$ s, and a rate of change of the period $dP/dt = 3 \times 10^{-6}$ s/yr. Find (a) the kinetic energy; (b) the rate at which the kinetic energy is decreasing; and (c) the lifetime of the pulsar if it loses energy at this rate.

Solution

We find the energy from equation (11.10), remembering that the moment of inertia for a sphere is $\frac{2}{5}MR^2$. We also use the fact that the frequency is $\omega = 2\pi/P$.

$$E = (\tfrac{1}{2})(\tfrac{2}{5})MR^2\omega^2$$

$$= \tfrac{1}{5}(4 \times 10^{33} \text{ g})[(1.5 \times 10^6 \text{ cm})(2\pi/0.1 \text{ s})]^2$$

$$= 7 \times 10^{48} \text{ erg}$$

To find the rate at which the energy is changing, we find the fractional rate at which the period is changing. We first convert dP/dt into more useful units:

$$dP/dt = \frac{3 \times 10^{-6} \text{ s/yr}}{3 \times 10^7 \text{ s/yr}}$$

$$= 1 \times 10^{-13}$$

The fractional rate of change is

$$(1/P)(dP/dt) = (1/0.1 \text{ s})(1 \times 10^{-13})$$

$$= 1 \times 10^{-12}/\text{s}$$

This means that each second the period increases by 10^{-12} of a period. We can use equation (11.15) to find the fractional rate of change of energy:

$$(1/E)(dE/dt) = -2(1/P)(dP/dt)$$

$$= -2 \times 10^{-12}/\text{s}$$

Multiplying both sides by E gives

$$dE/dt = (-2 \times 10^{-12}/\text{s})E$$

$$= -1.4 \times 10^{37} \text{ erg/s}$$

The lifetime of the pulsar T is just the energy, divided by the rate at which it is being lost:

$$T = E/(dE/dt)$$

$$= \frac{1}{2 \times 10^{-12}/\text{s}}$$

$$= 5 \times 10^{11} \text{ s}$$

$$= 1.7 \times 10^4 \text{ yr}$$

This means that pulsars have very short lifetimes by astronomical standards. (Actually, the rate is not constant, so the above calculation actually gives a lower limit to the lifetime.)

We should note that the rate at which the pulsar in the Crab Nebula is losing energy is equal to the rate at which the nebula itself is losing energy via synchrotron radiation. This solves a longstanding problem: It explains the source of energy for the nebula. As the pulsar loses energy, somehow that energy ends up in the nebula. The connection between the pulsar and the nebula is probably the strong magnetic field. When we see radiation from the Crab Nebula, it is being indirectly fueled by the slowing down of the Crab pulsar.

Pulsars show another variation in their periods. These are sudden decreases in the period, called *glitches* (Fig. 11.4). After each glitch, the normal slowdown resumes. If the angular momentum is conserved, the rotation of the star can only speed up if the size of the star decreases. We think that the fast rotation and extreme conditions at the surface put a tremendous strain on the solid crust. Periodically this strain is relieved in a quake, and the crust then settles into a more stable configuration. (Actually, this explanation of glitches may work for the Crab pulsar, which has small glitches, but it cannot explain the larger glitches in other pulsars. It is thought that the glitches may involve a transfer of angular momentum from the superfluid interior to the crust.)

We can relate the change in period to the change in radius. The angular momentum is given by

$$J = I\omega$$

$$= I(2\pi/P)$$

$$= \tfrac{2}{5}MR^2(2\pi/P) \tag{11.16}$$

$$= (4\pi/5)MR^2/P$$

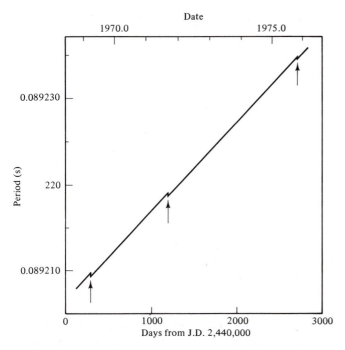

Figure 11.14   Glitches in the pulse period. The glitches are the sudden jumps in the period.

Solving for P gives

$$P = (4\pi/5)MR^2/J$$

Differentiating with respect to R gives

$$dP/dR = (4\pi/5)2MR/J$$

$$= (4\pi/5)(2MR/MR^2)(5/4\pi)P \qquad (11.17)$$

$$= 2(P/R)$$

This means that a small change in period dP will be caused by a small change in radius

$$dR/R = \tfrac{1}{2}(dP/P) \qquad (11.18)$$

The fractional change in radius is half the fractional change in the period.

EXAMPLE 11.5   Pulsar Glitch

For the pulsar in the above example find the change in radius for a period change of 10^{-7} s.

Solution

We can find dR by multiplying both sides of equation (11.18) by R, to give

$$dR = \tfrac{1}{2}(dP/P)R$$

$$= \tfrac{1}{2}(1 \times 10^{-7}/0.1)(1.5 \times 10^6 \text{ cm})$$

$$= 0.75 \text{ cm}$$

This is not a very large change.

Until relatively recently, two classes of pulsars had been found: radio pulsars and x-ray pulsars. The pulsars we have been discussing in this chapter are the radio pulsars. We will discuss x-ray pulsars in Chapter 12. It now seems that another class of pulsar has been found. These are pulsars with extremely short periods. Since the periods are a few milliseconds, they are called *millisecond pulsars*. An amazing feature of these objects is that they are not slowing down very much. For example, for one pulsar, with a period of about 1.5 ms, the rate of slowdown is 1.26×10^{-19} s/s. This is contrary to the expectation from normal pulsars that faster pulsars have the greatest slowdown rates. This means that there is a very small energy loss rate in the millisecond pulsars. Theoreticians are still working on this problem.

11.4   PULSARS AND DISPERSION

Let's look at what happens to pulsar signals as they travel through interstellar space. The speed of the radiation depends on the index of refraction n, which is a function of wavelength λ. The speed of the radiation is given by

$$c(n) = c/n \tag{11.19}$$

where c is the speed of light in a vacuum (where $n = 1$). We can write the speed as a function of wavelength

$$c(\lambda) = c/n(\lambda) \tag{11.20}$$

Figure 11.15 Dispersion. Pulses are emitted at two different wavelengths on the left. They are received at different times on the right.

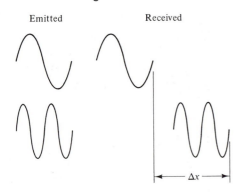

Emitted   Received

Suppose we have two signals at wavelengths λ_1 and λ_2. Their speeds will be different. Their times t_1 and t_2 to travel a distance d are

$$t_1 = dn(\lambda_1)/c$$
$$t_2 = dn(\lambda_2)/c \tag{11.21}$$

The difference between these two times is

$$\Delta t = (d/c)[n(\lambda_1) - n(\lambda_2)] \tag{11.22}$$

This time delay for signals of different wavelength is called *dispersion* (Fig. 11.15). If we know $n(\lambda)$, then we can measure Δt and find out the distance d that the signals traveled. If d is known, then measuring t tells us about the index of refraction.

For radio waves passing through interstellar space, the index of refraction results from the interaction of the radio waves with electrons, and is proportional to the interstellar electron density n_e. That is,

$$n(\lambda) - 1 \propto n_e \tag{11.23}$$

Then the time delay must be given by

$$\Delta t \propto dn_e \tag{11.24}$$

When we observe a pulsar at two wavelengths, the signals will not arrive at the same time. However, we know they left the pulsar at the same time. We can attribute the time difference to interstellar dispersion. By measuring the time delay Δt, we can derive the quantity dn_e. If we observe objects for which we have another distance estimate, we can derive the average interstellar electron density. This turns out to be $\frac{1}{30}$ cm^{-3} on average. Once we know this average n_e, we can measure Δt for other pulsars, which gives us dn_e. Since we know n_e, we can estimate d. This gives us a way of estimating distances to pulsars that we cannot otherwise see.

11.5 STELLAR BLACK HOLES

Just as special relativistic effects put a limit on the mass that can be supported by electron degeneracy, they put a limit on the mass that can be supported by neutron degeneracy. There is some uncertainty about that limit. Neutron degeneracy, alone, can support about 2 $M_\odot$. Neutron–neutron interactions raise this to about 3 $M_\odot$, and this is the number that has been regularly used. However, some calculations show that mass as large as 8 $M_\odot$ might be supported. Whatever the limit, if a neutron star is more massive than this limit, it cannot be stable. We know of no other source of pressure to halt the collapse. The star will eventually be small enough to be contained within its Schwarzschild radius. It will become a black hole, as described in Chapter 7. Therefore, unless all neutron stars are formed with less than about 3 $M_\odot$, black holes appear to be a normal stellar final state. Though this can happen for individual stars or stars in binary systems, it is not likely that we can detect an isolated stellar black hole. We have to deduce its presence by its gravitational effects on another star. We will therefore discuss attempts to detect stellar black holes in the next chapter, where we discuss close binary systems.

CHAPTER SUMMARY

In this chapter we saw what happens to stars of higher mass after they leave the main sequence. The higher mass of the core means a higher temperature, meaning that nuclear reactions can proceed farther in a high-mass star than in a low-mass star. A shell structure develops, with each layer closer to the center fusing heavier nuclei than the layer outside it.

When an iron core is built up, there is no longer a source of energy. The core begins to collapse. When a high enough density is reached, the star explodes in a supernova. We see the blown-off material as a supernova remnant.

The leftover core of the star becomes a neutron star, supported by neutron degeneracy pressure. Neutron stars are observed as pulsars. We discussed the accidental discovery of pulsars, and the chain of reasoning to make the connection between pulsars and neutron stars. The conditions on neutron stars are quite extreme, often involving enormous magnetic fields. These strong fields are involved in the pulse emission mechanism. As pulsars give off energy they slow down.

If the mass of a neutron star is greater than roughly 3 $M_\odot$, neutron degeneracy pressure will not support it, and it will become a black hole. Such objects must be detected in binary systems.

QUESTIONS AND PROBLEMS

11.1 Why is the stability of ^{56}Fe important in supernovae?

11.2 Suppose a supernova explosion throws off 5 $M_\odot$ of material at an initial speed of 10^3 km/s. (a) Calculate the initial energy of the shell and the sum of the magnitudes of the momenta of all of the pieces of the shell. (b) Suppose the shell slows down by conservation of momentum in sweeping up interstellar material. How much mass will be swept up before the shell slows to 10 km/s? (c) If the average density of interstellar material is 1 H atom/cm^3, what will the radius of the shell be when it reaches 10 km/s?

11.3 What are the distinguishing characteristics of synchrotron radiation?

11.4 At a given density, for equal numbers of electrons and neutrons, electrons will exert a greater degeneracy pressure. Why then are neutron stars supported by neutron degeneracy pressure?

11.5 Estimate the pressure in a 1.5-$M_\odot$ neutron star with a radius of 15 km.

11.6 We have already seen that hydrostatic equilibrium allows us to estimate the central pressure of a star as $P_C \approx (2)GM^2/R^4$. (a) Show that this provides the following mass–radius relation for neutron stars:

$$M^{1/3}R = (2)(1/4)^{5/3}(\hbar^2/G)(1/m_p)^{7/3}$$

(b) Use the expression to find the radius of a 1.5-$M_\odot$ neutron star.

11.7 (a) If $F(r)$ is the force on an object as a function of radius r, show that the "tidal" force on an object of length Δr is

$$\Delta F = (dF/dr)\Delta r$$

(b) Calculate the tidal force on your body near the surface of a neutron star. (Do the calculation for standing and prone positions.)

11.8 For a 1.5-$M_\odot$ neutron star with $R = 15$ km, rotating 100 times per second, compare the gravitational force on an object at the surface with the force required to produce the circular motion for that object (at the star's equator). In other words, compare the weight of an object with the centrifugal force on it.

11.9 A uniform density sphere of mass M has an initial radius r and an angular speed ω_0. It collapses under its own gravity to a radius r, conserving angular momentum. (a) How do the initial and final kinetic energies compare? (b) Account for any difference.

11.10 (a) What is the escape velocity from a 1.5-$M_\odot$ neutron star of radius 10 km? How does it compare with the speed of light?

11.11 Suppose we can measure the arrival time of pulses to within 10^{-8} s. (a) Explain how we can measure a pulsar period more accurately by timing a large number of pulses. (b) How many pulses would you have to measure to measure a period of 0.1 s with an accuracy of 10^{-10} s?

11.12 Explain why the average brightness of an object is approximately equal to its duty cycle multiplied by its peak brightness. Use the fact that if $b(t)$ is the brightness as a function of time, then the average of $b(t)$ over some time interval T is given by

$$\langle b \rangle = (1/T)\int_0^T b(t)\,dt$$

11.13 (a) Prove that the requirement that for a rotating

object the gravitational force must at least balance the centrifugal force produces an expression for the minimum radius like equation (11.9). (b) Using this result, what is the minimum rotation period for a white dwarf?

11.14 Suppose we approximate the rate of change of a magnetic field for a stationary observer near a neutron star as the field strength divided by the rotation period. (a) Taking the magnetic field to be 10^{12} gauss, calculate the magnitude of the induced electric field. (b) Over what distance will an electron have to travel in this field to reach a speed of 0.1 c?

11.15 (a) Suppose you have a drumhead vibrating at 10^3 Hz. You want to "freeze" its motion at different parts of the cycle. Explain how you could use a strobe light to obtain a sequence of photographs, each showing the drumhead at a slightly different point in the cycle. (b) How is this like the technique for studying the optical pulses from a pulsar?

11.16 Suppose that, as a pulsar slows down, the quantity $(1/P)(dP/dt)$ stays constant (say at a value of $-b$, where b is a positive number). (a) If the initial ($t = 0$) period of the pulsar is P_0, find an expression for $P(t)$, the period as a function of time. (b) If the initial rotational energy is E_0, find an expression for $E(t)$, the energy as a function of time. (c) If a pulsar is formed with $P = 10^{-3}$ s, how long will it take to reach $P = 3$ s?

11.17 (a) In a pulsar of mass M, radius R, and period P, a glitch in the period of size ΔP is observed. What is the corresponding change in energy ΔE? (b) Where did this energy come from or go to?

11.18 (a) For a pulsar of mass M, radius R, and period P, slowing at a rate dP/dt, at what rate is its angular momentum changing? (b) What provides the torque to change the angular momentum?

11.19 Find an expression, analogous to that for the Chandrasekhar limit discussed in Chapter 10, for the maximum star that can be supported by neutron degeneracy pressure.

11.20 Compare the radius of a 1.4-$M_\odot$ neutron star with its Schwarzschild radius.

11.21 Compare the quantity $(1/P)(dP/dt)$ for normal pulsars and millisecond pulsars, using quantities given in the chapter.

12

Evolution in Close Binaries

12.1 CLOSE BINARIES

If the two stars in a binary system are very close to each other, each star has the effect of altering the structure of the other star. When this occurs, we call the system a *close binary* system. The surface of a star can be distorted by the fact that the gravitational force that the companion exerts is stronger on the side closer to the campanion than it is on the side away from the companion. Any effect which depends on variations in the gravitational force from one position to another is called a *tidal effect*. (A similar situation applies as the Sun and Moon distort the Earth's ocean surface, raising the tides.)

The distortion of stars results in an internal dissipation of energy. As a star rotates, different material is incorporated in the bulge. Different layers of material rub against each other, in a fluid friction. This lost energy has to come from somewhere. It comes from both the orbital energy and the rotational energy of the star. As a result, eventually the orbits circularize and the two stars always keep the same sides toward each other. This is the lowest energy arrangement for the system. We say the spins are *synchronized*.

In certain situations, it is possible for material from one star to be pulled off the surface onto the other star. To see how this can happen, we look at a binary system from a coordinate system rotating with the same period as that of the orbits. If we look at the energy of a particle in this system, the rotation of the coordinate system introduces a term in addition to the gravitational potential. This term is equal to $L^2/2mr^2$, where L, m, and r are the angular momentum, mass, and distance, respectively, from the origin of some particle. (We can think of it as the term in the potential energy which gives rise to the "centrifugal force" in the rotating system.) When we add this term to the gravitational potential, we have an *effective potential* that can be used to describe the motion of the particle. We can draw surfaces of constant, effective potential, as in Fig. 12.1. The effective force (gravity plus "centrifugal") at any point on one of these surfaces is perpendicular to the surface.

There are five points, called *Lagrangian points*, where the effective gravitational force is zero. These points are designated L1, L2, L3, L4, L5. (Note that the "L5 Society" wants to place a space station at the L5 point for the Earth/Moon system.) The point L1 lies between the two stars, at the intersection point of the "figure eight" shaped surface. L1 is the dividing point between material being attracted to one star or the other. The two sides of the figure eight are called *Roche lobes* (Fig. 12.2).

The equipotential surfaces in a fluid must be surfaces of constant pressure. If they were not, there would be pressure differences forcing fluid along the surface, and these forces could not be balanced by any gravitational forces, which must be perpendicular to the surfaces. The equipotential surfaces must therefore also be surfaces of constant density.

In discussing the evolution of close binary systems, we divide them into three classes (Fig. 12.3): (1) *detached*, in which each star is totally contained within its own Roche lobe; (2) *semidetached*, in which the photosphere of

Figure 12.1 Surfaces of constant effective potential. The effective potential is the true gravitational potential plus that resulting from the centrifugal force in the coordinate system rotating with the orbiting system. In the upper figure we look from above and see the intersection of surfaces of constant potential and the plane of the orbit. The heavy figure "8" is the intersection of the Roche lobe and the plane of the orbit. A three-dimensional view of the Roche lobe is shown below.

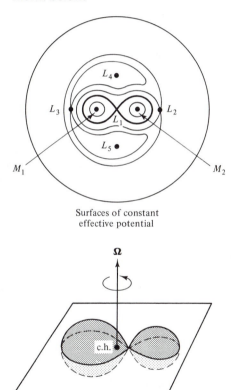

Surfaces of constant effective potential

Equatorial plane

Figure 12.2 The sizes of the two Roche lobes depends on the masses and separations of the orbiting systems. This figure shows the Roche lobes for three mass combinations. The masses are the best estimates for the masses in three observed close binary systems.

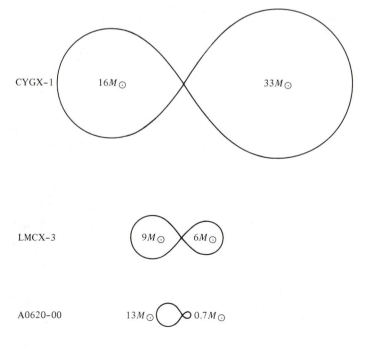

Figure 12.3 Classification of close binary systems. Shaded areas are filled with material.

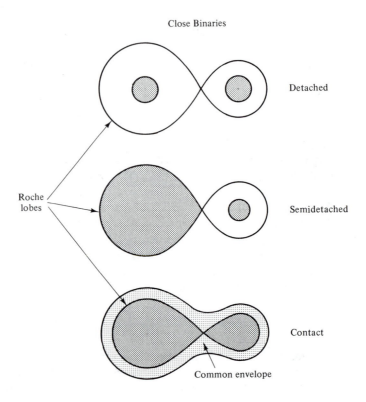

one star exactly fills its side of the Roche lobe; and (3) *contact binaries*, in which both stars are at or over the Roche lobe.

So far we have discussed the evolution of isolated stars. However, we have already seen that approximately half of all stars are in binary systems. If the binaries are completely detached, and there is no mass transfer, then the evolution will not be altered by the presence of the companion. However, mass transfer in semidetached systems can influence stellar evolution.

In general, the more massive star in a binary system will evolve off the main sequence first. When that star becomes a red giant, it may become large enough to fill its Roche lobe. In that case mass transfer to the companion will take place. This can alter the evolution of the companion. The degree to which it alters the evolution depends on the nature of the companion. As the more massive star continues to lose mass, its Roche lobe shrinks, but the Roche lobe for the companion grows. This means that mass transfer will take place until the masses of the stars are about equal. Some slow mass transfer may continue after that point.

At some point the star that was losing mass will become a white dwarf or some other collapsed object. In the following sections we look at examples of each of the type of collapsed objects we have discussed—white dwarf, neutron star, and black hole.

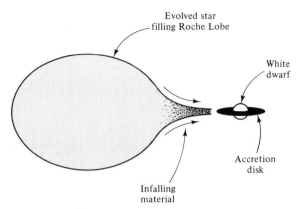

Evolved star
filling Roche Lobe

White
dwarf

Accretion
disk

Infalling
material

Figure 12.4 Matter from a normal star falling onto a white dwarf.

12.2 SYSTEMS WITH WHITE DWARFS

We first consider systems in which the first star to evolve off the main sequence becomes a white dwarf (Fig. 12.4). Eventually, the white dwarf's companion goes through its evolution off the main sequence. The companion becomes a red giant and fills its Roche lobe. At this point mass transfer starts back in the other direction from the original mass transfer. Now mass is falling in on the white dwarf. Not all of the infalling matter strikes the white dwarf. Because of its angular momentum, some of the material goes into orbit around the white dwarf. This orbiting material forms a disk, called an *accretion disk*. (The disk forms because material can fall parallel to the axis of rotation but not perpendicular to that axis.)

As material falls in, its potential energy decreases, so its kinetic energy increases. According to the virial theorem, half of the lost potential energy must be radiated away. This allows us to calculate the luminosity resulting from the mass transfer. If the mass starts at a distance r_1 from the white dwarf and ends up at a distance r_2, the luminosity is given by (see Problem 12.11)

$$L = \tfrac{1}{2}(dM/dt)(GM/r_2 - GM/r_1) \qquad (12.1a)$$

$$= (GM/2)(dM/dt)(1/r_2 - 1/r_1) \qquad (12.1b)$$

In this equation dM/dt is the rate at which mass is falling in, and M is the mass of the white dwarf.

EXAMPLE 12.1 Mass Accretion Luminosity

Calculate the luminosity for a mass infall rate of $10^{-8}\ M_\odot/\text{yr}$, onto a $1\text{-}M_\odot$ white dwarf. Assume that the material starts 1.0×10^{11} cm away from the white dwarf, and ends 1.0×10^9 cm away.

Solution

We first convert the mass loss rate into g/s:

$$dM/dt = \frac{(1.0 \times 10^{-8} M_\odot/\text{yr})(2.0 \times 10^{33}\text{g}/M_\odot)}{3.1 \times 10^7 \text{ s/yr}}$$

$$= 6.5 \times 10^{17} \text{ g/s}$$

The luminosity is then

$$L = (6.67 \times 10^{-8} \text{ dyn-cm}^2/\text{g}^2)(2.0 \times 10^{33} \text{ g})(\tfrac{1}{2})(6.5 \times 10^{17} \text{ g/s})$$
$$\times \{[1/(1.0 \times 10^9 \text{ cm})] - [1/(1.0 \times 10^{11} \text{ cm})]\}$$

$$= 4.3 \times 10^{34} \text{ erg/s}$$

This is approximately ten times the luminosity of the Sun.

(a)

(b)

Figure 12.5 Nova Cygni 1975. A before-and-after photograph. The arrow indicates the location of the nova. (a) This photograph shows the region before the nova outburst. (b) We see it after the outburst.

Occasionally we observe a star that suddenly brightens by 5 to 15 magnitudes. These objects are called *novae* (Fig. 12.5). The name suggests the appearance of a new star where one was not previously seen. Some of these novae appear to be recurrent, on time scales of weeks up to hundreds of years. There is evidence that mass is ejected in the process. In some cases, this material can be seen expanding away from the star. The amount of mass ejected is about $10^{-5} M_\odot$.

We think that novae are the result of thermonuclear explosions on the surface of white dwarfs with mass falling in from a companion. The mass falling in is from the envelope of a red giant, and therefore contains hydrogen. (Remember, the white dwarf has used up all of the hydrogen in the core, and has expelled the rest in its planetary nebula.) The surface of the white dwarf is hot enough for fusion of the hydrogen to take place. It takes place rapidly in a small explosion. The explosion probably stops the mass transfer for a while. When the transfer resumes, another explosion can take place.

We think that mass transfer onto a white dwarf can account for type I supernovae. Sometimes enough mass falls onto the white dwarf to make its mass greater than M_{CH}. In this case, electron degeneracy no longer supports the star, and it collapses. The energy from the collapse drives nuclear reactions which eventually build up ^{56}Ni (with an even number of ^{4}He nuclei). The ^{56}Ni beta decays to form ^{56}Co, which, in turn, beta decays to ^{56}Fe. Type I supernovae have light curves with a double exponential behavior. The time scales of the two exponentials turn out to be characteristic of these two beta decays. We think that this process accounts for most of the iron in the universe, since the iron created in more massive stars is destroyed in the type II supernova, as discussed in Chapter 11.

The nuclear energy released in these reactions is greater than the binding energy of the white dwarf, and the star is destroyed, leaving no remnant. This explanation accounts for the light curves of type II supernovae, their spectra and luminosities, and their occurrence in what are thought to be old systems (as we will discuss in Chapter 13).

12.3   NEUTRON STARS IN CLOSE BINARY SYSTEMS

In the last section we saw how mass transfer in semidetached systems can alter the evolution of a star. In this section we consider a neutron star and a normal star in orbit around their common center of mass (Fig. 12.6). As the normal star evolves towards a red giant, it fills its Roche lobe and matter starts to fall onto the neutron star.

At first it was thought that this situation could not develop. It was not clear how such a system could form. The problem is that for a neutron star to be present, there must have been a supernova explosion in the past. The first star to go supernova in a binary is the more massive star. The supernova explosion drives away most of the mass of the more massive star, meaning that more than 50% of the original mass of the system was blown away. If a system in virial equilibrium loses more than half of its mass suddenly, it

Figure 12.6   Model for an x-ray source with matter from a normal star falling onto a neutron star.

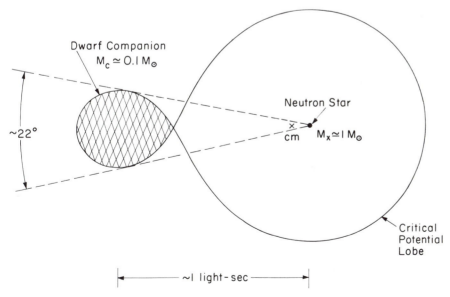

becomes unbound (see Problem 12.7). Therefore, the explosion should break up the binary system. This scenario explains the existence of ''runaway stars.'' These are individual stars moving through space with much higher than average speeds.

However, x-ray observations were made of systems in which it appeared that a neutron star was still in orbit with a normal star. This meant that there must be some way of forming such a system, and eventually theoreticians came up with a plausible scenario. Before the supernova explosion, the more massive star might have filled its Roche lobe and transferred mass to the less massive star. If enough mass is transferred before an explosion, the system can stay bound. An alternative explanation is that the compact star may have originally been a white dwarf, not a neutron star. However, mass transfer from the companion may have pushed the mass of the white dwarf beyond the $1.4\ M_\odot$ limit. The electron degeneracy pressure could no longer support the star and it would collapse until it became a neutron star. The neutron star would have formed without a supernova explosion. (An interesting sidelight to this is that a $1.44\text{-}M_\odot$ white dwarf that suddenly collapsed to form a neutron star would, according to general relativity, appear to exert the gravitational force of a $1.3\text{-}M_\odot$ star. Thus, there may be stars we think are white dwarfs, but which are really neutron stars.)

We now suppose we have a binary system with a neutron star and a normal star, with mass being transferred from the normal star to the neutron star. The mass falling in is heated and gives off irregular bursts of x-ray emission. To see how this works, we look at the case of a well-studied x-ray source, Her X-1 (Fig. 12.7). (The name implies the first x-ray source in the constellation Hercules.) It is also coincident with a variable star HZ Her. This star is a binary with a period of 1.7 days. The mass of the unseen companion is estimated to be in the range 0.4 to $2.2\ M_\odot$. The x-rays are observed to pulse with a period of 1.24 s. This period changes regularly throughout the 1.7-day cycle. This can be interpreted as a Doppler shift. We can think of the x-rays as a signal being emitted with a period of 1.24 s. When the source is moving away from us the period appears longer, and when the source is coming toward us the period appears shorter. The x-ray source also appears to be eclipsed every 1.7 days.

The mass transfer rate, estimated from the x-ray observations, is about $10^{-9}\ M_\odot/\text{yr}$. The luminosity is about 10^{37} erg/s. The temperature is estimated to be 10^8 K. At this temperature, we can estimate the frequency of a photon with energy kT. The frequency, ν, is kT/h, or $2\ \times\ 10^{18}$ Hz. This corresponds to a wavelength of 1.4 Å, clearly in the x-ray part of the spectrum.

Theoreticians have speculated on the future evolution of such a system. The mass transfer rate may become so large that the x-ray emission is quenched. The outgoing x-rays are effectively blocked by the infalling material. This system may eventually end up as two compact objects.

SAS-3 OBSERVATIONS OF RAPIDLY REPETITIVE
X-RAY BURSTS FROM MXB 1730-335

24-minute snapshots from 8 orbits on March 2/3, 1976

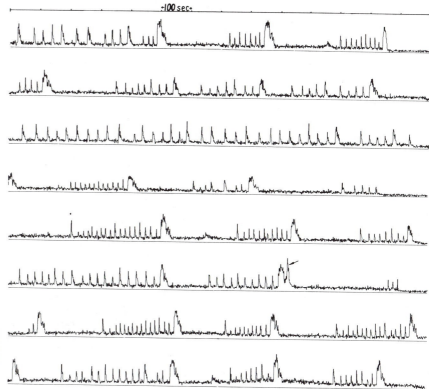

Figure 12.7 (a) X-ray emission from HZ Her. The intensity is plotted as a function of time into the period. (b) X-ray emission from a burster.

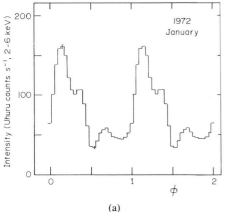

(a)

(b)

While discussing binary systems, we briefly mention one other interesting object. This is an interesting radio pulsar discovered in 1974 by Joseph Taylor and Russell Hulse, of the University of Massachusetts. This pulsar has a period of 0.059 s, but the period varies periodically, suggesting that the pulsar is in a binary system. The variation in the period is like a Doppler shift (Fig. 12.8), with the period appearing longer when the pulsar is moving away from us and shorter when it is coming toward us. This system provides us with an interesting test of the idea of gravitational radiation.

Since the gravitational radiation carries away energy, the total energy of the orbit should be decreasing. In hopes of seeing this, Taylor and his co-workers have monitored this binary pulsar for several years. Some results are shown in Fig. 12.9. They find that the orbital period is changing by -2.3×10^{-12} s/s. The change in energy of the orbit corresponds to the

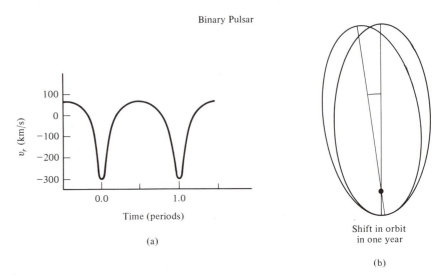

Binary Pulsar

(a)

Shift in orbit
in one year

(b)

Figure 12.8   Binary pulsar orbit. (a) The radial velocity as a function of time within a period, expressed as a fraction of the period. (b) The shift in the orbit over a 1-yr period.

energy that would be given off by gravitational radiation from the orbiting bodies. This may provide us with the first indirect observational confirmation of gravitational radiation. To date, a total of six binary pulsars have been found.

12.4 SYSTEMS WITH BLACK HOLES

In Chapter 11 we saw that neutron stars are supported by neutron degeneracy pressure. However, if the neutron star is too massive, neutron degeneracy pressure is insufficient to support the star. We think this limit is about $3 \, M_\odot$. We know of no other source of pressure that will stop the collapse of the

Figure 12.9   A comparison of theory and experiment for the binary pulsar. The plotted quantity keeps track of the energy lost in the orbit over the years that it has been observed. The prediction from general relativity includes gravitational radiation as the mechanism for energy loss.

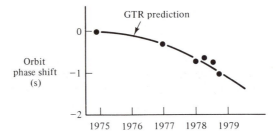

star. It will collapse right through the Schwarzschild radius for its mass, and will become a black hole. Black holes would be a normal state for the evolution of some stars.

How would we detect a stellar black hole? We obviously could not see it directly. We could not even see it in silhouette against a bright source, since the area blocked would be only a few kilometers across. We have to detect stellar black holes indirectly. We hope to see their gravitational effects on their surrounding environment. This is not a hopeless task, since we might expect to find a reasonable number in binary systems. By studying a binary with a suspected black hole, the problem would be to show the existence of a very small object (as inferred from small orbits) with a mass in excess of 3 $M_\odot$.

How do we find a candidate binary to study? In Section 12.3 we saw that neutron stars in binary systems can give rise to strong x-ray sources. The importance of the neutron star is that its radius is so small that infalling material acquires enough energy to give off x-rays. A stellar black hole would be smaller than a neutron star, so material falling in could also emit x-rays (before crossing the event horizon). We could start searching for stellar black holes by looking for irregular x-ray sources.

One interesting possibility is known as CygX-1 (Fig. 12.10), the brightest x-ray source in Cygnus. The Uhuru satellite showed this to have both short- and long-term variability. Until the Einstein observatory was launched in 1978 the positions of x-ray sources were not accurately determined. However, there is also a radio source associated with the x-ray source. We know that the x-ray and radio source are associated because they have the same pattern of variability. The position of the radio source is determined accurately from radio interferometry. Once the radio position was known, optical photographs were studied to see if there is an optical counterpart. This optical counterpart was found, and is shown in Fig. 12.11. It is the ninth-magnitude star HDE226868. A study of the star's spectrum shows that it is an 09.7Ib (blue supergiant) star. This places its mass at about 15 $M_\odot$.

This star is also a spectroscopic binary. Its orbital period is 5.6 days. The star also varies in brightness with this period. The amount of the variation is 0.07 mag. The source of the variation is thought to be a distortion in the shape of the blue supergiant due to the strong tidal effects of the unseen companion. The distortion results in the star appearing different sizes at different points in its orbit. From the amount of brightness variation, it is concluded that the inclination of the orbit is 30°.

We now look at what can be deduced about the companion. From the analysis of the Doppler shift data, the mass function (discussed in Chapter 5) can be found. Since we know the mass of the blue supergiant and have an estimate of the inclination angle of the orbit, we can derive the mass of the unseen companion from the mass function. The best estimate for this is 8 $M_\odot$. There is some uncertainty, but it seems very likely that the mass of the unseen companion is at least greater than 4 $M_\odot$.

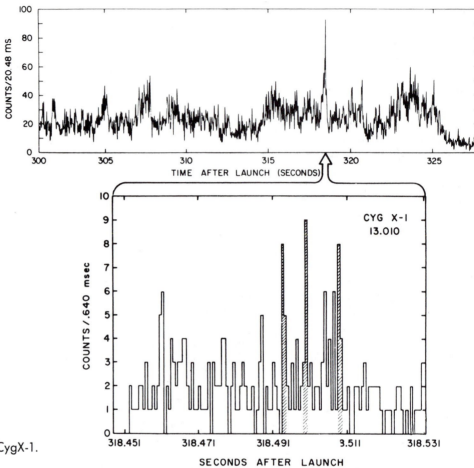

Figure 12.10 X-rays from CygX-1.

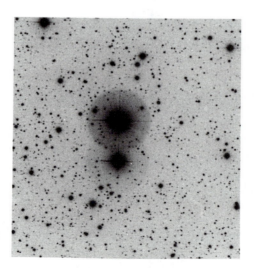

Figure 12.11 An optical photograph of HDE22868, the optical star identified with CygX-1.

The x-ray observations tell us that the companion must be very small. This is because the x-ray emission varies significantly in intensity on time scales of about 5 ms. This requires that the emitting region be less than 150 km in extent. Thus, we have an object that is definitely smaller than a normal star, but which has a mass greater than $4\ M_{\odot}$ (and probably closer to $8\ M_{\odot}$). It would therefore seem likely that the object is a black hole!

It is interesting that such an astonishing conclusion rests on the foundation of some "standard" observational techniques. These include using the spectrum of HDE226868 to find its mass, and the classical analysis of spectroscopic binary orbits, using the mass function.

There are a few other x-ray sources in which the evidence also points to a stellar black hole. A source in the Large Magellanic Cloud, one of our companion galaxies, is an even better black hole candidate than CygX-1. Though most astronomers probably accept the black hole explanation for CygX-1, the argument is not airtight. For example, our knowledge of the mass of the visible star comes from assuming that it is like any other 09.7Ib star. However, we know that it is in a close binary system, and is tidally distorted. We are only now beginning to understand stellar structure and evolution in close binary systems. It may be that the spectrum we classify as 09.7Ib is really produced by a star of a different mass. We are also not sure of the inclination of the orbit, and that enters into the mass function as $\sin^3 i$. It is also possible to avoid an $8\ M_{\odot}$ companion if we postulate the triple-star system, but there is no evidence to support this. (However, 10% of all stellar systems are triples, so it is not out of the question.)

It is amazing to contemplate how far astronomers have come when we can calmly say that the most likely explanation for some observed phenomenon is the presence of a black hole!

12.5 SS433

To get an idea of the fascinating range of phenomena encountered in close binary systems, we take a look at an object, best known by its designation in a particular catalogue, SS433. (It was catalogued long before its unusual nature was realized.) The object is a binary at the center of a supernova remnant. It is therefore not surprising that one of the members of the binary system is a neutron star or black hole. The period of the binary system is 13.087 days. The system is also a periodic x-ray source, as shown in Fig. 12.12.

Optical observations reveal absorption and emission lines with very large Doppler shifts. The required speeds are up to $0.26\ c$. At any time, both blue-shifted and red-shifted components are present. The magnitude of the Doppler shifts goes through a 164-day cycle, as shown in Fig. 12.12. There is an interesting asymmetry in the Doppler shifts. The red shift is always larger than the blue shift. The maximum red shift corresponds to about 50,000

Figure 12.12 Doppler shift data for SS433. (a) The vertical axis shows the red shift $z \equiv \Delta\lambda/\lambda$, relative to the average velocity of the system. The horizontal axis is in days. The dots represent the actual measurements. The smooth curves are fits of models to the data. (b) Actual spectra of SS433 at three different times showing the shifts in the velocities of the Hα emission. (The absorption features marked "$\oplus$" are from the Earth's atmosphere.)

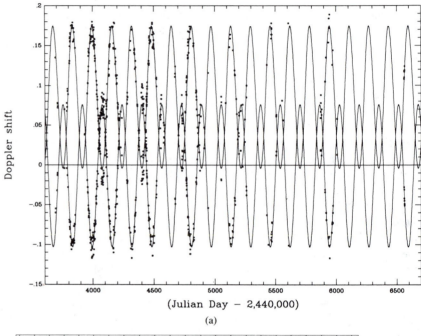

(a)

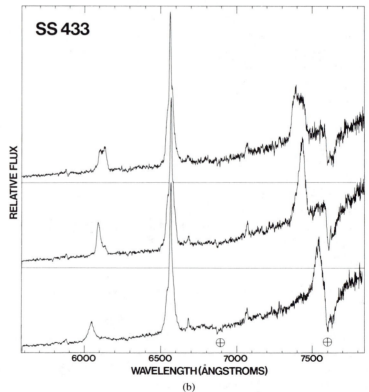

(b)

km/s, and the maximum red shift corresponds to about 30,000 km/s. If we take the average of the red-shifted velocity and blue-shifted velocity at any instant (remembering the blue shift corresponds to a negative radial velocity) we get a fairly constant value of about 12,000 km/s.

The basic model to explain this behavior is shown in Fig. 12.13. It involves a binary system in which one member is either a black hole or neutron star. The period of the binary is 13.087 days. The compact object is a source of two jets, moving in opposite directions. The jets are emitted in a cone with a half-angle of about 20°. The cone is inclined by 79° to the line of sight. The compact object precesses with a period of 164 days. (By "precession," we mean the changing of the orientation of the axis of rotation caused by an external torque.) This causes the projected angle of the jets to go through a 164-day cycle, giving the variations in the Doppler shifts.

In this model there is a natural explanation for the 12,000 km/s offset in the Doppler shifts. It comes from a transverse Doppler shift, arising from the fact that the jet is moving at an appreciable fraction of the speed of light. (We discussed the transverse Doppler effect in Chapter 7; it is essentially a time dilation effect.) At $v/c = 0.26$, $\gamma = 1.036$. According to equation (7.5) this produces a wavelength which is 1.036 times the rest wavelength. If we interpret this shift as arising from a normal Doppler shift, it corresponds to $v/c = 0.036$ or about 11,000 km/s. Since this is just a time dilation effect, this gets added in as a constant to any kinematic Doppler shift. Since it always increases the wavelengths, it makes the red shifts larger and the blue shifts smaller.

In this general model the x-rays come from material falling onto an accretion disk around the compact object. The binary is an eclipsing binary, so we can estimate the masses of the components. The best estimate for the mass of the companion is about $4\,M_{\odot}$. This would make it a black hole. However, there are enough uncertainties in the estimate to allow it to be either a black hole or neutron star. Theoreticians are still working on the mechanism for the collimation of the jets, and for generating the energy to get the jets moving so fast.

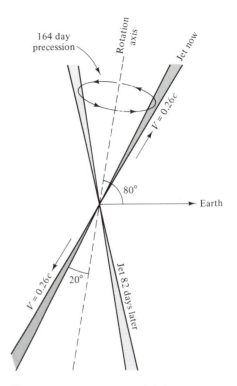

Figure 12.13 A model for SS433, showing the twin jet geometry.

CHAPTER SUMMARY

We saw in this chapter how the evolution of stars is altered by placing them in close binary systems. The changes in evolution arise mainly from strong tidal effects and from mass transfer.

If the component receiving mass is a compact object—white dwarf, neutron star, or black hole—infalling material can acquire enough energy to emit x-rays as it falls in to form an accretion disk.

We saw how mass transfer onto white dwarfs can account for novae. If the mass transfer is large enough, it can account for type I supernovae.

We saw how mass transfer onto neutron stars can produce strong x-ray emission, such as that from HZ Her. We also saw how studies of a binary pulsar have provided evidence for gravitational radiation.

We also saw how studies of CygX-1 have provided us with evidence for the existence of a stellar black hole.

QUESTIONS AND PROBLEMS

12.1   Show that, for a fixed total angular momentum, the "synchronized" spins situation is the lowest-energy state for an orbiting system. (*Hint*: Consider the sum of the orbital and rotational energies.)

12.2   A radial force F_r is related to a potential $V(r)$ by

$$F_r = -dV/dr$$

Show that the "centrifugal" force can be derived from the term $L^2/2mr^2$ in the effective potential.

12.3   Explain why surfaces of constant effective potential must also be surfaces of constant density.

12.4   Why is the binary pulsar so important?

12.5   Suppose we have a pulsar in orbit around another object. The pulse period, as emitted by the pulsar, is P_0. The orbit is circular with a speed v. We are observing in the plane of the orbit. Find an expression for the observed pulse period as a function of the position of the pulsar in its orbit. (*Hint*: Consider the advance and delay in the arrival times of pulses as the pulsar moves toward us and away from us.)

12.6   For the rate of change in the orbital period for the binary pulsar given in the chapter, what is the fractional change in the orbital energy $\Delta E/E$ per second?

12.7   Suppose we have a gravitationally bound system in virial equilibrium. Show that if more than half of the mass is lost with no change in the velocities of the remaining material, then the system will be unbound.

12.8   (a) Consider a pulsar in an eclipsing binary system. The pulsar is in a circular orbit of radius r and period T. The pulsar's period is P. Show that the observed pulse period will vary with the position in the orbit and find an expression for $P(t)$. Show that the expression is just like a Doppler shift. (b) For the system HZ Her, how large a shift in the pulsar period is oberved if the sum of the masses is $5.0 \, M_\odot$ and the orbital period is 1.7 days? The pulsar period is 1.24 s.

12.9   For HZ Her, assuming that material falls in from far away at the rate given, how close must it get to the star to provide the given x-ray luminosity?

12.10   Explain the steps leading to the conclusion that CygX-1 is a black hole? Which are the most suspect?

12.11   Derive equation (12.1a).

Clusters of Stars

When we look at the spatial distribution of stars in our galaxy, we find that most of the light is concentrated in a thin disk. We are inside this disk, so we see it as a band of light on the sky, called the Milky Way. We will discuss this further in Part III, but we will see in this chapter that location of stars in the galaxy can tell us something about those stars. In particular, some stars are confined to the thin disk of the Milky Way, while others form a more spherical distribution. In this chapter we will discuss groupings of stars, called clusters, and see how they vary in size, content, and galactic distribution.

13.1 TYPES OF CLUSTERS

We distinguish between two types of star clusters—galactic clusters and globular clusters.

Galactic clusters are named for their confinement to the galactic disk. A familiar galactic cluster, the Pleiades, is shown in Fig. 13.1. Note the open appearance in which individual stars can be seen. Because of this appearance, galactic clusters are sometimes called *open clusters*. Galactic clusters typically contain $\leq 10^3$ stars, and are less than ~ 10 pc across. In the photograph we see some starlight reflected by interstellar dust. Galactic clusters are sometimes associated with interstellar gas and dust.

Globular clusters are named for their compact spherical appearance (Fig. 13.2). They have 10^4–10^6 stars, and are 20 to 100 pc across. They seem to have no associated gas or dust. Globular clusters are not confined to the disk of the galaxy. Harlow Shapley used RR Lyrae stars and Cepheids to find the distances to globular clusters. This placed the globular clusters in three dimensions. It was found that the globular clusters form a spherical distribution with the Sun being about 10 kpc from the center. (This is still one of the best techniques for finding the distance to the galactic center.)

Before we look at the properties of clusters themselves, we will look at an important technique for determining distances to relatively nearby galactic clusters.

Figure 13.1 Two open clusters. (a) The Pleiades (M45), in Taurus. The nebulosity seen here is starlight reflected from interstellar dust. (b) NGC 457, in Cassiopeia.

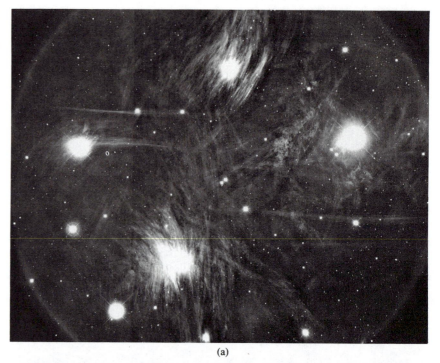

(a)

(b)

Figure 13.2   The globular cluster, M92.

13.2   DISTANCES TO MOVING CLUSTERS

Let's assume that we have a star (or cluster) moving through space with a velocity **v**. The velocity makes an angle A with the line of sight. We can break the velocity into components parallel to the line of sight and perpendicular to the line of sight. The component parallel to the line of sight is the radial velocity v_r and is responsible for the Doppler shift we observe. The component perpendicular to the line of sight is the *transverse velocity* v_T. It is responsible for the motion of the star across the sky, called the *proper motion*.

From the right triangle in Fig. 13.3, we can see that these quantities are related by

$$v^2 = v_r^2 + v_T^2 \tag{13.1}$$

$$v_r = v \cos A \tag{13.2}$$

$$v_T = v \sin A \tag{13.3}$$

$$\tan A = v_T/v_r \tag{13.4}$$

Figure 13.3   Space velocity. The velocity of the star is **v**, which makes an angle A with the line of sight. The radial and tangential components of the velocity are v_r and v_T, respectively.

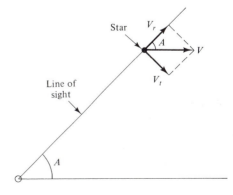

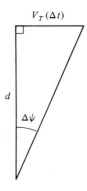

Figure 13.4 The dependance of the proper motion on distance d and tangential speed v_T. We follow the motion of the star for a time Δt.

The proper motion μ, expressed in radians per second, is just the transverse velocity, divided by the distance to the star d:

$$\mu \text{ (rad/s)} = v_T \text{ (km/s)}/d \text{ (km)} \qquad (13.5)$$

The greater the transverse velocity, the faster the star will appear to move across the sky. Also, the closer the star is, the greater is the motion across the sky. In general, proper motions are very small, amounting to a few arc seconds per year, or less. (The largest proper motion is 10.3 arc sec/yr for Barnard's star.) For this reason we would like to rewrite equation (13.5), expressing μ in arc seconds per year, v in kilometers per second, and d in parsecs. We can rewrite equation (13.5) as

$$v_T(\text{km/s}) = [\mu(\text{rad/s})d(\text{km})][2.063 \times 10^{5}{}''/\text{rad}][3.156 \times 10^7 \text{ s/yr}]$$
$$\times [3.086 \times 10^{13} \text{ km/pc}]$$

$$= 4.74 \; (''/\text{yr})d(\text{pc}) \qquad (13.6)$$

In general, we can measure the radial velocity (from the Doppler shift), and we can also measure the proper motion. If we know the transverse velocity, we can find the distance from

$$d = v_T/4.74\mu \qquad (13.7)$$

If, instead of v_T, we know A, then we can use equation (13.4) in equation (13.7) to give

$$d = v_T \tan A/4.74\mu \qquad (13.8)$$

With a cluster of stars, we can compare the proper motion with the rate at which the angular size of the cluster changes to find A. To see how this works, we consider the case of a cluster moving away from us (positive radial velocity). As the cluster moves farther away (Fig. 13.5), A gets smaller and

Figure 13.5 Convergent point. As the cluster gets farther away, the angle between **v** and the line of sight approaches zero. That is, $A_1 > A_2 > A_3$. As this happens, the cluster approaches one line of sight, the convergent point.

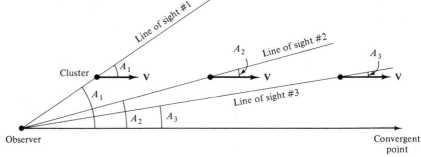

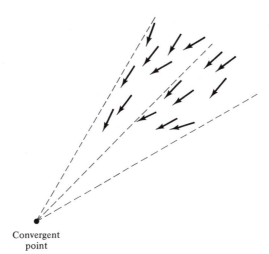

Convergent
point

Figure 13.6 Schematic representation of motions of stars in a cluster. Arrows represent proper motions.

approaches zero. As the cluster gets farther away, the proper motion approaches zero [by equation (13.5)]. Thus, the cluster appears to be heading toward a particular point in the sky. We call this point the *convergent point* of the cluster. We can see from the figure that the angle between the *current line of sight* and the line of sight to the convergent point is the *current value of A*. Similar reasoning applies to clusters coming toward us. They are moving away from their convergent point, so we find it by extrapolating their motion backward.

To apply these ideas, we take a series of photographs a number of years apart. From the proper motion, and change in angular size, we can find the convergent point, and therefore, we know A. We measure the radial velocity v_r and proper motion μ. Since we know A, v_r, and μ, we can find d from equation (13.8). The best determination of a distance to a moving cluster is for the Hyades (Fig. 13.6). This determination is an important cornerstone in our determination of distances to more distant objects in our galaxy and to other galaxies.

13.3   CLUSTERS AS DYNAMICAL ENTITIES

In this section we look at the internal dynamics of star clusters. If the gravitational forces between the stars are sufficient to keep the cluster together, we say that the cluster is gravitationally bound. However, gravity does more than assure the overall existence of the cluster. As stars move around within the cluster, pairs of stars will pass near each other. The gravitational attraction between the pair will not be enough to pull them together into a binary star, but it will alter the motion of each star. The momentum and energy of each star will change in this gravitational encounter. Thus, these encounters alter the distribution of speeds, the number of stars traveling at a given speed. If there has been sufficient time for many encounters to occur, the distribution of speeds will reach some equilibrium. For every star that suffers a collision, changing its speed from v_1 to v_2, there is another collision in which some other star has its speed changed from speed v_2 to v_1. (We refer to these gravitational encounters as collisions, even though the stars never actually get close enough for their surfaces to touch.) When a cluster has reached this state, we say that it is dynamically *relaxed*.

13.3.1 Energies

In a gravitationally bound, relaxed system, the various forms of energy (e.g., gravitational potential, kinetic) are related by the virial theorem, discussed in Section 8.2. The virial theorem tells us that

$$0 = 2\langle K \rangle + \langle U \rangle$$

where U is the gravitational potential energy, and K is the kinetic energy. The $\langle\ \rangle$ denote time averages. In Section 8.1, we saw that the gravitational potential energy of a constant density sphere of mass M and radius R is

$$U = -\tfrac{3}{5}GM^2/R$$

EXAMPLE 13.1   Potential Energy for a Globular Cluster

Find the gravitational potential energy for a spherical cluster of stars with 10^6 stars each of $0.5\ M_\odot$. The radius of the cluster core is 5 pc.

Solution

We use above equation to give

$$U = -\frac{(0.6)(6.67 \times 10^{-8}\ \text{dyn-cm}^2/\text{g}^2)[(0.5)(2 \times 10^{33}\ \text{g})(10^6)]^2}{(5\ \text{pc})(3.18 \times 10^{18}\ \text{cm/pc})}$$

$$= 2.5 \times 10^{51}\ \text{erg}$$

We now look at the kinetic energy. When we considered the collapse of interstellar clouds to form stars, the kinetic energy was in the random thermal motions of the individual atoms and molecules, and could be characterized by a temperature. In a cluster of stars, the kinetic energy is in the random motions of the stars. If the cluster has N stars, each of mass m, the kinetic energy is

$$K = \tfrac{1}{2} \sum_{i=1}^{N} m v_i^2$$

$$= \tfrac{1}{2} m \sum_{i=1}^{N} v_i^2$$

If N is the total number of particles, then $Nm = M$, the total mass; so

$$K = \tfrac{1}{2}(mN)(1/N) \sum_{i=1}^{N} v_i^2 \tag{13.9}$$

If we take the sum of N quantities and then divide by N, the result is the average of that quantity. Therefore, $(1/N) \Sigma\, v_i^2$ is the average of the quantity v^2. We write this average as $\langle v^2 \rangle$. Remembering that $mN = M$, equation (13.9) becomes

$$K = \tfrac{1}{2}M\langle v^2 \rangle \tag{13.10}$$

If we put this and the potential energy into the virial theorem, we find

$$M\langle v^2 \rangle = \tfrac{3}{5}GM^2/R \tag{13.11}$$

Dividing both sides by M gives

$$\langle v^2 \rangle = \tfrac{3}{5}GM/R \tag{13.12}$$

The quantity $\langle v^2 \rangle$ is the mean (average) of the square of the velocity. If we take the square root of this quantity, we have the *root mean square velocity*, or *rms velocity*. It is a measure of the internal motions in the cluster.

EXAMPLE 13.2   RMS Velocity in a Cluster

Find the rms velocity for the cluster used in the previous example.

Solution

We use equation (6.46) with the given quantities

$$\langle v^2 \rangle = \frac{(0.6)(6.67 \times 10^{-8} \text{ dyn-cm}^2/\text{g}^2)(10^6)(0.5)(2 \times 10^{33} \text{ g})}{(5 \text{ pc})(3.18 \times 10^{18} \text{ cm/pc})}$$

$$= 2.5 \times 10^{12} \text{ (cm/s)}^2$$

Taking the square root gives

$$v_{\text{rms}} = 1.6 \times 10^6 \text{ cm/s}$$

$$= 16 \text{ km/s}$$

We can relate the gravitational potential energy to the *escape velocity* v_e, the speed with which an object must be launched from the surface to escape permanently from the cluster. Consider a particle of mass m, moving outward from the surface at speed v_e. If the object escapes, it must get so far away that the potential energy is essentially zero. Since the kinetic energy is always greater than or equal to zero, the total energy of the object far away must be greater than or equal to zero. Since the total energy is conserved, the total energy for an escaping object must be zero or positive when it is launched. The kinetic energy of the particle is

$$\text{KE} = \tfrac{1}{2}mv_e^2 \tag{13.13}$$

Since it is at the surface of the sphere of mass M and radius R, its potential energy is just

$$\text{PE} = -GmM/R \tag{13.14}$$

For the total energy to be zero (the condition that the particle barely escapes), $\text{KE} = -\text{PE}$, giving

$$v^2 = 2GM/R \tag{13.15}$$

Note that the escape velocity is approximately twice the rms speed. For a gravitationally bound system, we would expect $v_e > v_{\text{rms}}$. If it were the other way around, many particles would have speeds greater than the escape velocity and would escape. The cluster would not be gravitationally bound.

13.3.2   Cluster Dynamics

In any given cluster, stars will be in orbits about the center of mass. Pairs of stars can exchange energy via gravitational encounters. As we have said, if there are enough collisions, an equilibrium velocity distribution will be reached. That distribution will be similar to the velocity distribution of the molecules of the gas in a room. Not all are moving at the average speed. Some move faster and others move slower. In a cluster, there will be some stars with speeds greater than v_e. They will escape. This alters the velocity distribution by removing the highest-velocity stars. The remaining stars must adjust, reestablishing the equilibrium velocity distribution. (This is equivalent to the evaporation of a puddle of water on the Earth. The highest-speed molecules escape, leaving the water a little cooler.)

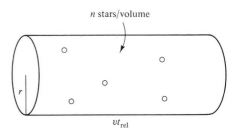

Figure 13.7   Calculation of relaxation time.

We call the time it takes to reestablish equilibrium the *relaxation time*. We can estimate the relaxation time t_{rel} by following a single star as it moves through the cluster (Fig. 13.7). We assume that there are n stars per unit volume in the cluster. We would like to know how long our star will go between collisions with other stars. That depends, in part, on how we define a collision. We would like to define a distance r and say that if two stars pass within this distance we will count it as a collision. We define r so that the potential energy of the stars is equal in magnitude to the kinetic energy of our star. If our star is moving with speed v, this means that r is defined by

$$Gm^2/r = mv^2/2 \qquad (13.16)$$

We can think of our star as sweeping out a cylinder in a given time t_{rel}. The radius of the cylinder is r, and the length is vt_{rel}. Therefore, the volume swept out is $\pi r^2 vt_{rel}$. The number of stars in this volume is n multiplied by the volume. If we define t_{rel} so that it is the time for one collision, we have the condition

$$n(\pi r^2 vt_{rel}) = 1 \qquad (13.17)$$

Solving for t_{rel} gives

$$t_{rel} = 1/n\pi r^2 v \qquad (13.18)$$

Substituting for r from equation (13.16) gives

$$t_{rel} = v^3/4\pi G^2 m^2 n \qquad (13.19)$$

The number of stars per unit volume is simply

$$n = \frac{N}{(4\pi/3)R^3}$$

$$= \frac{M/m}{(4\pi/3)R^3} \qquad (13.20)$$

Substituting into equation (13.19) gives

$$t_{rel} = 3v^3R^3/G^2mM$$

$$= \frac{(R/v)[3v^4R^2(M/m)]}{G^2M^2}$$

(13.21)

Using the virial theorem to eliminate two factors of v^2, and ignoring numerical factors that are close to 1 (since this is an estimate), this simplifies to

$$t_{rel} \cong (R/v)(M/m)$$

$$= NR/v$$

(13.22)

Equation (13.22) is just an estimate in which the effects of a few close encounters dominate. However, for every close encounter that a star has, it has many more distant encounters. This is because the effective area for an encounter at distance r is proportional to r^2. The effect of any one distant encounter is small. However, because there are so many of them, they actually dominate the relaxation process and speed it up. A more detailed calculation shows that the effect of many distant encounters is to reduce t_{rel} by a factor of $12 \ln(N/2)$.

EXAMPLE 13.3   Relaxation Time

Estimate the relaxation time for the cluster discussed in the two previous examples.

Solution

We use equation (13.22). For the speed v we use the calculated v_{rms}:

$$t_{rel} = \frac{(10^6)(5 \text{ pc})(3.18 \times 10^{18} \text{ cm/pc})}{1.6 \times 10^6 \text{ cm/s}}$$

$$= 1.0 \times 10^{19} \text{ s}$$

$$= 3 \times 10^{11} \text{ yr}$$

If we apply the correction for distant encounters, $12 \ln(N/2) = 1.6 \times 10^2$, t_{rel} is reduced to 2×10^9 yr.

We also can define an evaporation time, which is the time for a significant number of stars to leave the cluster. The evaporation time is approximately 100 times the relaxation time.

Once relaxation takes place, mostly by a large number of distant encounters, the cluster evolution is dominated by two-body collisions. This means that the velocity distribution will evolve toward a Maxwell–Boltzmann distribution. Calculations show that a core denser than the outer parts of the cluster will develop. It has been speculated that, at some point, the core can become so massive that it collapses to form a large black hole. Recent observations of some globular clusters have revealed the existence of a luminous extended object at the center. The sizes of these objects are in the 0.1 pc range. In each case the object is bluer than the rest of the cluster, meaning that it is not an unresolved group of red stars.

13.4   *H–R* DIAGRAMS FOR CLUSTERS

By studying the *H–R* digram for a cluster, we are studying a group of stars with a common distance. We can study their relative properties without knowing what their actual distance is. If we do know the distance to the cluster, we plot directly the absolute magnitudes on the *H–R* diagram. If we don't know the distance, we plot the diagram in apparent magnitudes. We then see how many magnitudes we would have to shift the diagram up or down to get the right absolute magnitudes for each spectral type. The amount of the shift gives us the distance modulus for the cluster, and therefore the distance. This procedure is known as *main sequence fitting*. It is like doing a spectroscopic parallax measurement, but it uses the information from all of the main sequence stars in the cluster. This is more accurate than studying a single star.

The *H–R* diagram for a group of galactic clusters is shown schematically in Fig. 13.8. Note that the lower (cooler or later) part of the main sequence is the same for all of the clusters shown. For each cluster, there is some point at which the main sequence stops. Beyond that point, no hotter stars are seen on the main sequence. The hotter stars all appear to be above the main sequence. The point at which this happens for a given cluster is called the *turn-off point*. Stars of earlier spectral type (hotter than) the turn-off point appear above the main sequence, meaning that they are more luminous, and therefore larger than main sequence stars of the same spectral type. Each cluster has its turn-off point at a different spectral type.

We interpret this behavior as representing stellar aging, in which stars use up their basic fuel supply, as described in Chapter 10. Hotter, more massive stars evolve faster than the cooler, lower-mass stars, and leave the main sequence sooner. We assume that the stars in a cluster were formed at

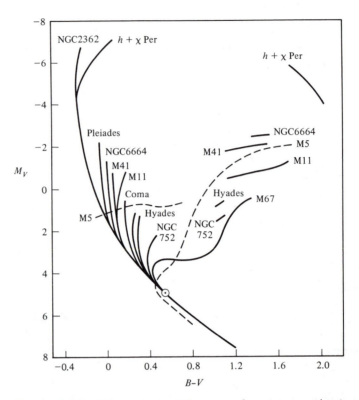

Figure 13.8 Schematic *H–R* diagrams for various galactic clusters.

approximately the same time. As a cluster ages, later and later spectral types evolve away from the main sequence. This means that the turn-off point shifts to later spectral types as the cluster ages. We can tell the relative age of two clusters by comparing their turn-off points. If we know how long different spectral type stars actually stay on the main sequence, we can tell the absolute age of a cluster from its turn-off point.

We note that there are some galactic clusters that are missing the lower (cooler) end of the main sequence. We think that these clusters are very young. The lower-mass stars are still in the process of collapse, and have not yet reached the main sequence. (We think that lower-mass stars take longer to collapse than higher-mass stars.)

Representative *H–R* diagrams for globular clusters are shown in Fig. 13.9. These appear to be different from the galactic cluster *H–R* diagrams. For globular clusters, only the lower (cooler) part of the main sequence is present. All earlier spectral types have turned off the main sequence. This tells us that globular clusters must be very old. Globular clusters contain a large number of red giants. In Chapter 10 we saw that the red giant state is symptomatic of old age in a star.

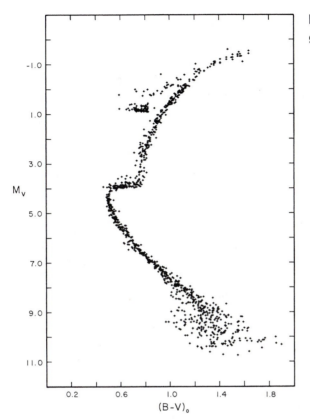

Figure 13.9 *H–R* diagram for the globular cluster, M3.

13.5 THE CONCEPT OF POPULATIONS

There is another important difference between the stars in galactic and globular clusters. It concerns the abundances of "metals," elements heavier than hydrogen and helium. Many globular clusters have stars with very low metal abundances, while galactic cluster stars are higher in metal abundance. We refer to high-metal stars as *population I* stars and low-metal stars as *population II* stars. We have the general sense that population I stars represent younger, more recently formed stars. We interpret the metal abundance differences as reflecting the conditions in our galaxy at the time each type of star was formed. When the older stars were formed, our galaxy had only hydrogen and helium. When the newer stars were formed, the galaxy had been enriched in the metals. This enriched material comes from nuclear processing in stars, followed by spreading into the interstellar medium, especially through supernova explosions.

The differences between galactic and globular clusters start us thinking about old and new material in our galaxy. The globular clusters are older, and form a spherical distribution, while the galactic clusters are confined to the galactic disk. This suggests that a long time ago, star formation took place in a large spherical volume, but now it only takes place in the disk. This is supported by the fact that globular clusters are free of interstellar gas and dust, the material out of which new stars can form, while galactic clusters are sometimes associated with gas and dust.

The concept of stellar populations is important in our understanding of the evolution of our galaxy. This will be discussed further in Chapters 14, 15, and 16.

CHAPTER SUMMARY

In this chapter we saw what could be learned from studying clusters of stars. We developed the idea of stellar populations, signifying old and new material in the galaxy. Globular clusters seem associated with the old material, and galactic (open) clusters seem associated with the newer material. Some of these differences are very evident in comparing H–R diagrams, as well as in comparing the metal content of the stars in the two types of clusters.

We looked at an important technique for determining the distances to nearby clusters. It is free of any assumptions about luminosities, and serves as an important cornerstone in our distance determination scheme.

We looked at the dynamical properties of clusters. Galactic clusters evaporate on a short time scale, while globular clusters are still around from the time the galaxy formed.

QUESTIONS AND PROBLEMS

13.1   The Hyades has a proper motion of 0.07 arc sec/yr and appears 26° from its convergent point. The radial velocity is 35 km/s. (a) How far away is the cluster? (b) What is the actual speed of the cluster?

13.2   Suppose we can detect proper motions down to 0.1″/yr. How far away can we detect a transverse velocity of 10 km/s?

13.3   Suppose you had two photographs of a cluster taken ten years apart. How would you use those photographs to find the convergent point of the cluster?

13.4   Suppose we have two photographs of a cluster, taken 10 years apart. In the second photograph the cluster has moved over by 2.0″. Two stars that were originally 20.0″ apart are now 19.5″ apart. What is the angle between the current line of sight to the cluster and the line of sight to the convergent point?

13.5   List the distance measurement techniques that we have encountered so far in this book, and estimate the distance range over which each is useful.

13.6   In Chapter 4 we discussed the resolving power of gratings. How is the minimum radial velocity shift we can measure related to this resolving power?

13.7   What is the relationship between main sequence fitting and spectroscopic parallax?

13.8   What is the significance of the main sequence turn-off point in a cluster?

13.9   Make a table, contrasting the properties of globular clusters and galactic clusters.

13.10   For the globular cluster treated in the examples in this chapter, estimate the average time between collisions in which the stars actually hit.

13.11 Compare the rms speeds in a typical globular cluster with those in a typical galactic cluster.

13.12 If a cluster is moving through space at 100 km/s, how should this motion be included in $\langle v^2 \rangle$, which appears in equation (13.12)?

13.13 In a cluster for which $v_e > v_{rms}$, some stars can still escape. How does this happen?

13.14 Verify that equation (13.22) can be obtained, as outlined, from equation (13.21).

13.15 The *crossing time* for a cluster is the average time for a star to get from one side of a cluster to the other. (a) What is the relationship between the crossing time and the relaxation time? (b) How would you explain that relationship?

13.16 Compare the evaporation times for the typical galactic and globular clusters discussed in this chapter. Include the correction for the effects of distant collisions.

PART III

THE MILKY WAY GALAXY

Most of the light we can see from our galaxy appears as a narrow band around the sky. From this appearance, we think that we are in the plane of a disk, and that this disk looks something like the Andromeda galaxy. However, our location within our own galaxy makes its structure very difficult to study. In this part we will see both how we learn about our galaxy and what we have learned about it so far.

Most of the light we see comes directly from stars. Among all of the objects we can see the stars provide most of the mass. Averaged over the whole galaxy, the gas and dust between the stars—the interstellar medium—contains only about 1% as much mass as the stars themselves. Of the interstellar medium, 99% of the mass is in the form of gas, and 1% is in the form of dust. However, this small amount of dust is very efficient at blocking light, making optical observations of distant objects difficult.

We expect that stars form out of interstellar material. Since most of the mass of the interstellar material is in the form of gas, it is the gas that will provide the gravitational attraction for the star formation process. In this part we will first look at the contents of the interstellar medium. We will then look at how stars are born. Finally, we will see how the stars and interstellar medium are arranged in the galaxy as a whole.

Contents of the Interstellar Medium

14.1 OVERVIEW

When we look at photographs of the Milky Way (see p. 355), we note large regions where no light is seen. We think that these are due to dust blocking the light between us and the stars. We can see the same effect on a smaller scale (Fig. 14.1). Note that there is a high density of stars near the edges of the photograph. As one gets closer to the center, the density of stars declines sharply. Near the center, no stars can be seen. This apparent hole in the distribution of stars is really caused by a small dust cloud, called a *globule*. The more dust there is in the globule, the fewer background stars we can see through the globule. We can use photographs like this to trace out the interstellar dust. We find that it is not uniformly distributed. Rather, it is mostly confined to concentrations, or interstellar clouds.

We detect the presence of the gas by observing absorption or emission lines from the gas. By tracing these lines, we find that the gas also has an irregular distribution. Often the gas appears along the same lines of sight as dust clouds. From this apparent coincidence we get the idea that the gas and dust are generally well-mixed together, with the gas having about 99% of the mass in a given cloud. In this chapter we will see how the masses of different types of clouds are determined.

One of the reasons the interstellar medium is so interesting is that it is the birthplace of stars. How do we know that stars are still forming in our galaxy? We have seen that stars are dying, and we know that there is still a large number of stars in the galaxy. We therefore presume that stars are being created at a rate that approximately offsets the rate at which they are dying. This is not an airtight argument, because it could be that many stars were formed early in the history of the galaxy and that we are just seeing the ones that haven't died yet. However, we know that an O star lives only about 10^7 years or less on the main sequence. Since we see O stars today, there must

Figure 14.1 A photograph of a globule. The globule is in the region with the fewest number of stars per unit area. The dust in the globule is blocking the light from the background stars.

have been O star formation in the last 10^7 years. We think that the galaxy is ten billion years old. Compared to this, 10^7 years ago is almost like yesterday. If the conditions were right for star formation in the last 10^7 years, they must be right for star formation now. The actual star formation process will be discussed in Chapter 15.

14.2   INTERSTELLAR EXTINCTION

If we want to see direct emission from the dust, we have to look in the infrared, as we will discuss in the next section. In the visible part of the spectrum the dust is generally evidenced by its blocking of starlight. The blocking arises from two processes, scattering and absorption. In scattering, the incoming photon is not destroyed, but its direction is changed. In absorption, the incoming photon is destroyed, with its energy remaining in the dust grain. The combined effects of scattering and absorption are called interstellar *extinction*. In Fig. 14.2 these two processes are depicted schematically. A dark nebula, in which background light is being blocked, and a *reflection nebula*, in which scattered starlight is being sent in our direction, are depicted in Color Plate 1.

14.2.1   The Effect of Extinction

We measure interstellar extinction as the number of magnitudes by which a cloud dims starlight passing through it. For example, if a particular star would have an apparent magnitude m without extinction, but its light passes through

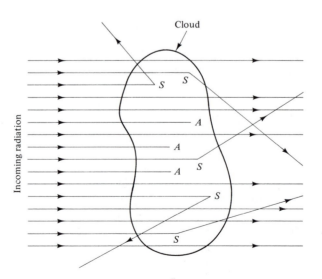

Figure 14.2 Scattering and absorption. Light is incident from the left. Light rays that are absorbed stop inside the cloud. The points of absorption are indicated by A. Light rays that are scattered change direction. The points of scattering are indicated by S.

a cloud with A magnitudes of extinction, then the star will be observed to have a magnitude $m' = m + A$. (Remember, extinction dims the starlight, so the magnitude increases.)

We can relate the extinction, in magnitudes, to the optical depth τ of the dust. This is useful, since it is the extinction in magnitudes that will be directly measurable, but it is the optical depth that is directly related to the dust properties. If we have light of incident intensity I_0 passing through the cloud of optical depth τ, and intensity I emerges, then these are related [as we saw in equation (16.18)] by

$$I = I_0 \exp(-\tau) \tag{14.1}$$

From the definition of extinction and the magnitude scale

$$A = m' - m$$
$$= 2.5 \log_{10}(I/I_0) \tag{14.2}$$

Using equation (14.1) this becomes

$$A = 2.5 \log_{10}[\exp(\tau)]$$
$$= 2.5\tau \log_{10}(e)$$
$$= (2.5)(0.4343)\tau \tag{14.3}$$
$$= (1.086)\tau$$

This means that one magnitude of extinction corresponds approximately to an optical depth of one.

If we have a star of known distance and spectral type, we can determine the extinction between the star and us. The spectral type gives us the absolute magnitude M. We can measure the apparent magnitude m. In the presence of A magnitudes of extinction, the star will appear A magnitudes fainter than without the extinction, so

$$m = M + 5 \log(r/10 \text{ pc}) + A \tag{14.4}$$

Since we know m, M, and r, we can find A. Obviously, the presence of interstellar extinction will affect distance measurements by spectroscopic parallax. If we don't correct for extinction, then a star will appear to be farther away than it actually is. You can see that if both r and A are unknown, then equation (14.4) only gives us one equation with two unknowns. We will see below that there is a way of getting additional information by observing at different wavelengths.

EXAMPLE 14.1 Interstellar Extinction

Suppose we observe a B5 ($M = -0.9$) star to have an apparent magnitude of 9.2. The star is in a cluster whose distance is known to be 400 pc. What is the extinction between us and the star?

Solution

We solve equation (14.4) for A to give

$$A = m - M - 5 \log(r/10 \text{ pc})$$

$$= 9.2 + 0.9 - 5 \log(40)$$

$$= 2.1 \text{ mag}$$

14.2.2 Star Counting

If we record an image of a field which has some interstellar extinction, fewer stars will appear than if the extinction were not present. This is because the light from each star is dimmed by the extinction. Some stars that would have appeared on the image if there were no extinction now are too dim to appear with extinction. We can estimate the extinction in a cloud by comparing the number of stars we can see through the cloud with the number we can see in an unobscured region of the same size. Suppose a photograph is exposed to a threshold magnitude m_0. All stars whose apparent magnitude is less than m_0 (that is, stars that are brighter than m_0) will appear on the photograph. If the light from each star is dimmed by A magnitudes, only stars that have undimmed magnitudes of $m_0 - A$ will appear.

There are two ways of applying this idea. In one we measure the number of stars per unit area in each magnitude range. In the more common way we measure just the total number of stars per unit area. We define the function $N'(m)$ such that $N'(m) \, dm$ is the number of stars per unit area with magnitudes between m and $m + dm$ in the absence of extinction. We measure $N'(m)$ for a region we think is partially obscured by dust and for a nearby region we think is unobscured. If we graph these two quantities, as shown in Fig. 14.3, we see that the two curves look like each other, except that one is shifted by a certain number of magnitudes. The amount of the shift is the extinction in the partially obscured region.

Often we don't have enough stars in each magnitude range to get a good measure of $N'(m)$. In that case we must use integrated star counts. We let $N(m)$ be the number of stars per unit area brighter than magnitude m. This is related to $N'(m)$ by

Figure 14.3 Effect of extinction of star counts. The plot gives the number of stars per magnitude interval, as a function of magnitude m. The distribution is such that when the vertical axis is logarithmic, the curve is close to a straight line. The effect of a cloud with A magnitudes of extinction is to make each star A magnitudes fainter, shifting the curve to the right by A magnitudes.

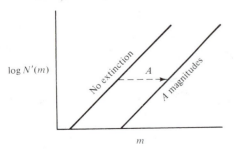

$$N(m) = \int_{-\infty}^{m} N'(m') \, dm' \qquad (14.5)$$

If a photographic plate has a limiting magnitude m_0, then the number of stars per unit area, without extinction, is

$$N(m_0) = \int_{-\infty}^{m_0} N'(m') \, dm' \qquad (14.6)$$

If we now look at a region with extinction A, only the stars that would have had magnitude $m_0 - A$ without extinction will show up. We therefore count

$$N(m_0 - A) = \int_{-\infty}^{m_0 - A} N'(m') \, dm' \qquad (14.7)$$

Therefore, if we know $N'(m)$, we can predict $N(m_0 - A)$ for various values of A. If we use plates with a limiting magnitude of 20, then we can generally get good star count data for A in the range 1–6 mag. For A much less than 1, the difference between an obscured and an unobscured region is hard to detect. For A much greater than 6, there are very few stars bright enough to shine through, and the obscured region will appear blank, a situation in which 6 mag is indistinguishable from 20 mag.

14.2.3 Reddening

When we measure interstellar extinction we find it is not the same at all wavelengths. In general, the shorter the wavelength is, the higher is the extinction. This means that blue light from a star is more efficiently blocked than red light. In the presence of extinction, the images of stars will therefore appear redder than normal. This is called *interstellar reddening*. (NOTE: You should not confuse reddening due to extinction with a red shift produced by the Doppler effect.)

Suppose we measure the magnitude of a star in two different wavelength ranges, say those corresponding to the B and V filters. Then, from equation (14.4) we have

$$m_V = M_V + 5 \log(r/10 \text{ pc}) + A_V \qquad (14.8a)$$

$$m_B = M_B + 5 \log(r/10 \text{ pc}) + A_B \qquad (14.8b)$$

If we take the difference $m_B - m_V$, the distance r drops out, giving

$$(m_B - m_V) = (M_B - M_V) + (A_B - A_V) \qquad (14.9)$$

In equation (14.9) the quantity on the left-hand side is directly observed. The first quantity on the right-hand side depends only on the spectral type of the star. It is simply the B $-$ V color of the star. We know it because we can observe the star's absorption line spectrum to determine its spectral type. Since the spectral type determination depends on the presence of certain spectral lines, it is not greatly influenced by interstellar extinction. We can therefore determine the quantity $A_B - A_V$.

Since both A_B and A_V are proportional to the total dust column density N_D, then their difference is also proportional to N_D. If we define a quantity,

$$R = A_V/(A_B - A_V) \tag{14.10}$$

it will not depend on N_D, since N_D appears in both numerator and denominator. We call this quantity the *ratio of total-to-selective extinction*. Extensive observational studies have shown that in almost all regions, R has a value very close to 3.1. (There are a few special regions where R is as high as 6.) This has a very important consequence. It means that if we can measure $A_B - A_V$, we need only multiply it by 3.1 to give A_V. We have already seen that the difference can be determined by knowing the spectral type of a star and measuring its B and V apparent magnitudes, and then using equation (14.9). Note that we have not made use of knowing the distance to the star r. We can still go back to equation (8.9a) and find the distance to the star. The method of spectroscopic parallax works even in the presence of interstellar extinction.

EXAMPLE 14.2   Spectroscopic Parallax with Extinction

Suppose we observe a B5 star ($M_V = -0.9$, B $-$ V $= 0.17$) to have $m_B = 11.0$ and $m_V = 10.0$. What is the visual extinction between us and the star, and how far away is the star?

Solution

From equation (14.9) we have

$$A_B - A_V = (m_B - m_V) - (M_B - M_V)$$

$$= 1.00 - 0.17$$

$$= 0.83$$

We can now use the ratio of total-to-selective extinction to convert this to A_V.

$$A_V = R(A_B - A_V)$$

$$= 2.6 \text{ mag}$$

We can now find the distance from

$$5 \log(r/10 \text{ pc}) = m_V - M_V - A_V$$

$$= 8.3$$

This gives

$$r = 460 \text{ pc}$$

14.2.4 Extinction Curves

If we study how interstellar extinction varies with wavelength, we can learn something about the properties of interstellar dust grains. We try to measure the $A(\lambda)$ in the directions of several stars, to see the degree to which the grain properties are the same in different directions. Since the dust column densities are different in various directions, we do not directly compare values of A. Instead, we divide by $A_B - A_V$ to get a quantity that is independent of the column density. It is conventional to plot the following function to represent *interstellar extinction curves*:

$$f(\lambda) = \frac{A(\lambda) - A_V}{A_B - A_V} \tag{14.11}$$

Some curves are shown in Fig. 14.4. One general feature is that in the visible part of the spectrum $f(\lambda)$ is roughly proportional to $1/\lambda$. In the ultraviolet

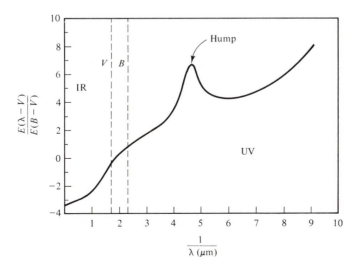

Figure 14.4 Average interstellar extinction curve. The vertical axis is just $[A(\lambda) - A(V)]/[A(B) - A(V)]$. This is plotted as a function of $1/\lambda$, since it is approximately linear in $1/\lambda$ near the visible. The wavelengths corresponding to the V and B filters are marked by vertical dashed lines.

there is a broad "hump" in the curve. The size of this hump varies from one line of sight to another. In the infrared there are absorption features of various strengths. We will see that these infrared features can tell us a lot about the grain composition.

14.2.5   Polarization

Sometimes the light we receive from celestial objects is polarized. We can detect this polarization by placing a polarized filter in front of our detector. Such a filter only passes radiation whose electric field vector is parallel to the polarization direction of the filter. As we rotate the filter, different polarizations of the incoming light are passed. If the incoming light is unpolarized, the amount of light coming through will not depend on the angle through which the filter has been rotated. If the incoming light is completely linearly polarized, there will be one position of the filter for which we see the image at full brightness and another position, 90° away, for which we see no image. If the incoming light is partially polarized, we will see a maximum brightness at one position and a minimum brightness when the filter is rotated by 90°. The greater the amount of polarization, the greater the contrast between the maximum and the minimum.

When unpolarized starlight passes through interstellar dust clouds, it can emerge with a slight degree of linear polarization. This means that the polarization must be caused by the dust itself. The amount of polarization is very small, only a few percent at most. We find that there is a weak wavelength dependence on the amount of polarization. We also find that the amount of polarization depends on the visual extinction A_V. When A_V is low, the polarization is low. When A_V is high, the polarization can be low or high. Dust is necessary for the polarization, but something else must be necessary: There must be a mechanism of aligning the nonspherical dust grains to produce the polarization. We will discuss this in the next section.

14.2.6   Scattering versus Absorption

We have said that extinction is the combined effect of scattering and absorption. The relative importance of the two effects depends on the physical properties of the grains and the wavelength of the incoming light. The fraction of the extinction that results from scattering is called the *albedo* a_λ of the dust grains. If a_λ is the fraction scattered, then $1 - a_\lambda$ must be the fraction absorbed.

The albedo is much harder to measure than the extinction. Studies of reflection nebulae are particularly useful, since they provide us with light that we know is reflected by the dust. It appears that the albedo is quite high, about 50–70%, at most wavelengths. (The albedo is lower in the range of the ultraviolet bump in the extinction curve, meaning that the bump is due to a strong absorption feature.) The high albedo means that a photon may be scattered a few times before it is actually absorbed.

If a photon is scattered by a dust grain, we would like to know the directions in which it is most likely to travel. Studies indicate that about half of the scattered photons move in almost the same direction as they were going when they struck the grain. The rest of the photons have almost an equal probability for being scattered in any direction.

14.3 PHYSICS OF DUST GRAINS

In this section we will see how we use observations and theory to deduce a number of grain properties. We will also see how grains interact with their environment. Some of the things we would like to know about interstellar dust are:

1. size and shape
2. alignment mechanism for polarization
3. composition
4. temperature
5. electric charge
6. formation and evolution

14.3.1 Size and Shape

We try to deduce grain sizes from the observed properties of the interstellar extinction curve, the variation of extinction with wavelength. If the grain size r is much greater than the wavelength λ, then we are close to the situation in which geometric optics applies. The wavelength is unimportant, and $A(\lambda)$ is roughly constant. If $r << \lambda$, the waves are too large to ''see'' the dust grains, and $A(\lambda)$ is very small (explaining the low extinction at long wavelengths). If r is comparable to λ, then diffraction effects in the scattering process are important. Hence, the wavelength dependence is strongest in this range.

We compare the observations with theoretical calculations of scattering and absorption by grains of different sizes and compositions to see which gives the best agreement. We find that interstellar grains are not all the same size. There is some spread about an average value (just as all people are not the same height, but have some spread of heights about some average value). In fact, the situation is even more complicated than that. Observations of extinction curves indicate that there are probably at least two different types of grains with distinctly different average sizes (just as men and women are

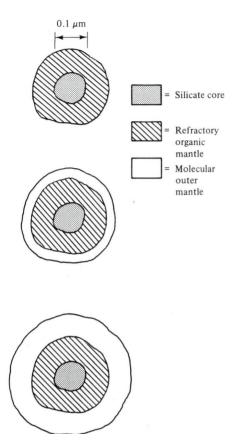

0.1 μm

□ = Silicate core

▨ = Refractory organic mantle

□ = Molecular outer mantle

Figure 14.5   Some models that have been proposed for grain compositions and sizes in different types of clouds.

different types of people with different average heights). The overall size distribution looks something like that shown in Fig. 14.5. The larger grains are responsible for the extinction in the visible and infrared, and the smaller grains are responsible for the extinction in the ultraviolet.

We can deduce something about the shapes of grains from their ability to polarize and unpolarized beams of light. If the grains were perfect spheres, there would be no preferred direction, and there would be no way of producing the polarization. Therefore, some significant fraction of the grains must either be elongated, like cigars, or flattened, like disks.

There must also be some mechanism for actually aligning the asymmetric grains. This mechanism probably involves the interstellar magnetic field. The grains are probably not ferromagnetic. This means that a collection of grains cannot be made into a permanent magnet. However, they might be paramagnetic. In a paramagnetic material the individual particles have magnetic moments. These can be aligned by a magnetic field. The tendency to align is offset by the random thermal motions of the grains. We think that a partial alignment arises from the combination of two effects: (1) the tendency for elongated grains, shaped like cigars, to rotate end-over-end rather than about the long axis, since less energy is required for the end-over-end motion; and (2) the tendency for the magnetic field to align with the rotation axes.

14.3.2   Composition

We can deduce the composition of the larger grains from infrared absorption features like those in Fig. 14.6. This is not as exact as using absorption spectra to tell us about the compositions of stellar atmospheres. Because the dust grains are solids, certain motions of atoms within the grains are inhibited by the close bonding to neighboring atoms, so the spectra consist of a few smeared-out features instead of many sharp spectral lines. We observe absorption features at 10 μm and 12 μm, which correspond to wavelengths for vibrational transition (stretching of bonds) in silicates (SiO and SiO_2) and water ice. Since silicates are an important component of normal dirt on the Earth, we sometimes talk of the grains as being ''dirty ice.'' However, the extinction in the ultraviolet (including the bump) cannot be explained by dirty ice. For that, carbon is probably needed. Therefore, interstellar grains are probably a combination of large dirty ice grains and small graphite grains. There are additional elements present in smaller quantities in the grains. We also think that it is possible for the grains to develop a thin layer of hydrogen on the surface.

14.3.3   Electric Charge

We can deduce the electric charge of grains from theoretical considerations. There are two ways for grains to acquire charge: (1) Charged particles (both positive and negative) from the gas can strike the grains and stick to the surface. (2) Photons striking the grain surface can eject electrons via the

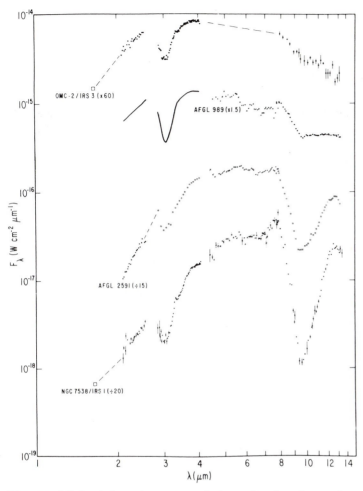

Figure 14.6 Infrared spectra of dust, showing the main spectral lines. The absorption spectra of some protostars.

photoelectric effect. The grain is left with one unit of positive charge for every electron ejected. In an equilibrium situation the net charge on the grains must be constant.

We first consider the situation in which the photoelectric effect is not important. This would be the case in regions of high extinction. For particles striking the grains the negative charges are mostly carried by electrons, and the positive charges are mostly carried by protons. At any given temperature, the average speed of the electrons, which have a lower mass, will be greater than that of the protons. Therefore, electrons will hit grains at a greater rate than protons or C^+. If the grains are initially neutral, this will tend to build up a negative charge. However, once the grains have a small negative charge, the electrons will be slowed down as they approach the grains, while the positive charges will be accelerated. Therefore, if the grains have a net negative charge, it is possible to have protons and electrons striking the grain at the same rate, keeping the charge on the grain constant. Note that, if the grains have a net negative charge, the gas must have a net positive charge if the interstellar medium as a whole is to be neutral.

EXAMPLE 14.3   Charge of Dust Grains

Estimate the electric charge (in multiples of e) required to keep the charge of a dust grain constant. Take the radius of the grain to be 10^{-5} cm and the gas kinetic temperature to be 100 K.

Solution

We estimate the grain charge for which the electric potential energy of an electron at the grain surface is equal in magnitude to the average kinetic energy of the electrons in the gas. If the net charge on the grain is $-Ne$, the potential energy for an electron at the surface, a distance r from the center, is

$$U = Ne^2/r$$

The average kinetic energy is $\frac{3}{2}kT$. Equating these and solving for N gives

$$N = \tfrac{3}{2}kTr/e^2$$

$$= \frac{(3)(1.38 \times 10^{-16} \text{ erg/K})(100 \text{ K})(10^{-5} \text{ cm})}{(4.8 \times 10^{-10} \text{ esu})^2}$$

$$= 0.9$$

This says that each grain should have a net charge of about $-e$. However, the actual charge is about a factor of 10 larger because we have only considered electrons of the average energy. Electrons moving faster than average contribute significantly to charge buildup. Also, the charge becomes more negative at higher temperatures.

If the photoelectric effect is dominant, the grains will have a positive charge. There must be a balance between the rate at which electrons are ejected and the rate at which they strike the grain. For positively charged grains, the electrons in the gas are attracted, meaning that the tendency for the electrons to strike the grains at a greater rate than the more positively charged particles is enhanced.

14.3.4 Temperature

The temperature of a large dust grain is determined by the fact that, on a time average, it must emit radiation at the same rate as it receives radiation. This keeps the temperature constant. The temperature of a dust grain will therefore depend on its environment. If it is very close to a star it will be hot. If it is far from any one star, it is cool, receiving energy only from the combined light of many distant stars.

Let's look at the case of a dust grain a distance d from a star whose radius and temperature are R_* and T_*. We will assume that the albedo is the same at all wavelengths. (If the albedo varies with wavelength, as in more realistic cases, the fraction of incoming radiation absorbed is different at different wavelengths, and the calculation is harder. See Problem 14.4.) The luminosity of the star is

$$L_* = 4\pi R_*^2 \sigma R_*^4 \tag{14.12}$$

The fraction of this power striking the grain is the projected area of the grain πr_g^2, divided by the area of a sphere of radius d. That is,

$$\text{Fraction striking grain} = \pi r_g^2/4\pi d^2 \tag{14.13}$$

If a is the albedo, then $1 - a$ is the fraction of incoming radiation absorbed by the grain. Therefore, the rate P at which energy is being absorbed by the dust grain is

$$
\begin{aligned}
P &= \frac{(1 - a)(4\pi R_*^2)(\sigma T_*^4)(\pi r_g^2)}{4\pi d^2} \\
&= \frac{(1 - a)\,\pi\sigma T_*^4 R_*^2 r_g^2}{d^2}
\end{aligned}
\tag{14.14}
$$

The quantity $\pi R_*^2/d^2$ is the solid angle subtended by the star as seen from the dust grain. We say that the star acts like a *dilute blackbody*. It has the spectrum of a blackbody at temperature T_*, but the intensity is down by a factor of (solid angle$/4\pi$).

We now look at the rate at which the grain radiates energy. Since it can only absorb $1 - a$ of the radiation striking it, it can only emit $1 - a$ of the radiation that a perfect blackbody would emit. (A perfect blackbody has an albedo of zero, by definition.) If the grain is at a temperature T_g, the power radiated is

$$P_{\text{rad}} = (1 - a)4\pi r_g^2 \sigma T_g^4 \tag{14.15}$$

Equating the power radiated and the power received, and solving for T, gives

$$T_g = T_*(R_*/2d)^{1/2} \qquad (14.16)$$

Note that the final result doesn't depend on the size of the grain or the albedo. That is because both enter into the emission and absorption processes. (This result is the same as that derived for a planet in Section 23.2.)

EXAMPLE 14.4   Temperature of a Dust Grain Near a Star

What is the temperature of a dust grain a distance 5000 stellar radii from a star whose temperature is 10^4 K?

Solution

Using equation (14.16), we have

$$T_g = (10^4 \text{ K}) \left[\frac{R_*}{(2)(5000)R_*} \right]^{1/2}$$

$$= 100 \text{ K}$$

When dust is sufficiently warm ($T_g > 20$ K), it is a good emitter in the infrared, and we can determine its temperature directly from infrared observations.

14.3.5 Evolution

We still know very little about the evolution of dust grains. The densities in interstellar clouds are probably too low for the grains to be formed directly where we see them. We think that most of the grains are formed in the envelopes of red giants undergoing mass loss. As material leaves the surface, it is hot enough to be gaseous. However, as it gets farther from the surface, it cools. When the temperature is low enough, about 1000–2000 K, many of the materials such as the silicates can no longer exist as a gas. They form small solid particles. These particles are blown into the interstellar medium either via the effects of stellar winds, or as part of a planetary nebula. This is another example of the cyclical processing of material between stars and the interstellar medium.

Once the grains are in clouds, they can collect particles from the gas and grow. There are some limits. For example, once a layer of molecular hydrogen (H_2) one molecule thick forms on the grains, no more hydrogen will stick. Grains can be destroyed or diminished in size by a number of processes. Sometimes molecules can simply *sublime* from the surface. (Sublimation is a phase change from the solid to the gas state.) Collisions with atoms in the gas can break up grains. Collisions between grains can also destroy the grains.

14.4 INTERSTELLAR GAS

14.4.1 Optical Studies

Early studies of cold interstellar gas utilized optical absorption lines. When light from a star passes through a cloud as shown in Fig. 14.7, some energy is removed at wavelengths corresponding to transitions in the atoms and molecules in the cloud. These studies revealed the existence of trace elements such as sodium and calcium. (These elements happen to have convenient spectral lines to study.) In addition to these atoms, some simple molecules were discovered: CH (in 1937), CN (in 1940), and CH^+ (in 1941). These simple molecules are not generally stable in the laboratory. CH^+ is charged and would combine with a negative ion or electron under laboratory conditions. CH and CN have an outer electronic shell with only one electron, making them chemically reactive. The presence of these unstable molecules in the interstellar gas suggests densities much lower than in the typical laboratory.

In these early studies no hydrogen absorption lines were observed. This is not because there is no hydrogen present. The temperatures in interstellar clouds are generally low, and most of the hydrogen is in the ground state. Therefore, the only H absorption lines that are possible are the Lyman lines in the ultraviolet. Now that ultraviolet observations are possible from satellites, astronomers can study these absorption lines.

You might wonder how we know that the absorption lines are coming from interstellar gas and not from the stars themselves. After all, we have already seen the large number of absorption lines present in stars. One distinguishing feature is that the interstellar lines are much narrower than the stellar absorption lines (Fig. 14.3). By narrower we mean they cover a smaller range of frequency (or wavelength). Interstellar lines are typically a few kilometers per second wide. If the Doppler broadening is produced by random thermal motions, this suggests a temperature of about 1000 K. Also, systematic studies show that, on the average, the absorption lines are stronger when detected in the light of more distant stars. The more distant a star is, the more interstellar material there is between us and the star. The narrow interstellar lines don't appear in the spectra of all stars. This suggests that the interstellar gas is clumpy, just as the interstellar dust is clumpy.

Figure 14.7 Arrangement for interstellar absorption lines. A star is observed through an interstellar cloud. Starlight is absorbed by the cloud at wavelengths corresponding to transitions in atoms and molecules in the cloud.

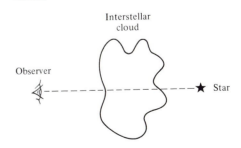

14.4.2 Radio Studies of Atomic Hydrogen

Much of what we now know about the interstellar medium comes from radio observations. We have already seen that the galactic center, supernova remnants, pulsars, and planetary nebulae are sources of radio emission. These are generally hot sources, or sources with high-energy electrons that produce a high radio luminosity. However, most of the interstellar gas is cool and does not produce a strong radio continuum emission. The cool interstellar gas must be observed via radio spectral lines.

The first interstellar radio line to be observed was from atomic hydrogen, but was not a transition in which an electron jumps from one orbit to another. Those transitions are in the visible and ultraviolet parts of the spectrum. For this radio transition, the hydrogen stays in the ground electronic state. Both the electron and proton have intrinsic angular momentum, called spin. We have already seen that this spin can have two possible orientations. We refer to them as "up" and "down." This means that the electron and proton spins can either be parallel or antiparallel. The relative orientation of the spins affects the magnetic force between the electron and the proton. The state with the spins parallel has slightly more energy than the state with the spins antiparallel. The atom can undergo transitions between these two states. The energy difference corresponds to a frequency of about 1400 MHz, or a wavelength of 21 cm. This is generally referred to as the *21-cm line* (Fig. 14.8).

If we take the energy of the transition $h\nu$ and divide by Boltzmann's constant, the quantity $h\nu/k$ gives us the temperature necessary to get collisional excitation of the hydrogen to the upper state. This is about 0.07 K. This means that even at the low temperatures of interstellar space there will be sufficient energy to excite transitions between these two states in hydrogen. The 21-cm line is easily observable under interstellar conditions. The possibility of detecting this line was discussed in Leiden (Netherlands) in the early 1940s by Henk van de Hulst. After that there was a race among Australian, Dutch, and American groups to detect the line. The first detection of the 21-cm line from interstellar hydrogen was by a group at Harvard, in the early 1950s.

Since that time there have been extensive observations of the 21-cm line by radio astronomers all over the world. It is probably fair to say that it was the dominant tool for studying the interstellar medium and galactic structure through the 1960s and continues to be very useful. In these studies, the line was observed in both emission and absorption. In order for a line to be in absorption, there must be a background continuum source whose brightness temperature at 21 cm is greater than the excitation temperature of the atoms in the particular cloud being observed. Under most conditions the excitation temperature of the 21-cm line is close to the kinetic temperature of the clouds (Fig. 14.9).

By studying absorption and emission lines in a given region it is possible to deduce both the excitation temperature and the optical depth. The exci-

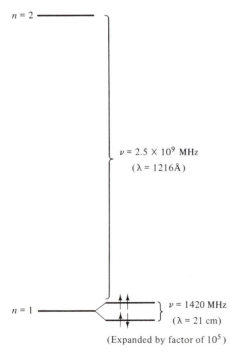

Figure 14.8 Origin of the 21-cm line. It arises from the splitting of the hydrogen ground state ($n = 1$). The splitting is greatly exaggerated in this figure. The splitting comes from a magnetic interaction which depends on the spin directions of the electron and proton. These spins are represented by the arrows. The energy is higher when the spins are parallel and lower when they are antiparallel.

$n = 2$

$\nu = 2.5 \times 10^9$ MHz
($\lambda = 1216\text{Å}$)

$n = 1$

$\nu = 1420$ MHz
($\lambda = 21$ cm)

(Expanded by factor of 10^5)

tation temperature enables us to calculate the kinetic temperature of the gas, an important quantity. The optical depth can be converted into a column density for atomic hydrogen. If we know the column density and the size of a cloud, we can also find the average local density of hydrogen. So you see that the 21-cm line observations provides radio astronomers with a powerful tool for studying the physical conditions in many interstellar clouds.

The importance of the radio observations is that interstellar dust is transparent at radio wavelengths. Therefore, we can use radio telescopes to detect objects clear across the galaxy, far beyond what we can see optically in the presence of dust. Since we can use it to observe clouds anywhere in the galaxy, the 21-cm line is a very useful tool for studying galactic structure. Also, since it is a spectral line, we can observe its Doppler shift and learn about motions throughout our galaxy. We will see how these studies are used in Chapter 16.

We have already seen (in Chapter 6) that some energy levels shift in the presence of a magnetic field (the Zeeman effect). The levels involved in the 21-cm line fall into this category. The stronger the magnetic field is, the greater is the shift. This means we can use the Zeeman shift in the 21-cm line to measure the strength of interstellar magnetic fields. The experiment is a difficult one because the Zeeman shift is much less than the width of a normal 21-cm line. However, opposite polarizations are shifted in opposite directions. Since we can detect different polarizations separately, we can subtract one polarization's spectrum from the other, leaving a very small signal (Fig. 14.10). The experiment is also difficult because a small difference in the response of the telescope to the two polarizations can mimic the effects of a Zeeman shift. Despite these difficulties, recent experiments have succeeded in measuring fields of the order of 10 μG in a few interstellar clouds. Fields of this strength may sound very weak, but they are strong enough to influence the evolution of these clouds.

By making maps of the 21-cm emission (Fig. 14.11) astronomers have been able to get a good picture of the cloud structure of the interstellar gas.

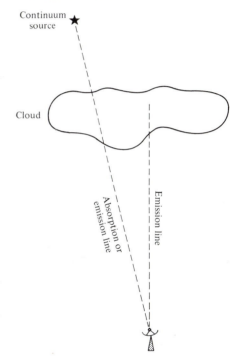

Figure 14.9 Conditions for radio absorption and emission lines. If a radio continuum source is viewed through an interstellar cloud, then radio absorption lines can be seen against the continuum source. (This also requires the continuum source to appear hotter than the cloud at the wavelength of observation.) If there is no background continuum source, then only emission lines can be seen.

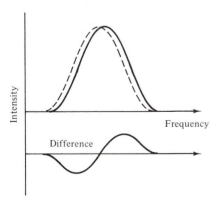

Figure 14.10 Zeeman effect in atomic hydrogen. The line shift is a very small fraction of the linewidth. However, as shown in the upper figure, opposite polarizations are shifted in opposite directions. When we subtract one polarization from the other, we are left with a very distinctive pattern. (Note that the lower curve is approximately the derivative of the upper curve. This follows directly from the definition of a derivative.)

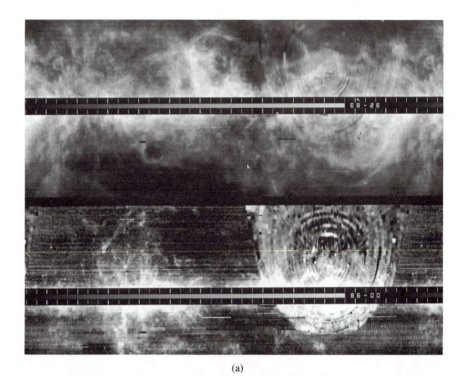

(a)

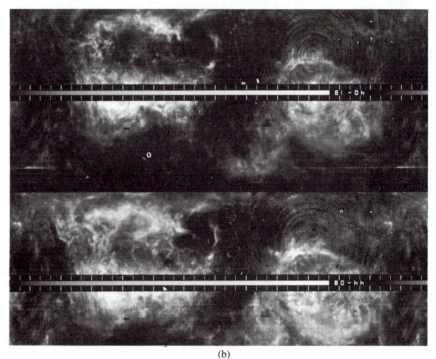

Figure 14.11   Images made from 21-cm observations, showing the large-scale structure, projected in a galactic coordinates system, with the plane of the Milky Way acting as the equator. (a) This map shows the emission of gas moving at all velocities. (b) This shows the gas in only one velocity range, revealing large loops in the gas distribution.

(b)

These maps show an irregular cloud structure, similar to that shown in the dust clouds. Typical clouds have the following physical parameters: temperature, 100 K; hydrogen density, $n_H \simeq 10$–100 cm^{-3}; lengths of tens of parsecs; hydrogen column densities up to $\sim 10^{21}$ cm^{-2}. The clouds fill about 5% of the volume of interstellar space, meaning that the average density of hydrogen in the interstellar medium is of the order of 0.1 cm^{-3}.

The regions between the clouds are not empty. Studies of the line profiles of the 21-cm line show very broad, faint wings. This is interpreted as coming from a small amount of very hot gas. Temperatures of about 10^4 K have been estimated for this low-density gas between the stars. We will see later in this chapter that the low density means it is very hard for the gas to lose energy and cool (a situation somewhat similar to the solar corona). It has been noted that if we compare the pressure within a cloud P_{cl} with the pressure in the intercloud medium P_{ic}, we find

$$P_{cl}/P_{ic} = n_{cl}T_{cl}/n_{ic}T_{ic}$$

$$= \frac{(10 \text{ cm}^{-3})(10^2 \text{ K})}{(0.1 \text{ cm}^{-3})(10^4 \text{ K})}$$

$$= 1$$

The pressure in the clouds and intercloud medium is approximately the same. Some theoreticians have proposed a picture of the interstellar medium, known as the *two-phase model*, in which this equality of pressures is not a coincidence, but follows from the ways in which the gas can cool. The two-phase model is now considered overly simplified and has been replaced by more dynamic pictures of the interstellar medium.

For clouds that are near enough to be seen optically, it was found that the 21-cm emission often follows the optical obscuration of the dust. This suggests that the gas and dust are well mixed. This idea was tested in detail by seeing the degree to which the hydrogen column density N_H correlates with the visual extinction, A_V. The results of these studies are shown in Fig. 14.12. There is some scatter in the data, but it is clear that N_H and A_V are related. The general ratio of A_V/N_H is approximately 1 mag/10^{21} atoms cm^{-2}.

This ratio was found to hold as long as the extinction is less than 1 mag. When the extinction gets higher, the relationship no longer holds. This was a mystery for several years. Two possible solutions were proposed. One was that with a lot of dust, very little radiation can penetrate to heat the cloud. It is possible that the hydrogen is so cold that the emission lines are just very weak. The other possibility is that under the higher extinction conditions, pairs of hydrogen atoms combine to form molecular hydrogen, H_2. Molecular hydrogen obviously has a very different structure than atomic hydrogen, and has no equivalent of the 21-cm line. In fact, cold molecular hydrogen has no

Figure 14.12 Hydrogen column density N_H as a function of visual extinction A_V. Though there is some scatter, there is a good correlation in the two quantities as long as A_V is less than one magnitude.

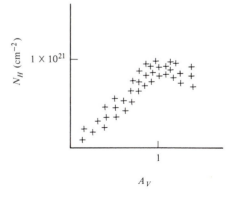

emission or absorption lines in the radio or visible part of the spectrum. The recent discovery of a large number of interstellar molecules, including H_2, tells us that the latter explanation is correct.

14.5   INTERSTELLAR MOLECULES

14.5.1   Discovery

The discovery of optical absorption lines of CH, CH^+, and CN raised the possibility that molecules might be an important constituent of the interstellar gas. However, it was thought that the densities were too low for chemistry to proceed very far. In fact, the existence of these three unstable species supported that general notion. If chemistry had proceeded very far, these species would have been incorporated into more complex molecules.

To see how these arguments worked we can estimate the rate at which molecules form in a cloud. Let's take the simple example of C and O coming together to form the simple molecule CO. The rate of formation of CO per unit volume is given by

$$R_{form} = n_C n_O v \sigma \qquad (14.17)$$

where n_C and n_O are the C and O densities, respectively, v is the relative speed of the atoms, and σ is the cross section for a collision. We take σ to be the geometric cross section (the approximate size of an atom, 10^{-16} cm^2), and v to be the average thermal speed at a temperature of 100 K (about 10^5 cm/s). Finally, since both C and O have cosmic abundances of about 10^{-3} that of H, we take each of their densities to be $10^{-3} n_H$. This will give us a factor of n_H^2 in the rate, explaining why the density is so important. If $n_H = 10$ cm^{-3}, the rate becomes 10^{-15} cm^{-3}/s.

We have to compare this with the rate at which the CO is destroyed. One destruction mechanism is photodissociation. An ultraviolet photon strikes the molecule with sufficient energy to break it apart. An unprotected CO molecule can live an average of 10^3 years (3×10^{10} s) in the interstellar radiation field. The dissociation rate per molecule is the inverse of the lifetime. The dissociation rate per unit volume is the density of CO molecules divided by the lifetime,

$$R_{dis} = n_{CO}/t \qquad (14.18)$$

If we equate the formation and destruction rates, we can solve for the equilibrium CO abundance,

$$n_{CO} = 1 \times 10^{-15} t$$

$$= 3 \times 10^{-5} \text{ cm}^{-3}$$

Since $n_H = 10$ cm^{-3} the fractional abundance of CO, n_{CO}/n_H, is about 3×10^{-6}. This is low enough that it did not raise the hopes of finding very complex molecules. We have even been optimistic by assuming that every collision between a C and an O leads to a CO molecule.

However, radio searches for small molecules were carried out. In the 1960s three simple molecules were found, OH (at a wavelength of 18 cm), H_2O (1 cm), and NH_3 (1 cm). The abundances of these molecules were surprisingly high, and astronomers were encouraged to carry out searches for other molecules. In 1969 one of the most important molecular discoveries took place. CO was found at a wavelength of 2.6 mm. This was the first molecule to be found at millimeter wavelengths. Remember, at shorter wavelengths we can get good angular resolution with modest-sized telescopes. (Of course, the telescope surfaces require greater precision.) The abundance of CO is also very high, with CO densities of about 1 cm^{-3}, much higher than our previous estimate. As we will see, the 2.6-mm line of CO has become the complement to the 21-cm line in studying the cool interstellar gas.

Following these initial discoveries, a large number of interstellar molecules were found. Over 50 have been discovered to date. They are listed in Table 14.1. There are many familiar molecules, such as formaldehyde (H_2CO), methyl alcohol (CH_3OH), and ethyl alcohol (CH_2CH_3OH). There are some unfamiliar molecules. Some of these are charged species, such as HCO^+, and others have unpaired electrons and are chemically active in the laboratory, such as CCH. Even carbon chain molecules of moderate length have been found, such as $HC_{11}N$. Many of these molecules were discovered by observations at millimeter wavelengths on telescopes such as that shown in Fig. 14.13.

Figure 14.13 The NRAO 12-m-diam telescope (originally 11-m until resurfacing), located on Kitt Peak, Arizona, at an altitude of 6000 ft. Its accurate surface allows it to work at wavelengths as short as 1 mm. The high altitude is particularly important at these short wavelengths, the atmospheric absorption is significant. The incoming radiation bounces off the parabolic dish and then off a secondary mirror. It passes through a hole in the center of the dish to a radio receiver. Signals are then amplified and sent to a computer in a control room, where they can be studied by the observer.

TABLE 14.1   Interstellar Molecules (Arranged by Number of Heavy Atoms)

0	1	2		3		4
H_2	$^*CH$	CO	CS	$HCOOH$		$^*C_3N$
	CH^+	$^*HCO$		NH_2HCO		$^*C_4H$
	$^*OH$	HCO^+	HCS^+	CH_3CN		HC_3N
	H_2O H_2S	H_2CO	H_2CS	CH_3C_2H		CH_3CH_2CN
	$^*CH_4$	CH_3OH	CH_3SH	CH_3CHO		CH_3OOCH
	NH_3	NO	NS	CH_3CH_2OH		CH_2CHCN
			SO	$(CH_3)_2O$		
		HNO		$HNCO$	$HNCS$	$\dfrac{6}{HC_5N}$
		$^*CN$	H_2CCO			
		HCN	NH_2CN			
		HNC			SO_2	$\dfrac{8}{HC_7N}$
		CH_2NH			OCS	
		CH_3NH_2		$HOCO^+$		$\dfrac{10}{HC_9N}$
		$^*C_2H$		$HOCN$		
		C_2H_2				$\dfrac{12}{HC_{11}N}$
		N_2H^+				
		SiO	SiS			
		CO^+				
		C_2				

*Radical

The discovery of so many interstellar molecules was obviously a surprise. How could the predictions that molecules could not form have been so wrong? The clouds in which molecules have been found are not the same clouds that were studied at 21 cm. They have higher densities and visual extinctions and lower temperatures. The higher densities mean that chemical reactions can take place faster. The higher visual extinctions provide shielding from the ultraviolet radiation that dissociates the molecules.

We don't see 21-cm emission from these clouds because the atomic hydrogen has been converted to molecular hydrogen. As we have already seen, the molecular hydrogen has no radio or optical spectrum. Since hydrogen is the most abundant element, we classify interstellar clouds by the form in which the hydrogen is found. For example, clouds in which the hydrogen is mostly atomic are called HI clouds. Clouds in which the hydrogen is mostly ionized are called HII regions (to be discussed in the next chapter). Clouds in which most of the hydrogen is molecular are called *molecular clouds*.

14.5.2 Chemistry

Since the discovery of so many interstellar molecules, considerable effort has gone toward a better understanding of interstellar chemistry. It appears that some of the chemical reactions take place on the grain surfaces. The grain surface provides a place for two atoms to migrate around until they find each other. They also provide a sink for the binding energy of a molecule. For example, we think that molecular hydrogen must be formed on a grain surface. If two H atoms combined in the gas phase, the particular properties of the H_2 molecule would keep it from radiating away the excess energy before the molecule flew apart. On a grain surface, the energy can efficiently be transferred to the grain, resulting in an increase in the grain temperature. The fact that the dust plays an important role in the formation of H_2 and in the protection of the H_2 once it is formed, results in an interesting situation. When a cloud has an extinction of less than 1 mag, almost all of the hydrogen is atomic. When the extinction is greater than 1 mag, almost all of the hydrogen is molecular. This explains the breakdown in the relationship between N_H and A_V.

Despite the important role that the dust plays in the formation of the most abundant molecule H_2, most of the interstellar chemistry cannot proceed in this way. Many of the molecules are formed in the gas. At the beginning of this section we calculated a very low rate for two atoms to collide in the gas to form a molecule. However, the densities in molecular clouds are at least 10^3 times those we used for our estimate, and the reaction rate goes as the square of the density. Therefore, the reaction rates in molecular clouds are much faster than our initial calculation suggests. There is also another factor that increases the cross section for collisions if one of the reactants is an ion and the other is neutral. Such a reaction is called an *ion–molecule reaction*.

To see how the rate is enhanced, let's consider the case of a positive ion. (We have already said that the grains must be negatively charged, so the gas must be positively charged. In addition, Table 14.1 shows that many positive ions have been detected.) The neutral atom can still have an electric dipole moment even though it has no net charge. The dipole will tend to line up with the electric field of the ion (Fig. 14.14). Since the ion is positive, the negative end of the dipole will end up closer to the ion. The negative end of the dipole will therefore feel an attractive force which is slightly greater than the repulsive force felt by the positive end which is farther away. The dipole will feel a net attractive force. This attractive force increases the effective cross section of the reactants and speeds the reactions.

Figure 14.14 Dipole in electric field. In this case, the electric field is provided by the positive charge, and gets weaker with distance from that charge. The negative end of the dipole is closer to the positive charge, so an attractive force felt by the negative end is greater than the repulsive force felt by the positive end. The dipole is attracted to the charge. (The same thing would happen with a negative charge.)

Theoreticians have tried to identify the chemical reactions that might be important in the interstellar medium. They then carry out model calculations in which they calculate the equilibrium abundances of various molecules. These are the abundances for which the rates of destruction and formation are equal. These theories have been quite successful at predicting the abundances of most of the simpler (especially two- and three-atom) molecules. More work is still needed to understand the heavier molecules. In addition, it may be that many interstellar clouds are not old enough to have reached an equilibrium situation. If that is the case, abundances should still be changing.

14.5.3   Observing Molecules

When we observe interstellar molecules, we are not observing transitions in which electrons jump from one level to another. Such transitions do exist for molecules, as they do for atoms. However, they require energies of the order of at least a few electron volts and are in the visible part of the spectrum. These transitions are not easily excited in the cool interstellar medium. Another type of transition in molecules, involving lower energies, is vibrational. We can think of a molecule as consisting of a number of balls, connected by springs. The springs can stretch and bend. However, according to quantum mechanics, they can only stretch and bend at certain frequencies, with certain energies. Transitions between vibrational states are possible. The energies associated with vibrational transitions usually place the resulting photons in the infrared. This is still too energetic for the cool clouds.

There is another type of transition, with even lower energies. They involve the rotation of the molecule. The rotational motion is also quantized, and transitions among rotational states can take place. The photons associated with these transitions are generally in the radio part of the spectrum. To see what the energy levels look like in this case, we consider a diatomic molecule (like CO), rotating end-over-end about its center of mass. The moment of inertia for this motion is I. If the molecule is rotating with an angular speed ω, the energy is given by

$$E = \tfrac{1}{2}I\omega^2 \tag{14.19}$$

The angular frequency can be expressed in terms of the angular momentum L as

$$\omega = L/I \tag{14.20}$$

Using this, equation (14.19) becomes

$$E = \tfrac{1}{2}L^2/I \tag{14.21}$$

The condition for the quantization of angular momentum is different than the one we saw for electrons in an atom. If J is an integer, called the *rotational quantum number*, then L^2 is related to J by

$$L^2 = J(J + 1)(\hbar)^2 \qquad (14.22)$$

(NOTE: For large values of J this is not very different from the condition $L = J\hbar$). If we put equation (14.22) into equation (14.21), the energy becomes

$$E = \frac{J(J + 1)\hbar^2}{2I} \qquad (14.23)$$

For any given molecule the energy levels are determined by the moment of inertia. If I is large, the energy levels will be close together. If I is small, the energy levels will be farther apart. The energy levels for CO are shown in Fig. 14.15. Note that the closest spacing is for the first two energy levels ($J = 0$ and $J = 1$). As we go to higher values of J the energy levels get farther apart. This means that at low temperatures only a few energy levels are populated. For example, the 2.6-mm transition in which CO is most commonly observed is the $J = 1 \rightarrow 0$ transition. The moments of inertia for many simple molecules are such that the lowest transitions lie in the millimeter part of the radio spectrum. This is why so many molecules were discovered at millimeter wavelengths.

If we want to look for a new interstellar molecule, we need to know the wavelengths at which it can emit. For the most part we rely on accurate laboratory measurements of molecular spectra. Once the wavelengths of a few transitions have been measured, those of other transitions can be calculated. There are some molecules that have been found in interstellar space without prior laboratory study. These were found accidentally, in the course of searches for other molecules. In some cases the interstellar medium provides us with a unique opportunity to study molecules that are not stable in the laboratory.

The most important feature of interstellar molecules is that they provide us with a way of obtaining information about the physical conditions in the molecular clouds. If we take the energy corresponding to a 2.6-mm photon and divide by k, we find an equivalent temperature of 5.5 K. This means that rotational transitions in molecules are excited even at low temperatures. Also, the factor $\exp[-E/kT]$ in the Boltzmann equation is most sensitive to changes in temperature and density when E is of the order of kT.

To see how we use molecules to probe the physical conditions in interstellar clouds, we must first know the ways in which a molecule can get from one rotational state to another. One set of processes involves interaction with radiation, either by emission or absorption of photons. Radiative processes

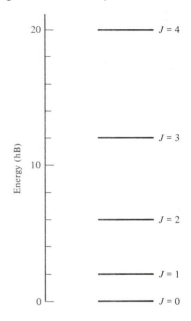

Figure 14.15 Rotational energy levels of a linear molecule. The energy is in units of Planck's constant multiplied by the molecule's rotation constant *B*. The states are designated by the total angular momentum quantum number *J*.

Figure 14.16   Types of interactions between radiation and matter. (a) Absorption. A photon is absorbed, leaving the atom or molecule in an excited state. In this figure the excited state is denoted by a larger symbol for the atom or molecule. (b) Emission. The atom or molecule starts in the excited state, and spontaneously makes a transition to the lower state, giving off a photon of the appropriate energy. (c) Stimulated emission. The atom or molecule is in the excited state. A photon whose energy would be right to produce an absorption if the atom or molecule were in the lower state, strikes the atom or molecule, inducing a downward transition and the emission of a photon. There are now two photons. They are at the same frequency, traveling in the same direction and in phase.

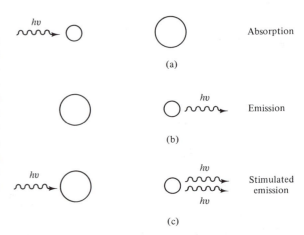

are illustrated in Fig. 14.16. We are already familiar with absorption and emission of photons. For absorption, the photon must have the energy corresponding to a transition from the molecule's initial state to a higher energy state. For emission the molecule goes from a higher-energy state to a lower energy state. The emitted photon has an energy equal to the energy difference between the states.

There is an additional radiative process that is important—*stimulated emission*. This is emission of a photon, stimulated by the presence of another photon. Suppose we consider only two energy states. The molecule starts in the higher-energy state. When the molecule is struck by a photon whose energy is equal to the energy difference between the states, the molecule cannot absorb the photon, since the molecule is already in the higher state. However, the presence of that photon can cause the molecule to drop to the lower state, emitting a second photon. In the process of stimulated emission, one photon comes in and two photons go out. The two photons have the same wavelength, are in phase with each other, and travel in the same direction. As we will see below, it is stimulated emission that is responsible for the amplification in masers and lasers.

Molecules can also be induced to make transitions by collisions with other particles. In a molecular cloud most of the matter is in H_2, which we don't usually directly observe. However, this H_2 makes its presence felt by forcing transitions in other molecules. The process works in both directions. An H_2 molecule can strike a CO molecule, for example. In the process, the H_2 can lose kinetic energy, while the CO is excited to a higher-energy state, or the H_2 can gain kinetic energy with the CO going to a lower-energy state.

In order to understand the excitation of interstellar molecules, we must be able to calculate the rates at which these various processes occur under different conditions. It is then necessary to carry out large calculations to model the conditions in a cloud. In these models we require that the population of each level stay constant. The rate at which molecules reach any energy state must equal the rate at which they leave that state. We use the model calculations to predict the strengths of various molecular lines. We then compare those predictions with observations. The models are adjusted until agreement is found. The model is then used to predict the results of new observations, and the process continues.

In studying molecular lines we can also learn about the velocities within a cloud by studying the line profiles. An interesting feature of molecular lines is that they are almost always wider (cover a larger frequency range) than would be expected for a line in which the only Doppler broadening is from random thermal motions. This implies that we are seeing the effects of other internal motions in the cloud. These motions can include collapse, expansion or rotation. We may also be seeing the effects of turbulent stirring of the gas. This stirring may be driven by mass loss from both old and young stars.

Probably the most usefully studied interstellar molecule is carbon monoxide, CO. It is very abundant. Its abundance is about 10^{-4} to 10^{-5} times that of hydrogen, while most other molecules have abundances that are about 10^{-9} times that of hydrogen, or less. The CO is easy to excite and to observe. It is particularly useful in tracing out the extent of molecular clouds. Also, observations of CO allow us to estimate cloud masses and kinetic temperatures. Some other molecules that are very useful are carbon sulfide (CS) and formaldehyde (H_2CO). These molecules are rarer and harder to excite than CO. We only see them in very dense parts of clouds. Therefore, we can use these molecules to tell us about the densities in clouds.

Another interesting aspect of interstellar molecules involves the substitution of various isotopes. For example, the most common form of carbon is ^{12}C, and most of the CO has this form. However, some CO is formed with the rarer species ^{13}C. Making such a substitution changes the moment of inertia of the molecule, shifting the spectral lines. The shifts are quite large, and are easy to detect. We can therefore measure the amounts of different isotopes in the interstellar medium. Since all of the heavy elements come from stars, these measurements can tell us about the ways in which earlier generations of stars have enriched the interstellar medium. It also turns out that changing isotopes changes chemical reaction rates. Therefore, observing the same molecule with different isotopic substitutions can tell us about how that molecule is made.

14.6 THERMODYNAMICS OF THE INTERSTELLAR MEDIUM

The temperature of any object is determined by the balance between heating and cooling. There is generally some temperature for which the rates of heating and cooling will be the same, allowing the temperature to stay constant. In this section we will look at the heating and cooling processes. When we talk about a heating process we mean one that tends to increase the kinetic energy of the gas. When we talk about a cooling process we mean one that tends to decrease the kinetic energy of the gas.

One way for an interstellar cloud to be heated is by the absorption of photons. These photons can come from a nearby star, or can be from the combined light of many distant stars. A photon entering a cloud is not, by itself, a mechanism for heating the cloud. We must have a way of converting the energy of the photon into kinetic energy in the gas. The most important methods for this are as follows:

1. *Heating the dust*. A photon strikes a dust grain with the photon energy going toward increasing the grain temperature. The hot grain is then struck by an atom or molecule in the gas, and it transfers some of its energy to that atom or molecule.

2. *Excitation of atoms or molecules*. A photon strikes an atom or molecule, leaving it in an excited state. The excited atom or molecule undergoes a collision with another atom or molecule. The first atom or molecule drops back to the lower energy state, and the energy shows up as an increased kinetic energy for the second atom or molecule. It is important that the collision take place before the first atom or molecule has had time to simply emit a photon and drop to the lower state, since the photon can escape leaving the cloud with no additional energy.

3. *Ionization*. The incoming photon strikes an atom or molecule, ejecting an electron. The electron can then transfer its kinetic energy to the rest of the gas through collisions. These collisions must take place before the electron recombines with an ion, releasing a photon.

4. *Photoelectric effect*. In this process, an incoming photon strikes a grain surface, causing the ejection of an electron. The electron's kinetic energy is then available to heat the gas.

The above processes also work for heating by streams of high-energy particles, known as cosmic rays, which permeate the interstellar medium. The sources of cosmic rays will be discussed in Chapter 19.

The interstellar medium can also be heated by the direct injection of mechanical energy from high-velocity flows. For example, a supernova remnant, traveling at high speeds, will transfer some of its kinetic energy to material it overruns. Stellar winds can accomplish the same thing. When we look at large-scale maps of interstellar clouds, we see evidence for many loops and ''bubbles.'' These suggest that the interstellar medium is constantly being stirred up by processes such as supernova explosions.

We now look at cooling processes. We must remember that a cooling process must take kinetic energy from the gas and remove that energy from the cloud. These processes are just the inverse of the heating process:

1. *Emission from grains*. An atom or molecule strikes a grain, with the atom or molecule losing kinetic energy and the grain getting hotter. The

grain can then radiate this excess energy away. It must radiate away before it is struck by another atom or molecule that might take back the energy.

2. *Excitation*. One atom or molecule strikes another, with the first losing kinetic energy and the second being driven into an excited state. The second one then emits a photon and drops back to its lower state. Of course, the emission of the photon must take place before another collision forces the second one back to the lower state.

3. *Ionization*. One atom or molecule strikes another, with the second being ionized. The electron then recombines, accompanied by the emission of a photon, before it can collide with another particle in the gas.

In the heating and cooling processes, different atoms and molecules play important roles in different density and temperature regimes. For example, in the cool molecular clouds, much of the cooling comes from radiation by CO. In very hot regions the cooling can come from unusual emission lines in certain ions (discussed in Chapter 15). This multitude of processes allows us to have a wide variety of temperatures in the interstellar medium, ranging from the cool (10 K) high-density regions to the hot (10^4 K) low-density regions.

CHAPTER SUMMARY

In this chapter we looked at the various components of the interstellar medium. We saw how they are observed, and we looked at the physical processes that are important in their current state and evolution.

Though only 1% of the interstellar mass, the dust is the most easily visible part of the interstellar medium. We detect dust by its blocking of starlight, known as extinction. Warm dust can be detected in the infrared. The extinction consists of both scattering and absorption. The extinction is wavelength dependent, producing interstellar reddening. We saw what could be deduced about grain sizes from the extinction curves. We get information on grain composition from infrared observations. We saw that the equilibrium temperatures of grains are determined by a balance between radiation absorbed and radiation emitted.

The interstellar gas is observed in the radio part of the spectrum. Extensive studies have been made using the 21-cm line of hydrogen. However, the clouds revealed by these studies do not have high enough densities for them to be the likely sites of star formation.

Star formation probably takes place in the cooler, denser molecular clouds. In these clouds, most of the hydrogen is in the form of H_2. We cannot observe the H_2 from the ground, except when it is heated in a few small regions. Instead, we study molecular clouds by emission from trace molecules, such as CO. At these low temperatures (10–50 K), we are usually observing transitions from one rotational state to another. At these low temperatures and densities, most of the chemical reactions are probably between ions and neutral species.

We also looked at how interstellar clouds are heated, and how they can cool. In heating, any energy input must eventually be converted into kinetic energy in the gas. In cooling, the kinetic energy of the gas must be converted into energy, such as radiation, that can leave the cloud.

QUESTIONS AND PROBLEMS

14.1   Prove that, for the same angular momentum, end-over-end rotation for a cigar-shaped object has a lower energy than rotation about the long axis.

14.2   Show that in a gas, the number of particles hitting a surface per second per unit surface area is nv, where n is the number density (particles per volume), and v is the speed of the particles.

14.3   How would the equilibrium temperature of a dust grain change if the albedo is less in the infrared than in the visible?

14.4   Suppose a dust grain has an albedo $a(\lambda)$. It is near a star whose spectrum is that of a blackbody of temperature T_* and whose radius is R_*. The grain is a distance d from the star. Derive an expression for the dust temperature. [You will have to leave your answer in terms of an integral, since the form of $a(\lambda)$ is not given.]

14.5   What is the range of distances from a B0V star for which the dust temperature is between 50 and 1000 K?

14.6   In our discussion of the temperature of a dust grain near a star, we did not account for the fact that dust grains near the star would block some light from reaching dust grains far from the star. Show how this effect modifies the results.

14.7   What frequency resolution would be needed to observe the 21-cm line with a velocity resolution of 0.1 km/s?

14.8   For an excitation temperature of 100 K, what is the ratio of populations for the two levels in the 21-cm transition? (Take the statistical weight of the lower level to be 1 and the upper level to be 3.)

14.9   How does the angular resolution of a 100-m-diam telescope at 21 cm compare with that of a 12-m-diam telescope at 2.6 mm? What is the significance of these numbers?

14.10   Estimate the rate at which two H atoms can form an H_2 molecule on a grain surface. Assume that all atoms hitting a grain stick, and that as soon as two H atoms are on the grain surface they immediately form a molecule.

14.11   (a) Suppose we have an electric dipole made up of two charges, $+q$ and $-q$, a distance d apart. The dipole is placed a distance r from a charge $+Q$. Find an expression for the net force on the dipole. (b) How is this related to chemical reactions in the interstellar medium?

14.12   Find the value of T (in terms of E and k) for which the expression $\exp[-E/kT]$ is most sensitive to changes in temperature. What does this have to do with finding the right atom or molecule to tell you about physical conditions in a given region?

14.13   For scattering of light by interstellar grains we define the *phase function* as the average value of $\cos \theta$, where θ is the angle through which the light is scattered, with $\theta = 0$ for forward scattering. Find the value of the phase function in the following limiting cases: (a) all forward scattering; (b) all rearward scattering; (c) random scattering with all directions equally likely.

14.14   If we are far from any individual star, we can treat the interstellar radiation field as a dilute blackbody of temperature 10^4 K, with a dilution factor of 10^{-14}. Use this to estimate the temperature of a dust grain.

PLATE 1
(Above) The Horsehead nebula, in Orion's belt
is formed by dust blocking the light from the
glowing gas in the background.

PLATE 2
(Left) The Orion nebula, an HII
region in Orion's sword.

PLATE 3
False color far-infrared image of the
Orion region, taken with IRAS.

PLATE 4
False color far-infrared image of the Andromeda
galaxy, M31, taken with IRAS.

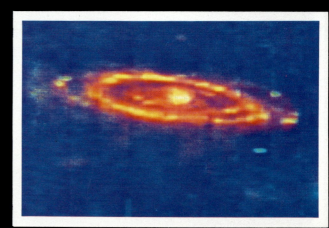

PLATE 5
Optical image of the Andromeda galaxy, M31.

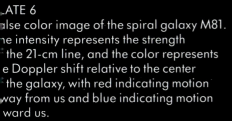

PLATE 6
False color image of the spiral galaxy M81.
The intensity represents the strength
of the 21-cm line, and the color represents
the Doppler shift relative to the center
of the galaxy, with red indicating motion
away from us and blue indicating motion
toward us.

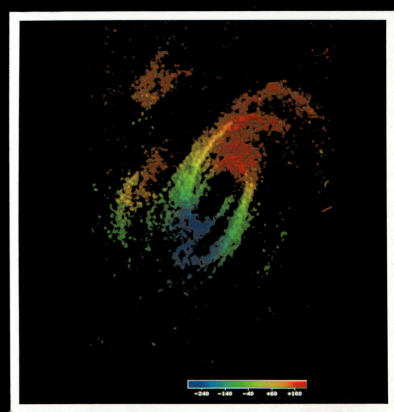

-240 -140 -40 +60 +160

PLATE 7
Jupiter, as seen from Voyager.

PLATE 8
The four largest moons of Jupiter, as seen from Voyager

PLATE 9
The surface of Mars, as seen from Viking.

Star Formation

In Chapter 14 we discussed the contents of the interstellar medium. It is this material out of which new stars must be formed. In this chapter we will identify those parts of the interstellar medium that are involved in star formation, and see what we know, and what we have to learn, about the star formation process. We begin by looking at questions that come to the front when astronomers discuss star formation.

15.1 PROBLEMS IN STAR FORMATION

We would like to know the conditions under which stars will form. We would like to know which types of interstellar clouds are the most likely to form stars, and which locations within the clouds are the most likely sites of star formation. We would also like to know whether star formation is spontaneous or whether it needs some outside *trigger*. When we say a trigger is necessary, we mean that the conditions in a cloud are right for star formation, but something is necessary to compress the cloud somewhat to get the process started. Once started, it continues on its own. Some sources for triggering star formation that have been suggested are the passage of a supernova remnant shock front, or the compression caused by a stellar wind. (In Chapter 18 we will also discuss density waves associated with galactic spiral structure. These might also induce star formation.)

Once the collapse to form stars starts, we would like to know how it proceeds. We would like to know what fraction of the cloud mass ends up in stars. This is sometimes referred to as the *efficiency of star formation*. We would also like to know how much of the mass that goes into stars goes into stars of various masses. This is called the *initial mass function*. By ''initial'' we mean the distribution of stellar masses at the time that a cloud gives birth to stars. The actual mass distribution in the galaxy is altered by the fact that stars of different masses have different lifetimes.

An important problem in understanding the evolution of star-forming clouds comes from the angular momentum of the cloud. In Chapter 8 we saw that the collapse can be slowed down or even stopped in a rotating cloud. How a cloud distributes and loses its angular momentum probably affects the efficiency of star formation and the initial mass function. In addition, it may account for the high fraction of multiple star systems and for the formation of planetary systems.

An interesting set of problems is posed by groupings of stars called *OB associations*. These are groups of stars in which it has been suggested all O stars form. We refer to these groupings as associations rather than clusters because the associations are not gravitationally bound. They are expanding, and eventually dissolve into the background of stars. We would like to know how an initially bound cloud can give birth to an unbound grouping of stars. In Chapter 12 we saw that, if a system in virial equilibrium loses more than half its mass without the velocity distribution changing, then the system becomes unbound. It is clear then that clusters must have lost more than half their mass, but we don't know how. Another interesting feature of OB associations is the existence of *subgroups*. Some associations can have as many as three or four distinct groupings of stars. The subgroups have different ages, as determined from their *H–R* diagrams. Also, the older subgroups seem to be larger, which makes sense if they are always expanding. A major question in star formation is explaining what appears to be a sequential wave of star formation through an association. It is in this context that triggers have been discussed the most actively.

15.2 MAGNETIC EFFECTS AND STAR FORMATION

Astronomers are becoming increasingly aware of the fact that magnetic fields can have an important effect on star formation in an interstellar cloud. Work in this area has been slow for two reasons: (1) As we have seen, measurements of interstellar magnetic fields are very difficult. Until we have a good idea of field strengths, it is hard to estimate their effects. (2) Theories that include magnetic fields are much harder to work out than those that don't. Even computer simulations of gravitational collapse in clouds with substantial magnetic fields take large amounts of time on the fastest computers. However, we can still get a feel for the types of effects that can be expected.

We would expect magnetic effects to be important when the energy associated with the presence of the magnetic field is comparable to the gravitational energy in magnitude. In cgs units, the energy density (erg/cm^3) associated with a magnetic field B is

$$u_B = B^2/8\pi$$

EXAMPLE 15.1   Magnetic Energy

For what magnetic field strength B does the magnetic energy of a cloud equal the absolute value of the gravitational energy? Assume a spherical cloud with radius $R = 10$ pc, and a density of molecular hydrogen $n_{H_2} = 300$ cm^{-3}.

Solution

The magnetic energy U_B is the energy density, multiplied by the volume of the cloud:

$$U_B = (B^2/8\pi)(4\pi/3)R^3$$

$$= B^2R^3/6$$

The magnitude of the (negative) gravitational potential energy is

$$-U_G = \tfrac{3}{5}(GM^2/R)$$

$$= \tfrac{3}{5}G[(4\pi/3)R^3(n_{H_2})(2m_p)]^2/R$$

$$= (16\pi^2/15)R^5(2m_p n_{H_2})^2$$

$$= 10GR^5(2m_p n_{H_2})^2$$

Equating these, and solving for B, we have

$$B = (60G)^{1/2}(2n_{H_2}m_p R)$$

$$= 6 \times 10^{-5} \text{ G}$$

$$= 60 \text{ }\mu\text{G}$$

This is comparable to the fields that have been measured in molecular clouds, implying that magnetic effects should be important.

As an interstellar cloud collapses, the magnetic field strength will increase. This is because of the flux freezing, discussed in Chapter 11. (Remember, Faraday's law requires that the flux through a surface be constant.) This only takes place if the cloud is a good conductor. Most interstellar clouds have sufficient ionization for this to be the case. The ionization in cold clouds probably results from cosmic rays. Most of the mass of the cloud is in the form of neutral atoms or molecules. However, as the cloud collapses, these neutral particles carry the charged particles along with them. The charged particles, in turn, provide the conductivity to insure the flux freezing. This process allows the magnetic field to effectively exert a pressure which can inhibit the collapse.

EXAMPLE 15.2   Flux Freezing

For a spherical cloud with the magnetic flux constant as the cloud collapses, find how the magnetic energy varies with the cloud radius R. Compare this with the gravitational energy.

Solution

For a uniform cloud the magnetic flux is the product of the field B and the projected area of the cloud πR^2. This means that the flux is proportional to BR^2. If the flux is constant, then BR^2 must be constant. This means that

$$B \propto 1/R^2$$

From the previous example we see that the magnetic energy density U_B is proportional to $B^2 R^3$. This means that

$$U_B \propto (1/R^2)^2 R^3$$

$$\propto 1/R$$

The gravitational potential energy is

$$U_G \propto GM^2/R$$

Since the mass of the cloud stays constant as it collapses,

$$U_G \propto 1/R$$

Therefore the magnetic and gravitational energies have the same dependence on R as the cloud collapses. If the magnetic field cannot prevent the initial collapse, then it cannot prevent a further collapse. However, if the magnetic field is important in the initial collapse, it will continue to be important.

As a cloud evolves, the ions and neutrals do not always stay perfectly mixed. The ions can drift with respect to the neutrals. If this happens, some of the magnetic flux can escape from the cloud, meaning that the field is not as high as one would calculate from flux freezing. The process, called *ambipolar diffusion*, has another effect. As the ions move past the neutrals, some collisions occur. This converts some of the drift motion into random motions of the neutrals, meaning an increase in the cloud temperature. Therefore, ambipolar diffusion can serve as a general heat source in a cloud.

15.3 MOLECULAR CLOUDS AND STAR FORMATION

We discussed the properties of molecular clouds in Chapter 14. They are important for star formation because they are both cool and dense, relative to the rest of the interstellar medium. In Chapter 8 we discussed the conditions under which an interstellar cloud is gravitationally bound, and expressed the result as a Jeans length. For a cloud of a given temperature T and number density n, the Jeans length is the minimum size of a gravitationally bound cloud. We approximated the Jeans length [equation (8.9)] as

$$R_J = [kT/Gm^2n]^{1/2}$$

EXAMPLE 15.3 Jeans Length for Atomic and Molecular Clouds

Compare the Jeans length for an atomic cloud ($T = 100$ K, $n = 1$ cm^{-3}) and a molecular cloud ($T = 10$ K, $n = 10^3$ cm^{-3}).

Solution

We simply take the ratios, noting also that the mass per particle in molecular clouds is twice that in atomic clouds.

$$R_J \text{ (at)}/R_J \text{ (mol)} = [(10^3)(10)(4)]^{1/2}$$

$$= 200$$

This means that a much smaller piece of a molecular cloud can become gravitationally bound than for an atomic cloud. It is therefore much easier to get a bound section in a molecular cloud than in an HI cloud.

We find a number of different types of molecular clouds. Their basic properties are summarized in Table 15.1. The simplest are the *globules*, like that shown in Fig. 14.1. They are sometimes called *Bok globules*, after Bart

TABLE 15.1   Interstellar Molecular Clouds

Type	T_k (K)	n_{H_2} (cm^{-3})	R (pc)	M ($M_\odot$)	A_v (mag)	Probes
Dark cloud	10	$\sim 10^3$	5–10	10^4	1–6	CO
Globule	10	$\sim 10^3$	1	10^2–10^3	5–15	CO, CS
GMC (envelope)	15	300	50	10^5	1–5	CO
Hot core	30–100	10^5–10^6	1	10^3–10^4	100	CO, CS, H$_2$CO, CH$_3$OH, NH$_3$, . . .
Protostellar cores						OH, H$_2$O maser
Energetic flows						CO
Envelope of evolved stars						CO, HC$_3$N SiO masers

Figure 15.1 Dark clouds blocking the light from background stars. Note the intricate shapes.

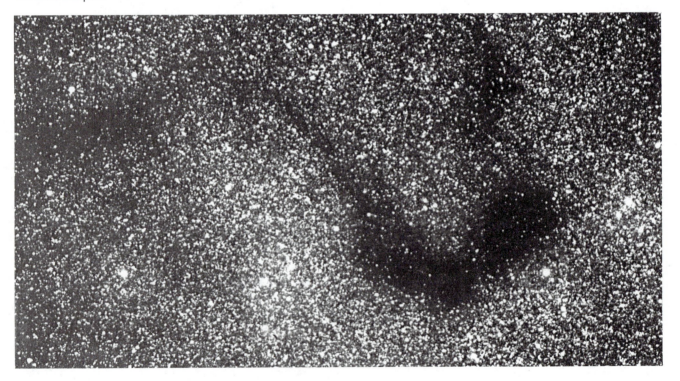

Bok, who suggested that they are potential sites of star formation. Globules are typically a few parsecs across. They generally have a simple, round appearance. This simplicity makes them attractive to study. Their visual extinctions fall in the range 1–10 mag, which can be determined by star counting. From CO observations, we find that their kinetic temperatures are about 10 K. From observations of CO and CS, we estimate their densities at 10^3 cm^{-3}, and mass in the range of 10 to 100 $M_\odot$. We think that they are mostly in a state of slow gravitational contraction.

The *dark clouds*, like those shown in Fig. 15.1, have local conditions (density, temperature) similar to those in globules, but the dark clouds are larger. Typical sizes for dark clouds are in the 10 pc range. Often a size is hard to define because dark clouds appear to consist of a number of small clouds in an irregular arrangement. There is evidence that they contain low-mass stars.

The largest molecular clouds are called *giant molecular clouds*, or *GMCs*. They are generally elongated, with a length of about 50 pc. Their densities are about 300 cm^{-3}, a little lower than for globules or dark clouds. They are also warmer, with $T = 15$ K. Their extent can be traced using CO observations, like those shown in Fig. 15.2. By observing the CO in nearby GMCs, where we can see the dust, we gain confidence in the fact that the CO tells

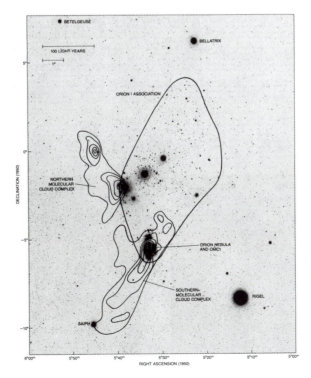

Figure 15.2 Molecular clouds in the Orion Region. To give an idea of the scale, the whole constellation of Orion is shown. The single outline shows the full extent of the Orion OB Association. There are two giant molecular clouds, indicated by their contours of emission from the CO molecule at 2.6 mm. The Northern Cloud Complex is associated with the stars in the Belt region. The Southern Cloud Complex is associated with the stars in the Sword region. This cloud contains the large HII region, the Orion Nebula.

us where the dust (and hydrogen) is. We therefore use the CO to trace out GMCs that are so far away that foreground dust blocks our view of them. GMCs have masses of $10^5 \, M_\odot$, and seem to come in complexes of clouds whose masses exceed $10^6 \, M_\odot$. These complexes are among the most massive entities in the galaxy. There also seems to be a close connection between giant molecular clouds and OB associations. It therefore appears that O and B stars form in GMCs. (GMCs also have lower-mass stars in addition to the O and B stars.)

Within giant molecular clouds we find denser regions, called *dense cloud cores*. These are warmer than the surrounding cloud, with temperatures above 50 K. Their densities, determined from studies of a number of different molecules, are in the range 10^5–$10^6 \, \text{cm}^{-3}$. (Even though we call these clouds dense, these densities are comparable to the best vacuums we can obtain in a laboratory!) These cores are very small, only a few tenths of a parsec across, and have masses of a few hundred $M_\odot$. Our ability to study them is limited by the angular resolution of our telescopes. We think these are the places in the GMCs where the star formation is actually taking place.

15.4 REGIONS OF RECENT STAR FORMATION

When we study star formation we find that there are some very obvious signposts of recent or ongoing star formation. In this section we will look at some of the most prominent: (a) HII regions, (b) masers, and (c) energetic flows. In each case, the object becomes prominent either because of the unique conditions that accompany star formation or because of the effect of newly formed stars on the cloud out of which they were born.

15.4.1 HII Regions

When a massive star forms it gives off visible and ultraviolet photons. Photons with wavelengths shorter than 912 Å (in the ultraviolet), have enough energy (≥ 13.6 eV) to ionize hydrogen. The stars that give off sufficient ultraviolet radiation to cause significant ionization are the O and early B stars. When most of the hydrogen is ionized, we call the resulting part of the cloud an *HII region* (Fig. 15.3).

In equilibrium in an HII region there is a balance between ionizations and recombinations. Free electrons and protons collide, forming neutral hydrogen atoms. However, the ultraviolet photons from the star are continuously breaking up atoms to form proton–electron pairs. The balance between these two processes determines how large a particular HII region can be. Within the HII region, almost all of the hydrogen is ionized. There is a rapid transition at the edge, from almost entirely ionized gas to almost entirely

Figure 15.3 The Carina Nebula, a large HII region.

neutral gas. The theoretical reasons for this sharp transition were first demonstrated by the astrophysicist, Bengt Stromgren. For this reason, HII regions are often referred to as *Stromgren spheres*, and the radius of an HII region is called the *Stromgren radius*.

We can see how the balance between ionizations and recombinations determines the Stromgren radius r_s. If N_{uv} is the number of ultraviolet photons per second given off by the star capable of ionizing hydrogen, then this is the number of hydrogen atoms per second that can be ionized. That is, the rate of ionizations R_i is given by

$$R_i = N_{uv} \qquad (15.1)$$

The higher the density of protons and electrons is, the greater is the rate of recombinations. The recombination rate is given by

$$R_r = \alpha n_e n_p V \qquad (15.2)$$

TABLE 15.2 Rates of H-
Ionizing Photons for MS Stars

Spectral type	Photons/s
O5	51×10^{48}
O6	17.4
O7	7.2
O8	3.9
O9	2.1
B0	0.43
B1	0.0033

where V is the volume of the HII region and α is a coefficient (which depends on temperature in a known way). For the volume, we can substitute the volume of a sphere of radius r_s. If the only ionization is of hydrogen, the number density of electrons must equal that of protons, since both come from ionizations of hydrogen. Equation (15.2) then becomes

$$R_r = \alpha n_p^2 (\tfrac{4}{3} \pi r_s^3) \tag{15.3}$$

Equating the ionization and recombination rates gives

$$N_{uv} = (4\pi/3) n_p^2 r_s^3 \tag{15.4}$$

Solving for r_s gives

$$r_s = (4\pi \alpha N_{uv}/3)^{-1/3} n_p^{-2/3} \tag{15.5}$$

From equation (15.5) we see that the size of an HII region depends on the rate at which the star gives off ionizing photons and the density of the gas. If the gas density is high, the ionizing photons do not get very far before reaching their quota of atoms that can be ionized. The rate at which hydrogen ionizing photons are given off changes rapidly with spectral type, as indicated in Table 15.2, so the HII region around an O7 star is very different from that around a BO star. Often, O and early B stars are found in small groupings. In these groupings, the HII regions from the various stars overlap, and the region appears as one large HII region.

The ultraviolet radiation from stars can also ionize other elements. For example, after hydrogen, the next most abundant element is helium. However, the ionization energy of helium is so large that only the hottest stars produce large numbers of photons capable of ionizing helium. On the other hand, the ionization energy of carbon (for removing one electron) is less than that of hydrogen. There are many photons that are capable of ionizing carbon that will not ionize hydrogen. This, combined with the lower abundance of C relative to H, means that CII regions are generally much larger than HII regions (see Problem 15.26).

There are actually two conditions under which a boundary for an HII region can exist. One is that which we have already discussed. The cloud continues beyond that range of the hydrogen-ionizing photons. When this happens, we say that the HII region is *ionization bounded*. The other possibility is that the cloud itself comes to an end while there is still hydrogen-ionizing radiation. In this case we say that the HII region is *density bounded*, since its boundary is determined by the place where the density is so low that we no longer think of the cloud as existing. When an HII region is density bounded, hydrogen-ionizing radiation can slip out into the general interstellar radiation field.

The termperature of HII regions is quite high—about 10^4 K. HII regions are heated by the ionization of hydrogen. When an ultraviolet photon causes an ionization, some of the photon's energy shows up as the kinetic energy of the free proton and electron. Cooling in an HII region is inefficient, since there are no hydrogen atoms and no molecules. Cooling can only take place through trace constituents, such as oxygen. Transitions within these constituents are excited by collisions with protons and electrons. The collisions transfer kinetic energy from the gas to the internal energy of the oxygen. The oxygen then radiates that energy away. Since the heating is efficient and the cooling is inefficient, the temperature is high.

HII regions can give off continuous radiation, which can be detected in the radio part of the spectrum. This radiation results from collisions between electrons and protons in which the two do not recombine. Instead, the electron scatters off the proton. In the process the electron changes its velocity. When a charged particle changes its velocity it can emit or absorb a photon. This radiation is called *Bremsstrahlung* (from the German for "stopping radiation"). It is also called *free–free radiation*, because the electron is free (not bound to the proton) both before and after the collision. The spectrum for free–free radiation is characterized by the temperature of a gas (Fig. 15.4). The spectrum is not that of a blackbody because the gas is not optically thick. The spectrum is a blackbody curve multiplied by a frequency-dependent opacity. (Because the radiation can be described by the gas temperature, it is also known as *thermal radiation*.) This radiation is strongest in the radio part of the spectrum. Therefore, we can use radio continuum observations to see HII regions anywhere in our galaxy. A map of continuum emission from an HII region is shown in Fig. 15.5.

HII regions also give off spectral line radiation, called *recombination line* radiation. When an electron and proton recombine to form a hydrogen atom, the electron often ends up in a very high state. The electron then starts to drop down. It usually falls one level at a time. Larger jumps are also possible, but less frequent. With each jump, a photon is emitted at a frequency corresponding to the energy difference for the particular jump. For very high states, the energy levels are close together and the radiation is in the radio part of the spectrum. As the electron gets to lower states the lines pass through the infrared and into the visible. Generally, the electron can get all the way down to the ground state before the atom is re-ionized. We even see Hα emission as part of this recombination line series. This gives HII regions a red glow. (This red glow allows us to distinguish HII regions from reflection nebulae which appear blue.)

HII regions expand with time. When an HII region first forms, it must grow to its equilibrium radius. Even after it reaches this equilibrium size, it will continue to expand. This is because the pressure in the HII region is greater than that in the expanding cloud. The higher pressure results from the higher temperature in the HII region. Remember, the temperature of an HII region is about 10^4 K, while that of the surrounding cloud is less than 100 K. (The densities in the HII region and surrounding cloud are similar).

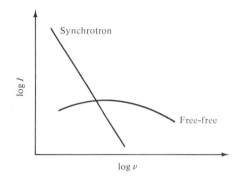

Figure 15.4 Schematic spectra of synchrotron radiation and free–free emission. The vertical axis is intensity and the horizontal axis is frequency, both on a logarithmic scale. The log–log representation emphasizes the power-law behavior of the synchrotron radiation.

Figure 15.5 Radio map of free–free emission from an HII region.

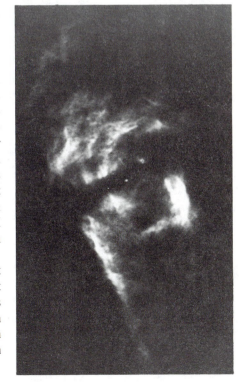

As the HII region expands, it can compress the material in the surrounding cloud, possibly initiating a new wave of star formation. This is one possibility that has been discussed for the triggering of star formation. The gas compressed by an expanding HII region will not automatically form stars. This is because the gas will be heated as it is compressed. If that heat is not lost, the temperature of the cloud will increase. The pressure will increase, and the gas will expand again. This re-expansion of the compressed gas can only be avoided if the gas can cool as it is compressed. Radiation from molecules like CO in the surrounding cloud can help with this cooling process.

15.4.2 Masers

We have already seen how the process of stimulated emission can lead to a multiplication—or amplification—in the number of photons passing through a material. In the stimulated emission process, one photon strikes an atom or molecule, and two photons emerge. The two photons are in phase and are traveling in the same direction. The fact that they are in phase means that their intensities add constructively. Stimulated emission can only take place if the incoming photon has an energy corresponding the difference between two levels in the atom or molecule, and the atom or molecule is in the upper of the two levels.

If only a few atoms or molecules are in the correct state, there will not be a significant increase in the number of photons. Suppose we designate the two states in the transition as 1 and 2. The population of the lower state is n_1, and the population of the upper state is n_2. The requirement for amplification is that n_2/n_1 be greater than g_2/g_1, where the g's are the statistical weights. This situation is called a *population inversion*, since it is the opposite of the normal situation. Formally, it corresponds to a negative temperature in the Boltzmann equation (see Problem 15.20). This is clearly not an equilibrium situation. The population inversion in a particular pair of levels must be produced by a process, called a pump. The pump may involve both radiation and collisions. The net effect of the pump process is to put energy into the collection of atoms or molecules so that some of that energy can come out in the form of an intense, monochromatic, coherent (in phase) beam of radiation.

This was first realized in the laboratory in the 1950s. Since microwaves are being amplified in the process, the device was called a *maser* (Fig. 15.6), an acronym for *microwave amplification by stimulation emission of radiation*. Subsequently, *lasers* were developed for the amplification of visible light. In any laser or maser, two things are necessary: (1) a pump to provide the population inversion; and (2) a sufficient path length to provide significant amplification.

Shortly after the development of laboratory masers, an interstellar maser was discovered. It involved the molecule OH. Four emission lines of OH

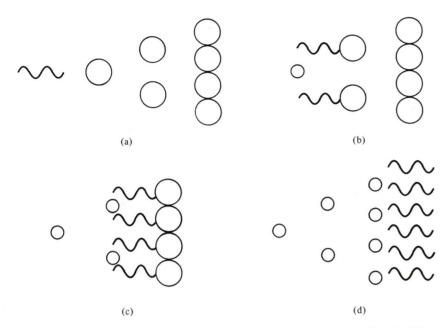

(a)

(b)

(c)

(d)

Figure 15.6   Maser. In each frame, a molecule in the upper level of the maser transition is indicated by a large circle, while one in the lower level is indicated by a small circle. (a) All of the molecules are in the upper state, and a photon is incident from the left. (b) The photon stimulates emission from the first molecule, so there are now two photons, in phase. (c) These photons stimulate emission from the next two molecules, resulting in four photons. (d) The process continues with another doubling of the number of photons.

were observed, but their relative intensities were wrong for a molecule in equilibrium. As radio telescopes were developed with better angular resolution, the emission appeared stronger and stronger. This meant that the emission is probably very intense, but coming from a very small area. This behavior was all suggestive of an interstellar maser. The next maser discovered was in the water molecule (H_2O), at a wavelength of 1 cm. As observations with better resolution became possible, it was clear that the objects were giving off as much energy as a 10^{15} K blackbody over that narrow wavelength range in which the emission was taking place.

A small size for these sources was also deduced from rapid variations in their intensity. Suppose we have a sphere of radius R, as shown in Fig. 15.7. If the sphere were suddenly to become luminous, then the first photons to leave each point on the surface would not reach us simultaneously. The photons from the edge of the sphere have to travel an extra distance R farther than the photons from the nearest point. These photons will arrive a time, $\Delta t = R/c$, later than the first photons. Therefore, it will take this time for

Figure 15.7   Time variability and source size. The signal from the farthest point the eye can see must travel an extra distance R over that from the nearest point the eye can see.

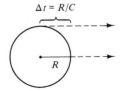

$\Delta t = R/C$

R

the light we see to rise from its initial low level to the final high value. A similar analysis holds for the time it would take for us to see the light turning off.

The above analysis tells us that an object's brightness cannot vary on a time scale faster than the size of the emitting region, divided by c. (This is actually true only as long as objects are not moving close to the speed of light.) If we see variations in intensity over a timescale of a year, the source cannot be larger than a light year across. Interstellar masers were found to be varying in intensity on an even shorter timescale, of the order of a month, indicating an even smaller size.

EXAMPLE 15.4 Maser Size

Estimate the maximum size of a maser that varies on the timescale of a month. What is the angular size of this object at a distance of 500 pc?

Solution

The timescale for the variations is

$$t = (24 \text{ h/day})(3600 \text{ s/h})(30 \text{ day})$$

$$= 2.6 \times 10^6 \text{ s}$$

This corresponds to a size of

$$R = c/\Delta t$$

$$= 7.8 \times 10^{16} \text{ cm}$$

$$= 5.2 \times 10^3 \text{ AU}$$

The angular size (in arc seconds) is related to R (in astronomical units), and the distance d (in parsecs) by

$$\theta \, ('') = R \, (\text{AU})/d \, (\text{pc})$$

$$= 5.2 \times 10^3/500$$

$$= 10''$$

In fact, masers are even smaller than this size, and have angular extents much less than one arc second. We have already seen that typical radio telescopes have resolutions of only one arc minute. This means that normal radio techniques cannot be used to measure the size of masers. We have to use combinations of radio telescopes, called *interferometers*, to improve our angular resolution, as we discussed in Chapter 4. The largest interferometers, used to study interstellar masers, involve telescopes all around the Earth.

When we try to understand interstellar masers we must explain both the pump and the path length for the gain. Many of the theories require very high densities. For example, we think that the presence of water masers suggests densities in excess of 10^8 cm^{-3}. This is much denser than even the dense cores that we normally see in molecular clouds. We therefore think that masers are associated with objects collapsing to become protostars. We take the presence of H_2O or OH masers in a region to indicate the presence of ongoing star formation.

When we observe masers, we often see them in clusters, like that depicted in Fig. 15.8. With radio interferometry, we can measure the positions of the masers very accurately. We can even measure their proper motions. We can use Doppler shifts to measure their radial velocities. However, we expect the motions in a cluster to be random, so the average radial velocity should equal the average transverse velocity v_T. From equation (6.6) we see that the distance is related to the proper motion and transverse velocity by

$$d(\text{pc}) = \frac{v_T \; (\text{km/s})}{4.74 \; \mu \; ('' / \text{yr})}$$

Therefore, an accurate study of the motions of masers allows us to determine the distance to a cluster of masers. It is hoped that this will develop into a very powerful distance measurement technique.

Maser emission is also observed in the molecule SiO (silicon monoxide). From the regions in which it is observed, it seems that the SiO maser emission is associated with mass outflow from evolved red giant stars.

15.4.3   Energetic Flows

A major recent discovery is that many regions of star formation seem to be characterized by strong outflows of material. One piece of evidence for such flows comes from the observation of very broad wings on the emission lines of CO (Fig. 15.9). The widths of these wings range from 10 to 200 km/s. The broad wings are usually seen only over a small region where the CO emission is the strongest. A peculiar feature of this emission is that the red-shifted wing and blue-shifted wing seem to be coming from different parts of the cloud (Fig. 15.10). This suggests that we are seeing two jets of gas, one coming partially toward us and the other going partially away from us. Because of this structure, we call these objects sources of *bipolar flows*.

Actually, we could also envision a model in which we are seeing infall rather than outflow. However, there is evidence we are seeing the effects of a wind striking the surrounding cloud, heating a small region. These small

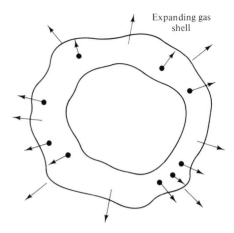

Figure 15.8   Cluster of masers in an expanding shell.

Figure 15.9 Spectrum of the CO line in the direction of the Orion Molecular Cloud, behind the Orion Nebula. The horizontal axis is wavelength, expressed as the velocity that would be necessary to Doppler shift the CO line to that wavelength.

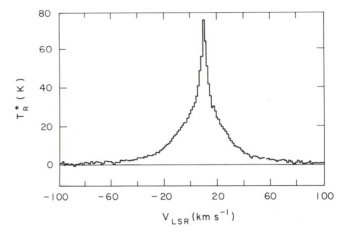

Figure 15.10 A bipolar flow source. (a) A negative image optical of the region, containing a number of Herbig Haro objects. The solid contours are of CO emission that is blue-shifted with respect to the average velocity. The dashed contours show the red-shifted CO. The plus (+) marked *B* is the

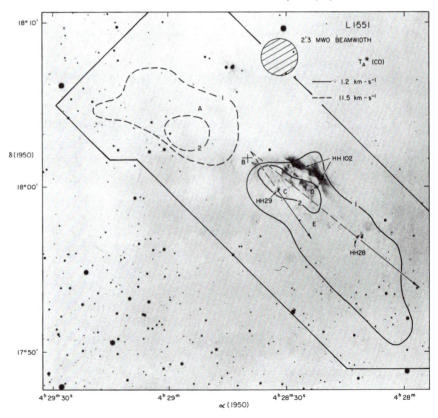

(a)

heated regions show emission in the infrared from H_2, requiring temperatures of about 10^3 K. Current theories of these sources involve strong stellar winds. The star is also surrounded by a dense disk of material. This disk blocks the wind in most directions, but allows it to escape along the axis, explaining the bipolar appearance. Remember, when we discussed the collapse of protostars in Chapter 8 we saw that disks are likely to form around the collapsing star.

These flows are seen around a variety of objects associated with star formation. For example, they are seen around T Tauri stars. In Chapter 8 we saw that T Tauri stars have not yet reached the main sequence. There is also evidence for strong winds from their optical spectra. We also see flows near

Figure 15.10 (Continued)   location of the suspected source of the flow. (b) A model for sources like this. The stellar wind comes out in all directions, but is blocked in most directions by a disk around the star. The wind emerges mostly at the poles of the disk. This drives material in the surrounding cloud away. Below, the effects of the motion on the CO line profiles are shown, assuming that the wind to the upper left also moves away from the observer and the wind to the lower right moves toward the observer.

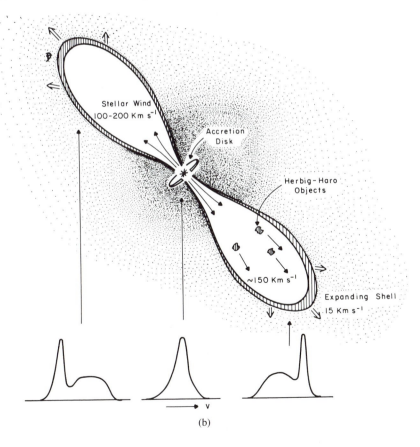

(b)

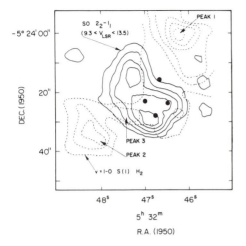

Figure 15.11 Maps of molecular infrared emission, superimposed on a photograph of the Orion nebula. The molecular and infrared sources are actually behind the HII region. The dashed lines are contours of infrared emission from molecular hydrogen. The solid contours are from the dense molecular gas.

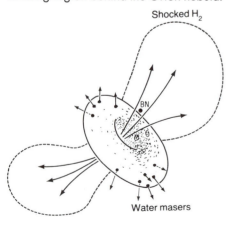

Figure 15.12 A model for the star forming region behind the Orion nebula.

Herbig–Haro objects. We have seen, in Chapter 8, that these objects are also thought to be associated with strong stellar winds. Regions of massive star formation also seem to have very strong flows.

These observations indicate that winds are an important feature of protostellar evolution for most stars. For low-mass stars like the Sun, this wind is relatively gentle, and can clear some of the debris from around the forming star leaving any planetary systems intact. For massive (O and B) stars the strong winds drive away a large mass. It has been suggested that the combined effects of the winds in OB associations can drive off enough mass to unbind the association, explaining why such associations are not gravitationally bound. Winds can also carry away some of the angular momentum in a cloud, allowing the collapse of the remaining material to continue.

15.5 PICTURE OF A STAR-FORMING REGION: ORION

The Orion region is one of the most extensively studied star-forming regions. It is relatively nearby, only about 500 pc from the Sun. It is away from the plane of the galaxy, so there is little confusion with foreground and background stars. There is also an interesting variety of activity in this region.

The region contains a large OB association, shown in Fig. 15.2. There are four distinct subgroups. The two oldest are near Orion's belt, and the two youngest are near the Orion Nebula in Orion's sword. The Orion Nebula is an HII region powered by the brightest stars in the youngest subgroup.

The region also contains two giant molecular cloud complexes, also shown in Fig. 15.2. The northern complex is associated with the belt region, and the southern complex is associated with the sword region. The southern complex contains the Orion Nebula and several smaller HII regions. The Orion Nebula is actually on the front side of the molecular cloud. Behind the HII region is a dense molecular core. It is totally invisible in the optical part of the spectrum because of the foreground material. However, it can be studied in detail using radio observations of molecules. In addition, it is a source of infrared emission (Fig. 15.11).

The general picture that has emerged has the HII region expanding into the molecular cloud, compressing it. This compression probably triggered a new generation of star formation. This new cluster of stars now appears as a series of small infrared sources and masers (Fig. 15.12). There is also evidence for an energetic flow. We see very broad (over 100 km/s) lines in CO and other molecules. These wings have the characteristics of the bipolar flows discussed in the last section. We also see evidence for small regions of gas heated to high temperatures in the regions where the wind strikes the surrounding cloud. Infrared line emission from H_2 is seen from this 2000 K gas. We can also study the proper motion of the masers and see that the region is expanding. (The best measurement of the distance to this region comes from studying the proper motions of these masers, as discussed in the last section.) It is likely that this dense region will appear as another OB subgroup when there has been sufficient time to clear the interstellar material away.

CHAPTER SUMMARY

In this chapter we saw how the various components of the interstellar medium, discussed in Chapter 14, are involved in the process of star formation. Current problems in star formation include how the collapse is actually initiated, what fraction of the mass is converted into stars, how planetary systems form, and how OB associations form.

We saw how the magnetic field in a cloud can affect its collapse if the magnetic energy is comparable to the gravitational potential energy. As a cloud collapses we expect that flux freezing will lead to a increase in the magnetic field strength within the cloud.

We saw how the molecular clouds, being cool and dense, are the most likely sites of star formation. The most massive stars seem to be born in the giant molecular clouds. These clouds have masses in excess of $10^5 M_\odot$.

Once an O or early B star forms in a molecular cloud, the gas around the cloud is ionized, producing an HII region. The size of the HII region is determined by a balance between ionizations and recombinations. The continuous radiation from HII regions comes from free–free scattering of electrons. Recombinations lead to electrons passing through many energy levels, giving off recombination line radiation.

Other indicators of recent star formation are masers and energetic flows.

QUESTIONS AND PROBLEMS

15.1 Compare the Jeans mass and radius for typical HI and molecular clouds.

15.2 Suppose we made an interferometer with one telescope on the Earth and the other on the Moon. What would its angular resolution be at a wavelength of 1 cm?

15.3 From observation of masers in Orion, we find an average radial velocity of __ km/s, and a proper motion of __ "/yr. How far away are these masers?

15.4 What do we mean by a "trigger" for star formation?

15.5 How is it possible for a gravitationally bound cloud to give birth to a gravitationally unbound association?

15.6 Suppose that a cloud contracts to one-tenth of its initial size. How do the Jeans length and mass compare with that in the original cloud?

15.7 Compare the density and pressure of a dense interstellar cloud with that for the gas in your room.

15.8 (a) Find the Jeans length for a cloud with density 10^6 H_2 molecules/cm^3 and $T = 100$ K. (b) What angle does this subtend at a distance of 1 kpc from the Earth?

15.9 (a) For a star of radius R, emitting like a blackbody of temperature T, find an expression for the number of photons per second emitted capable of ionizing hydrogen. (b) For $h\nu \gg kT$, evaluate the integral. (c) Why is the approximation in part (b) valid for an ordinary star?

15.10 (a) What is the difference between ionization- and density-bounded HII regions? (b) Why are density-bounded HII regions important?

15.11 Find the radius of an HII region around an O7 star if the density is $n_H = 10^3$ cm^{-3}. (Use Table 15.1.)

15.12 Explain why the CII zone around a star is much larger than the HII region. Consider (a) the relative ionization energies, and (b) the relative abundances.

15.13 Explain why almost all of the carbon ionizations result from photons with energies between the ionization energy of carbon and that of hydrogen, even though photons that can ionize hydrogen can also ionize carbon.

15.14 Why is the temperature of an HII region so high?

15.15 What is the Doppler width of the H110α line at a wavelength of 6 cm, for a temperature of 10^4 K?

15.16 What is the shift from the H110α line to the C110α line wavelength, if it just depends on the nucleus–electron reduced mass?

15.17 Suppose an HII region is formed in a cloud with a density of 10^4 H molecules/cm^3. If the temperature in the HII region is 10^4 K and the temperature in the surrounding molecular cloud is 10^2 K, by how much will the HII region expand before the pressures equalize?

15.18 (a) How can an expanding HII region trigger star formation? (b) Why is the rate at which the gas can cool important to the process?

15.19 Why is coherence important for a maser?

15.20 For a maser, a negative excitation temperature gives a negative optical depth. Show from the radiative transfer equation that it implies amplification.

15.21 How do we know that masers are small?

15.22 If the Sun turns off, how long will it take for the light to go from full intensity to zero?

15.23 (a) If an object dims in a day, how large can it be? (b) What angle would it subtend at a distance of 1 kpc?

15.24 Calculate the rate at which energy is delivered to a cloud by a $10^{-8}\, M_\odot$/yr flow with a speed of 100 km/s.

15.25 For the flow in the previous question, what is the rate at which interstellar material is being swept up if the sweeping occurs as long as the wind speed is greater than 5 km/s?

15.26 (a) Derive an expression, analogous to equation (15.5), for the radius $R_s(\text{He})$ of an ionized helium region around a star in terms of $N_{uv}(\text{He})$, the number of photons per second emitted by the star capable of ionizing helium. Give your answer in terms of n_p and n_{He}. Assume that α is the same for H and He. Assume that anywhere that He is ionized, all of the H is ionized, so that $n_e = n_p$. (b) Repeat the calculation for carbon, giving your answer in terms of $N_{uv}(\text{C})$, n_p, and n_C. For carbon, assume that it is ionized over a much larger region than H because of its lower ionization energy, so that $n_e = n_C$.

The Milky Way Galaxy

16.1 OVERVIEW

Throughout this book so far we have discussed the components of our galaxy: stars, clusters of stars, interstellar gas, and dust. We now look at how these components are arranged in the galaxy. The study of the large-scale structure of our galaxy is difficult from our particular viewing point. We are in the plane of the galaxy, so all we see is a band of light (Fig. 16.1). The interstellar dust prevents us from seeing very far into the galaxy. We get a distorted view.

The first evidence on our true position in the galaxy came from the work of Harlow Shapley who studied the distribution of globular clusters (Fig. 16.2). He found the distances to the clusters from observations of Cepheids and RR Lyrae stars. Shapley found that the globular clusters form a spherical distribution. The center of this distribution is some 10 kpc from the Sun. Presumably, the center of the globular cluster distribution is the center of the galaxy. This means that we are about 10 kpc from the galactic center, near the edge of the galaxy.

Figure 16.1 The Milky Way. From within, it is difficult to determine the structure.

355

Figure 16.2 Two-dimensional projections of the distribution of globular clusters in the galaxy. The frame marked "F" corresponds to metal poor clusters and that marked "G" corresponds to metal-rich clusters.

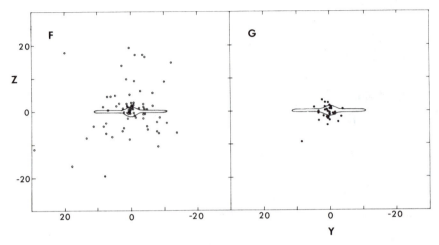

In Chapter 13, when we studied *H–R* diagrams of clusters, we introduced the concept of stellar populations, I and II. The distribution of these populations in the galaxy can help us understand how the galaxy has evolved. Population I material is loosely thought of as being the young material in the galaxy. Population I stars are found in galactic clusters, and are characterized by high metalicity. Some are also associated with gas and dust, suggesting that they are young enough to still have some of their parent cloud around them. Population I stars are confined to the galactic plane.

Population II stars are thought of as being the "old" component of the galaxy. They are found in globular clusters and are characterized by low metalicity. They have no gas and dust around them. Their galactic distribution is very different from that of population I stars. The population II stars form a spherical distribution, as opposed to a disk. This spherical distribution is sometime called the *halo*. When we talk about a spherical distribution, we do not mean just a spherical shell around the galaxy. Instead, we mean the spherically symmetric distribution whose density falls off with increasing distance from the galactic center. Population II objects also seem to have a larger velocity spread in their motions than do population I objects. Table 16.1 shows the characteristic thicknesses and velocity dispersions for some components of the galaxy.

16.2 DIFFERENTIAL GALACTIC ROTATION

All of the material in the galaxy orbits the galactic center. If the galaxy were a rigid body, all of the gas and dust would orbit with the same period. However, material closer to the galactic center orbits with a shorter period than material farther from the galactic center. This is not an unusual situation.

After all, the planets in the solar system exhibit the same behavior: Mercury takes less time to orbit the Sun than does the Earth, and so on. When the orbital period depends on the distance from the center, we say that the material is exhibiting *differential rotation*.

16.2.1 Rotation and Mass Distribution

The orbital period of any particle will depend on the mass about which it is orbiting. Just as we used the period and size of the Earth's orbit to tell us the mass of the Sun, we can use the orbital periods at different distances from the galactic center to tell us about the distribution of mass in the galaxy.

To see how this works, we assume that all matter follows circular orbits. At a distance R from the center, the orbital speed is $v(R)$, and the angular speed is $\Omega(R)$. The mass interior to radius R is

$$M(R) = \int_0^R \rho(r)\, dV \qquad (16.1)$$

where $\rho(r)$ is the density at radius r, and dV is a volume element. For a spherical mass distribution the motion of an object at R only depends on $M(R)$. Furthermore, the mass $M(R)$ behaves as if it were all concentrated at the center. (This also works for particles in the plane of a thin disk.)

For a particle of mass m orbiting at a radius R, the gravitational force is $GM(R)m/R^2$, and this must provide the acceleration for circular motion, so

$$GM(R)m/R^2 = mv(R)^2/R \qquad (16.2)$$

Solving for $M(R)$ gives

$$M(R) = v(R)^2 R/G \qquad (16.3)$$

Therefore, if we can measure $v(R)$, we can deduce $M(R)$, the mass distribution in the galaxy. Equivalently, we can use $\Omega(R)$, since

$$\Omega(R) = v(R)/R \qquad (16.4)$$

Substituting into equation (16.3) gives

$$M(R) = \Omega(R)^2 R^3/G \qquad (16.5)$$

If all of the mass is, indeed, concentrated at the center of the galaxy, then $M(R)$ is a constant, so equation (16.5) gives Kepler's third law, mentioned in Chapter 5. (We speak of the orbits as being "Keplerian.")

The function $v(R)$ [or $\Omega(R)$] is called the *rotation curve* for the galaxy.

If we know the rotation curve, we have a reliable way of determining the mass distribution in the galaxy. It is much more reliable than counting the visible mass, since it relies on the gravitational force, in which all matter participates. Therefore, if we want to know the mass distribution in a galaxy, we only need to find the rotation curve. However, because of our location in our galaxy, the measurement of the rotation curve is not easy.

16.2.2   Rotation Curve and Doppler Shift

In determining the rotation curve for our galaxy, it is convenient to introduce a set of coordinates known as *galactic coordinates,* measured from the Sun. The galactic latitude b measures the distance above or below the galactic plane of an object. The galactic longitude l is an angle measured around the galactic plane, starting in the direction of the galactic center.

We also define a convenient reference frame for measuring velocities, called the *local standard of rest,* or *lsr.* If the Sun were moving in a circular orbit, then the local standard of rest would coincide with the Sun. However, the Sun has a small motion superimposed on its circular motion, so it is not a convenient reference point for velocities. There are actually two ways of defining the local standards of rest:

1. *Dynamical.* The origin of the coordinate system orbits at a distance R_0 from the galactic center, where R_0 is the distance of the Sun from the galactic center. The coordinate system moves with the velocity, $v(R_0) = v_0$, or $\Omega(R_0) = \Omega_0$, appropriate for a circular orbit at R_0. Defined this way, the motion of the lsr depends only on $M(R_0)$.

2. *Kinematic.* The origin of the coordinate system moves with the average velocity of all of the stars in the vicinity of the Sun. This averages out the effects of the random motions of these stars.

The two definitions should result in the same coordinate system. However, there are small differences, which we will ignore. (The difference tells us about the dynamical properties of the galaxy.) With respect to the lsr, the Sun is moving at about 20 km/s toward a right ascension of 18 hr and a declination of 30°. (In galactic coordinates, this is $l = 56°$, $l = 23°$.) Once we know the Sun's motion when we measure the Doppler shift of any object in the galaxy, we can correct it to give us the radial velocity of the object with respect to the lsr.

We now look at the Doppler shifts we will observe for material some distance R from the galactic center, moving in a circular orbit with a speed $v(R)$. The situation is shown in Fig. 16.3 for $R < R_0$, but the result holds for $R > R_0$ (see Problem 16.1). The relative radial velocity is given by

$$v_r = v(R) \cos(90° - \theta) - v_0 \cos(90° - l) \qquad (16.6)$$

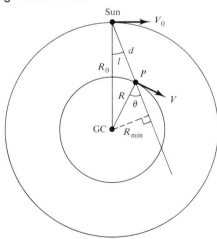

Figure 16.3 Differential rotation and radial velocities. The Sun is a distance R_0 from the galactic center. We observe an object at P, along a line making an angle ℓ with the line from us to the galactic center. P is a distance R from the galactic center.

$$= v(R) \sin \theta - v_0 \sin l$$

$$= R\Omega(R) \sin \theta - R_0\Omega_0 \sin l \tag{16.7}$$

We can measure l, but not θ, so we must eliminate it using the law of sines:

$$\sin(180° - \theta)/R_0 = \sin l/R \tag{16.8}$$

Simplifying the left-hand side gives

$$\sin \theta/R_0 = \sin l/R \tag{16.9}$$

Substituting into equation (16.7) gives

$$v_r = R_0\Omega_0(R) \sin l - R_0\Omega_0 \sin l \tag{16.10}$$

Factoring out the $R_0 \sin l$ gives

$$v_r = [\Omega(R) - \Omega_0]R_0 \sin l \tag{16.11}$$

Let's look along a line of sight at some galactic longitude l and see how the radial velocity changes with increasing distance d from the Sun. We first consider the case $l < 90°$ (the first quadrant of the galaxy). As we look at material closer to the galactic center, the quantity $\Omega - \Omega_0$ gets larger. This means that v_r gets larger. We see that this line of sight has a point of closest approach to the galactic center. This point is called the *subcentral point*. Of all the material along this line of sight, the material at the subcentral point produces the largest v_r. As we go beyond the subcentral point, we are re-crossing orbits, and v_r gets smaller. Eventually, v_r reaches zero, when the Sun's orbit is crossed again. For points beyond the Sun's orbit, $\Omega(R) < \Omega(R_0)$. This means that v_r is negative and increases in absolute value as d increases (Fig. 16.4).

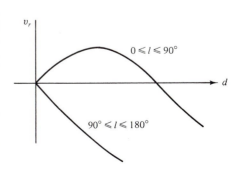

Figure 16.4 Radial velocity as a function of distance from the Sun, d, along a given line of sight. The upper curve is typical of ℓ between 0 and 90°. The maximum v_r corresponds to the point at which the line of sight passes closest to the galactic center. The close point with $v_r = 0$ corresponds to local material, and the far point with $v_r = 0$ corresponds to our line of sight recrossing the Sun's orbit. Inside the Sun's orbit, each v_r (except for the maximum) occurs twice, since the line of sight crosses each circle twice. Outside the Sun's orbit, each circle is crossed only once, and each v_r is reached only once. The lower curve corresponds to values of ℓ between 90° and 180°. All of these points are outside the Sun's orbit, so each circle is crossed only once, and each v_r is reached only once.

For $l > 270°$ (the fourth quadrant), the behavior of v_r is similar to that in the first quadrant, except that when v_r is positive in the first quadrant it is negative in the fourth quadrant, and when it is negative in the first quadrant it is positive in the fourth quadrant.

In the second quadrant ($90° < l < 180°$), all lines of sight pass only through material outside the Sun's orbit. There is no maximum v_r; it just keeps increasing with d. The behavior in the third quadrant is the negative of that in the second quadrant.

We can also find an expression for the relative transverse velocity. This will produce proper motions. The relative velocity is given by

$$v_T = v(R) \sin(90° - \theta) - v_0 \sin(90° - l)$$

$$= v(R) \cos \theta - v_0 \cos l \tag{16.12}$$

$$= R\Omega(R) \cos \theta - R_0\Omega_0 \cos l$$

From Fig. 16.3 we see that

$$R_0 \cos l = d + R \cos \theta \tag{16.13}$$

This gives

$$R \cos \theta = R_0 \cos l - d \tag{16.14}$$

Substituting this into equation (16.12) gives

$$v_T = \Omega(R)(R_0 \cos l - d) - R_0 \Omega \cos l$$

Grouping the terms with $\cos l$ gives

$$v_T = [\Omega(R) - \Omega_0]R_0 \cos l - \Omega(R)d \tag{16.15}$$

16.2.3   Oort Formulae

Early studies of the Milky Way were carried out in the visible part of the spectrum. Interstellar extinction limited these studies to nearby objects. It was therefore useful to find simplified approximations for equations (16.11) and (16.15), valid for $d << R_0$. These equations were derived by the Dutch astronomer, Jan Oort, and are known as the *Oort formulae*.

We use the fact that for R close to R_0, we can approximate $\Omega(R)$ by the first two terms in a Taylor series expansion about R_0. This allows us to write

$$[\Omega(R) - \Omega_0] = (d\Omega/dR)_{R_0}(R - R_0) \tag{16.16}$$

Since $\Omega(R) = v(R)/R$, we find the derivative

$$d\Omega/dR = (1/R)(dv/dR) - v(R)/R^2 \qquad (16.17)$$

Evaluating this at R_0 gives

$$(d\Omega/dR)_{R_0} = (1/R_0)(dv/dR)_{R_0} - v_0/R_0^2 \qquad (16.18)$$

We first apply this to the radial velocity [equation (16.11)], giving

$$v_r = [(dv/dR)_{R_0} - v_0/R_0](R - R_0) \sin l \qquad (16.19)$$

We can express $R - R_0$ in terms of d by using the law of cosines:

$$R^2 = R_0^2 + d^2 - 2R_0 d \cos l$$

Subtracting R_0^2 from both sides gives

$$R^2 - R_0^2 = d^2 - 2R_0 d \cos l$$

The left-hand side can be simplified as

$$R^2 - R_0^2 = (R - R_0)(R + R_0)$$

$$= (R - R_0)(2R_0)$$

where we have used the fact that $R + R_0 = 2R_0$ when $R \cong R_0$. We therefore have, ignoring the d^2 compared to the other terms,

$$(R_0 - R)(2R_0) = -2R_0 d \cos l$$

which simplifies to

$$R_0 - R = d \cos l \qquad (16.20)$$

We now substitute this into the radial velocity expression:

$$v_r = [v_0/R_0 - (dv/dR)_{R_0}]d \sin l \cos l \qquad (16.21)$$

Using the trigonometric identity

$$\tfrac{1}{2} \sin(2l) = \sin l \cos l$$

and defining

$$A = \tfrac{1}{2}[v_0/R_0 - (dv/dR)_{R_0}] \qquad (16.22)$$

equation (16.21) becomes

$$v_r = Ad \sin(2l) \qquad (16.23)$$

The quantity A is known as the *Oort A constant*. An interesting consequence of equation (16.23) is that v_r goes through two full cycles as l goes through one cycle. The observation of this distinctive pattern in stellar radial velocities was an early demonstration of the differential rotation of our galaxy.

We now look at the transverse velocities. This time we approximate

$$\Omega(R) - \Omega_0 = [(dv/dR)_{R_0} - v_0/R_0](R - R_0)/R_0 \qquad (16.24)$$

The quantity $\Omega(R)d$ is given by

$$\Omega(R)d = [\Omega_0 + (d\Omega/dR)_{R_0}(R - R_0)]$$

Using equation (16.20), this becomes

$$\Omega(R)d = \Omega_0 d - (d\Omega/dR)_{R_0}d^2 \cos l \qquad (16.25)$$

Since $d \ll R_0$, we can ignore the d^2 term in equation (16.25), leaving the simple result

$$\Omega(R)d = \Omega_0 d \qquad (16.26)$$

We now substitute this into the transverse velocity equation to give

$$
\begin{aligned}
v_T &= [(dv/dR)_{R_0} - v_0/R_0](R - R_0) \cos l - \Omega_0 d \\
&= [v_0/R_0 - (dv/dR)_{R_0}]d \cos^2 l - v_0 d/R_0
\end{aligned}
\qquad (16.27)
$$

We use the identity

$$\cos^2 l = \tfrac{1}{2}[1 + \cos(2l)]$$

to give

$$v_T = \tfrac{1}{2}[v_0/R_0 - (dv/dR)_{R_0}]d \cos l - \tfrac{1}{2}[v_0/R_0 - (dv/dR)_{R_0}]d \qquad (16.28)$$

We define the *Oort B constant* as

$$B = -\tfrac{1}{2}[v_0/R_0 + (dv/dR)_{R_0}] \qquad (16.29)$$

which allows us to simplify equation (16.28) to

$$v_T = d[A \cos(2l) + B] \qquad (16.30)$$

Using the relationship between proper motion and transverse velocity, this becomes

$$\mu = [A \cos(2l) + B]/4.74 \qquad (16.31)$$

From equations (16.22) and (16.29) we see that the sum and difference of the A and B constants provide us with interesting quantities (see Problem 16.3):

$$A - B = \Omega_0 \qquad (16.32)$$

$$A + B = -(dv/dR)_{R_0} \qquad (16.33)$$

In addition, for nearby material, the maximum radial velocity along any particular line of sight is (see Problem 16.4)

$$v_{max} = 2AR_0(1 - \sin l) \qquad (16.34)$$

By measuring v_{max} along various lines of sight we can use equation (16.24) to get an estimate of the quantity AR_0.

The values of the Oort constants have been measured by extensive optical observations. The quantity A is determined mostly from radial velocity measurements, and B is determined from proper motion measurements (and a knowledge of A). The currently accepted values are (assuming no large radial motions near the Sun)

$$A = 15 \text{ km s}^{-1}/\text{kpc}$$

$$B = -10 \text{ km s}^{-1}/\text{kpc}$$

The quantity R_0, our distance from the galactic center, is determined from studies of the distribution of globular clusters. For the past 20 years the generally accepted value was 10.0 kpc. However, data accumulated during that time points to a smaller value. As of the 1985 International Astronomical Union meeting, the recommended value was

$$R_0 = 8.5 \text{ kpc}$$

By using an agreed upon value, astronomers can be sure that they are using the same values when they compare their studies of various aspects of galactic structure.

Our orbital speed about the galactic center v_0 is determined from the quantity $A - B$, using equation (34.32) and the fact that $v_0 = \Omega_0 R_0$. There are additional ways of determining v_0. The adopted value had been 250 km/s, but the newly recommended value is 220 km/s.

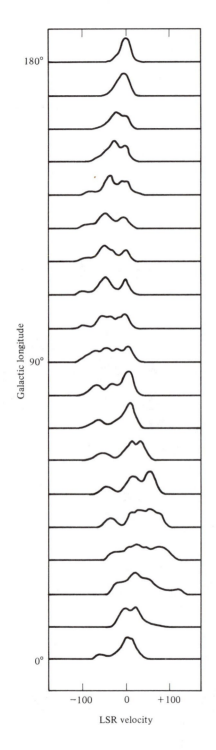

LSR velocity

EXAMPLE 16.1   Galactic Rotation

For the values of the galactic rotation constants, find the time it takes the Sun to orbit the galactic center and the mass interior to the Sun's orbit.

Solution

The time for the Sun to orbit is simply the circumference, $2\pi R_0$, divided by the speed v_0:

$$t_0 = 2\pi R_0/v_0$$

$$= (2\pi)(8.5 \times 10^3 \text{ pc})(3.1 \times 10^{13} \text{ km/pc})/(220 \text{ km/s})$$

$$= 7.5 \times 10^{15} \text{ s}$$

$$= 2.4 \times 10^8 \text{ yr}$$

The Sun orbits the galactic center in 240 million years. We find the mass interior to R_0 from equation (16.3).

$$M(R_0) = v_0^2 R_0/G$$

$$= \frac{(240 \times 10^5 \text{ cm/s})^2(8.5 \times 10^3 \text{ pc})(3.1 \times 10^{18} \text{ cm/pc})}{6.67 \times 10^{-8} \text{ dyn-cm}^2/\text{g}^2}$$

$$= 2.3 \times 10^{44} \text{ g}$$

$$= 1.1 \times 10^{11} \, M_\odot$$

16.3   DETERMINATION OF THE ROTATION CURVE

The rotation curve for material within the Sun's orbit can be determined reasonably well from 21-cm-line observations, like those shown in Fig. 16.5. In determining the rotation curve, we make two important assumptions: (1) the orbits are circular. This means that we need to determine $v(R)$ at only

Figure 16.5   Sample 21-cm profiles at 10° intervals of galactic longitude. In each spectrum, intensity is plotted as a function of Doppler shift velocity.

one point on the orbit for each R. (2) There is some atomic hydrogen all positive along any given line of sight. It is especially important that there be some hydrogen at the subcentral point of each line of sight.

The method takes advantage of the fact that, for lines of sight through the part of the galaxy interior to the Sun's orbit, there is a maximum Doppler shift. It is easy to inspect the 21-cm spectrum at each longitude and determine the maximum Doppler shift. We then assign that Doppler shift to material at the subcentral point (the point of closest approach to the galactic center) for that particular longitude. We can see from Fig. 16.3 that the distance of the subcentral point R_{min} from the galactic center is

$$R_{min} = R \sin l \qquad (16.35)$$

From equation (16.11), we see that if v_{max} is the maximum radial velocity along a given line of sight, then the angular speed $\Omega(R_{min})$ for that line of sight is given by

$$\Omega(R_0 \sin l) = (v_{max}/R_0 \sin l) + \Omega_0 \qquad (16.36)$$

By studying lines of sight with longitudes ranging from 0° to 90°, the corresponding values of R_{min} will range from 0 to R_0. This means that we measure $\Omega(R)$ once for each value of R from 0 to R_0. However, we have already said that if the material is moving in circular orbits, one measurement per orbit is sufficient to determine the rotation curve.

There are some limitations to this technique. We have already seen that the distribution of interstellar gas is irregular. If there happens to be no atomic hydrogen at the subcentral point for some line of sight we will see a v_{max} which is less than the value that we would see if there were material at the subcentral point. There are also problems arising from noncircular orbits. The effect of both of these problems can be reduced by repeating the procedure for the fourth quadrant. Because of the inclination of the galactic plane relative to the celestial equator, the fourth-quadrant studies must be done from the southern hemisphere, and have been done in Australia. For a number of years it seemed that there were disagreements between the first- and fourth-quadrant rotation curves, but we now think we understand them in terms of noncircular motions. There is further evidence for such motions in the large radial velocities observed for l close to 0° and 180°.

In addition, we cannot really cover the full range from 0 to R_0. For l close to zero, $\sin l$ is close to zero, and the Doppler shift is very small. Random motions of clouds are much larger than the radial velocity due to galactic rotation. Similarly, for l near 90°, Ω is close to Ω_0, providing a small radial velocity from galactic rotation. When we eliminate these ends, a reasonable rotation curve has been derived for R in the range 3–8 kpc. This is shown as part of Fig. 16.6.

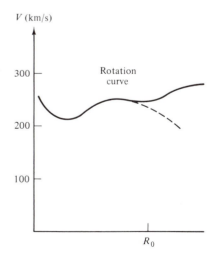

Figure 16.6 Rotation curve for the Milky Way. The curve inside the Sun's orbit is well determined from 21-cm studies. Outside the Sun's orbit, it is more difficult to determine.

The determination of the rotation curve outside the Sun's orbit is more difficult. There is no maximum Doppler shift along any line of sight. It is therefore necessary to independently measure v_0 and d, the distance from the Sun to the material being studied. From d and l, we can deduce R (see Problem 16.8). Until recently there was no reliable rotation curve for $R > R_0$. The best astronomers could do was to derive the mass distribution for $R < R_0$ from the rotation curve for $R < R_0$, and then make some assumptions on how the mass distribution would continue for $R > R_0$. From the assumed mass distribution, a rotation curve could be derived. It was assumed that there was relatively little mass outside the Sun's orbit, so the rotation curve was characterized by a falloff in $v(R)$ that was close to that predicted by Kepler's third law. This rotation curve is depicted as a dashed line in Fig. 16.6.

However, recent observations of molecular clouds have provided a direct method of measuring the rotation curve outside the Sun's orbit. Molecular clouds associated with HII regions were studied. The radial velocities were determined from observations of carbon monoxide (CO) emission from the clouds. The distances to the stars exciting the HII regions were determined by spectroscopic parallax. This gives a reliable rotation curve at least out to about 16 kpc. The results are shown in Fig. 16.6, and you can see that there is a substantial difference between the measured curve and the assumed curve. There is no falloff in $v(R)$ out to 16 kpc, and there may even be a slight rise. This means that there is much more mass outside the Sun's orbit than previously thought!

We can see from equation (16.3) that if $v(R)$ is constant from 10 to 20 kpc, $M(20\ \text{kpc})$ will be twice $M(10\ \text{kpc})$. This means that there is as much mass between 10 and 20 kpc as there is out to 10 kpc. However, the luminosity of our galaxy is falling very fast as R increases. Since the luminous matter is mostly in the disk, it would seem that this extra mass cannot be part of the disk. Current thinking places the extra mass in the halo (spherical distribution) of the galaxy. We still have little idea on the form of this matter. It has been suggested that it can be in the form of faint red stars, but recent results make this seem unlikely. We will see that this problem exists for other galaxies, and we will discuss it farther in Chapter 17.

Once we have a rotation curve for our galaxy, it is possible to use measured Doppler shifts to determine distances to objects. Since these distances are determined from the motions of the objects, they are called *kinematic distances*. For any particular object, we measure v_r and l. We then determine its orbital angular speed Ω from

$$\Omega = (v_0/R \sin l) + \Omega_0 \tag{16.37}$$

It is assumed we know R_0 and Ω_0. Once we know Ω we use the known rotation curve to find the value of R to which the Ω corresponds.

There are a few limitations to this technique. It doesn't work for material whose radial velocity due to galactic rotation is less than that due to the random motions of the clouds. This rules out material near longitudes of 0° and 180°, as well as material very close to the Sun.

Another problem arises for material inside the Sun's orbit. There are two points along the same line of sight that produce the same radial velocity. (The one exception is the subcentral point.) Both of these points are the same distance from the galactic center, but they are different distances from us. This problem is called the *distance ambiguity*. We can use the rotation curve to say that the object is in either of two places, and we must then use other techniques to resolve the ambiguity.

16.4   AVERAGE GAS DISTRIBUTION

To understand star formation on a galactic scale we must know how the interstellar gas, out of which the stars will be formed, is distributed in the galaxy. We are interested in the average distributions of various constituents. By "average" we mean that we are interested only in the large-scale structure. We would like to know the radial distribution of the interstellar gas. [Remember, this is not the same as $M(R)$, which includes mass in all forms.] We would also like to know the degree to which the gas is confined to the plane. We can express this as a thickness of the plane as determined from various constituents. We would also like to know whether the thickness is constant, or whether it varies with position in the galaxy. Finally, we would like to know if the plane of the galaxy is truly flat, or if it has some large-scale bumps and wiggles.

The radial distributions of various constituents are shown in Fig. 16.7. We first look at the HI. The amount of HI doesn't fall off very quickly as one goes to larger R. For example, the mass of HI interior to R_0 is about $1 \times 10^9 \, M_\odot$, and the mass exterior to R_0 is about $2 \times 10^9 \, M_\odot$. (Of course, this larger mass is spread out over a larger volume. See Problem 16.7.) Note that the mass of HI is only about 1% of the total mass interior to a given radius. This means that the gas does not provide most of the gravitational force in the galaxy. It just responds to the force of the stars (and whatever "dark" matter is in the halo).

We generally express the thickness of the plane by finding the separation between the two points, one above and one below, at which the HI density falls to half of its value in the middle of the plane. This is called the full-width at half-maximum, or *FWHM*. At the orbit of the Sun, the thickness of the HI layer is about 300 pc. At $R = 15$ kpc, the thickness is about 1 kpc (Fig. 16.8). This means that the plane gets thicker as one gets farther from the center of the galaxy. This is called the "flaring" of the plane. In addition, we find that the plane isn't flat. It has a warp to it, like the brim of a hat (Fig. 16.8). The warp is most prominent outside the Sun's orbit. The bend

Figure 16.7 Radial distribution of HI and H_2, with the H_2 deduced from CO observations. The H_2 is more closely concentrated toward the center of the galaxy.

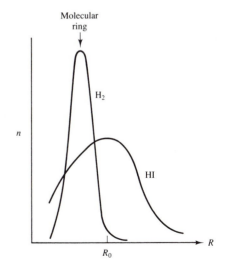

Figure 16.8 Height of the plane of the Milky Way, as determined from 21-cm observations. The solid contours are heights (in kpc) above the plane, and dashed contours are heights below the plane.

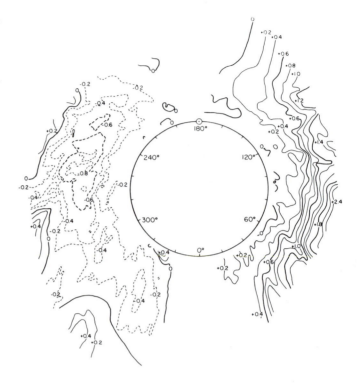

is upward in the first and second quadrants and downward in the third and fourth quadrants.

The distribution of molecular hydrogen H_2 is generally deduced indirectly from observations of CO. There are still disagreements over how to derive the H_2 abundance from the intensity of the CO emission. The derived radial distribution of H_2, in comparison to that of HI, is shown in Fig. 16.8. The abundance of H_2 falls off more rapidly with R than does that of HI. Inside the Sun's orbit, the mass of H_2 is approximately equal to that of HI, about $1 \times 10^9\ M_\odot$. Outside the Sun's orbit the mass of H_2 is about $5 \times 10^8\ M_\odot$, about one-quarter that of HI. There appears to be a peak in the H_2 distribution at about 6 kpc from the galactic center. This is sometimes called the *molecular ring*. It has been suggested that most of the H_2 seems to be concentrated into a few thousand giant molecular clouds, rather than into a larger number of small clouds.

Molecular hydrogen is more closely confined to the galactic plane than is atomic hydrogen. The thickness of the plane in H_2 is about half that in HI. We have seen that one feature of population I material is its confinement to the galactic plane. We therefore think of molecular clouds as representing a more extreme population I than that represented by the HI clouds. This might indicate that the molecular clouds, as a whole, are more recently formed than the HI clouds. Finally, the H_2 shows the same flaring and warp as the HI.

Additional information on the total gas distribution, H plus H_2, comes from observations of gamma rays. These gamma rays are created when cosmic rays strike protons. It doesn't matter what types of atoms or molecules the protons are in. The results are still somewhat uncertain, but provide an additional constraint. The general conclusion from such observations supports those conclusions from the CO observations.

16.5 SPIRAL STRUCTURE IN THE MILKY WAY

Early photographic surveys revealed a number of nebulae with a spiral appearance. In Chapter 17 we will discuss the reasons for believing that these spiral nebulae are distant galaxies, distinct from our own, rather than part of the interstellar medium in our galaxy. Some examples of spiral galaxies are shown in Fig. 17.3. Given that many other galaxies seem to have a spiral appearance, it is reasonable to wonder whether our galaxy is also a spiral. Unfortunately, from our vantage point within the galaxy, it is very difficult to see the overall pattern.

However, we can detect certain similarities between our galaxy and spiral galaxies. In particular, spiral galaxies have a significant amount of gas and dust. The amount of gas and dust in our galaxy is comparable to that in other spirals. For this reason, astronomers have been encouraged in trying to unravel the spiral structure of our galaxy.

16.5.1 Optical Tracers of Spiral Structure

When we look at spiral galaxies we see that the spiral arms are not continuous bands of light. Rather, they appear to contain knots of bright stars and glowing gas. For example, it appears that HII regions and OB associations trace out spiral arms in other galaxies. For this reason, we have tried to see if the HII regions we can see optically in our galaxy form any distinct pattern. By using optical observations we can rely on distances determined by spectroscopic parallax for the stars exciting the HII regions. Similarly, we can also look at the distribution of OB associations. A drawback is that, with optical observations, we cannot see very far along the plane of the galaxy.

The distribution of HII regions and OB associations is shown in Fig. 16.9. It is clear that the placement is not entirely random. We seem to see at least pieces of connected chains of HII regions and OB associations. These pieces have been identified as a series of named "arms," identified by the constellation in which they are most prominent. This is all a tantalizing hint of spiral structure, but not a definitive picture.

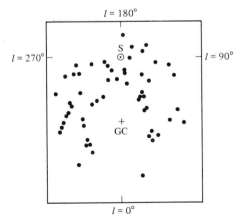

Figure 16.9 The locations of HII regions in the Milky Way. Some efforts have been made to connect them into spiral arms.

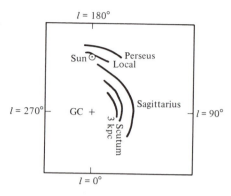

Figure 16.10 Arms of the galaxy as determined from 21-cm observations. The Perseus, Sagittarius, and Scutum arms are named after constellations in which the arms are prominent. The 4-kpc arm is approximately that distance from the galactic center.

Figure 16.11 Spiral structure of the galaxy as determined from giant molecular cloud complexes. The sizes of the circles indicate the masses of the complexes, as indicated in the upper right. The 4-kpc and Scutum arms are drawn in from the 21-cm maps. The Sagittarius arm is drawn in as it would best fit the CO data.

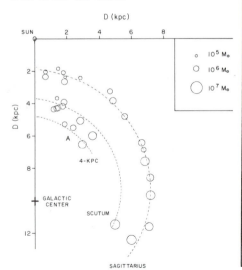

16.5.2 Radio Tracers of Spiral Structure

We can get a view on a larger scale by using radio observations to look at the distribution of interstellar gas. This allows us to see across the whole galaxy. We can utilize kinematic distances, but we must still deal with the distance ambiguity. Initial radio studies of the interstellar gas and spiral structure involved the 21-cm line. Some results are shown in Fig. 16.10, and are inconclusive, though, again, long connected features have been identified.

With the discovery of molecular clouds, it was hoped they would reveal the spiral structure of our galaxy. This is because the optical tracers of spiral structure we see in other galaxies—OB associations, HII regions, dust lanes—are all associated with giant molecular clouds. A number of groups have carried out large-scale surveys of emission from CO throughout the galaxy.

Most of the work has involved material inside the Sun's orbit. The problem is that the distance ambiguity makes it difficult to uniquely place all of the emitting regions. One approach has been to take specific models for

Figure 16.12 Molecular clouds outside the solar circle and spiral structure. The cloud masses are denoted by the symbols, as shown at the lower left. The circle at 13 kpc is drawn in for reference.

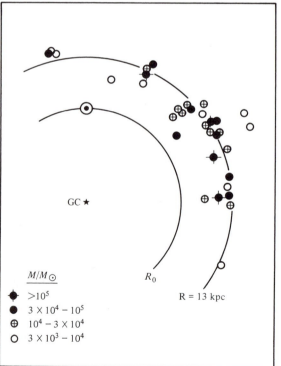

spiral arms and predict the outcome of observations. Again, pieces of arms have been identified (Fig. 16.11). Outside the sun's orbit, the approach is more direct. Since we have a rotation curve for the outer part of the galaxy, and there is no distance ambiguity, it is easier to trace out the large-scale structure. It is in the region outside the Sun's orbit that we see the best evidence for spiral structure, with some features being traced over at least a quarter of the galaxy (Fig. 16.12). This work is still going on.

Much of our understanding of spiral structure comes from comparing our galaxy to other galaxies. Therefore, we will leave further explanation of spiral structure until Chapter 17, when we look at other galaxies.

16.6   THE GALACTIC CENTER

Ever since it was realized that the galactic center is someplace else from where we are, astronomers have wondered about its nature. Is it simply a geometric location, or is it the site of unusual activity? Visual extinction in the galactic plane makes optical studies of the galactic center virtually impossible. We are able to observe the galactic center in the radio part of the spectrum, detecting continuum emission from ionized gas and line emission from molecular clouds. Recently, infrared observations have been used. Continuum observations tell us about the dust temperature and opacity, and spectral line observations tell us about the neutral and ionized gas. Because of the great distance to the galactic center, many observations are limited by poor angular resolution. Maps of infrared and radio emission are shown in Figs. 16.13 and 16.14.

Figure 16.13   Maps of the galactic center. (a) Far infrared map, made at a wavelength of 10.6 μm. Individual sources are numbered. (b) A map of radio emission at a wavelength of 6 cm. Much of this pattern is viewed as coming from a tilted ring of ionized material.

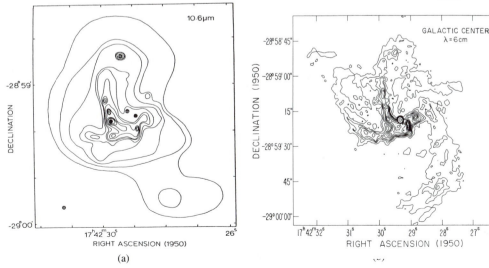

(a)

(b)

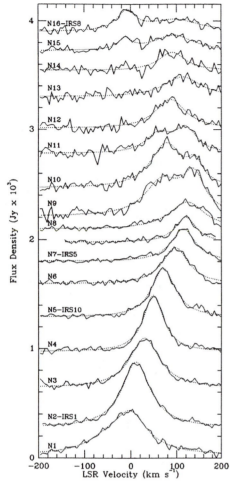

Figure 16.14 Distribution of an infrared spectral line. The spectra were taken at a sequence of positions along a single line. At each position the spectrum is shown as a plot of intensity vs. velocity needed to produce a given Doppler shift. Positive velocities correspond to material moving away, and negative velocities correspond to material moving toward us, relative to the average velocity of the material. The shift in velocity from position to position is interpreted as being due to the orbital motion of the material near the galactic center. This orbital motion can be used to determine the mass of the material at the galactic center.

Studies of the ionized gas show a bent arc of emission perpendicular to the galactic plane. This structure is about 15 arc min in extent. It shows a filamentary structure. This feature is also seen in the infrared. The emission seems to be a combination of thermal (free–free) radiation and nonthermal (synchrotron) radiation.

The neutral gas is concentrated into a disk which is tilted with respect to the galactic plane. The outer part of this disk is a 2 pc radius ring of gas and dust. From the orbital speed of the material in this ring, the mass interior to the ring is deduced to be 2–5 million $M_\odot$. (See Problem 16.26.) The material inside the ring is at least partially ionized gas, and shows up in radio maps as well as in the infrared. High-resolution maps, like that shown in Fig. 16.14, show a curved feature that is sometime called the "spiral arm," because of its appearance. It is not related to the spiral arm in the disk of the galaxy.

There has been considerable speculation on the nature of the central object. It has been suggested that it might be a few million $M_\odot$ black hole. Currently, observations do not limit the size of the object to the point where we can say that it must be smaller than the Schwarzschild radius for this mass. However, it has been argued from the dynamics of the material near the center that the central mass must be highly concentrated.

There is indirect evidence for explosive activity in the galactic center region. For example, there is an armlike feature in our galaxy, some 3 to 4 kpc from the galactic center, called the *3-kpc arm*, which appears to be expanding at about 50 km/s. Speculation is that this expansion is due to some explosion in the relatively recent past (see Problem 16.9). We see other similar features closer in, suggesting that this activity has taken place on a

TABLE 16.1   Scale Heights and Velocity Dispersions

Constituent	Scale height (pc)	Velocity dispersion (km/s)
O Stars	50	5
GMCs	60	7
Galactic clusters	80	
HI clouds	120	
F stars	190	20
Planetary nebulae	260	20
M stars	350	
RR Lyrae stars		
Short period	900	25
Long period	2000	30
Globular clusters	3000	

continuing basis. We will see in Chapter 17 that the activity in our galactic center is small compared to that in many galaxies. However, it gives us our best opportunity to study such activity "close up." For this reason, study of the galactic center is a very active field.

CHAPTER SUMMARY

We saw in this chapter how many of the parts of the Milky Way, stars, and interstellar medium are arranged in the galaxy.

We saw how the rotation curve tells us about the mass distribution in the galaxy, and how the rotation curve is determined. Different techniques are needed for material inside and outside the Sun's orbit about the galactic center. Once the rotation curve is known, velocities of objects can be used to help estimate the distances.

In looking at the average gas distribution, we found the HI is extended beyond the Sun's orbit, while the number of molecular clouds falls off more sharply with dis-

tance from the center of the galaxy. The molecular clouds are also more tightly concentrated toward the galactic plane than the HI clouds. This concentration toward the plane suggests that the molecular clouds are younger.

We discussed evidence for spiral structure in the Milky Way and the difficulties in tracing out spiral arms in our galaxy. We saw that tracers for spiral structure include HII regions, OB associations, and molecular clouds.

In looking at the galactic center, we found that it contains a small, active region. The center has a mass concentration which could be a black hole of a few million $M_{\odot}$.

QUESTIONS AND PROBLEMS

16.1 Show that the radial velocity equation (16.11), derived for the first quadrant of the galaxy, holds for the second quadrant.

16.2 For material observed at a radial velocity v_r along a line of sight at galactic longitude l, find an expression for the separation between the near and far points producing that v_r.

16.3 Derive the relationship between the Oort constants, equations (16.32) and (16.33).

16.4 Show that the maximum v_r along a line of sight for l slightly less than 90° (so that $d << R_0$) is given by equation (16.34).

16.5 Calculate $\Omega(R)$ and $v(R)$ for the following density models: (a) all the mass M at the center of the galaxy; (b) a constant density, adding up to a mass $M(R_0)$ at the Sun's orbit, and no mass beyond.

16.6 For what mass distribution is $v(R)$ constant?

16.7 Convert the HI and H_2 masses given in the chapter into: (a) average volume densities and (b) average surface densities for the regions $R < R_0$ and $R_0 < R < 2R_0$ (assuming all of the mass outside R_0 is between R_0 and $2R_0$).

16.8 For material outside the Sun's orbit, derive an expression to convert observed v_r and d into $\Omega(R)$ and R.

16.9 (a) Estimate the age of the 3-kpc arm from its radius and expansion speed. (b) Is this an upper or lower limit to the age? Explain.

16.10 How can we be sure that $v(R)$ tells us the *total* mass interior to R?

16.11 Why should the kinematic and dynamical definitions of the local standard of rest be the same?

16.12 Draw a diagram showing two points with the same radial velocity along the same line of sight. Show that the component of their velocity along the line of sight is the same.

16.13   Why is it sufficient to measure the rotation curve only for the subcentral points for $R < R_0$?

16.14   Show that the radial velocity of a point in the fourth quadrant is the negative of that of the corresponding point in the first quadrant.

16.15   We can only measure kinematic distances for points with $v_r > 10$ km/s. What range of R can we study?

16.16   Why can't we use transverse velocities to measure the rotation curve of our galaxy?

16.17   Contrast the method for measuring the rotation curve inside the Sun's orbit with that for measuring it outside the Sun's orbit.

16.18   For $l = 45°$, we observe $v_r = +30$ km/s. What are R and d?

16.19   If $v(R) = 220$ km/s for $R_0 < R < 2R_0$, find expressions for v_r as a function of (l, R).

16.20   For the constants given, what is $(dV/dR)_{R_0}$? How does this compare with the idea of a flat rotation curve?

16.21   Describe the difficulties in studying spiral structure in our galaxy.

16.22   Compare using GMCs and HII regions to study spiral structure.

16.23   Why is it easier to study spiral structure in the outer part of our galaxy than in the inner part?

16.24   What is the Schwarzschild radius for a 5×10^6 $M_\odot$ black hole? How does this compare with the sizes of the structures found in the galactic center?

16.25   When the galactic constants R_0 and v_0 changed from 10.0 kpc and 250 km/s to 8.5 kpc and 220 km/s, respectively, by what factor did the mass inside the Sun's orbit change?

16.26   The mass distribution for material near the galactic center can be determined from studies of the rotation curve close to the center. When the rotation velocity is plotted vs. log R for the inner 10 pc, the result is approximately a straight line. $V(R)$ is 200 km/s at $R = 2$ pc and 70 km/s at $R = 10$ pc. (a) Use this data to find $M(R)$ for 2 pc $< R <$ 10 pc. (b) Assuming a spherical distribution, find $\rho(R)$.

Normal Galaxies

Our study of the Milky Way has been aided greatly by studies of other galaxies. However, for a long time it wasn't clear that the spiral nebulae we see in the sky are really other galaxies, far from the Milky Way, and comparable to the Milky Way in size and mass. From their appearance, it might just as well be assumed that these nebulae are small, nearby objects, just as HII regions are part of our galaxy.

The issues were crystallized in 1920 in a debate between Harlow Shapley and Heber D. Curtis. Curtis argued that spiral nebulae were really other galaxies. His argument was based on some erroneous assumptions. First, he thought that our galaxy is smaller than it actually is. Also, he confused novae in our galaxy with supernovae in other galaxies. Shapley thought that the spiral nebulae were part of our galaxy, partly based on an erroneous report of a measurable proper motion for some of the nebulae.

The issue was settled in 1924 by the observational astronomer Edwin Hubble (after whom the Space Telescope was named). Hubble studied Cepheids in three spiral nebulae (including the Andromeda galaxy), and clearly established their distance as being large compared to the size of the Milky Way. There is still some problem with Hubble's analysis, involving the type I vs. type II Cepheids. However, even this factor of 2 error in distance was not enough to alter the basic conclusion that spiral nebulae are not part of our galaxy. Following this work, Hubble made a number of pioneering studies of other galaxies, essentially opening up the whole field of extragalactic astronomy.

17.1 TYPES OF GALAXIES

In his studies Hubble realized immediately that not all spiral galaxies have the same appearance. Furthermore, he found galaxies that do not have a spiral structure. Hubble classified the galaxies he studied according to their basic appearance. It was originally thought that the different types of galaxies

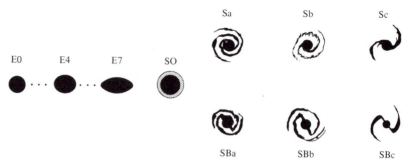

Figure 17.1   Hubble classification of galaxies. Ellipticals range from E0 (round) to E7 (the most oblate). The regular spirals are divided according to the relative size of the nucleus and disk, and the tightness of the spiral arms. The Sa's have the largest nuclei, and the tightest arms, while the Sc's have the smallest nuclei and the most open arms. The barred spirals SB follow the same classification as the normal spirals. S0's have nuclei and very small disks, but no spiral arms.

represented different stages of galactic evolution. (Similarly, some astronomers thought that different spectral-type stars along the main sequence were different evolutionary states of the same star.) We now know that this is not the case. However, Hubble's classification scheme, depicted in Fig. 17.1, is still quite useful.

17.1.1   Elliptical Galaxies

Elliptical galaxies, as their name suggests, have a simple elliptical appearance. Some examples are shown in Fig. 17.2. The ellipticals are classified according to their degree of eccentricity. The ones that look spherical (zero eccentricity) are called E0, and the most eccentric are called E7.

The most common types of elliptical galaxies are called *dwarf ellipticals*, since they are also the smallest. Their sizes are typically a few kiloparsecs and their masses are only a few million solar masses. More spectacular are the *giant ellipticals*, with extents up to 100 kpc and masses of about 10^{12} $M_{\odot}$, with some masses up to a factor of 10 higher.

The gas content of ellipticals is low. Studies of HI, using 21-cm emission, as well as IRAS observations of the weak emission from the dust, suggest that the mass of the interstellar medium may be up to about 1% of the mass of the stars we see. This low gas content rules out the possibility that ellipticals eventually flatten to form spirals. The continuing process of star formation in a galaxy depletes the supply of interstellar matter, so if spirals are merely evolved ellipticals, we have no way of understanding why spirals have more gas and dust than ellipticals.

Ellipticals generally contain an evolved stellar population, with no O and B stars. However, their metal abundances are not low. Giant ellipticals have metal abundances that are quite high, about twice the solar value.

(a)   (b)

Figure 17.2   Elliptical galaxies. (a) M87, in Virgo, which is a giant elliptical, type E0. (b) M32, in Andromeda, which is a dwarf elliptical companion to the Andromeda galaxy (M31) and is type E2.

Some ellipticals are rotating very slowly. They have a higher ratio of random velocities to rotational velocities than do spirals. We think that their slow rotation means they could collapse without much flattening. Remember, when we discussed collapsing interstellar clouds, we found that rapid rotation retards collapse perpendicular to the axis of rotation, resulting in the formation of a disk.

We can also use photometry to study the brightness distribution in ellipticals. Since we see the galaxy projected as a two-dimensional object on the sky, it is convenient to speak of the luminosity per unit surface area $L(r)$, where r is the projected distance from the center of the elliptical. Studies show that the light from most ellipticals can be described well by the simple relationship (known as de Vaucouleur's law)

$$L(r) = L(0)\exp[-(r/r_0)^{1/4}] \qquad (17.1)$$

In this expression, $L(0)$ and r_0 are constants. The values of $L(0)$ are found not to vary by very much, with a typical value of about $2 \times 10^5 \, L_\odot/\mathrm{pc}^2$. The values of r_0, however, show a very large spread.

Figure 17.3 Various types of spiral galaxies. (a) NGC 2811, type Sa. (b) NGC 4622, type Sb. (c) NGC 5364, type Sc. (d) NGC 4565, type Sb, viewed edge on.

17.1.2 Spiral Galaxies

Spirals make up about two-thirds of all bright galaxies. They are subdivided into classes Sa, Sb, and Sc. The two important features in the classification are: (1) the openness or tightness of the wrapping of the spiral pattern; and (2) the relative importance of the central bulge and the disk of the galaxy. Sa galaxies have the largest bulges and the tightest wound arms. Sc have the smallest bulges and the most open arms. (We think that the Milky Way is between an Sb and Sc.) Different types of spirals are shown in Fig. 17.3.

Some spiral galaxies have a bright bar running through their center, out to a point where the arms appear to start. These are called *barred spirals*. Some examples are shown in Fig. 17.4. The barred spirals are also subclassified into SBa, SBb, and SBc, according to the same criteria as the Sa, Sb, and Sc. In general, barred spirals have grand design spiral patterns.

Spirals of a given type may show a clear-cut, continuous spiral pattern running through the galaxy. These are referred to as *grand design spirals*. Other spirals have a less-organized appearance. These are called *flocculent spirals*. A comparison is shown in Fig. 17.5.

It has recently been realized that spirals of the same type can have different luminosities. This point is important in trying to determine the distances to galaxies that are so far away we cannot distinguish individual stars or HII regions. At the suggestion of Sydney van den Bergh, astronomers have recently taken to adding a *luminosity class* to the spiral classification. This is indicated with a I through V following the Hubble classification, with I being the brightest (just as for stars). Efforts are still underway to find other properties of a spirals that correlate with luminosity class. In this way the luminosity of a galaxy can be determined without our needing to know its distance. (Similarly, the luminosity class of a star can be determined from the shapes of certain spectral lines, allowing us to know the absolute magnitude of a star without knowing its distance.) Once the absolute magnitude of a galaxy is known, and its apparent magnitude is observed, its distance can be determined.

An important feature of spirals is the obvious presence of an interstellar medium—gas and dust. Even when a spiral is seen edge-on, we can tell it is a spiral by the presence of a lane of obscuring dust in the disk of the galaxy. The light from spirals contains an important contribution from a relatively small number of young blue stars, suggesting that star formation is still taking place in spirals. Where galaxies are found in clusters, to be discussed in Chapter 18, approximately 80% of the galaxies are ellipticals. Outside of clusters, approximately 80% are spiral. Typical radii for the luminous part of the disk in spirals are about 10 to 30 kpc. Stellar masses of the galaxies that we can see are in the range 10^7–10^{11} $M_\odot$.

From optical studies, we can get a good idea of the distribution of young stars as we go farther out in the disks of spiral galaxies. A typical result is shown in Fig. 17.6. The luminosity of the disk falls off very sharply with r,

(a)

(b)

(c)

Figure 17.4 Various types of barred spirals. (a) NGC 175, type SBab. (b) NGC 1530, type SBb. (c) NGC 1073, type SBc.

(a)   (b)

Figure 17.5 Grand design and flocculent spirals. (a) A grand design galaxy, M81, type Sb. (b) A flocculent galaxy, M94, also type Sb.

the distance from the center. We can approximately fit the observed falloff with an exponential expression. That is, if L is the luminosity at the center, then $L(r)$, the luminosity at radius r, is given by

$$L(r) = L_0 \exp(-r/D) \qquad (17.2)$$

In this expression, D is called the *luminosity scale length* and gives a measure of the characteristic radius of the galaxy as seen in visible light. Typical values of D are about 5 kpc. This means that the luminosity of the disk of a galaxy fall to $1/e$ of its peak value at $r = 5$ kpc.

Figure 17.6 Blue luminosity vs. distance from center, in a typical spiral. At each radius, an average luminosity is taken around a full circle, so the effects of spiral arms are smoothed out.

17.1.3 Other Types of Galaxies

There is an additional type of galaxy that has certain features in common with spirals, but does not show spiral arms. This type is called *S0* (Fig. 17.7). The bulge in an S0 is almost as large as the rest of the disk, giving the galaxy

TABLE 17.1   Properties of Spirals and Ellipticals

Property	Spirals	Ellipticals
Gas	Yes	None
Dust	Yes	None
Young stars	Yes	None
Shape	Flat	Round
Stellar motions	Circular rotation	Random
Color	Blue	Red

an almost spherical appearance. Some S0's also contain gas and dust, suggesting that they belong in the spiral classification. However, most S0's have no detectable gas. The role of S0's is still not well-understood.

Some galaxies have no regular pattern in their appearance. These are called *irregular galaxies*. The Magellenic clouds, shown in Fig. 17.8 are irregular companions to our own galaxy. Irregulars make up a few percent of all galaxies. We distinguish between two types of irregulars: Irr I galaxies

(a)

(b)

Figure 17.7   Examples of S0 galaxies. (a) M84 (NGC 4374), type S0. (b) NGC 2859, type SB0.

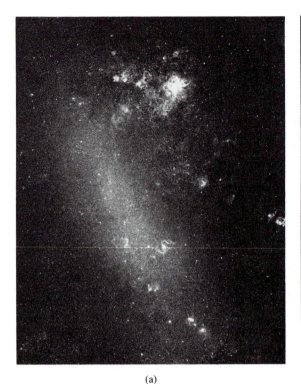

(a)

(b)

Figure 17.8   The Magellenic clouds. (a) Large cloud. (b) Small cloud.

are resolved into stars and nebulae; Irr II galaxies just have a general amorphous appearance.

Peculiar galaxies have a general overall pattern, but also have some irregular structure indicative of unusual activity in the galaxy. An example is shown in Fig. 17.9.

There are two types of galaxies that are characterized by a very bright nucleus. *Seyfert galaxies* are named after their discoverer, Carl Seyfert, who reported their existence in 1943. These are spiral galaxies with a bright, small nucleus. The spectra show broad emission lines, indicating a very hot or energetic gas, Seyferts make up about 2 to 5% of all spiral galaxies. An example of a similar phenomenon in ellipticals is found in *N galaxies* (where the N stands for ''nucleus''). There is also a class of galaxies that give off strong radio emission, *radio galaxies*. These *active galaxies* will be discussed in Chapter 19.

17.2   STAR FORMATION IN SPIRAL GALAXIES

The existence of O stars in spiral galaxies, detected either directly or via the HII regions they illuminate, tells us that massive star formation is still taking place. We discussed the basic ideas of star formation in the interstellar me-

Figure 17.9 This galaxy, M82, is the scene of very unusual activity. At first it was thought that this galaxy was exploding. However, it just seems to have undergone a rapid wave of star formation. Galaxies like this are known as starburst galaxies.

dium in Chapters 14–16. We would like to apply the ideas we developed when studying star formation in our galaxy to help us understand other galaxies. In turn, our understanding of other galaxies will help in our analysis of our galaxy. We can ask a number of questions about star formation in galaxies:

1. What is the large-scale distribution of star-forming material? How does it vary in the disk with distance from the galactic center? How does it vary with distance from the central plane of the disk?
2. Is the interstellar medium concentrated into the spiral arms?
3. How do the sizes of molecular clouds compare with those in our galaxy? Are the physical conditions within the star-forming clouds the same?

As with our galaxy, astronomers use a variety of observational techniques to study the interstellar medium in other galaxies. Optical images

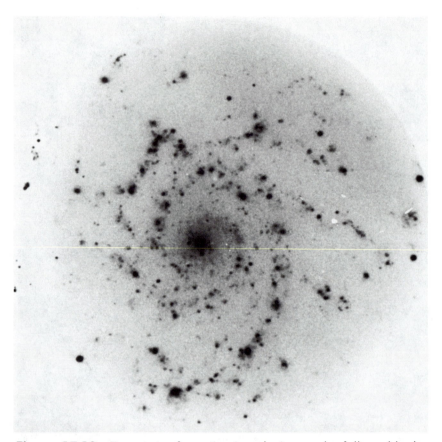

Figure 17.10   Recent star formation in galaxies can be followed by locating the O stars. We can do this indirectly by looking for the HII regions excited by the O stars. HII regions are strong sources of Hα emission. Photographs like this one, taken through an Hα filter show the locations of the HII regions.

reveal the positions of large concentrations of dust, and show us the distribution of bright stars and HII regions. For studies of the HII regions, it is helpful to look at a galaxy through a filter that only passes Hα light (Fig. 17.10). The recent development of optical imaging systems, such as the CCDs discussed in Chapter 4, has made more detailed optical studies of galaxies possible (Fig. 17.11).

In studying the interstellar medium of our galaxy or any other, radio observations play an important role. Except for the nearest galaxies, single-dish radio observations do not provide much spatial detail. However, with the extensive use of interferometers we have now obtained very detailed maps of many galaxies (Fig. 17.12). Continuum observations can be used to study the positions of HII regions and young supernova remnants, both signs of relatively recent star formation.

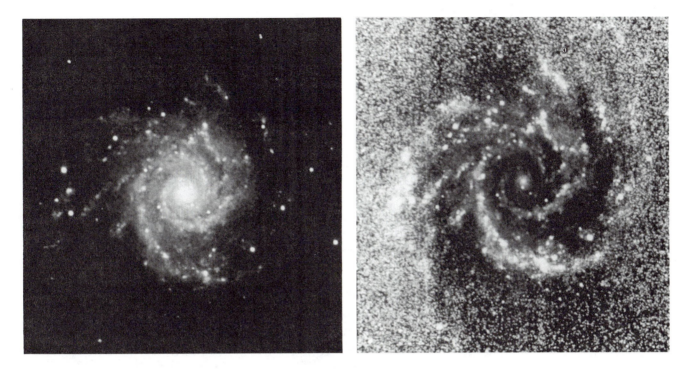

Figure 17.11 This shows how we can process a CCD image of a spiral to emphasize the spiral structure. The unprocessed image is on the left. In the image on the right, the average brightness at each distance from the center is subtracted out, highlighting things that stand out from the average, such as spiral arms.

Figure 17.12 21-cm maps of the Andromeda Galaxy, M31

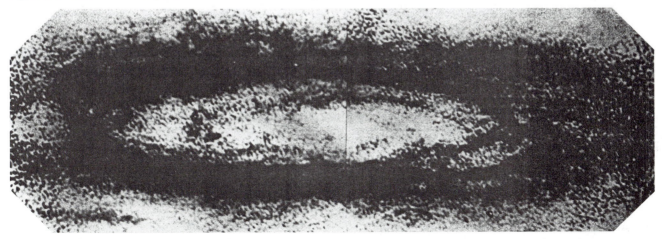

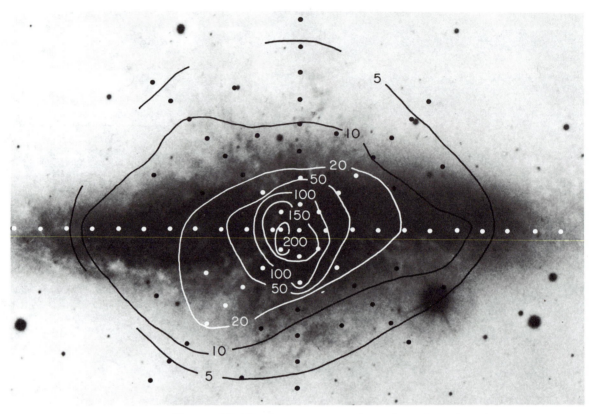

Figure 17.13 CO emission in the galaxy M82. The CO emission is rep-resent by contours of differented emission levels.

In our galaxy, we have found that star formation takes place in molecular clouds, as opposed to HI clouds. These molecular clouds must be observed through their trace molecular constituents, such as CO. Most of the CO studies (Fig. 17.13) of galaxies to date have been with single dishes, since interferometry at millimeter wavelengths is still being developed. Recently, millimeter interferometers have been used to study CO emission from the centers of a few galaxies.

The single-dish CO observations of spirals reveal a sharp falloff in brightness with radius, similar to that of the visible light. The falloff in CO emission may indicate the true gas distribution (Fig. 17.14). However, it may be due to the fact that the gas gets colder, and therefore radiates less strongly, where there are fewer stars to heat it. There may still be significant amounts of molecular gas in the outer parts of the spiral galaxies.

In studies of several spirals it is found that the relative amounts of molecular and atomic hydrogen vary significantly. These variations occur both within a galaxy and from galaxy to galaxy. Within a galaxy the general

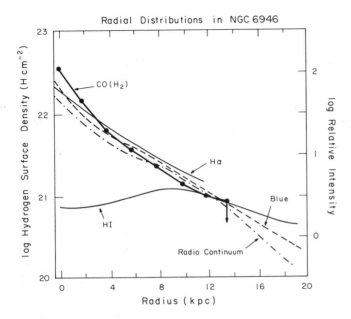

Figure 17.14   The CO brightness is plotted as a function of distance from the center of the galaxy, NGC 6946. For comparison, the Hα luminosity, blue light, and radio continuum distributions are shown.

trend is to have the molecular hydrogen abundance fall off faster with radius than the atomic hydrogen abundance. We find some galaxies in which molecular hydrogen makes up over half of the interstellar medium, and others in which it seems to be less than 10%, as determined from a deficiency of CO. In galaxies that seem to be deficient in molecular hydrogen, we still don't have observations with sufficient resolution to tell us whether this is because they contain fewer molecular clouds than other galaxies do, or whether the clouds are less dense. Also, even galaxies that appear to be deficient in molecular clouds have O stars. This tells us that we still do not understand fully the connection between molecular clouds and massive star formation.

Studies of spiral structure have been limited by poor angular resolution for single-dish studies. However, sufficient resolution is available to study nearby galaxies. Figure 17.15 shows a CO map of a section of M31 superimposed on an optical photograph. The CO emission is seen in the direction of the arms, but not in between. It therefore appears that there are at least some galaxies for which the molecular clouds are confined to the spiral arms.

In Chapter 16 we saw that once star formation starts in a cloud, the heated dust glows in the infrared. Therefore, infrared observations can also tell us about the distribution of star formation in a galaxy. Until recently very few positions per galaxy could be observed. However, the recent IRAS ob-

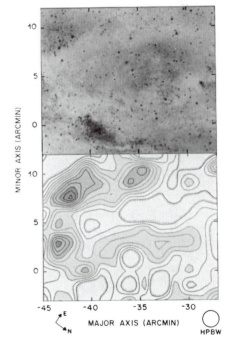

Figure 17.15   This figure shows that the CO emission follows the well-defined spiral arms in M31. In the upper photograph we have a negative optical image of a small piece of a spiral arm. In the contour map below we see the intensity of CO emission.

Figure 17.16 An optical photograph of Arp 220, an example that is not very bright optically, but is very luminous in the far infrared, as determined by IRAS observations. It emits 100 times as much energy in the infrared as it does in the visible.

servations have provided us with extensive maps of many galaxies (Fig. 17.16). The one thing lacking is good angular resolution. Follow-up studies are currently going on to see the details of the infrared emission in these galaxies.

These various studies have revealed a number of galaxies that seem to have undergone excessive star formation for short periods of time. These galaxies are called *starburst galaxies*. There is some speculation that this may be the way that most spiral galaxies work. In this scenario, most of the star formation in a given galaxy occurs during the burst. The burst is followed by a quiet period, and then another burst, and so on.

17.3 EXPLANATIONS OF SPIRAL STRUCTURE

It is actually quite surprising that we see any spiral structure in galaxies. The differential rotation of a galaxy should smear out any pattern on a timescale comparable to the orbital period. For the Sun the orbital period is about 240 million years. We think that the age of the galaxy is about 10 billion years. This means that any initial spiral pattern would have had ample time to smear out. Therefore, spiral arms must be temporary or there must be a way of perpetuating them.

A scenario for temporary spiral structure is shown in Fig. 17.17. For one reason or another, star formation starts in one region. It may even spread via some of the triggering mechanisms discussed in Chapter 16. The region with new stars is then stretched out by the differential rotation into a piece of a spiral pattern. This may explain why we see pieces of spiral features but no complete arms in our galaxy. In this scenario each ''arm'' lives for a short time, and new ones are always forming.

If spiral patterns persist for many revolutions of the galaxy, then it may be that the pattern and the matter itself are moving at different speeds. At first this may seem strange, but we can use an analogy to see how it might work (Fig. 17.18). Suppose we have cars moving along a two-lane highway, and you are looking from above, in a traffic helicopter. A truck breaks down in the right lane, causing a traffic jam. If you look from above, you see cars backed up for some distance behind the truck. The density of cars is higher for this region. Far behind the truck, the cars are still moving at their normal speed, and after the cars squeeze past the truck, they resume their normal speed. If you come back a few minutes later, you will see the same pattern of cars. However, the specific cars involved in the buildup will be different. The cars that you saw originally will be far down the road. In this case, the cars are moving along, but the pattern stays in the same place because the truck stays in the same place.

Now suppose the truck is moving along at a slow speed. Again, there will be a buildup behind the truck as cars squeeze into one lane to get past the truck. As in the case of the stationary truck, you see the cars moving at their normal speed. However, now the pattern moves. The speed of the pattern is not related to the speed of the cars—it is determined by the speed of the truck. The truck is responsible for the pattern. The cars simply respond to the presence of the truck. This is the type of situation in which the pattern (the traffic jam) can move at one speed, and the matter (the cars) at another.

There is a theory that the same type of situation can occur in spiral galaxies. Since the matter moves at a different speed from a density buildup, the theory is called the *density wave* theory. In a galaxy the dynamics is controlled by the halo, which contains most of the mass. The bright spiral arms contain a small fraction of the mass of the galaxy, and represent material which is orbiting at its normal speed, but responding to the gravitational effects of the asymmetric distribution of the stars in the halo. The mathematician C. C. Lin has shown that once a spiral pattern is established in a galaxy, it can sustain itself for a long time in this type of wave. Eventually, the wave may die out and a new one must be generated.

In the density wave picture, the visible arms are a result of a gathering of interstellar matter. When high enough densities are reached, star formation

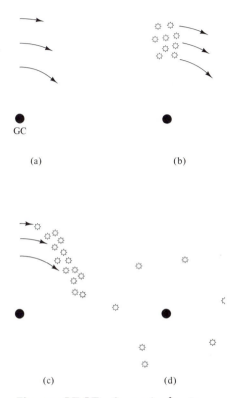

Figure 17.17 Scenario for temporary spiral structure. (a) The dark area is the center of the galaxy, and the arrows indicate the orbital speeds of material at different distances from the center. (b) A large-scale burst of star formation takes place. (c) The differential rotation stretches the stars out, producing part of a spiral arm. (d) After a few rotations the arm is stretched out so much that we can no longer detect its presence.

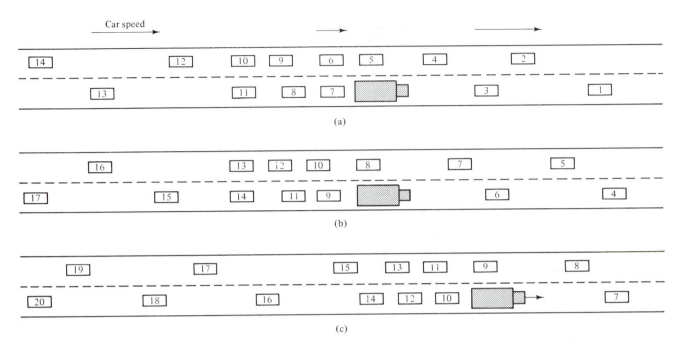

Figure 17.18 Car analogy to density wave. (a) A truck (shaded) is broken down in the right lane. Far in front and behind the truck, the cars have their normal speed and low density. Just behind the truck, the density of cars goes up, and their speed goes down. (b) As time goes on, cars slowly squeeze by the truck. The basic pattern is retained. However, as the numbers on the cars show, different cars are stuck behind the truck than in the earlier frame. We therefore have a density concentration along the highway while the individual cars are not permanently attached to the concentration. (c) This concentration can even move. Instead of being stuck, suppose the truck is moving slowly. The pattern moves along with the truck, while the individual cars move at a higher speed.

may take place. One scenario for this is illustrated in Fig. 17.19. A large HI cloud, or a group of small clouds, approaches an arm at a speed of about 100 km/s relative to the arm. (In this case, the arm may be moving at 100 km/s, and the matter overtaking it at 200 km/s.) The arm acts like a gravitational potential well, causing material to take more time to traverse the arm than a similar distance between arms. The matter entering an arm will leave its circular path, and have some motion along the arm before finally emerging. It should be noted that, even if the density waves don't cause strong visible spiral arms, they alter the orbits, resulting in noncircular motions. Some of the results of their critical calculations of density waves are shown in Fig. 19.20.

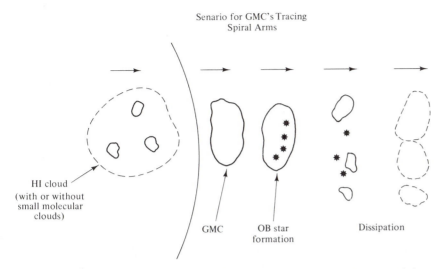

Figure 17.19 Scenario for molecular clouds tracing spiral arms. If the gas is circulating faster than the spiral pattern, as with a density wave picture, the gas can overtake the spiral arms from behind. Before reaching the arms, the gas is in the form of low-density HI clouds, possibly containing some small molecular clouds. The entry into the arm slows the HI clouds down and compresses them. It may also gather together small molecular clouds. In any event, giant molecular clouds form. Those clouds give birth to O and B stars (as well as to all the other types). The radiation from the O and B stars disrupts the clouds.

As the front of the cloud enters the arm, it slows down. The only way the back of the cloud can know that this has happened is for a pressure wave (sound wave) to travel from the front of the cloud to the back. However, the speed of the cloud is greatly in excess of the speed of sound within the cloud. The back of the cloud doesn't get the message to slow down until it has almost overtaken the front. The cloud has been compressed. (This may also gather small molecular clouds into giant molecular clouds.) At this point, we see the cloud as part of a dust lane, explaining why dust lanes appear at the backs of spiral arms. Since the cloud has been compressed, star formation is initiated. Massive (as well as low-mass) star formation takes place. The bright stars don't live very long, so we see them only over a small range, forming the bright chains that mark the front of the arms.

The massive stars also have the effect of driving the clouds apart. This can be through the effect of stellar winds, expanding HII regions, and supernova explosions. The clouds dissipate. The material again resembles that which originally entered the back of the arm. It remains this way until it overtakes the next arm. According to this picture, giant molecular clouds

Figure 17.20 The results of theoretical calculations demonstrating the ability of galaxies to amplify small perturbations into well-defined spiral structure.

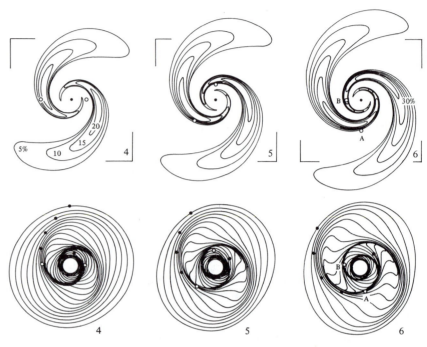

should be seen almost exclusively in spiral arms. It is a good observational test of the theory, but, as we have seen above, the observations are difficult.

We have already mentioned that spirals seem to fall into two categories: grand design and flocculent. It may be that the underlying cause of the two types of spiral structure is different. One observation to support this is that grand design spirals seem to occur in galaxies with internal bars or with nearby neighbors. It is suspected that the tidal interactions between the galaxy and the neighbors set up the spiral density wave in the mass distribution of the galaxy. As the gas streams through the density wave it is compressed into giant molecular clouds, which give birth to stars and then dissipate, as shown in Fig. 17.17. In flocculent galaxies, the spiral structure may be a series of temporary patterns, in which the results of local bursts of star formation are drawn into a spiral by the differential rotation of the galaxy.

The density wave theory seems to be best applied to grand design galaxies. We will briefly look at two that have been studied in detail, M51 (the ''Whirlpool'') and M81. These are shown in Fig. 17.21.

In M51 we can see clearly the location of the dust lanes with respect to the bright arms. The lanes are on the inside edges of the bright arms. Remember, in the density wave picture, the interstellar gas is swept up, forming giant molecular clouds. These giant molecular clouds are visible as the dust patches. Eventually, these clouds give birth to O stars, which illuminate HII

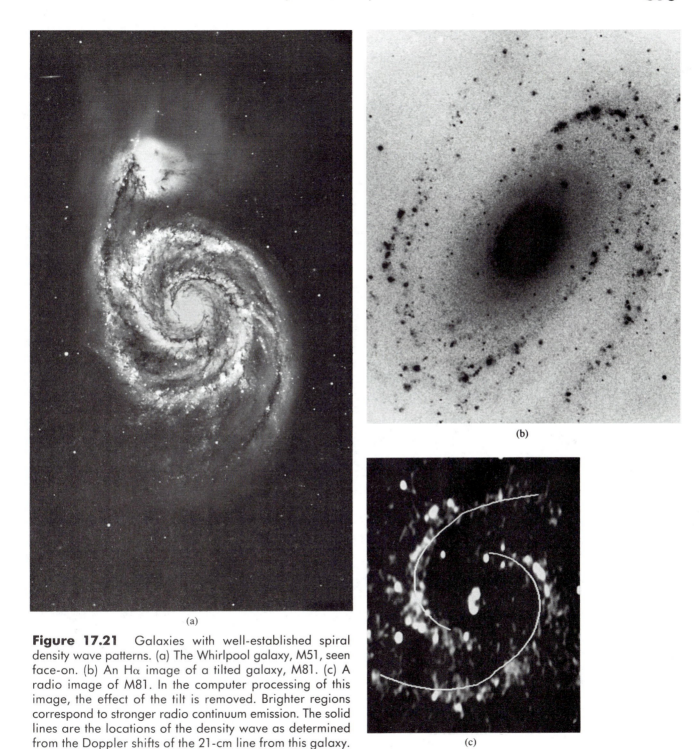

Figure 17.21   Galaxies with well-established spiral density wave patterns. (a) The Whirlpool galaxy, M51, seen face-on. (b) An Hα image of a tilted galaxy, M81. (c) A radio image of M81. In the computer processing of this image, the effect of the tilt is removed. Brighter regions correspond to stronger radio continuum emission. The solid lines are the locations of the density wave as determined from the Doppler shifts of the 21-cm line from this galaxy.

(a)

(b)

(c)

regions. The O stars and HII regions trace out the bright arms. The density wave picture therefore makes a specific prediction on the relative positions of the dust and bright arms. Radio continuum maps of M51 show that the synchrotron emission is strongest in the direction of the dust lanes. At first, you might expect the synchrotron emission to be strongest where there are the most supernovae, on the bright side of the arm. However, the compression of the interstellar medium on the dark dust side of the arms produces relatively strong magnetic fields, and the synchrotron emission gets stronger when the magnetic field gets stronger.

The face-on appearance of M51 means that we can easily see the relative placement of features. However, we get very little velocity information, since all of the galactic rotation is perpendicular to our line of sight. We see the largest Doppler shifts in edge-on galaxies, but we cannot make out any spatial structure from their edge-on appearance. To get a good idea of the velocity distribution in a galaxy, we need an inclined one, such as M81. In Fig. 17.21b we see an Hα image of that galaxy, so that the arms are clearly delineated. In Fig. 17.21c, we see a representation of the continuum emission, with the locations of the spiral density waves as determined from HI velocities, shown as solid lines. When the gas crosses the arms, the paths are very different from circular. In the arms, the gas has an inward motion along the arm. The density wave model for M81 has been successful at predicting many of the features of the galaxy.

17.4   MASS-TO-LIGHT RATIOS IN GALAXIES

When we look at a galaxy in visible light we obviously see the most luminous objects. However, some of the mass may not be luminous. It could be there, but hard to detect. The only sure way to trace out the total mass, whether it is bright or dark, is to study its gravitational effects. In a galaxy, the easiest way to study the gravitational forces is to measure the rotation curve. We have already discussed the rotation curve for our galaxy in Chapter 16.

Rotation curves can be determined from measurement of Doppler shifts in spectral lines. This can be done with optical lines, such as Hα. However, we only see the optical lines out to radii where the luminosity is high. We can study radio emission lines, such as the 21-cm line, out to larger radii. Typical results are shown in Fig. 17.22. We find that $v(r)$ stays roughly constant out to radii well beyond the most luminous material. This immediately tells us that the mass doesn't fall off as fast as the luminosity does. The masses that are being found are as high as $2 \times 10^{12}\ M_\odot$. In many galaxies, no edge has been found yet. The rotation curves are still flat out to radii where the interstellar medium can no longer be detected.

Where can this matter reside? One possibility is that it is part of the disk. However, theoretical models show that such a large mass would gather

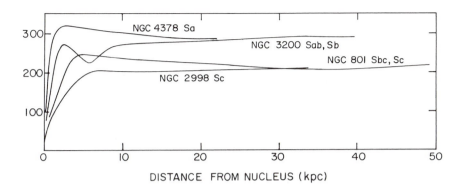

Figure 17.22 Rotation velocity determined from 21-cm observations as a function of distance from the center of a galaxy. The fact that the rotation velocity is large at great distances from the center of the galaxy is taken as evidence for the existence of dark matter.

the disk into a bar. The disks that we see would not be stable. The galaxies are more stable if the dark mass has a spherical distribution. We call this spherical part of a spiral galaxy the *halo*. Remember, a halo is not a ring around a galaxy. It is a spherically symmetric mass distribution.

We can use the rotation curve to give us the mass distribution in the halo. If $M(r)$ is the mass interior to radius r, then $v(r)$ is related to it by the fact that the acceleration of gravity must provide the acceleration for a circular orbit (v^2/r), so

$$GM(r)/r^2 = v(r)^2 r$$

Solving for $M(r)$ gives

$$M(r) = rv(r)^2/G \tag{17.3}$$

We can relate $M(r)$ to the density distribution $\rho(r)$ by the equation of mass continuity [equation (9.23)], which was one of the equations of stellar structure:

$$dM(r)/dr = 4\pi r^2 \rho(r) \tag{17.4}$$

Solving for $\rho(r)$ gives

$$\rho(r) = (1/4\pi r^2)[dM(r)/dr] \tag{17.5}$$

If we take $v(r) = v_0$, a constant, then differentiating equation (17.3) with respect to r gives

$$dM(r)/dr = v_0^2/G \tag{17.6}$$

Finally, substituting equation (17.6) into equation (17.5) gives

$$\rho(r) = v_0^2/4\pi G r^2 \tag{17.7}$$

The density in the halo therefore falls off as $1/r^2$. This is not nearly as fast as the exponential falloff in the light of the disk.

With a $1/r^2$ falloff in density, it might seem that there is not much matter very far out. However, if we divide the galaxy into spherical shells, each of thickness dr, the volume of each shell is $4\pi r^2\, dr$. This means that the mass of each shell is constant! As far out as the rotation curve remains flat, we are adding significant amounts of matter to the galaxy. This is particularly important, since the rotation curves seem flat as far out as we can measure. It may be that there is still a significant amount of matter beyond those points.

What is the dark matter in the halo? We can get a clue by looking at the *mass-to-light* ratios of various objects. This is the ratio of the mass, expressed in solar masses, to the luminosity, expressed in solar luminosities. By definition, the mass-to-light ratio of the Sun is 1. The mass-to-light ratios of main sequence stars are given in the mass–luminosity relationship discussed in Section 5.5. If we know the mass-to-light ratio of a galaxy, we can see what types of objects have similar mass-to-light ratios. For spiral galaxies, the mass-to-light ratio is 1:3 near the center. This means that most of the mass in the center probably comes from normal stars. However, near the edge of the visible disk, the ratio climbs to 20, and is above 100 for the farthest points to which rotation curves have been measured.

It has been suggested that the halos might consist of faint, old red stars. These stars would have masses less than $1\ M_\odot$. There is even some direct observational evidence for such stars in the halos of nearby galaxies. The mass-to-light ratio for such stars is about 20. They might therefore provide the dark matter out to the edge of the visible disk, but something else is needed beyond that. More recent observations seem to rule out all nuclear burning material as a significant part of the halo. Some astronomers have suggested that a lot of mass could be hidden in Jupiter-sized objects, which are obviously not very luminous.

It has been suggested that the dark matter could be in the form of neutrinos, if neutrinos have a small rest mass. Experiments on the nature of the neutrino are continuing in many laboratories. If electron neutrinos have a mass, it is less than 10^{-4} of the electron mass. This is small enough to have been overlooked in previous experiments. However, there are so many neutrinos in the universe that even a small mass for each neutrino can add a lot of mass to a galaxy. The idea of neutrinos as the dark matter in galaxies is still not generally accepted.

This is our first encounter with dark matter. We know it is there, because we can measure its gravitational effects, but we cannot see it. When astronomers were not so sure the matter was there, this was called the ''missing mass problem.'' However, the mass really isn't missing. It is just nonluminous. As we go to larger scales in the universe in subsequent chapters, we will find that there is evidence that the dark matter becomes more and more the dominant matter in the universe.

CHAPTER SUMMARY

In this chapter we looked at the properties of various types of galaxies.

Elliptical galaxies have no evidence for recent star formation. However, the metal abundances are high. Ellipticals are classified according to the eccentricity of their appearance. Spirals have an evident interstellar medium, as well as O and B stars, meaning that star formation is still taking place. Spirals are classified according to the tightness of the spiral arms and the relative sizes of the nucleus and disk.

We applied some of the ideas of star formation, discussed in previous chapters, to look at star formation in spirals. We also looked at how the density wave theory might explain the spiral structure.

In studying the rotation curves of galaxies we found that the masses of galaxies are greater than would be determined from the luminous material. It has been suggested that most of the matter is distributed in a spherically symmetric halo.

QUESTIONS AND PROBLEMS

17.1 If we see a spiral galaxy edge on, how can we tell that it is a spiral?

17.2 (a) Why is it not likely that single spirals formed from single ellipticals? (b) Why is it not likely that ellipticals formed from spirals? (*Hint:* Think of the effects of rotation.)

17.3 Compare the properties of dwarf elliptical galaxies with the properties of globular clusters in our own galaxy.

17.4 What features of S0 galaxies make them similar to (a) spirals? (b) ellipticals?

17.5 Assume there is some way, either by spectroscopy or by some aspect of the shape of the galaxy, to determine the luminosity class for a galaxy. How can this information be used, along with other observations, to determine the distance to a galaxy?

17.6 How do the relative abundances of atomic and molecular hydrogen vary within a spiral galaxy?

17.7 Which parts of the interstellar medium would you expect to best trace out spiral arms? Explain your answer.

17.8 Given the luminosity profile in equation (17.1), for a given galaxy, how far out would you have to go before you have 95% of the galactic luminosity? Express your answer as a multiple of D.

17.9 What are the differences between grand design and flocculent spiral galaxies (a) in their appearance and (b) in the scenario by which the spiral arms are formed and maintained?

17.10 Given the luminosity profile in equation (17.1) and the density profile in equation (17.7), find an expression for the mass-to-light ratio as a function of distance from the galactic center r. Assume that the ratio is 1 at $r = D$.

17.11 Calculate the mass of a galaxy with a flat rotation curve, with $v = 300$ km/s, out to $r = 20$ kpc. (Express your answer in $M_\odot$.)

17.12 In equation (17.4) we used the relationship between mass and density for a spherically symmetric system. However, spiral galaxies have a disklike appearance. Why is the use of equation (17.4) valid?

17.13 If neutrinos have a rest mass equal to 10^{-4} of the electron rest mass, how many neutrinos are needed to give $10^{12} M_\odot$?

17.14 Use the mass–luminosity relationship discussed in Chapter 5 to graph the mass-to-light ratio for main sequence stars as a function of spectral type.

17.15 How does the density wave theory help us understand spiral structure?

PART IV

THE UNIVERSE AT LARGE

To this point we have been studying the stellar life cycle and how stars and other material are arranged into galaxies. We will now turn to studies on a much larger scale. When we talk about how the universe is put together, each galaxy has only as much importance as a single molecule of oxygen has in describing the gas in a room.

In this part we will start by studying how galaxies are distributed on the sky and how they move relative to one another. We will also see how the problem of dark matter becomes more important as we go to larger and larger scales.

As we go to larger scales, increasing the number of galaxies that we observe, we also find an increasing variety of interesting phenomena associated with galaxies. In Chapter 19 we will discuss the various aspects of galactic activity, particularly as evidenced by radio galaxies and quasars.

In Chapters 20 and 21 we will turn to cosmology, the study of the universe on the largest scales. This also includes the study of the past and future evolution of the universe. It is in the study of the past that we will encounter one of the most fascinating aspects of modern astrophysics research, the merging of physics on the smallest (elementary particles) and largest (structure of the universe) scales.

Clusters of Galaxies

18.1 DISTRIBUTION OF GALAXIES

If we look at the distribution of galaxies, such as that shown in Fig. 18.1, we see that the galaxies are not randomly arranged on the sky. Among the patterns we see are distinct groupings, called *clusters of galaxies*.

Clusters are interesting for a number of reasons. They may provide us with clues on the formation of galaxies themselves. This is especially true if, as many think, cluster-sized objects formed first and then broke into

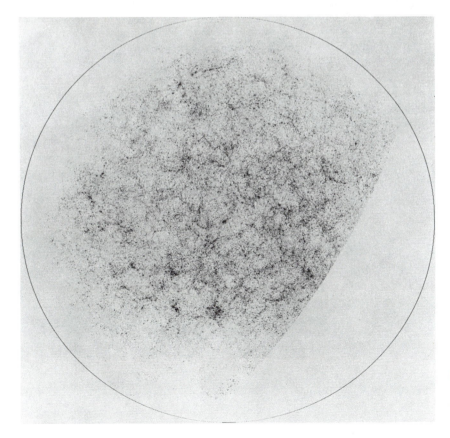

Figure 18.1 Distribution of galaxies. This is a two-dimensional view as seen from the Earth.

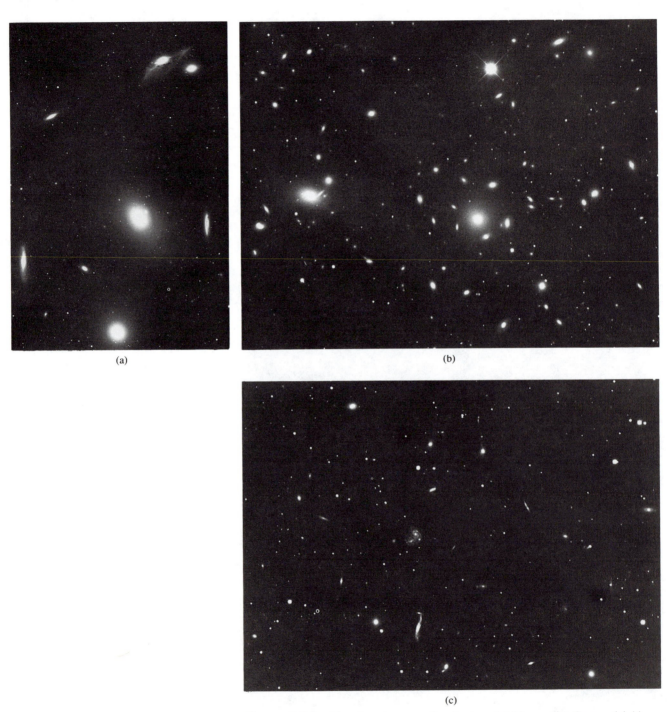

Figure 18.2 Nearby clusters of galaxies. (a) Virgo. (b) Coma. (c) Hercules.

galaxy-sized objects. (The alternative view is that galaxies formed first and then gathered into clusters.) Clusters also pose us with interesting dynamical problems, including a dark-matter problem of their own. Finally, when we reach the scale of clusters of galaxies, we are beginning to reach a scale which has some significance in the overall structure of the universe.

The cluster of galaxies to which the Milky Way belongs is called *the local group*. As clusters go, it is not a very rich one. Besides the Milky Way, it contains several irregulars, including our companions, the Large and Small Magellenic Clouds, the spiral galaxies M31 and M33, and a number of dwarf ellipticals. Other nearby clusters are named by the constellation in which they are centered. For example, the Virgo, Coma, and Hercules clusters are shown in Fig. 18.2.

18.2 CLUSTER DYNAMICS

Just as with clusters of stars, clusters of galaxies may be isolated collections of masses interacting gravitationally. As such, they are interesting systems to understand. In addition, by studying the gravitational interactions, we learn more about the masses of the individual galaxies and of the cluster as a whole. It has been found that the number of galaxies per unit area in a cluster falls off approximately as $\exp[-(r/r_0)^{1/4}]$. This is the same as the surface brightness in elliptical galaxies. We know that elliptical galaxies are a dynamically relaxed system. If clusters and ellipticals have similar density distributions, then it may be that some clusters are dynamically relaxed also.

EXAMPLE 18.1 Crossing Time for a Cluster of Galaxies

The time for a galaxy to cross from one side of a cluster to the other is called the crossing time. Find the crossing time for a cluster of galaxies with an extent of 1 Mpc, and galaxies moving at 10^3 km/s.

Solution

The time for a galaxy to cross is the diameter divided by the speed, so

$$t_{\text{cross}} = (10^6 \text{ pc})(3 \times 10^{18} \text{ cm/pc})/(10^8 \text{ cm/s})$$

$$= 3 \times 10^{16} \text{ s}$$

$$= 1 \times 10^9 \text{ yr}$$

Figure 18.3 This Einstein observatory image shows X-ray emission from the hot gas between the galaxies in a cluster A1367. The contours of x-ray intensity are superimposed on a photograph of the cluster. Some of the x-ray emission is associated with individual galaxies, while some come from the gas between the galaxies.

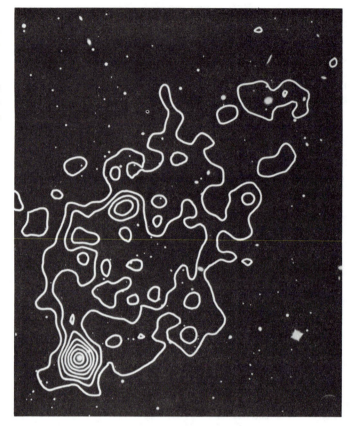

We think that clusters of galaxies have been around for over 10^{10} years (the age of our galaxy as determined from globular cluster H–R diagrams). If they were not gravitationally bound, they would have had many crossing times to evaporate. We therefore think that clusters are gravitationally bound. They have also had sufficient time to become relaxed, so we can apply the virial theorem to analyze the internal motions. In its simplest form, the virial theorem tells us that

$$2K + U = 0 \qquad\qquad (18.1)$$

where K is the cluster kinetic energy, and U is the potential energy. If we treat the cluster as a spherically symmetric collection of N galaxies, each of mass m, and the total mass of the cluster $M = mN$, the cluster potential energy is (ignoring factors of order one)

$$U \cong -GM^2/R$$

where R is the radius of the cluster. The average kinetic energy per galaxy is $\frac{1}{2}m\langle v^2\rangle$, so the total kinetic energy is N times this, or

$$K = \tfrac{1}{2}mN\langle v^2\rangle$$

$$= \tfrac{1}{2}M\langle v^2\rangle$$

Substituting these into the virial theorem gives

$$M\langle v^2\rangle = GM^2/R \qquad (18.2)$$

Solving for M, we have

$$M = \langle v^2\rangle R/G \qquad (18.3)$$

The quantity $\langle v^2\rangle$ is the average of the squares of the galaxy velocities with respect to the center of mass of the cluster. The best way for us to measure the velocities of the individual galaxies is through their Doppler shifts. However, this only gives us the component of the velocity along the line of sight. This means that we are measuring $\langle v_r^2\rangle$ rather than $\langle v^2\rangle$. However, if the motions of the galaxies are random, we can relate these two quantities.

Suppose we resolve the motion of any galaxy into its components in an (x, y, z) coordinate system. The velocity can then be written in terms of its components as

$$\mathbf{v} = v_x\hat{\mathbf{x}} + v_y\hat{\mathbf{y}} + v_z\hat{\mathbf{z}} \qquad (18.4)$$

where $\hat{\mathbf{x}}$, $\hat{\mathbf{y}}$, and $\hat{\mathbf{z}}$ are unit vectors in the three directions, respectively. To be definite, we can let the x-direction correspond to the line of sight. The square of v, which is $\mathbf{v} \cdot \mathbf{v}$, is simply the sum of the squares of the components,

$$v^2 = v_x^2 + v_y^2 + v_z^2 \qquad (18.5)$$

If we then take the average of both sides of the equation, we have

$$\langle v^2\rangle = \langle v_x^2\rangle + \langle v_y^2\rangle + \langle v_z^2\rangle$$

However, if the motions are random, the averages of the squares of the components should be the same for all directions. This means that

$$\langle v_x^2\rangle = \langle v_y^2\rangle = \langle v_z^2\rangle \qquad (18.6)$$

Using this, equation (18.5) becomes

$$\langle v^2 \rangle = 3\langle v_x^2 \rangle$$

and since the x-direction is the one corresponding to the line of sight, $v_r = v_x$, so

$$\langle v^2 \rangle = 3\langle v_r^2 \rangle \tag{18.7}$$

If we substitute this into the virial theorem [equation (18.3)], the mass is given by

$$M = 3\langle v_r^2 \rangle R / G \tag{18.8}$$

EXAMPLE 18.2   Virial Mass of a Cluster

For the Coma cluster we have an average radial velocity of 860 km/s and a cluster radius of 6.1 Mpc. Find the virial mass of the cluster.

Solution

From equation (18.8) we have

$$M = \frac{(3)(8.6 \times 10^7 \text{ cm/s})^2 (6.1 \text{ Mpc})(3.18 \times 10^{24} \text{ cm/Mpc})}{(6.67 \times 10^{-8} \text{ dyn-cm}^2/\text{s}^2)}$$

$$= 6 \times 10^{48} \text{ g}$$

$$= 3 \times 10^{15} \, M_\odot$$

When we add up the mass that we can see in the cluster, we find that it doesn't add up to the amount required by the virial theorem. This was originally done using just the mass of the luminous matter in the galaxy. However, we saw in Chapter 17 that the halos of galaxies may contain dark matter. Even if we don't know what the dark matter is, we know it is there, and can add its mass to that of the luminous matter in each galaxy. However, clusters have mainly ellipticals and SO's, which may not have massive halos. We should only add the dark matter that we know is there, so we only add enough to account for the observed rotation curves in different types of galaxies. (We suspect that there may be more dark matter beyond the points where the rotation curves have been measured, because there is no evidence of the rotational velocities beginning to fall off.) Even this amount of dark matter is not enough to gravitationally bind the cluster.

Some of the mass may be in the form of low-density gas within the cluster, but between the galaxies. This gas has either been ejected from the galaxies, or has fallen into the cluster. In either case we would expect this gas to be very hot, about 10^7 K. It should be hot enough to give off faint x-ray emission. In fact, such emission is observed. Figure 18.3, an Einstein image, gives one example. The hot gas contributes a significant amount of mass, but still doesn't completely solve the problem.

There are still two possible solutions. One is that the individual galaxies have, as some have suspected, halos that go out even farther than the rotation curves can be measured. There is some evidence to support this in studies of the interactions in binary galaxies. The advantage of a binary galaxy over the rotation curve studies is that the galaxies in a binary system are far enough apart to sample the full mass of each other. The other possibility is that clusters contain their own dark matter. This matter may be the same as that in the halos of galaxies, but there just may be additional amounts in the cluster, not bound to any one galaxy. If a rich cluster has the mass implied by the virial theorem, then the mass-to-light ratio is about 200. This would be consistent with either the extended halos on individual galaxies or the generally distributed dark matter.

Another interesting feature about clusters of galaxies is that giant elliptical galaxies are found near the centers of some clusters (Fig. 18.4). These

Figure 18.4 (a) Optical image of M87, with a shorter exposure in the lower right corner, allowing us to see the jet emerging from the nucleus. (b) An X-ray image of the same region, using the Einstein observatory.

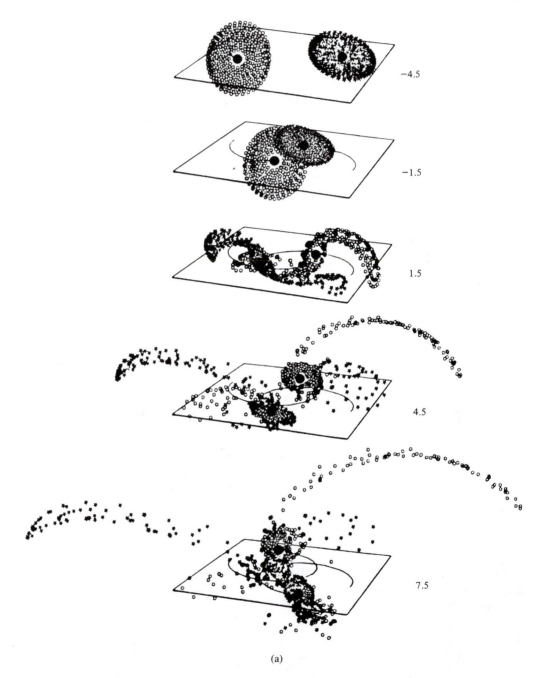

−4.5

−1.5

1.5

4.5

7.5

(a)

Figure 18.5   Interacting galaxies. (a) Steps in a computer simulation. Note that the tidal effects, tending to stretch structures, are very important. Also, in the encounters, individual stars never actually touch. (b) a pair of galaxies, NGC 4038 and 4039, which have an appearance similar to the results of the simulation.

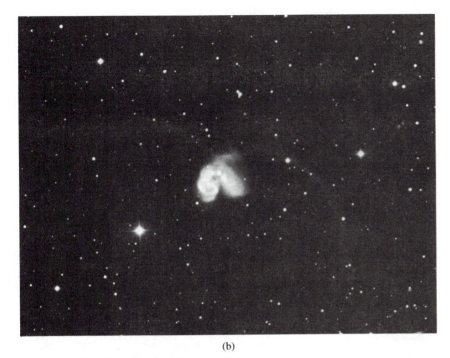

(b)

Figure 18.5 (Continued)

galaxies are called *central dominant*, or *cD*, galaxies. Some cD galaxies also seem to have multiple nuclei. It has been noted that the center of a cluster is the most likely place for galaxies to pass by each other. Some galaxy collisions will result in galaxy mergers. Once a few galaxies have merged, they can "swallow" galaxies that pass too close. This process, still under study, is called *galactic cannibalism*.

The whole subject of galactic encounters has been under active study. Numerical simulations have been carried out to find out what happens to the stars and gas in each of two colliding galaxies, both for very close encounters and for direct collisions. The result of one such calculation is shown in Fig. 18.5. As you can see, the calculations produce results that look like objects that are actually observed.

18.3 EXPANSION OF THE UNIVERSE

18.3.1 Hubble's Law

In Hubble's study of galaxies, he found that all galaxies have red-shifted spectral lines. This is indicated in Fig. 18.6. This means that they are all moving away from us. Furthermore, the rate at which any galaxy is receding

Figure 18.6 Distance and red shift. On the left are images of a galaxy in the indicated cluster. On the right are spectra for galaxies in the same clusters. The smaller images result from more distant galaxies. The spectra show the larger red shifts for these galaxies. In each spectrum, the band on top and bottom is for calibration; the actual spectrum is in the middle. The arrow shows the shift of a particular absorption line.

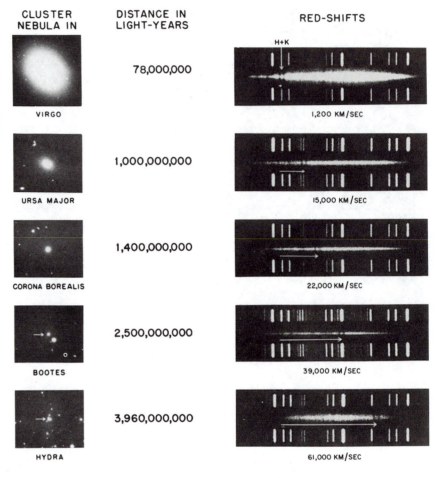

CLUSTER NEBULA IN	DISTANCE IN LIGHT-YEARS	RED-SHIFTS
VIRGO	78,000,000	1,200 KM/SEC
URSA MAJOR	1,000,000,000	15,000 KM/SEC
CORONA BOREALIS	1,400,000,000	22,000 KM/SEC
BOOTES	2,500,000,000	39,000 KM/SEC
HYDRA	3,960,000,000	61,000 KM/SEC

from us is proportional to its distance from us. We can write this in the simple form

$$v = H_0 d \tag{18.9}$$

where v is the speed of the galaxy, d is the distance to the galaxy, and H_0 is a constant, called the *Hubble constant*. (The subscript zero on the H indicates that this is the current value. As we will see in Chapter 20, H is a constant in a sense that it is the same at every place, but it can change with time.) The relationship given in equation (18.9) is called *Hubble's law*. Results of more recent studies are shown in Fig. 18.7.

At first it might seem unusual that we are in some special part of the universe, so that all things are moving away from us in a very particular way. However, we interpret Hubble's law as telling us that all galaxies are moving away from each other. This motion represents the overall expansion

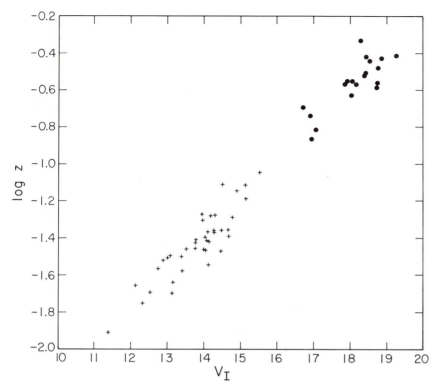

Figure 18.7 Hubble's law. The red shift is plotted on the vertical axis, and the apparent magnitude, a rough distance indicator, on the horizontal axis.

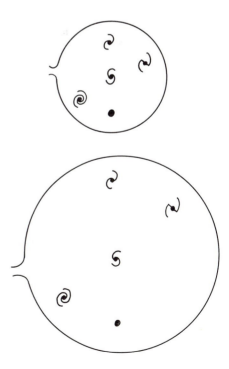

Figure 18.8 The universe as an expanding balloon. The galaxies are painted on the surface of the balloon. As the balloon expands, each galaxy moves away from every other galaxy.

of the universe. To visualize this situation, we imagine the galaxies as being dots on the surface of a balloon, as shown in Fig. 18.8. As the balloon is blown up, all of the dots move away from all of the other dots. In some time Δt, the balloon expands so that all distances are multiplied by a factor $1 + f$. If two objects were initially a distance d apart, their distance at the end of the interval is $(1 + f)d$. The change in the distance between the two objects is fd, so the average relative velocity of the two objects is $fd/\Delta t$. This is the same form as Hubble's law.

Suppose the rate of expansion has been constant over the age of the universe. If all objects started very close together at $t = 0$ (whatever that time means), then the current distance between any two objects would be

$$d = vt$$

where t is the current age of the universe. Solving for v gives

$$v = (1/t)d \qquad (18.10)$$

This is the same as Hubble's law, if we make the identification

$$H_0 = 1/t \qquad\qquad (18.11)$$

Therefore, $1/H_0$, called the *Hubble time*, is the age of the universe if the expansion has been constant. Actually, the expansion has been slowing down, so the age of the universe is less than the Hubble time.

The value that Hubble got for H_0 was 500 km s^{-1}/Mpc. Note that the units of the Hubble constant may seem strange, but there is a distance in the numerator (km) and in the denominator (Mpc), so the units really work out to 1/time. We use km s/Mpc because with it we can express both distance and velocity in convenient units. A value of 500 km s^{-1}/Mpc works out to a Hubble time of 2×10^9 yr (see Problem 18.11). This was a cause of concern, since our understanding of stellar evolution and the *H–R* diagrams for globular clusters in our galaxy tell us that these clusters are about 10×10^9 yr old. (In fact, radioactive dating places the age of the solar system at over 4×10^9 yr.) There is, however, an immediate error in Hubble's value due to the confusion between type I and type II Cepheids as a distance indicator. Over the years, other refinements have been made. As we will see below, the currently accepted value for the Hubble constant is in the range 50–100 km s^{-1}/Mpc.

Apart from telling us something interesting about the universe, Hubble's law is also of great value when determining distances to distant objects. It is important that this only be used for objects that are far enough away that their velocities relative to us are dominated by the expansion of the universe. We say that objects must be far enough away to be in the *Hubble flow*. Objects within our own cluster of galaxies are not in the Hubble flow. Their motions are dominated by the dynamics of our cluster. Even nearby clusters have random velocities relative to us that are a significant fraction of their Hubble's law velocity.

EXAMPLE 18.3   Hubble's Law and Distance

For some given cluster, we measure $v = 10^3$ km/s. What is the distance? Take $H_0 = 75$ km s^{-1}/Mpc.

Solution

We find the distance from

$$d = v/H_0$$

$$= \frac{10^3 \text{ km/s}}{75 \text{ km s}^{-1}/\text{Mpc}}$$

$$= 13.3 \text{ Mpc}$$

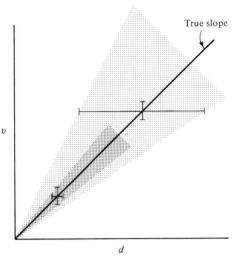

Figure 18.9 Sources of error in measuring the Hubble constant. For both nearby and distant galaxies, uncertainties are indicated schematically by error bars. Straight lines from the origin are then drawn at the widest angle that can still lie within one of the error bars. The error bar that determines this widest angle is the one that generates the largest uncertainty in H, since H would be determined by the slope of a line from our measured point to the origin.

18.3.2 Determining the Hubble Constant

Of course, if we are going to apply Hubble's law to determine distances, we need an accurate value for the Hubble constant. This means that we need an independent way of measuring distances to objects that are far enough away to be in the Hubble flow. The importance of it being far away can be seen by the following: Suppose an object has a velocity $H_0 d$ from the Hubble flow, and Δv from other sources. This Δv probably results from random motions of the galaxies, just as the gas in the room has random motions superimposed on any regular flow. Note that Δv can be positive or negative. The actual radial velocity that we measure will be

$$v_r = H_0 d + \Delta v \qquad (18.12)$$

In general, Δv is independent of d. Thus, for more distant objects, $H_0 d$ gets larger while Δv stays the same. For more distant objects, Δv represents a smaller fraction of $H_0 d$, and introduces a smaller fractional error into the determination of H_0 (Fig. 18.9).

It would therefore seem simple to get around this problem. All we have to do is use the most distant objects we can observe. Unfortunately, our distance indicators work best for nearby galaxies, where we can still see individual stars such as Cepheids. Therein lies the problem. We can measure distances most accurately for nearby objects, but we are not sure if they are in the Hubble flow. Arguments over the proper value for H_0 center around these two points: (1) What are the correct distance indicators. (2) Where does the Hubble flow start?

We now turn to the problem of measuring distances to distant clusters of galaxies. The procedure involves using our most secure distance indicators to get the distances to nearby galaxies, and then build up a series of distance indicators, useful to greater and greater distances. The problem can be quite involved, as shown in Table 18.1.

We start the process by using Cepheids to find the distances to nearby galaxies. Of course, we must calibrate the Cepheid period–luminosity relationship within our own galaxy. This involves starting with trigonometric parallaxes for nearby stars, and moving cluster observations for nearby cluster, to get a calibrated H–R diagrams. The calibration H–R diagram allows the use of spectroscopic parallax for individual stars, and for main sequence

TABLE 18.1   Distance Indicators

Method	Distance range (Mpc)
Cepheids	0–6
Novae	0–20
RR Lyraes	0–0.2
W Virginis stars	0–1
Eclipsing binaries	0–1
Red giants	0–1
Globular clusters	0–20
Supergiants	0–1
Stellar luminosity function	0–1
HII region diameters	0–25
HII region luminosities	0–100
HII loop diameters	0–4
Brightest blue star	0–25
Brightest red stars	0–15
Supernovae	0–200
21-cm linewidths	0–25
Disk luminosity gradients	0–100
U–B Colors	0–100
Luminosity classification	0–100
Brightest elliptical in cluster	50–5000

fitting for globular clusters containing Cepheids. This gives us a calibrated period–luminosity relation for Cepheids as well as for RR Lyrae stars. We then can use the Cepheids and RR Lyrae stars as distance indicators for galaxies that are close enough for us to see these stars individually.

For galaxies that are somewhat farther away, we can still use individual objects within the galaxy, but they have to be brighter than Cepheids. (Cepheids have been studied in galaxies as far away as M101, which is about 6 Mpc away.) We can, for example, measure the angular sizes of HII regions. Since we think we know what their linear sizes should be, this gives us a measure of distance. One promising technique is the use of supernovae. From the light curve we can tell whether a supernova is type I or type II. Type I supernovae appear to have similar peak luminosities. By measuring their apparent magnitude at peak brightness, we can find the distances.

Eventually, we reach a point where individual objects within galaxies cannot be measured. We must rely on being able to know the total luminosity of the galaxy. If we know the absolute magnitude of a given type of galaxy, and we measure the apparent magnitude, we have a measure of the distance.

The problem is to come up with independent indicators of galactic luminosity. Hubble made the simple assumption that all galaxies have the same visual magnitude. We know this is not the case. Instead, it has been suggested that the brightest ellipticals in each cluster have the same absolute magnitude. The luminosities of the brightest ellipticals seem to vary, however. For this reason, rather than using the brightest elliptical in a cluster, we use the second or third brightest elliptical in a cluster.

With the recent recognition that spiral galaxies have luminosity classes, much effort has gone into finding luminosity class indicators. In this way, we can observe a galaxy, determine its morphological type (Sa, Sb, etc.), and then have some indicator of its luminosity class. If we know absolute magnitude as a function of morphological type and luminosity class, and we measure the apparent magnitude, we can convert the difference into a distance. Another recent discovery, indicated in Figure 18.10, is that the width of the 21-cm line in a galaxy seems to correlate with its absolute infrared luminosity. Therefore, we can observe the 21-cm line, measure its width, and know the absolute magnitude of the galaxy. Of course, this relation must be calibrated, also. The calibration is complicated by the fact that the relationship may depend on type of galaxy.

Even with these methods there is another nagging problem. When we look at a distant galaxy, we are seeing it as it was a long time ago. However,

Figure 18.10 Fisher–Tully relation. The 21-cm line-width is plotted as a function of absolute magnitude for galaxies whose distances are determined by other methods.

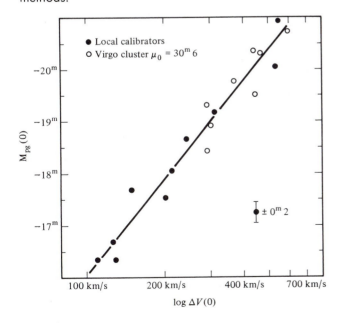

we know that galaxies evolve. As they evolve, their luminosity changes. Luminosity calibrations on nearby galaxies might not apply to distant galaxies.

So you see, there are problems at every step of the determination of the Hubble constant. For this reason, no one indicator is used at any step along the way. For example, when the distance to the Hyades, as determined from the moving cluster method, was recently revised by 10%, it only affected the extragalactic distance scale by 5%.

Currently accepted values of the Hubble constant range from 50 to 100 km s/Mpc. The corresponding Hubble times are 20×10^9 to 10×10^9 yr, the latter time being small enough to be somewhat worrisome. In a recent review of the problem of determining H_0, one astronomer noted that, given the problems in determining the Hubble constant, it is amazing that we seem to know it to within a factor of two.

18.4   SUPERCLUSTERS

Now that we have seen that galaxies are gathered into clusters, we might ask if the clusters are gathered into larger groupings, called *superclusters*. The answer is that they are. The first supercluster identified (in the 1950s) is the one in which we live, called the *local supercluster*. The Virgo cluster of galaxies is near the center of the local supercluster. The local group, our cluster of galaxies, is near the edge. The local supercluster contains 10^6 galaxies in a volume of about 10^{23} cubic light years!

Studies of more distant superclusters have been made difficult by a lack of extensive data on distances to galaxies. After all, we only see two dimensions projected on the sky. We can get a better idea of clustering if we also know the distances to galaxies. We get these distances from measuring the red shifts of clusters. Studies of nearby galaxies have provided sufficient information to identify additional nearby superclusters.

More recently, it has become possible to measure large numbers of red shifts in reasonable amounts of time. This now provides us with a large data base for examining clustering. We will look at the results of some of these studies in Chapter 21, when we discuss galaxy formation.

It might seem that we can determine the mass of a supercluster by using the virial theorem, in much the same way as we did for clusters of galaxies. However, the crossing time for a cluster in a supercluster is greater than the current age of the universe. This means that superclusters are not dynamically relaxed, and the virial theorem should not apply. It is even possible that superclusters are not gravitationally bound.

In addition to superclusters, there also appear to be *voids*, places with no (or at least very few) galaxies over a large volume. Several appeared in early surveys with no red shift information. Now that red shift data is avail-

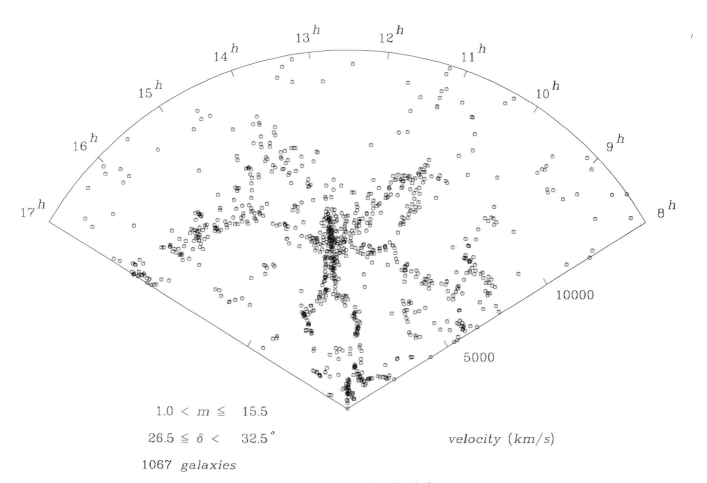

Figure 18.11   A 3-dimensional picture of clustering, with the third dimension provieded by red shifts. This is generated from a survey of red shifts of a large number of galaxies.

able, one void has been identified quite clearly. Its volume is 30×10^{24} cubic light years, or 300 times the volume of our local supercluster. It occurs for red shifts between 12,000 and 18,000 km/s. Red shift surveys have now been used to identify other voids of comparable size (Fig. 18.11).

The whole question of clustering is related to how galaxies formed in the first place. One might wonder whether there is any evidence for super-superclusters of galaxies. To date, it appears that such entities do not exist. However, we don't know what will be found as surveys are carried out to higher and higher red shifts.

CHAPTER SUMMARY

In this chapter we began to look at the universe on a large scale. We first looked at the distribution of galaxies, finding that they fall into clusters. We looked at the properties of these clusters. Finally, we looked at the large-scale motions of the clusters, relative to each other.

We find that clusters should be gravitationally bound, if they were formed over 10 billion years ago and are still around today. By applying the virial theorem, we find the masses necessary to keep them bound. The visible mass is insufficient for this, and, as with the halos of galaxies, we are confronted with a dark-matter problem.

All distant galaxies are moving away from us with a speed proportional to their distance, as described by Hubble's law. We interpret this as resulting from the expansion of the universe. We discussed the use of Hubble's law as a distance indicator, and saw the difficulties in measuring the Hubble constant.

Finally, we looked at structures that are even larger than clusters. These are the superclusters, or clusters of clusters. In addition to the superclusters, there are also very large voids, with relatively few clusters.

QUESTIONS AND PROBLEMS

18.1   What is the evidence for the existence of clusters and superclusters of galaxies?

18.2   Why do we expect a cluster of galaxies to obey the virial theorem and not a supercluster?

18.3   For some cluster of galaxies, the radius is 500 kpc, and the rms deviation in the radial velocities is 300 km/s. What is the mass of the cluster?

18.4   Rewrite equation (18.8) so that if velocities are entered in kilometers per second and distances in megaparsecs, the mass result in a solar mass.

18.5   Suppose that half of the mass of the cluster in Example 18.2 is in the form of hot intergalactic gas, spread uniformly over the cluster. What density of gas would this require?

18.6   In adding up the "visible" mass in clusters, what is the problem in accounting for the masses of the galaxies we can see?

18.7   Why would we expect intergalactic gas to be able to emit x-rays?

18.8   What are the current possibilities for the dark matter in clusters?

18.9   If we cannot see dark matter in clusters of galaxies, how do we know it is there?

18.10   Explain how we might determine the mass of binary galaxy system. Why is this important, since we already have masses determined from rotation curves?

18.11   Find a relationship between the Hubble constant, expressed in kilometers per second per megaparsec, and the Hubble time, expressed in years. Use this to find the Hubble times corresponding to Hubble constants of 50, 100, and 500 km s^{-1}/Mpc.

18.12   Suppose we have a universe whose size increases by 1% in 10^9 yr. Show that the average rate of separation in that time interval between any two points is proportional to their distance, and find the proportionality constant.

18.13   Why do we say that the Hubble time is an "upper limit" to the age of the universe?

18.14   What are the main problems in the accurate determination of the Hubble constant?

18.15   What do we mean by "Hubble flow"?

18.16   Can we use Hubble's law to determine distances within our galaxy?

18.17   What is the density of galaxies (galaxies/ly^3) in the local supercluster?

18.18   Why are red-shift surveys important in studying clustering of galaxies?

Active Galaxies and Quasars

In this chapter we will look at galaxies with unusual activity within and around them. As we go through the chapter, we will encounter phenomena of increasing energy.

19.1 RADIO GALAXIES

Many strong radio sources are not objects in our own galaxy. When a radio source is found, it is interesting to see if there is an identifiable optical object at the same position. In the past this could not always be done because the poor angular resolution in single-dish observations left radio astronomers with uncertain positions for their sources. With the development of interferometers, with better angular resolution, the finding of optical counterparts has been easier. Many of the strong radio sources appear in the directions of other galaxies. Galaxies with strong radio emission are called *radio galaxies*.

19.1.1 Properties of Radio Galaxies

The radio energy output of the radio galaxies is enormous, typically 10^6 times the total output of a normal galaxy. The radio spectrum of a typical radio galaxy is shown in Fig 19.1. The shape is the power law characteristic of synchrotron radiation. The radiation is also found to be polarized, another signature of synchrotron radiation. Synchrotron radiation requires a strong magnetic field and high-energy particles, most likely electrons.

The actual synchrotron spectrum depends on the energy distribution of the electrons, as shown in Fig. 19.2. The greater is the proportion of the highest-energy electrons, the greater is the proportion of high-energy photons. As the electrons radiate, they lose energy. Therefore, the synchrotron spec-

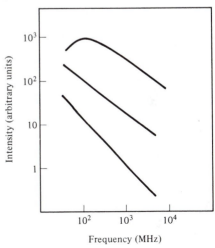

Figure 19.1 Schematic sample radio galaxy spectra.

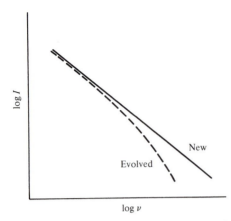

Figure 19.2 Schematic evolution of a radio galaxy spectrum. As the more energetic electrons give off energy, they become less energetic, so the evolved spectrum has proportionately less energy at higher frequencies.

trum evolves over time. The most energetic electrons lose their energy the fastest, so the proportion of high-energy electrons decreases with time. This means that the spectrum evolves, with more and more of the radiation coming out at long wavelengths. By studying the synchrotron spectrum of a radio galaxy, we can actually estimate how long a given region has been radiating.

Maps of radio galaxies, like those shown in Fig. 19.3, show us what truly amazing objects they are. (In order to determine the linear size, we must know the distance to the radio galaxy. This is found from the red shift of the visible associated galaxy, using Hubble's law.) A typical radio galaxy has a small source near the center of the galaxy, and then two large sources far beyond the optical limits of the galaxy. The optical galaxy is generally a giant elliptical. The two sources, or *radio lobes*, may be separated by up to 10^7 pc, and may be 10^6 pc wide. Sometimes, multiple pairs of lobes are seen. The structure of these sources is suggestive of matter being ejected from the galaxy. We sometimes see bending of the lobes, resulting from the motion of the central galaxy through the intergalactic medium. Higher-resolution observations show narrow structures, called *radio jets*, pointing from the center of the galaxy to the larger radio lobes.

Figure 19.3 Contour maps of radio emission from radio galaxies. These maps were made with the Very Large Array. (a) Hercules A, (b) Cygnus A, (c) 3C8318.

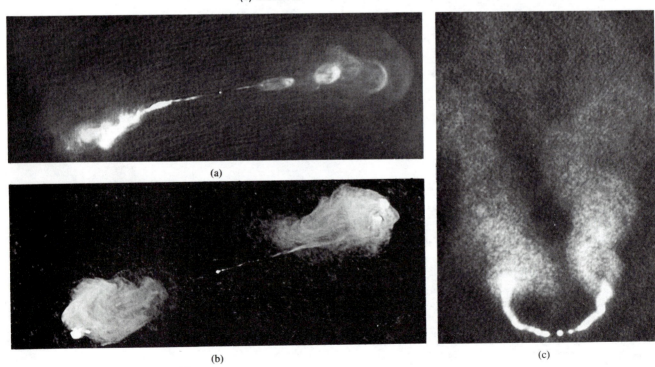

(a)

(b)

(c)

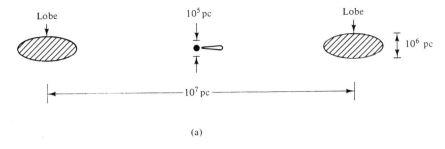

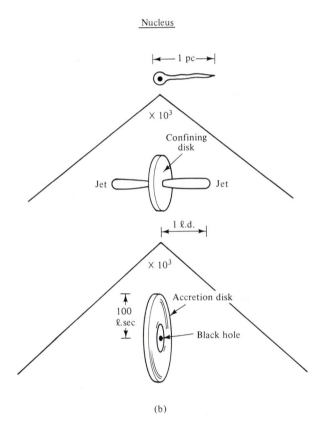

Figure 19.4 Schematic structure of radio galaxies. (a) Large scale. The shaded areas are in the large lobes. Closer to the center is the jet, pointing to one lobe or the other. (b) Small scale. A series of three frames, each blown up by a factor of 1000.

19.1.2 Model for Radio Galaxies

The generally accepted picture of what is going on in radio galaxies is shown in Fig. 19.4. We have interesting phenomena on a wide variety of length scales. On the largest scale we have the radio lobes themselves. The density in these lobes is very low, 10^{-4} to 10^{-3} electrons/cm^3. Because of this low density, collisions are infrequent. Therefore, when energy is deposited into this gas, it takes a long time for collisions to establish an equilibrium (Max-

Figure 19.5 Jet collimation mechanism. Material flows out of the small source in the center. It is confined by the ring of material, forming the nozzles for the jets. The shading in the confining material indicates that the density is higher on the inner side. The confining material is stabilized by material falling in from even farther out.

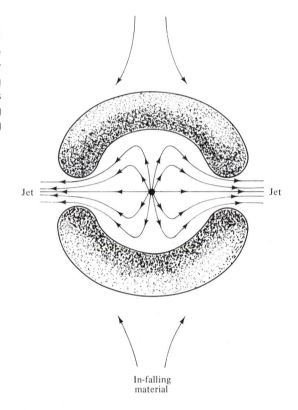

Jet   Jet

In-falling
material

well–Boltzmann) velocity distribution. The time is longer than the age of the universe, so it never happens. This is how we can have an unusually large number of high-energy electrons. The high-energy electrons lose their energy by synchrotron radiation, rather than by collisions with lower-energy electrons. The magnetic fields in the lobes are thought to be in the range 10–100 μG. The lobes contain smaller-scale structures with extents of 10^3 pc and smaller.

The jets are highly collimated, being some 10^5 pc long by 10^4 pc wide. The flow velocities in the jets are hard to measure, but estimates range from 10^3 to 10^5 km/s. (Some jets are moving close to the speed of light, as discussed in Section 19.1.3.) The densities are about 10^{-2} electrons/cm^3. The magnetic fields are comparable to those in the lobes.

Figure 19.5 shows one mechanism by which the jets can be collimated. We assume that there is material flowing outward in all directions. However, the source of the flow is surrounded by denser gas. This denser gas has a hole in it, and the outflowing material just follows the path of least resistance. Effectively, the material forms itself into a nozzle. One problem with this picture is that we would expect the outflowing material to drive the confining material outward. The confining material may be held in place by material falling in from even farther out, as shown in the figure.

It is possible to have two openings on opposite sides. However, high-resolution observations show that clumps on opposite sides of the center were not ejected at the same time. We also see single jets, but two lobes, in many radio galaxies. It therefore seems that only one nozzle operates at a time. Some flip-flopping of the nozzles is needed to produce the two lobes.

We now turn to the source of the energy in the nucleus of the galaxy. The energy requirements are enormous, since any of the energy ultimately given off in radio waves has its source in the galactic nucleus. At first we might think that nuclear reactions are the most efficient possible energy source. After all, they convert about 0.7% of the available mass into energy. How-ever, if we look at the mass available in the nucleus of a radio galaxy, a much higher fraction is being converted to energy. What energy source can be more efficient (by a factor of almost 100) than nuclear reactions?

The answer is mass falling into a black hole. A black hole is important because it allows us to have a large mass in a small radius, and hence strong gravitational forces. We can estimate the amount of energy available in drop-ping a particle of mass m from far away to the Schwarzschild radius. (Once the mass has passed the Schwarzschild radius, we can no longer get any energy out.) This energy gained by its mass is the negative of the potential energy at R_s, so the maximum energy we can get out is

$$E_{max} = GMm/R \tag{19.1}$$

Substituting $R_s = 2GM/c^2$ gives

$$E_{max} = \tfrac{1}{2}mc^2 \tag{19.2}$$

This tells us that we can get out up to half the rest energy of the infalling mass. The rate at which energy is generated then depends on the rate at which mass falls into the black hole, dm/dt. The maximum luminosity dE/dt is given by

$$L_{max} = \tfrac{1}{2}(dm/dt)c^2 \tag{19.3}$$

EXAMPLE 19.1   Luminosity for Mass Falling into a Black Hole

Calculate the energy generation rate for mass falling into a black hole at the rate of 1 $M_\odot$/yr.

Solution

Using equation (19.3) gives

$$L = \tfrac{1}{2}[(2 \times 10^{33} \text{ g})/(3 \times 10^7 \text{ s})](3 \times 10^{10} \text{ cm/s})^2$$

$$= 3 \times 10^{46} \text{ erg/s}$$

Remembering that the luminosity of the Sun is 4×10^{33} erg/s, this is almost 10^{13} solar luminosities!

However, extracting energy is not as simple as dropping mass into any black hole. If the mass is dropped straight in, most of the energy will be sucked into the black hole. In order to have most of the energy escape, it is necessary for the infalling matter to be in orbit around the black hole, slowly spiraling in. In this case, approximately 40% of mc^2 is available to power the galaxy. This 40% is very close to the limit of one-half that we found in our simple calculation [equation (19.3)].

In equation (19.3), we see that the luminosity doesn't depend on the mass of the black hole. However, when we take into consideration the spiraling trajectory for extracting the most energy, the mass of the black hole becomes important. The more massive the black hole is, the faster we can drop mass into it. We can simply think of a more massive black hole as having a larger surface areas. Calculations show that, in order to produce the luminosities we see in radio galaxies, black holes with masses of about 10^7 $M_\odot$ are needed!

19.1.3   The Problem of Superluminal Expansion

An interesting problem with some radio sources is that they appear to have small components that are moving faster than the speed of light! This is called the problem of *superluminal expansion*. Of course, we do not actually observe the velocities of these components. We observe the rate of change of the angular separation from the center of the source $d\theta/dt$, as shown in Fig. 19.6. We can convert this into a velocity only if we know the distance d to the source. If θ is measured in radians, then the speed is

$$v = (d\theta/dt)d$$

If our derived velocity is greater than c, then either (1) the sources are much closer than Hubble's law suggests, or (2) the apparent velocity doesn't represent a true physical velocity.

One explanation is based on the premise that we are not really seeing any one source moving. Instead, we are seeing a series of sources moving.

Figure 19.6   Superluminal expansion. (a) Radio maps of the source 3C273, taken at different times, show the expansion. (b) The geometry of the problem. We measure a change in angular position $\Delta\theta$ and relate that to a tangential velocity v by knowing the distance d.

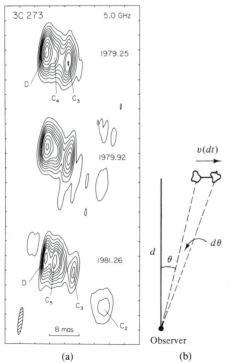

Each source "turns on" as the previous one fades. This creates the illusion of motion, just as do the lights on a movie marquee. Unfortunately, this doesn't solve the problem. It is unlikely that the individual sources will turn on in sequence by chance. Some siganl must be coordinating the time when each turns on. In order for us to see the superluminal expansion, this coordinating signal must be traveling faster than the speed of light. Having a signal travel faster than the speed of light is just as bad as having an object travel faster than c.

There is an alternative expansion, involving a special relativistic effect. The situation is illustrated in Fig. 19.7. Suppose we have an object starting at point O, and traveling to P, a distance r away. The object has a speed v, making an angle θ with the line of sight. In this arrangement, we take the x-direction to be along the line of sight. The y-direction is perpendicular to the line of sight, and the motion along the y-direction will be detected as proper motion. The distances x and y traveled along these two directions, are

$$x = r \cos \theta$$
$$y = r \sin \theta \tag{19.4}$$

The time for the object to move the distance r is

$$t = r/v \tag{19.5}$$

However, a light or radio wave emitted from P has to travel a shorter distance than one emitted from O before reaching the observer. The path from P to the observer is shorter than the path from O to the observer by a distance x. This means that a light wave emitted from P takes x/c less time to reach us than one emitted at O. Therefore, from the point of view of the observer, the apparent time t_{app} for the object to travel from O to P, is

$$t_{app} = t - x/c$$

Substituting from equations (19.4) and (19.5), we have

$$t_{app} = (r/v) - (r/c) \cos \theta$$
$$= (r/v)(1 - \beta \cos \theta) \tag{19.6}$$

where we have set $\beta = v/c$. The apparent velocity across the sky v_{app} is then

$$v_{app} = y/t_{app}$$

$$= \frac{r \sin \theta}{(r/v)(1 - \beta \cos \theta)}$$

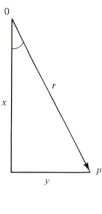

To observer

Figure 19.7 Apparent superluminal expansion as a clump of material moves from O to P.

Eliminating r gives

$$v_{app} = \frac{v \sin \theta}{1 - \beta \cos \theta} \qquad (19.7)$$

For $v \ll c$, β is close to zero, and $v_{app} \cong v \sin \theta$, the expected result. However, for v close to c, the quantity in square brackets can be greater than 1. It can be sufficiently greater than 1 if v_{app} is greater than c.

To see this, we can find the angle that gives the maximum v_{app} for a given v. Taking equation (19.7), dividing by v, differentiating with respect to θ, and setting the result equal to zero, gives

$$\left(\frac{\cos \theta}{1 - \beta \cos \theta}\right) - \left[\frac{\beta \sin^2 \theta}{(1 - \beta \cos \theta)^2}\right] = 0$$

Multiplying through by $(1 - \beta \cos \theta)^2$ gives

$$\cos \theta - \cos^2 \theta - \beta \sin^2 \theta = 0$$

Remembering that $\sin^2 \theta + \cos^2 \theta = 1$, this simplifies to

$$\cos \theta - \beta = 0$$

or

$$\cos \theta = \beta$$

We find $\sin \theta$ from

$$\sin \theta = (1 - \cos^2 \theta)^{1/2}$$

$$= (1 - \beta^2)^{1/2}$$

Substituting back into equation (19.7), we have

$$\left(\frac{v_{app}}{v}\right)_{max} = \frac{1}{(1 - \beta^2)^{1/2}} \qquad (19.8)$$

This is just the quantity γ that appears in the Lorentz transformations. We know that this quantity can get quite large as v approaches c.

EXAMPLE 19.2   Superluminal Expansion

For an object moving away from the nucleus of the galaxy at $v = 0.95\, c$, find the maximum value of v_{app} and the angle at which it must be moving to get this maximum.

Solution

We have

$$\cos \theta = \beta$$

so $\theta = 18.2°$.

From equation (19.8) we have

$$v_{\mathrm{app}} = \frac{0.95 \, c}{(1 - 0.95^2)^{1/2}}$$

$$= 3 \, c$$

There is a way to test this explanation. For a given speed and direction of motion, there should be a specific Doppler shift for radiation from the moving object. Unfortunately, these radio sources do not have any lines in their spectra. The Doppler shift does alter the shape of the synchrotron spectrum, but the interpretation is difficult. Studies of the spectra of these sources are still continuing.

19.2 SEYFERT AND N GALAXIES

Seyfert and N galaxies are characterized by having nuclei that strongly dominate the total light from the galaxy. On short exposure, they both look like stars. On longer exposures, they look like stars with a fuzzy patch around them (Fig. 19.8). For comparison, a spiral galaxy never looks like a star, even in a short exposure. This suggests that we are not resolving the nuclei

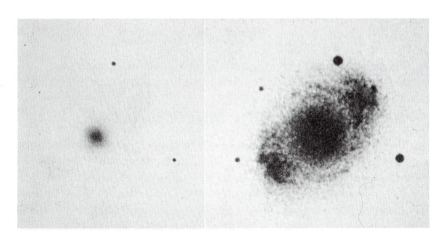

Figure 19.8 Different exposures of a Seyfert galaxy NGC4151, showing a quasar-like appearance with short exposures, and a spiral-like appearance with long exposures.

Figure 19.9 Schematic spectra of a quasar, a Seyfert galaxy, and a normal galaxy. The intensity scale is arbitrary and the vertical effect of the spectra is not meaningful.

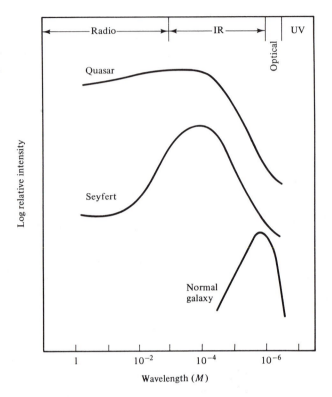

of Seyferts in most of our photographs of them. (There are a few Seyferts for which the source of narrow emission lines has been resolved.) When we can study the fuzzy patch around the nucleus, it seems like the disk of a spiral galaxy. This suggests that we are seeing spiral galaxies with unusually bright nuclei.

The optical spectra are characterized by strong, broad emission lines. The lines are more than 10^3 km/s wide. This is more than ten times the width of lines in normal galaxies. If the broadening is thermal Doppler broadening, this would imply a temperature in excess of 10^7 K (see Problem 19.23).

There are some differences among Seyferts in the appearance of lines called "forbidden lines." These are spectral lines that are not strong under normal circumstances. In some Seyferts, the forbidden lines are broad, and in others they are narrow. The ones with narrow lines are called type I, and the others are called type II. In addition, infrared emission from type I Seyferts is nonthermal, while that from type II Seyferts is thermal emission from dust. Also, type I Seyferts are weak radio sources, while half of the type II's have moderate radio emission. Type I's also have strong x-ray emission, with a correlation between optical and x-ray luminosity. In Fig. 19.9 we compare the schematic spectrum of a Seyfert with that of a normal galaxy and a quasar (to be discussed in the next section). It can be seen that the shape of the Seyfert spectrum is much closer to that of a quasar than the normal galaxy.

The brightest Seyfert galaxy is shown in Fig. 19.8. It is an 11th magnitude spiral in Coma Venatici. Ultraviolet emission lines from this galaxy have been studied using the International Ultraviolet Explorer satellite (IUE). The ultraviolet lines show rapid variations in strength and width. To explain these phenomena, a nucleus with a black hole of mass $10^9 \, M_\odot$ has been proposed.

19.3 QUASARS

19.3.1 Discovery of Quasars

In our discussion of radio galaxies we mentioned the importance of getting accurate radio positions for the purposes of finding optical counterparts. Before the use of interferometry, some sources were studied by lunar occultation. In such an experiment, the source is observed as the moon passes in front of the source. Since we know the position of the limb of the Moon very accurately, we can determine the location of a radio source that is occulted by noting the times at which the source disappears and appears. As Fig. 19.10 shows, the disappearance gives us an arc on the sky on which the source can be. The reappearance gives us another arc, intersecting the first arc in two places. The source must be at one of those two locations. A subsequent occultation will also give two positions, one identical to one of the positions in the first pair, and a different one. The position that is repeated in both occultations is the true position of the source.

This technique was used at the Parkes radio telescope, in Australia, to study the radio source 3C273. (The designation means that it is the 273rd

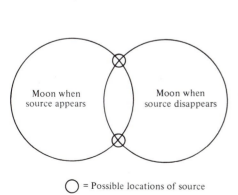

Figure 19.10 Lunar occultation of radio sources. The source must lie along the position of the leading edge of the Moon when the source disappears and the trailing edge when the source appears. These two points are indicated by the dark circles.

Figure 19.11   An optical image of the quasar 3C273.

source in the 3rd Cambridge catalog of radio sources.) When photographs of the area around the radio source were examined, a faint (13th magnitude) star was noted. This was an interesting discovery, since no radio stars were known at the time. Closer observation, producing photographs like that in Fig. 19.11, have shown that 3C273 is not really a star. It looks like a star with a fuzzy patch around it. The most detailed photographs show a jet extending from the core, just as in radio galaxies.

The optical spectrum of 3C273, shown in Fig. 19.12, is quite unusual. The spectrum puzzled astronomers for some time. It was finally noted that a

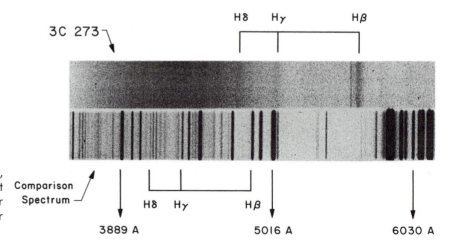

Figure 19.12   Spectrum of 3C273, showing the large shifts in the strongest Balmer lines. In the spectrum the lower frame is for comparison and the upper frame is the quasar.

series of lines looked like the Balmer series, but with a very large red shift. For example, the Hα line, whose rest wavelength is 6563 Å, was observed at about 7600 Å.

EXAMPLE 19.3 Red Shift of 3C273

In Chapter 21 we will see that Hubble's law is not a useful description for finding the distance modulus to an object with a very large red shift. This is because it is not proper to think of the red shift as being a Doppler shift. Rather, it is associated with the expansion of the universe, and distances can only be found by using a model for that expansion. However, it is still interesting to get a feel for the large distance involved by estimating $m - M$, using Hubble's law. Using the observed wavelength of the Hα line, find the velocity of 3C273. If its red shift follows Hubble's law (with $H_0 = 75$ km s^{-1}/Mpc), what are its distance and distance modulus?

Solution

The red shift is so large that we must use the relativistic Doppler shift formula, equation (12.8). Squaring it gives

$$\frac{1 + v/c}{1 - v/c} = \left(\frac{\lambda'}{\lambda}\right)^2$$

$$= \left(\frac{7600}{6563}\right)^2$$

$$= 1.34$$

Multiplying through by $(1 - v/c)$ gives

$$(1 + v/c) = 1.34 - (1.34)(v/c)$$

Solving for v/c gives

$$v/c = 0.15$$

This means that $v = 4.5 \times 10^4$ km/s. Using Hubble's law, we have the distance,

$$d = \frac{v}{H}$$

$$= \frac{4.5 \times 10^4 \text{ km/s}}{75 \text{ km s}^{-1}/\text{Mpc}}$$

$$= 600 \text{ Mpc}$$

We find the distance modulus, $m - M$, from

$$m - M = 5 \log_{10} (d/10 \text{ pc})$$

$$= (5)\log_{10}(6.0 \times 10^7)$$

$$= 38.9$$

Since the apparent magnitude is 13, the absolute magnitude is -26. For comparison, the absolute visual magnitude of the Sun is approximately $+5$. A difference of 31 magnitudes corresponds to a brightness ratio of a little more than 10^{12}. This means that 3C273 gives off 10^{12} times as much visible light as does the Sun. What makes this even more remarkable is that 3C273 gives off much more energy in the radio part of the spectrum than in the visible!

The proposal that the spectrum of 3C273 could be explained with a large red shift was made in 1963. At that time, the existence of a similar object, 3C48, was known. It had been noted in 1960 that there is a possible correspondence between the radio source and a 16th magnitude star. The spectrum of this star showed lines with an even greater red shift than in 3C273, corresponding to a speed of 0.37 c. These objects were given the name *quasi-stellar radio sources*, because of the starlike appearance on short-exposure photographs. This name was shortened to *QSR*, or *quasar*. Further studies have revealed a class of objects that are like QSRs in optical photographs, and also have large red shifts, but don't have any radio emission. These are called *quasi-stellar objects*, or *QSO*s. We now loosely call both types of objects quasars.

19.3.2 Properties of Quasars

The spectra of many quasars show a large ultraviolet excess. This means that they are much brighter in the ultraviolet than one would expect from just knowing their visual brightness. This provides us with a way of searching for quasars. We cannot take spectra of all stars to see if they have large red shifts. Since quasars are so faint, it takes a long time to observe a spectrum. We can study radio sources, but not all quasars are radio sources. However, we can compare visible, blue, and (now with space observations) ultraviolet images of large fields to find objects with a large ultraviolet excess. Spectra of these objects can be taken to see if they have large red shifts.

Some quasars are quite variable in their energy output (Fig. 19.13). We

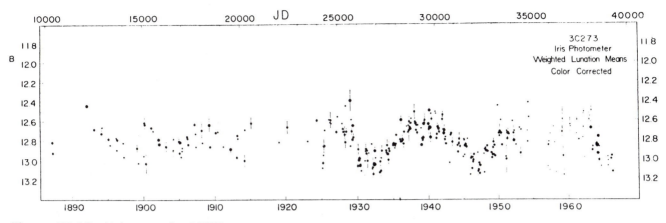

Figure 19.13 Light curve for 3C273.

have good records of the visible and radio variability of quasars for the past 20 years as a result of specific studies. However, we have optical records going even farther back, since observatories save photographic plates. A quasar may be on a plate exposed for an entirely different purpose. Once a quasar is discovered, an astronomer can go back through plate archives and find its image as far back as 100 years ago.

An important feature of the variability is that it allows us to place an upper limit on the size of the emitting region. We discussed this idea in our study of pulsars, in Chapter 11. If a significant fraction of the total power varies on a time scale t, then the emitting region can be no longer than ct (as long as motions close to the speed of light are not involved). In the case of quasars, variations on a time scale of a few months, limit the size of the emitting region to about 10^{12} km. (This is only about 10^4 AU.)

The spectra of quasars show both emission lines and absorption lines. Generally, all of the emission lines can be explained by a single red shift z, but a few groups of absorption lines appear with different red shifts. Over the past several years, extensive absorption line surveys have been carried out, providing good statistical information on absorption line properties. The absorption lines are generally narrow, less than 300 km/s wide. They also are mostly from the ground states of atoms, indicating a low temperature. The absorption line red shifts are usually less than or equal to the emission line red shift. This is taken as indicating that the absorption lines arise in material between us and the quasar.

As of the time of this writing, the largest observed quasar red shift was 3.53. The light from the most distant quasars has been traveling to reach us for over ten billion years. This means that we are seeing the universe as it was around the time our galaxy formed. For this reason, if their red shifts are cosmological, quasars provide us with a very important link to our past.

19.3.3 Energy–Red Shift Problem

The immediate problem that astronomers recognized with quasars was explaining their enormous energy output. What makes the problem even more difficult is the fact that the energy has to be generated in a small volume. One way out of the problem is to say that quasars are not as far away as we think they are. After all, we only observe their apparent brightness. We infer their absolute brightness by knowing their distances, determined from their red shifts and Hubble's law. (It should be noted that even a factor of 2 uncertainty in the Hubble constant has no real bearing on this particular problem.) If we are saying that the distances are wrong, we are saying that quasars do not obey Hubble's law. If that is the case, we must explain the large red shifts. This is the basic problem. Either we have to explain the large energy output, or we have to come up with another red shift mechanism. For this reason, we can think of this as the *energy–red shift problem*.

One possible source of a red shift is gravitational. We have already seen that photons are red-shifted as they leave the surface of any object. One problem with this explanation arises from the limited range of red shifts seen in the emission lines. This tells us that the emitting gas would have to be in a thin shell around some massive object. An analysis of such systems shows that quasars would have to be so close or so massive that our local part of the galaxy would be greatly affected by their presence. Even with a mass of $10^{11} M_\odot$, the objects would still have to be within our galaxy.

There is an additional problem. For the larger red shifts, we need black holes. Even with a black hole, to get $z > 1$ the photons must be emitted from very close to the Schwarzschild radius. For example, for $z = 2$ (see Problem 19.17), photons must be emitted from a surface whose radius is only 9/8 of R_s. We cannot think of any way to come up with a hot radiating gas that close to R_s, especially given the narrow range of red shifts in a given quasar. The accretion disks responsible for the x-ray emission around objects like CygX-1 are generally outside the photon sphere, which is at $1.5\,R_s$.

Another possible source of red shift is called kinematic. This means that we are observing a high velocity due to something other than the expansion of the universe. For example, they might be shot out of galaxies. However, if this were the case, we might expect to see at least a few quasars coming toward us. Quasars coming toward us would appear highly blue-shifted. To date, no objects have been found with blue shifts comparable to the red shifts seen in quasars. We might think that they have been ejected from our galaxy, but the kinetic energy for this becomes quite large. Also, if they were nearby objects moving at high speeds, we would expect to see some proper motion, but we don't.

In an effort to see if some kinematic explanation is possible, some effort has been made to search for galaxies and associated quasars, where the galaxy and quasar have very different red shifts. If it can be shown that a galaxy and quasar are directly connected, then they must be at approximately the

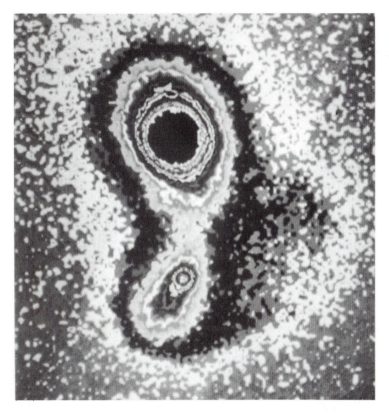

Figure 19.14 A galaxy and quasar that appear near each other on the sky, but have very different red shifts.

same distance from us. Presumably that distance is accurately indicated by the red shift of the galaxy. If the quasar has a different red shift, then it doesn't obey Hubble's law.

Examples have been found of galaxies and quasars that appear near each other on the sky, but which have different red shifts. An example is shown in Fig. 19.14. The problem, however, is in proving that the quasar and galaxy are associated. Just because two objects appear along almost the same line of sight doesn't mean that they are at the same distance. A galaxy and quasar don't have to be at the same distance any more than two stars in the same constellation have to be at the same distance.

Sometimes arguments over whether galaxies and quasars that appear near each other are associated come down to questions of probability and statistics. For example, we might ask, given a quasar, what is the probability that there will be an unassociated galaxy along the same line of sight? Arguments such as this are difficult to frame, and also can lead to fallacious conclusions, akin to the person who takes a bomb on board an airplane to safeguard his passage, on the assumption that if the probability of one person having a bomb on a plane is low, then the probability of two people having a bomb is even lower. To date, the statistical arguments seem to allow all of the cases of galaxies and quasars in the same direction as being coincidences, with the two being unassociated.

Some effort has also gone to trying to observe direct connections between the galaxy and quasar in such situations. Connections would be of the form of a detectable trail of luminous material between the galaxy and quasar. Even this problem is not easy, because there are photographic artifacts which can resemble such a bridge between two objects that appear close but are actually unrelated. To date, no convincing evidence has been presented to demonstrate a bridge between a quasar and a galaxy with different red shifts.

In addition, there are direct pieces of evidence that quasars are at cosmological distances. (By ''cosmological,'' we mean that their distances are given by Hubble's law.) The absorption lines in quasar spectra are one example. As we have said there are generally a few different red shifts in a given spectrum. The absorption line red shifts are less than or equal to the emission line red shifts. The most natural explanation is that the absorption lines are formed in material between us and the quasar. If the intervening material obeys Hubble's law, and the quasars are beyond the intervening material, then the quasars must be at cosmological distances. Also, the fact that Seyferts appear to be galaxies supports the belief that quasars are also galactic in scale and far away, rather than small and nearby.

Much of the concern over the energy–red shift problem has recently faded. Astrophysicists now feel that the energy generation requirements in quasars are not as outrageous as they seemed 20 years ago. This is partly because of the number of high-energy phenomena that have been observed and understood, at least partially. More specifically, our understanding of radio galaxies has reached the point where we think that quasars may be a different manifestation of the same phenomenon. There appears to be a natural progression, from normal galaxies, through the active galaxies—radio galaxies, Seyferts, and N galaxies—to quasars. The energy requirements are large, but not much larger than those for the most luminous radio galaxies.

One theory is that radio-quiet quasars are simply active galaxies in which the optical galaxy is a spiral. Remember, the radio galaxies discussed in Section 19.1 are giant ellipticals. The difference between the conditions near the centers of spirals and ellipticals could be sufficient to make the engine discussed for radio galaxies slightly more luminous in spirals.

The quasar problem is in many ways typical of problems in astronomy. On the observational side, it provides an excellent example of the interplay among observations in various parts of the spectrum. Radio and optical observations were important in the discovery of the phenomenon. Radio, optical, infrared, ultraviolet, and x-ray observations have been important in the understanding of the phenomenon. In addition, our understanding of quasars has required the results of extensive systematic surveys of quasar properties. Such surveys don't receive the same ''glory'' as the initial discovery, but they provide the data on which the solution can ultimately be based.

The quasar problem is also typical in another way. Many problems in astronomy come on the scene in a spectacular fashion, often with the unexpected observation of a new phenomenon. The solutions, however, do not come in such a single, spectacular step. For our understanding of quasars, it has been important to have the problem around long enough for astrophysicists to become more comfortable with the energies required. Part of the process is the discovery of other high-energy phenomena. For example, when some suggested twenty years ago that quasars might involve some form of black hole, the idea was not taken seriously by most. After all, black holes were theoretical playthings. However, the observations of objects like CygX-1 have made the existence of black holes seem more likely. It is a large jump from a few $M_\odot$ black hole to a 10^8-$M_\odot$ one, but for many it is not as big a jump as from no black holes to a few $M_\odot$ black hole. In addition, the success in explaining many details of radio galaxies has made astrophysicists more confident of the model that places a massive black hole at the center of a galaxy. (We may even have one at the center of our galaxy, as noted in Chapter 16.)

19.3.4 Gravitationally Lensed Quasar

Physicists have realized that if the path of light can be bent by a strong gravitational field, then, under the right conditions, that bending can lead to the formation of an image, just as does the bending by a piece of glass. When an image is formed through the gravitational bending of light, we say that we have a *gravitational lens*. It was even proposed that quasars might be gravitationally lensed objects. The idea was that the lensing effect might be making quasars appear much brighter than they actually are. There were, however, a number of difficulties with this theory.

However, the presence of a massive galaxy between us and a quasar would at least distort the appearance of the quasar, or could produce a multiple image, as indicated in Fig. 19.15. In the figure, light passing both above and below the galaxy can be bent to reach us. If we looked above and below the galaxy, we would see an image of the quasar.

Recent observations have revealed the existence of several such objects. Initially, two quasars were observed near each other, with identical spectra. The gravitational lens theory was advanced to explain the phenomenon. A test of this theory was to search for the intervening galaxy, which was not

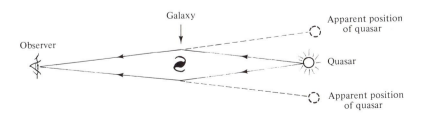

Figure 19.15 Gravitationally lensed quasar. The galaxy bends the path of the light from the quasar to the observer. The observer sees an image of the quasar back along the directions of the rays that reach the observer.

originally detected. This galaxy has now been found, confirming the basic theory. A number of additional lensed systems have also been found.

One important consequence of the gravitationally lensed quasar is that the geometry requires that the quasar be well beyond the intervening galaxy. This is another piece of evidence placing quasars at cosmological distances. Another use of the lensed systems comes from fact that we can use them to estimate the mass of the intervening galaxy.

19.3.5 BL Lacertae Objects

An interesting object, related to quasars, is the variable ''star'' BL Lacertae. It fluctuates between 14th and 16th magnitude. The optical ''star'' is surrounded by a diffuse optical emission; the optical spectrum of the star shows no emission or absorption lines. However, when the light from the star is blocked out, a spectrum of the nebulae can be taken. It shows lines with a red shift, $z = 0.07$, so it is not very far from us (see Problem 19.20). It is associated with a source of rapidly varying luminosity and polarization. The source shows superluminal expansion, with components appearing to move apart at $6\,c$.

Several similar objects have been found. The class is named after the prototype, so they are called *BL Lac objects*. They sometimes appear in the nuclei of galaxies. These objects always have nuclei, and some have nebulosity. The nebulosity looks like the radiation from late-type stars, like an elliptical galaxy. This is in contrast to the nebulosity in Seyferts, which looks like the disk of a spiral galaxy. BL Lac objects have been detected in the infrared. The timescale for variations in the luminosity is about 10 days. They are usually not associated with radio sources.

These objects have many similarities with quasars. A major exception is that emission and absorption lines are weaker in BL Lac objects than in quasars. Their continua are slightly different from those of quasars, and the timescale for variations is somewhat faster than that in quasars.

CHAPTER SUMMARY

In this chapter we looked at the unusual properties of active galaxies.

We first looked at radio galaxies. We discussed the large radio lobes and the jets leading out to them. We also discussed possible energy sources. The most likely sources involve mass falling into a small massive object, possibly a black hole.

We then looked at quasars as more extreme examples of a similar phenomenon. Many quasars are very far away, and are telling us about an earlier time in the universe.

We discussed the energy–red shift problem in quasars, and looked at alternative explanations for their large red shift. However, the best explanations seems to be that the quasars are at the distances indicated by their red shifts, and have central engines driven by material falling into a black hole. These black holes may be as massive as $10^8\,M_\odot$.

QUESTIONS AND PROBLEMS

19.1 (a) To the extent that we can ascribe a temperature to the large lobes of radio galaxies, it is about 10^6 K. For the numbers given in this chapter, estimate the thermal energy stored in one of these lobes. (b) The magnetic energy density is $B^2/8\pi$, where B is in gauss, and the energy density is in ergs per second. Compare the magnetic energy in the lobes with the answer in part (a).

19.2 For the sizes, densities, and speeds given in Section 19.12, estimate the kinetic energy in the flow of the jets in radio galaxies.

19.3 A radio galaxy has an angular extent of $2°$, and the associated optical galaxy has a red shift of $z = 0.5$. What is the diameter of the radio galaxy? (Take $H_0 = 75$ km s^{-1}/Mpc.)

19.4 What are the Schwarzschild radius and density for a 10^7-$M_\odot$ black hole?

19.5 At what rate must mass be falling into the central black hole to produce the radio galaxy luminosities discussed in this chapter?

19.7 In calculating the energy that we can get by dropping matter into a black hole, we only considered the energy gained in falling to the Schwarzschild radius. However, matter will still accelerate after crossing R_s. Why don't we add on this extra energy to what we can extract?

19.8 For the superluminal source discussed in Example 19.2 (with $\beta = 0.95$), what is the range of angles about $18.2°$ for which v_{app} will still be greater than c?

19.9 What β is required, assuming motion at the optimal angle, to produce a superluminal source with $v_{app} = 10\ c$?

19.10 What are the similarities between Seyfert and N galaxies and quasars?

19.11 Why were lunar occultations important in the discovery of quasars?

19.12 How would carry out a search for radio-quiet quasars?

19.13 If $v = 0.37\ c$ for 3C48, (a) at what wavelength is the Hα line observed, and (b) how far away is it, if its distance is cosmological and $H_0 = 75$ km s^{-1}/Mpc? (Assume that Hubble's law gives an adequate estimate of the distance.)

19.14 A quasar is observed with its Hα line at 8000 Å. What speed does this correspond to for a Doppler shift?

19.15 What would the red shift be for a quasar whose light has been traveling thirteen billion years to reach us? (Assume that Hubble's law gives a good estimate of the distance and take $H_0 = 75$ km s^{-1}/Mpc.)

19.16 If a quasar varies on a time scale of four months, what is the most masive black hole that can be in the center?

19.17 (a) If we observe the gravitational red shift of material coming to us from near a black hole, the ratio of observed to emitted wavelength is $1/(1 - 2GM/rc^2)^{1/2}$, where r is the distance from the center at which the photon is emitted. Rewrite this expression, expressing the mass in terms of the Schwarzschild radius R_s. (b) Show that to get a red shift $z = 2$, a photon must be emitted at a distance $r = \frac{9}{8} R_s$. (c) Why is this important for the gravitational red shift interpretation of quasar red shifts?

19.18 What are the similarities between quasars and radio galaxies?

19.19 What is meant by the "energy–red shift problem?"

19.20 What is the distance to BL Lac (assuming that $H_0 = 75$ km s^{-1}/Mpc)?

19.21 What does the gravitationally lensed quasar tell us about quasar red shifts?

19.22 What are the similarities between BL Lac objects and quasars?

19.23 For the 10^3 km/s wide emission lines in Seyferts, what temperature would be required for thermal Doppler broadening?

19.24 If the range of wavelength for a particular emission line at wavelength λ is $\Delta\lambda$ in a quasar, find an expression for the fractional range of radius, $\Delta R/R$, from which the emission could come from gravitational red shift caused by an object of mass M.

20 Cosmology

Einstein once said that the most incomprehensible thing about the universe is that it is comprehensible. It is amazing that we can apparently describe the universe with what are very simple theories. We can ask the truly fundamental questions of where we have come from and where we are going, and expect scientific answers. In this chapter and in the next, we will be studying *cosmology,* the large-scale structure of the universe. We can learn a great deal using only the physics that we have already introduced in this book. With the introduction of some more physics, namely, elementary particle physics, we will see that even more fascinating concepts are within our grasp.

20.1 THE SCALE OF THE UNIVERSE

When we study the gas in a room, we must deal with it as a collection of molecules. We don't care about the fact that the molecules are made up of atoms; or that the atoms are made up of protons, neutrons, and electrons; or that the protons and neutrons are made up of other particles. All we care about is how the molecules interact with one another, and how that affects the large-scale properties of the gas. When we study the universe, we also treat it as a gas. The molecules of this gas are galaxies. In the big picture, stars, planets, etc., don't matter. Of course, these smaller objects can still contain some hidden clues for us to learn about the larger structure. They simply don't affect the larger structure themselves.

How do we study cosmology? On the theoretical side, we look at the universe as a large-scale fluid and ignore the lumps. The only force that currently affects the large-scale structure is gravity. We can apply gravity as described by general relativity, though for many things Newtonian gravitation will be a sufficiently accurate approximation. The electromagnetic force is important in that electromagnetic radiation carries information, but it doesn't affect the large-scale structure. We will see in the next chapter that there was a time when all of the forces we know had an important effect on the structure of the universe.

Until recently, we have had very few observational clues about the large-scale structure of the universe. We will see that recent experiments, some characterized by great difficulty and resourcefulness, have greatly added to our knowledge. The field of observational cosmology is a growing one. In studying cosmology, as with other fields of astrophysics, we combine theory and observations to increase our insight into what is happening.

In making theoretical models of the universe, we start with an assumption, called the *cosmological principle*. It says that *on the largest scales, the universe is both homogeneous and isotropic*. By homogeneous we mean that, at any instant, the general properties, such as density, are the same everywhere. By isotropic we mean that, at any instant, the universe appears the same in all directions. It is important to remember that we are talking about the largest scales. We know that our everyday world is neither homogeneous nor isotropic, but for the universe on the scales of many superclusters of galaxies, this is a very good description.

There is another assumption that we might be tempted to make, namely, that the universe is also the same at all times. Cosmological theories which incorporate this assumption are called *steady-state theories*. Until relatively recently, there was a vigorous debate about whether or not the universe is steady state. The theory was favored by many because it had a certain philosophical simplicity. If the universe had no beginning, then there is no need to worry about what went ''before'' the beginning.

However, a long chain of observational evidence has been amassed against the steady-state theory, and very few hold to it today. For example, we will see that the relative abundance of hydrogen and helium can be easily explained in non-steady-state theories, but not in steady-state theories. In addition, the existence of many quasars some ten billion years ago, but very few now argues for the conditions in the universe changing over the last ten billion years.

You might think that the steady-state theory would have died with the observation of the expansion of the universe (Hubble's law). If the universe is expanding, its density must be decreasing. If the density is not constant with time, we cannot have a steady-state universe. However, the proponents of the steady-state theory pointed out that it might be possible to create matter out of nothing. We do not mean the creation of mass from energy, but from nothing. This theory, called *continuous creation*, calls for the violation of conservation of energy. (This violation would be permanent, not for the short times for temporary particle–antiparticle creation discussed in Chapter 7.) However, it requires a violation at a level well below that to which conservation of energy has been verified experimentally. We know of no mechanism to create this matter, but proponents of this idea say that there is no experimental evidence to rule it out.

EXAMPLE 20.1 Continuous Creation

Estimate the rate of continuous creation required by the current expansion of the universe. Take the current density of the universe to be 10^{-30} g/cm^3.

Solution

We start by considering a box with side x. The volume of the box is x^3. If all sides increase at the same rate, the rate of change of the volume is

$$dV/dt = 3x^2(dx/dt)$$

From Hubble's law, $(dx/dt) = H_0 x$, so

$$dV/dt = 3H_0 V$$

Dividing both sides by V gives

$$(1/V)(dV/dt) = 3H_0$$

The quantity $(1/V)(dV/dt)$ is the fractional rate of change of the volume. If we take $H_0 = 1/(15 \times 10^9)$ yr, and multiply through by dt,

$$(dV/V) = dt/5 \times 10^9 \text{ yr}$$

If we take $dt = 1$ yr, then

$$dV/V = 2 \times 10^{-10}$$

This means that the volume of any box increases by this fractional amount in one year. The density will decrease by this fractional amount unless we create 2×10^{-10} protons for every proton already in the universe. The current density of the universe, 10^{-30} g/cm^3, corresponds to approximately 10^{-6} protons/cm^3. This means that we need to create 2×10^{-16} protons per cubed centimeter per year. This corresponds to a proton per year in $\frac{1}{2} \times 10^{16}$ cm^3. This would be a box approximately 10^5 cm (1 km) on a side. Therefore, to keep the density of the universe constant, we must create one proton per year in each cubic kilometer. This would be essentially impossible to detect if it were taking place.

The big bang theory finally took the upper hand with the discovery, in 1965, of the *cosmic background radiation*. This radiation is the relic of a time when the universe was much hotter and denser than it is now. We will discuss this radiation in detail in Chapter 21, For now, we note that this discovery ended, for most astronomers, the debate over whether the universe is steady state.

20.2 EXPANSION OF THE UNIVERSE

20.2.1 Olbers's Paradox

One of the most important observed facts about the universe is that it is expanding. The direct evidence for the expansion is Hubble's law. However, there is a very simple observation that we make every day that also tells us that the universe is expanding. This simple observation is that the sky is dark at night. This mystery in this observation was noted by Heinrich Olbers in 1823. We refer to the problem as *Olbers's paradox*.

Qualitatively, Olbers said that, for a finite universe, every line of sight should eventually end on the surface of a star. Therefore, the whole night sky should look like the surface of a star. In fact, the whole day sky should appear the same, since the Sun would just be one of the large number of stars.

It may be easier to understand in terms of the quantitative argument, illustrated in Fig. 20.1. Suppose we divide the universe into concentric shells, centered on the Earth, each of thickness dr. The volume of each shell is then

$$dV = 4\pi r^2 \, dr$$

If there are n stars/volume, the number of stars in a shell is

$$N = 4\pi r^2 n \, dr \qquad (20.1)$$

The number of stars per shell goes up as r^2. However, the brightness that we see for each star falls off as $1/r^2$. Therefore, the r^2 and $1/r^2$ will cancel, and the brightness we see for each shell is the same. If there are an infinite number of shells, the sky should appear infinitely bright after we add up the contributions for each shell. Actually, this isn't quite the case. The stars have some extent, and, eventually, the nearer stars will block the distant stars. However, this will not happen until the whole sky looks as if it is covered with stars, with no gaps in between.

There may appear to be some obvious ways out of this. You might say that our galaxy doesn't go on forever. Most lines of sight will leave the galaxy before they strike a star. Unfortunately, the argument can be recast in terms of galaxies instead of stars, and the same problem applies. Another possible solution is to invoke the absorption of distant starlight by interstellar dust. However, if the universe has been here forever (or even a very long time), the dust will have absorbed enough energy to increase its temperature to the same as that of the surface of a star. (If the dust got any hotter than that, the dust would cool by giving off radiation that the stars would absorb.) Even if the sky wasn't bright from the light of stars, it would be bright from the light of the dust.

The expansion of the universe helps resolve the problem. There are

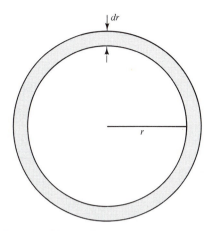

Figure 20.1 Olbers's paradox. We consider the contribution to the night sky from a shell of radius r and thickness, dr.

actually two aspects of the expansion that help. The first is the red shift. The energy of each proton is reduced in proportion of the distance that it travels before we detect it. This adds an additional factor of $1/r$ to the apparent brightness of each shell, meaning that the brightness per shell falls off as $1/r$, instead of being constant. Note that this doesn't completely solve the problem. If we add up the contributions from all of the shells, we have to take an integral of dr/r (see Problem 20.20). This gives $\ln(r)$, so if r can be arbitrarily large, the brightness can also.

However, for the models that we are discussing, expansion implies a finite age for the universe t_0. We can only see stars that are close enough for their light to have reached us in this time. That is, we can cut r off at ct_0. There may also be a cutoff due to our looking back far enough to where galaxies and stars have not yet formed. In either case, there is a finite cutoff to the number of shells that can contribute to the sky brightness, and the problem is solved. It should be noted that the same type of cutoff can be obtained if the universe has a finite size. This resolution of Olbers's paradox was not realized until the expansion of the universe had already been discovered. However, it is amazing that this simple observation—that the night sky is dark—could have led to the conclusion that the observable universe has a finite size or age.

20.2.2 Keeping Track of Expansion

We can show that Hubble's law follows from the assumption of homogeneity. In Fig. 20.2, suppose that P observes two positions, and O and O', with distance vectors from P being $\mathbf{r}$ and $\mathbf{r}'$, respectively. The vector from O' to O is $\mathbf{a}$, so that

$$\mathbf{a} = \mathbf{r} - \mathbf{r}' \tag{20.2}$$

We let $\mathbf{v}$ be a function to give the rate of change of length vectors ending at O and $\mathbf{v}'$ be the corresponding function for vectors ending at O'. We then have

$$\mathbf{v}'(\mathbf{r}') = \mathbf{v}(\mathbf{r}) - \mathbf{v}(\mathbf{a}) \tag{20.3}$$

Using equation (20.2) to eliminate $\mathbf{r}'$, this becomes

$$\mathbf{v}'(\mathbf{r} - \mathbf{a}) = \mathbf{v}(\mathbf{r}) - \mathbf{v}(\mathbf{a}) \tag{20.4}$$

The homogeneity of the universe means that the functional form of $\mathbf{v}$ and $\mathbf{v}'$ must be the same. (The functional form of the velocity cannot depend on where you are in the universe.) This means that

$$\mathbf{v}'(\mathbf{r} - \mathbf{a}) = \mathbf{v}(\mathbf{r} - \mathbf{a}) \tag{20.5}$$

Using this, equation (20.4) becomes

Figure 20.2 Vectors for locating objects in the universe.

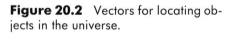

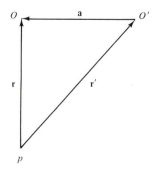

$$\mathbf{v}(\mathbf{r} - \mathbf{a}) = \mathbf{v}(\mathbf{r}) - \mathbf{v}(\mathbf{a}) \qquad (20.6)$$

This means that $\mathbf{v}(\mathbf{r})$ is a linear function of $\mathbf{r}$. The only velocity law that satisfies this relationship is

$$\mathbf{v}(r) = H(t)\mathbf{r} \qquad (20.7)$$

Note that we haven't required an expansion. (H could be zero.) However, if there is an expansion, it must follow this law, if the cosmological principle is correct.

When we want to keep track of the expansion of the universe, it is not convenient to think about the size of the universe. Instead, we introduce a *scale factor* with which we keep track of ratios of distances. We let t_0 be the age of the universe at some reference epoch. (It doesn't matter how this reference is chosen.) We let $\mathbf{r}(t)$ be the distance between two particular points as a function of time. (The points must be far enough apart so that their separation is cosmologically significant.) We define

$$\mathbf{r}_0 = \mathbf{r}(t_0) \qquad (20.8)$$

The scale factor $R(t)$ is a scalar, defined from

$$R(t) = \mathbf{r}(t)/\mathbf{r}(t_0)$$
$$= \mathbf{r}(t)/\mathbf{r}_0 \qquad (20.9)$$

Note from this definition that $R(t_0) = 1$. If the universe is always expanding, $R < 1$ for $t < t_0$, and $R > 1$ for $t > t_0$.

We can rewrite Hubble's law in terms of the scale factor. We start by writing Hubble's law as

$$d\mathbf{r}/dt = H(t)\mathbf{r} \qquad (20.10)$$

Using $\mathbf{r}(t) = R(t)\mathbf{r}_0$ [equation (20.9)] makes this

$$\mathbf{r}_0(dR/dt) = H(t)R(t)\mathbf{r}_0$$

Dividing through by $\mathbf{r}_0$ gives

$$dR/dt = H(t)R(t) \qquad (20.11)$$

Note that we now only have to deal with a scalar equation, instead of a vector equation. We can solve equation (20.11) for $H(t)$ to give the Hubble constant in terms of the scale factor,

$$H(t) = [1/R(t)](dR/dt) \qquad (20.12)$$

20.3 COSMOLOGY WITH NEWTONIAN GRAVITATION

We can learn a lot about the evolution of an expanding universe by applying Newtonian gravitation. In the next section, we will see how the Newtonian results are modified by general relativity.

The assumption of isotropy is equivalent to saying that the universe appears to be spherically symmetric from any point. This means that any spherical volume evolves only under its own influence. The gravitational forces exerted on the volume by material outside the volume sum (vectorially) to zero. If the volume in question has a radius r, and mass $M(r)$, the equation of motion for a particle of mass m on the surface of the sphere, as position **r,** is

$$m\ddot{\mathbf{r}} = -GM(r)m\hat{r}/r^2 \tag{20.13}$$

In this, we have let $\hat{r}$ be a unit vector in the r-direction. The assumption of homogeneity means that the density ρ is the same everywhere (though it can change with time). The mass $M(r)$ is given by

$$M(r) = (4\pi/3)r^3\rho$$

Substituting into equation (20.13) gives

$$\ddot{\mathbf{r}} = -(4\pi/3)G\rho\mathbf{r} \tag{20.14}$$

We use equation (20.9) to eliminate r by using the scale factor, to get

$$\ddot{R} = -(4\pi/3)G\rho R \tag{20.15}$$

As the universe expands, any given amount of mass occupies a larger volume. The density goes as 1/volume. The volume is proportional to R^3, so we have

$$\rho(t) = \rho_0 R_0^3/R^3(t)$$

We have used $\rho_0 = \rho(t_0)$. Also, since $R_0 = 1$, we have

$$\rho = \rho_0/R^3(t) \tag{20.16}$$

Substituting this into equation (20.15), we have

$$\ddot{R} = \frac{-(4\pi/3)G\rho_0}{R^2} \tag{20.17}$$

Note that if ρ_0 is not zero, then $\ddot{R}$ cannot be zero. A universe with matter cannot be static. It must be expanding or contracting.

To integrate the equation of motion [equation (20.17)], we first multiply both sides by $\dot{R}$, to give

$$\dot{R}\ddot{R} + \frac{(4\pi/3)G\rho_0\dot{R}}{R^2} = 0$$

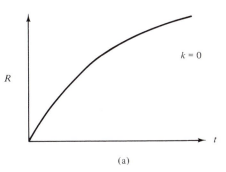

(a)

Noting that $d(\dot{R}^2)/dt = 2\dot{R}\ddot{R}$, this becomes

$$\frac{\frac{1}{2}d(\dot{R}^2)}{dt} + \left(\frac{4\pi}{3}\right)\left(\frac{G\rho_0}{R^2}\right)\left(\frac{dR}{dt}\right) = 0$$

Multiplying through by two, and using the fact that $(1/R^2)(dR/dt) = -d(1/R)/dt$, we have

$$\left(\frac{d}{dt}\right)\left[\dot{R}^2 - \frac{(8\pi G\rho_0/3)}{R}\right] = 0 \qquad (20.18)$$

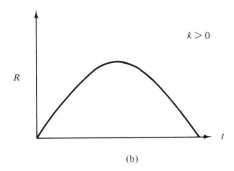

(b)

Since its time derivative is zero, the quantity in square brackets must be a constant. We can set it equal to some arbitrary constant k, so that

$$\dot{R}^2 = \frac{8\pi G\rho_0/3}{R} - k \qquad (20.19)$$

Further integration of equation (20.19) depends on whether k is zero, positive, or negative. We will consider each case separately. The schematic behavior of $R(t)$ for each case is shown in Fig. 20.3.

We first look at $k = 0$. Equation (20.19) then becomes

$$\dot{R}^2 = \frac{8\pi G\rho_0/3}{R} \qquad (20.20)$$

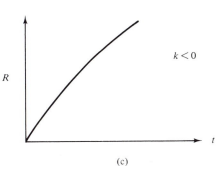

(c)

Figure 20.3 The scale factor R as a function of time for borderline, closed, and open universes.

It should be noted that $\dot{R}$ is always positive, but approaches zero as R approaches infinity. Equation (20.20) can be written so that we can integrate and get an explicit function for $R(t)$. We take the square root of both sides of the equation, giving

$$R^{1/2}dR = \left(\frac{8\pi G\rho_0}{3}\right)^{1/2} dt$$

We integrate the left side from zero to R, and the right side from O to t, to give

$$\tfrac{2}{3}R^{3/2} = \left(\frac{8\pi G\rho_0}{3}\right)^{1/2} t \qquad (20.21)$$

This means that R is proportional to $t^{2/3}$. The universe always expands, but the rate of expansion gets smaller and smaller.

We now look at the case $k > 0$. Since the first term on the right-hand side of equation (20.19) gets smaller when R gets larger, a point will be reached eventually, at some finite R, where $\dot{R} = 0$. The expansion stops at some maximum scale factor, R_{max}. We can find R_{max} as the value of R which makes R equal to zero in equation (20.19). That is,

$$0 = \frac{8\pi G\rho_0/3}{R_{max}} - k$$

Solving for R_{max},

$$R_{max} = \left(\frac{8\pi}{3}\right)\left(\frac{G\rho_0}{k}\right) \qquad (20.22)$$

After R_{max} is reached, the universe starts to collapse. We say that the universe is *closed*.

We next look at the case $k < 0$. If k is negative, then $-k$ is positive, and the right-hand side of equation (20.19) is always positive. As R gets very large, the first term on the right-hand side approaches zero, and $\dot{R}^2$ approaches $-k$. (Remember, $-k$ is a positive number.) This means that $\dot{R}$ approaches $(-k)^{1/2}$. The expansion continues forever, and we say that the universe is *open*.

We can think of the analogous situation of throwing a ball up in the air. If the total energy is negative, the ball will return to Earth. If the total energy is positive, the ball will escape, and its speed will remain positive. If the total energy is zero, the ball will reach infinity, but its speed will approach zero. We can therefore think of k as being related to the energy of the spherical region of the universe that we have been following.

Having established that the universe is expanding, we would now like to ask whether that expansion will continue forever. In other words, is the universe open or closed? We would like to have some quantity that we can measure to tell us. If we go back to our analogy of the ball thrown up in the air, if we know the position and velocity of the ball at some time, we also need to know its acceleration to know if the ball has sufficient energy to escape. Since the ball is slowing down, we might say that we want to know the deceleration of the ball.

For the universe, we define a *deceleration parameter*, whose value will tell us whether the universe is open or closed. We would like to define this parameter so that it is dimensionless (just as the scale factor is dimensionless), and so that it is independent of what time t_0 we choose for our reference epoch. The latter requirement says that our parameter should depend on quantities like $(\dot{R}/R)$ and $(\ddot{R}/R)$. With these ideas in mind, we define the deceleration parameter as

$$q = -\left(\frac{\ddot{R}}{R}\right)\left(\frac{R}{\dot{R}}\right)^2$$

$$= -\frac{R\ddot{R}}{\dot{R}^2} \qquad (20.23)$$

Since the expansion is slowing down, $\ddot{R}$ is negative, and q is positive. If we use equations (20.11), (20.16), and (20.17) to eliminate $\ddot{R}$ and $\dot{R}$, this becomes (see Problem 20.11)

$$q = \left(\frac{4\pi}{3}\right)\frac{G\rho}{H^2} \qquad (20.24)$$

We can look at the ranges of q for the three cases of (1) k being zero, (2) positive, or (3) negative:

1. $k = 0$. We combine equations (20.11) and (20.20) to give

$$H^2R^2 = \frac{(8\pi/3)G\rho_0}{R}$$

or

$$H^2 = \frac{(8\pi/3)G\rho_0}{R^3}$$

$$= \left(\frac{8\pi}{3}\right)G\rho \qquad (20.25)$$

where we used equation (20.16) in the last step. Substituting equation (20.25) into the equation for q, equation (20.23), we have

$$q = \tfrac{1}{2}$$

So, the case $k = 0$ corresponds to a specific value of q, namely, $\tfrac{1}{2}$.

2. $k > 0$. In this case, q can get arbitrarily large. It even approaches infinity as R approaches R_{max} (since $\dot{R} = 0$ at R_{max}). This means that any value of q in the range

$$q > \tfrac{1}{2}$$

will produce a closed universe.

3. $k < 0$. We have already said that q must be greater than zero, so the range of a q given by

$$0 < q < \tfrac{1}{2}$$

will produce an open universe.

The deceleration will depend on the density of matter in the universe. We can define a *critical density*, ρ_{crit}, such that the universe is closed if $\rho > \rho_{crit}$ and open if $\rho < \rho_{crit}$. If $\rho = \rho_{crit}$, we will have $k = 0$. This last point allows us to find ρ_{crit}, since it is the density for $k = 0$, or $q = \tfrac{1}{2}$. If we set $q = \tfrac{1}{2}$ in equation (20.24), and solve for ρ, we have

$$\rho_{crit} = (3/8\pi)(H^2/G) \qquad (20.26)$$

It is convenient to define a parameter Ω, which is the ratio of the true density to the critical density. That is,

$$\Omega = \rho/\rho_{crit} \qquad (20.27)$$

We can easily show (see Problem 20.12) that $\Omega = 2q$. If $H = 75$ km s^{-1}/Mpc, then ρ_{crit} is 1×10^{-29} g/cm^3. In the last section of this chapter, we will discuss observations to determine that actual value of Ω.

20.4 COSMOLOGY AND GENERAL RELATIVITY

When Einstein developed the general theory of relativity, he realized that it should provide a correct description of the universe as a whole. Einstein was immediately confronted with a result equivalent to equation (20.17), which says that if the density is not zero, the universe must be expanding or collapsing. This was before Hubble's work, and most believed in a static universe.

To get around this problem, Einstein introduced a constant, called the *cosmological constant*, into general relativity. It has no measurable effect on small scales, but altered results on cosmological scales. For example, in equation (20.17), the effect of the cosmological constant Λ would be to replace the density ρ by $\rho - \Lambda/4\pi G$. This makes it possible to have a nonzero density, but a zero value for $\dot{R}$. It should be emphasized that there is no evidence for the existence of this constant, and it doesn't affect any of the tests of relativity. Einstein withdrew the cosmological constant when he heard of Hubble's work. Some theoreticians still keep it in the theory, since there is nothing to say it cannot be there, and then set it equal to zero.

Figure 20.4 Albert Einstein.

Following Einstein's work (Fig. 20.4), a number of people worked out cosmological theories, using different simplifying assumptions. The models are generally named after the people who developed them. The de Sitter models are characterized by $k = 0$ and a nonzero (positive) cosmological constant; the Friedmann models have a zero cosmological constant and also zero pressure (a good approximation to very low density); Lemaitre models have nonzero density and a cosmological constant. As a result of his work, Lemaitre noted that there must have been a phase in its early history when the universe was very dense. This phase is called the *big bang*.

Many of the general relativistic results are similar to those we obtained in the previous section. This is because both depend on the fact that in a spherically symmetric distribution, matter outside a sphere has no effect on the evolution of the matter inside the sphere. One modification is the replacement of ρ by $\rho - \Lambda/4\pi G$ if you want a nonzero cosmological constant.

However, the general relativistic approach gives us a deeper insight by providing a geometric interpretation of the results. For example, the space-time interval, in spherical coordinates, becomes

$$(ds)^2 = (cdt)^2 - \left[\frac{R(t)}{1 + (1/4)kr^2}\right]^2$$

$$\times \{(dr)^2 + r^2[(d\theta)^2 + \sin^2\theta(d\psi)^2]\}$$

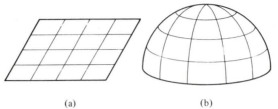

(a)   (b)

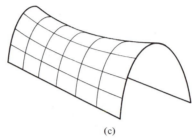

(c)

Figure 20.5   Schematic representations of universes with (a) flat, (b) positive, and (c) negative curvature.

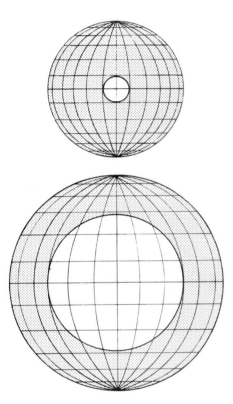

Figure 20.6   The universe as an expanding sphere. As the sphere expands, the coordinate system expands, and the radius of curvature changes. The dark circle with the lighter interior represents the part of the universe that we can see, with us at the center of the circle. A photon could have traveled from the circle to the center in the age of the universe. As the universe expands and gets older, our horizon expands, also. Since our horizon expands at the speed of light, new objects are always coming over the horizon.

In this equation, $R(t)$ has the same meaning as before. However, k now tells us something about the curvature of space-time (Fig. 20.5). If $k = 0$, space-time is flat (Euclidean). In this case space is infinite. If $k > 0$, space-time has positive curvature, like the surface of a sphere. Space must be finite, just as the surface of a sphere is finite. If space were infinite, it would look flat over any finite region. Finally, if $k < 0$, we say that space-time has negative curvature. In this case, space is infinite. (This equation for $(ds)^2$ is presented just to give you a feel for how k enters into the geometry.) The properties of various cosmological models are summarized in Table 20.1.

We can get a feel for the geometry of the universe by considering a two-dimensional analogy. We consider our universe to be confined to the surface of an expanding sphere, as shown in Fig. 20.6. One concept that we can now visualize is that we have a horizon due to the finite age of the universe t. We can only see light emitted toward us within a distance equal

TABLE 20.1   Parameters of Various Cosmological Models

Type	k	q	Ω	Curvature	Extent
Flat	0	$\frac{1}{2}$	1	Flat	Infinite
Closed	>0	$>\frac{1}{2}$	>1	Positive	Finite
Open	<0	$<\frac{1}{2}$	<1	Negative	Infinite
		>0	>0		

to ct. This horizon is growing. We have seen that over small distances even the surface of a sphere appears flat. The curvature becomes apparent as you can survey larger areas. This means that, as our horizon grows, the curvature might become more apparent.

We can also use our analogy to see that it is meaningless to talk about the radius of the universe. In three dimensions, our sphere has a radius, but in two dimensions we can only talk about the surface. This is one reason why the scale factor $R(t)$ is a better way of keeping track of the expansion. Even though we cannot talk about the radius of the universe in a meaningful way, we can talk about the curvature of our surface.

Our expanding-sphere analogy also tells us that it is not very meaningful to talk about the center of the universe. The sphere has a center, but it is not in the universe, which is the surface only. There is nothing special about any of the points on the surface of the sphere.

We can also see that the red shift (Hubble's law) fits in as a natural consequence of the expansion (Fig. 20.7). As the universe expands, the wavelengths of all photons expand in proportion to cosmological distances. That is, they expand in proportion to the scale factor. If radiation is emitted at wavelength λ_1, at epoch t_1, and detected at wavelength λ_2, at epoch t_2, then

$$(\lambda_2/\lambda_1) = R(t_2)/R(t_1) \tag{20.28}$$

If we let t_1 be some arbitrary time t, and t_2 be the reference epoch t_0 (for which $R = 1$), this becomes

$$\lambda_0/\lambda = 1/R(t)$$

Remembering that λ_0/λ is $1 + z$, where z is the red shift, we have

$$1 + z = 1/R(t) \tag{20.29}$$

Remember, since the radiation is emitted before the reference epoch, $R(t) < 1$, so $z > 0$.

We can derive an approximate expression for the red shift for radiation emitted some time Δt in the recent past, where $\Delta t \ll t_0$. Using a Taylor series, we have (see Problem 20.16)

$$1/R(t_0 - \Delta t) = R(t_0) + \Delta t \dot{R}(t_0)$$

$$= 1 + \Delta t \dot{R}(t_0) \tag{20.30}$$

Combining this with equation (20.29) gives

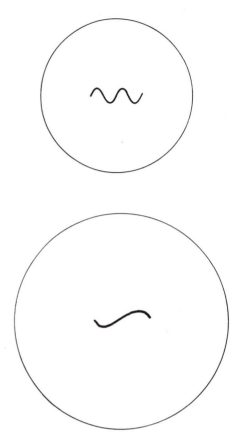

Figure 20.7 Cosmological red shift. As the universe expands, represented by the expanding sphere, the wavelengths of all photons expand in proportion to the scale increase.

$$1 + z = 1 + \Delta t \dot{R}(t_0)$$

Setting $t = t_0$ in equation (20.12) gives us $\dot{R}(t_0) = H_0$, so

$$z = H_0 \Delta t \qquad (20.31)$$

If a photon takes Δt to reach us, it must have been emitted from a distance $d = c \, \Delta t$. Using this to eliminate Δt in equation (20.31) gives

$$cz = H_0 d \qquad (20.32)$$

which is Hubble's law. (Remember, this approximation applies for small Δt.)

It is important to note that equation (20.29) tells us how to interpret the red shifts of distant galaxies. It is tempting to say that these galaxies are moving relative to us, and their radiation is therefore Doppler-shifted. However, in computing a relative velocity for a Doppler shift, we take the difference between the velocities of the two galaxies. These two velocities must be with respect to some coordinate system. In doing cosmology, it is important to use what is known as a co-moving coordinate system. This system expands with the universe. Therefore, apart from its peculiar random motion, each galaxy's velocity is zero with respect to this coordinate system. Therefore, strictly speaking, there is no Doppler shift due to the galaxies moving apart. The red shift arises as a result of the increase in the wavelengths of all photons moving through an expanding universe.

As a consequence of this, we should not directly interpret the red shift of a galaxy as giving us a particular distance. The amount of the red shift just tells us the amount by which the scale factor has changed between the time the photon was emitted and the time it was detected. For this reason, we often talk about the red shift of a particular galaxy, and don't bother to convert it to a distance. To convert a red shift to a distance we need a particular model for how $R(t)$ has evolved.

There are also different ways in which we could define the distance, since we are dealing with objects whose separation changes between the time a photon is emitted at one galaxy and received by another. A convenient definition of distance in this case is that which we would associate with a distance modulus, $m - M$. This would tell us how to convert apparent brightnesses (or magnitudes) into absolute luminosities (or magnitudes). If k is close to zero, then the relationship between distance modulus and the red shift is given approximately by

$$m - M = 25 + 5 \log(cz/H_0) + 1.086z(1 - q_0) \qquad (20.33)$$

where q_0 is defined in the next section. In this expression, cz/H_0 is in megaparsecs, accounting for the 25 in front (see Problems 20.21 and 20.22).

20.5   IS THE UNIVERSE OPEN OR CLOSED?

In this section we look at the evidence that might allow us to decide whether the universe is open or closed. It is impressive that we can even ask such a question and hope to get an answer.

The basic question is whether the actual density is less than or greater than the critical density. We could start then by adding up the density of all of the matter we can see, to find out if it gives us $\rho > \rho_{crit}$. However, we already know that there is a problem with dark matter, so if we only include the visible matter, we may be missing a significant amount of matter. Of course, if the visible matter is sufficient to close the universe, then we don't have to worry about the dark matter. If the visible matter is insufficient to close the universe, then we have to account for the dark matter. It turns out that the density of visible matter is about 1% of the critical density.

If the universe is to be closed, the dark matter must do it. From Table 20.2, we see that the amount of dark matter required to close the universe is greater than the amount of dark matter in clusters of galaxies. There also appears to be a trend toward more dark matter on larger scales. Therefore, we would not be surprised if there is enough dark matter to close the universe. However, in our attempt to see if the universe is open or closed, we can only include the dark matter that we know is present. We can therefore include the dark matter in clusters of galaxies, since we can detect its gravitational effects. This still leaves us a factor of 5 short of closing the universe.

We have said that the best way to measure the mass of an object is to measure its gravitational effect on something. If we want to determine the mass of the Earth, we measure the acceleration of gravity at the surface. Therefore, instead of trying to find all of the matter needed to close the universe, we can look for its gravitational effects. We can try to measure the

TABLE 20.2   Mass-to-Light Ratios
for Different Scales

Scale	M/L (solar units)
Milky Way to Sun	3
Spiral galaxy disk	10
Elliptical galaxy	30
Halo of giant elliptical	40
Rich cluster of galaxies	200
To close universe	1200

actual slowing down of the expansion of the universe and see if $-\ddot{R}$ is large enough to stop the expansion. When we do this, we are determining the current value of the deceleration parameter from its original definition [equation (20.23)]. Using the fact that $\dot{R}(t_0) = H_0$ [equation (20.12)], it becomes

$$q_0 = -\ddot{R}(t_0)/H_0^2 \tag{20.34}$$

We don't directly measure $\ddot{R}(t_0)$. What we try to measure is the current rate of change of the Hubble constant, $\dot{H}(t_0)$. We would therefore like to express $\ddot{R}(t_0)$ in terms of H_0 and $\dot{H}(t_0)$. We start with equation (20.11):

$$\dot{R}(t) = H(t)R(t)$$

Differentiating both sides with respect to t gives

$$\ddot{R}(t) = H(t)\dot{R}(t) + \dot{H}(t)R(t)$$

Setting $t = t_0$, and remembering that $\dot{R}(t_0) = H_0$, this becomes

$$\ddot{R}(t_0) = H_0^2 + \dot{H}(t_0) \tag{20.35}$$

Substituting this into equation (20.34), we have

$$q_0 = -[\dot{H}(t_0)/H_0^2 + 1] \tag{20.36}$$

Equation (20.36) tells us that if we can measure the rate of change of the Hubble constant, we can determine q_0. Unfortunately, measuring $\dot{H}(t_0)$ is not easy. It shouldn't surprise you, since measuring H_0 is not easy. In principle, we can measure $\dot{H}(t_0)$ by taking advantage of the fact that we see more distant objects as they were a long time ago. If we can determine H for objects that are five billion light years away, then we are really determining the value of H five billion years ago. If we include near and distant objects in a plot of Hubble's law, we should see deviations from a straight line as we look farther back in time. A sample result is shown in Fig. 20.8.

The difficulty comes in the methods used for measuring distances to distant objects. In our discussion of the extragalactic distance scale, we saw that for the most distant galaxies we can see, we cannot look at individual stars, such as Cepheids, within a galaxy. Instead, we must look at the total luminosity of a galaxy. We already know that the luminosities of galaxies change as they evolve. Galactic cannibalism provides us with the most spectacular example of this, but even normal galaxies change in luminosity with time. Therefore, if we calibrate the distance scale using the luminosities of nearby galaxies, we cannot apply this to more distant galaxies, precisely because we are seeing them as they were in the past. Before we can interpret a diagram like that in Fig. 20.8, we must apply a theoretical evolutionary

Figure 20.8 Hubble's law with deviations for distant objects. On the horizontal axis, we plot the brightness of a galaxy and the vertical axis the red shift. The dashed line is a continuation of a straight line extrapolation of Hubble's law. The theoretical predictions of deviations at large distances, for different geometries of the universe are indicated. The dark region is open; the lightly shaded region is closed, with the solid line in that region corresponding to one particular model.

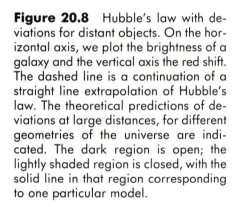
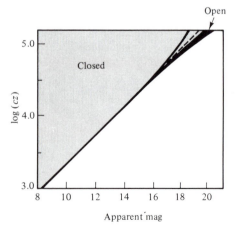

correction. These corrections can be so large that they can make an apparently closed universe open or an apparently open universe closed.

An alternative approach is to measure the constant k by actually surveying the universe on a large scale. Such surveying can determine the curvature of space-time, and therefore give us k. One way of carrying out this surveying is by *radio source counts*. We divide the universe into shells, as for our discussion of Olbers's paradox earlier in this chapter. We then count the number of radio sources in each shell. (We use radio sources because we can see them far away.) If the geometry of space-time is flat (Euclidean), the number of sources per shell will go up as r^2. If the geometry of space-time is curved, the curvature will become more apparent as we look farther away. Therefore, as we look farther away, we should see deviations from the r^2 dependence. What is actually varying is the relationship between r and the surface area. Of course, as we look far enough to see such deviations, we are also looking far back in time, and we are seeing the radio sources as they were. Again, evolutionary corrections are necessary.

There are also more indirect methods that have proved fruitful. These involve an understanding of the formation of elements in the big bang, and will be discussed in the next chapter. The results so far on these support a universe which is open, but not by a very large margin. In addition, they only give information on the density of material that can participate in nuclear reactions, and may not include the dark matter.

One of the interesting aspects of this whole problem is that we should be so close to the boundary. Of all the possible values for the density, ranging over many orders of magnitude, we seem to be tantalizingly close to the critical density. Cosmologists have wondered about whether this is accidental, or whether it is telling us something very significant about the universe.

There is a final point to consider if the universe turns out to be closed. After the expansion stops, a collapse will start. Eventually, all of the matter will come together into a dense, hot state for the first time since the big bang. Some people have taken to calling this event the "big crunch." It is natural to ask about what will happen after the big crunch. It has been suggested that the universe might reach a high density and then bounce back, starting a new expansion phase. If this can happen, it might happen forever into the future, and might have been happening for all of the past, as indicated in Fig. 20.9. Such a universe is called an *oscillating universe*.

If our universe turns out to be closed, can we tell if it is oscillating? Some theoreticians have argued that the big crunch/big bang in an oscillating universe strips everything down to elementary particles, and therefore destroys any information of what has come before. Others have argued that there are certain thermodynamic properties of the universe that might tell us if it is oscillating. And, others have taken a wait-and-see attitude, pointing out that we will need a quantum-mechanical theory of gravity to understand the densest state that is reached.

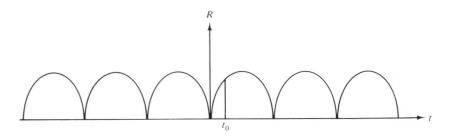

Figure 20.9   Oscillating universe. In this picture, t_0 is the current time. The universe we see is the single cycle that contains t_0. If the universe is oscillating, then, after all of the material comes back together, the expansion starts again.

CHAPTER SUMMARY

In this chapter we began to look at the large-scale structure of the universe.

We saw that on the largest scales, we could treat the universe as a homogeneous, isotropic fluid. We looked at the evolution of that fluid as described by Newtonian mechanics, and then as described by general relativity.

In keeping track of the evolution we found it useful to introduce a scale factor. The evolution of the scale factor depends on the density of the universe relative to a critical density. If the density is greater than the critical density, the universe is closed. If the density is less than the critical density, the universe is open, and will expand forever.

We saw the difficulties in trying to determine whether the universe is open or closed. Direct measurements, via the deceleration parameter, require corrections for the evolution of the luminosities of galaxies. Indirect methods suggest that the universe is close to the boundary. If it is closed or flat, it appears that dark matter is required.

QUESTIONS AND PROBLEMS

20.1   When we talk about the universe being homogeneous and isotropic, what is the smallest linear scale on which we mean this? How does this scale compare with the distance over which light could have reached us in the age of the universe?

20.2   For the rate of continuous creation, calculated in Example 20.1, how long would you have to wait for the creation of one proton in a particular cubic centimeter?

20.3   Why is a finite age or size of the universe necessary for the resolution of Olbers's paradox?

20.4   Suppose that we were trying to invoke interstellar dust as a way out of Olbers's paradox by saying that it is the scattering by the dust that blocks out the distant light, not absorption. Therefore the dust will not heat. Why doesn't this argument help?

20.5   List the basic observational facts about the universe covered in this chapter.

20.6   Show that if we have a universe with an infinite age, then the red shift is not enough to get us out of Olbers's paradox.

20.7   What are the advantages of using the scale factor $R(t)$ to keep track of the expansion of the universe?

20.8 Why is the density of the universe proportional to $1/R(t)^3$?

20.9 For the case $k = 0$, find an expression for dR/dt, and graph your result.

20.10 For the case $k < 0$, find an expression for $R(t)$, valid for large R. What are the limits on R for your expression to be valid?

20.11 Show that equation (20.24) follows from equations (20.11), (20.16), and (20.17).

20.12 Show that the density parameter Ω is twice the deceleration parameter q.

20.13 Rewrite equations (20.17) and (20.19) in terms of the density parameter, substituting for the critical density from equation (20.26).

20.14 If the current density of the universe is 1×10^{-29} g/cm^3, what value would be needed for the cosmological constant Λ in order for the universe to be static?

20.15 What are the various interpretations of the quantity k discussed in this chapter?

20.16 Show that equation (20.30) can be obtained by the appropriate use of a Taylor series.

20.17 What values of $\dot{H}(t_0)$ would be required to make q_0 equal to (a) 0, (b) $\frac{1}{2}$, (c) 1? (Take $H_0 = 75$ km s^{-1}/Mpc.)

20.18 What are the problems with each of the following ways of determining whether the universe is open or closed: (a) adding up the density of visible matter; (b) measuring $\dot{H}(t_0)$?

20.19 Derive the expression for the critical density [equation (28.26)] without introducing the deceleration parameter q_0.

20.20 In our discussion of Olbers's paradox we said that even with a cosmological red shift, an infinitely large and old universe leads to an infinite sky brightness. Show why this is true.

20.21 Show that if distances are given by Hubble's law, then the distance modulus is given by equation (20.33) without the last term on the right.

20.22 Compare the distances obtained using Hubble's law and equation (20.33) with $q = \frac{1}{2}$ for objects with $z = $ (a) 0.1, (b) 1.0, (c) 3.5. (Take $H_0 = 75$ km s^{-1}/Mpc.)

21

The Big Bang

In the last chapter we saw that Lemaitre first pointed out that an expanding universe must have had a stage in its past when it was very dense. We call that stage the big bang. In this chapter we will see what we can learn about the conditions in the big bang, and what the relationship is between those conditions and the current state of the universe.

21.1 THE COSMIC BACKGROUND RADIATION

Following the idea that the early universe was very hot and dense, George Gamow suggested, in 1946, that when the universe was less than about 200 seconds old, the temperature was greater than 10^9 K, hot enough for nuclear reactions to take place rapidly. In 1948, Ralph Alpher, Hans Bethe, and Gamow showed that these nuclear reactions could explain the current abundance of helium in the universe. (We will discuss the synthesis of elements in the next section.) In the course of following up this work, in 1948 Alpher and Robert Herman found that the early universe should have been filled with radiation, and that the remnant of that radiation should still be detectable as a low-intensity background of microwaves.

21.1.1 Origin of the Cosmic Background Radiation

When the universe was young enough to have its temperature higher than 3000 K, the atoms were all ionized. The universe was a plasma of nuclei and electrons. The free electrons provided a continuum opacity for any radiation present. The radiation would constantly scatter. The radiation therefore stayed in equilibrium with the matter. The spectrum of the radiation was that of a blackbody at the temperature of the matter. As the universe expanded, the density decreased and the temperature decreased. As the matter cooled, the radiation also cooled.

Then the point was reached at which the temperature dropped below 3000 K, and the electrons and protons recombined to make atoms. Except for a few wavelengths, corresponding to spectral lines and wavelengths with sufficient energy to ionize hydrogen, atoms are transparent to radiation. We say that the radiation and matter *decoupled*. The last photons emitted just before the decoupling should still be running around the universe. (A relatively small number have bumped into galaxies and been absorbed.) These photons should be visible, but they should be red-shifted.

We can calculate what this red-shifted blackbody radiation will look like. For a blackbody of temperature T, the energy density in photons with frequencies between ν and $\nu + d\nu$ is given by the Planck function,

$$U(\nu, T)d\nu = \left(\frac{8\pi h\nu^3}{c^3}\right)\left(\frac{1}{\exp(h\nu/kT) - 1}\right)d\nu \qquad (21.1)$$

To find the number of photons per unit volume with frequencies between ν and $\nu + d\nu$, we take the energy density and divide by the energy per photon, $h\nu$. This gives

$$n(\nu, T)d\nu = \left(\frac{8\pi \nu^2}{c^3}\right)\left(\frac{1}{\exp(h\nu/kT) - 1}\right)d\nu \qquad (21.2)$$

We assume that the radiation is emitted at some epoch t, with scale factor R, and that we detect the radiation at the reference epoch t_0. The observed wavelength λ_0, is related to the emitted wavelength λ by equation (28.29), as

$$\lambda_0/\lambda = 1/R$$

Since $\lambda = c/\nu$, the observed frequency ν_0 is related to the emitted frequency ν by

$$\nu/\nu_0 = 1/R$$

This means that

$$\nu_0 = R\nu \qquad (21.3)$$

In addition, photons emitted in the frequency range $d\nu$ will be observed in the frequency range $d\nu_0$, given by

$$d\nu_0 = R \, d\nu \qquad (21.4)$$

The photons emitted between ν and $\nu + d\nu$ are now observed between ν_0 and $\nu_0 + d\nu_0$. Also, all volumes are increased by a factor of $1/R^3$.

Combining all of these, we now have the number of observed photons per unit volume with frequencies between ν_0 and $\nu_0 + d\nu_0$ as

$$n(\nu_0, T)d\nu_0 = R^3\left(\frac{8\pi}{c^3}\right)\left(\frac{\nu_0}{R}\right)^2\left(\frac{1}{\exp(h\nu_0/kRT) - 1}\right)\left(\frac{d\nu_0}{R}\right)$$

$$= \left(\frac{8\pi\nu_0^2}{c^3}\right)\left(\frac{1}{\exp(h\nu_0/kT) - 1}\right)d\nu_0$$

Multiplying by $h\nu_0$ gives us back the energy density for radiation detected between ν_0 and $\nu_0 + d\nu_0$.

$$U(\nu_0, T)\,d\nu_0 = (8\pi h\nu_0^3/c^3)\{1/[\exp(h\nu_0/kRT) - 1]\}d\nu_0 \qquad (21.5)$$

This looks just like a blackbody spectrum at temperature $T_0 = RT$. Therefore, the radiation will still have a blackbody spectrum, but it will appear cooler by a factor of R.

The red shift of the background radiation has an interesting consequence on the evolution of the universe. The energy density of the matter in the universe is $\rho_{mat}c^2$, where ρ_{mat} is the density of matter (Fig. 21.1). We have already seen, in Chapter 20, that the density is proportional to $1/R^3$. The number density of photons in the universe is also proportional to $1/R^3$. However, the red shift means that the energy per photon is proportional to $1/R$, so the energy density of radiation in the universe is proportional to $1/R^4$. This means that the energy density of radiation drops more quickly than the energy density of matter. In the early universe, the energy density of radiation was greater than that of matter. We say that the universe was *radiation-*

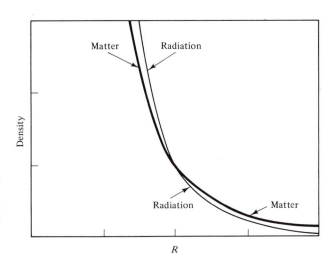

Figure 21.1 Density of matter and radiation as a function of the scale factor R. The energy density of matter falls as $1/R^3$ and that of radiation falls as $1/R^4$, with the extra factor of $1/R$ coming from the cosmological red shift.

dominated at that time. Now the opposite is true. We live in a *matter-dominated* universe. The time when the radiation drops below the matter energy density is called the *crossover time*.

To understand the distribution of this radiation in the sky, we go back to our expanding surface analogy (Fig. 21.2). At the instant before the electrons and protons recombined, making the universe transparent, matter at all points is emitting photons in all directions. Since the matter becomes transparent, these photons continue running around in all directions. We see a steady stream of photons, and not a brief flash as might be expected if we were looking at a localized explosion. The fact that the radiation is moving in all directions means that we see it coming from all directions. The radiation should appear isotropic. Also, cosmic background photons reaching us today are coming from one light day farther away than those that reached us yesterday.

21.1.2 Observations of the Cosmic Background Radiation

Alpher and Herman had predicted (in 1948) that the background radiation should currently have a temperature of about 5 K. This seemed to be to weak for the radio receivers then available, and no search was made. Unaware of the paper of Alpher and Herman, a group at Princeton rederived the result in the early 1960s, predicting a slightly higher temperature. The Princeton group also set up the equipment to try to detect the background radiation. While their work was going on, two physicists at the Bell Telephone Laboratory, Arno A. Penzias and Robert W. Wilson, accidentally detected this radiation, and reported their results in 1965.

Penzias and Wilson were unaware of the work of either Alpher and Herman or the Princeton group. They were using a very accurate radio telescope, depicted in Fig. 21.3, for both communications and radio astronomy. To have accurately calibrated results, they had to understand all sources of noise in their system. They found an unaccounted for source of noise at a very low level. The noise seemed to be coming either from within their system or from every direction in the sky. (This is a particular problem of observing the cosmic background. Radio astronomers can make their most accurate measurements when they can compare observations of a source with a nearby patch of empty sky. The cosmic background radiation is everywhere, so there are no empty patches for comparison.) After carefully analyzing their system, they were confident that the noise was coming from everywhere in the sky. They only had a measurement at one wavelength, so they could not confirm the shape of the spectrum. However, they found that the intensity corresponds to a blackbody at a temperature of about 3 K.

Still with no interpretation of the result, Penzias and Wilson prepared a paper describing it. A copy of the paper was seen by someone familiar with

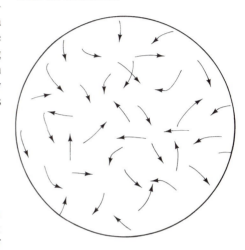

Figure 21.2 Isotropy of the background radiation. Using our balloon analogy we can think of the photons as running in all directions over the surface. An observer at any point on the surface would see radiation coming from all directions.

Figure 21.3 Arno A. Penzias and Robert W. Wilson, in front of the horn reflector radio telescope with which they discovered the cosmic background radiation.

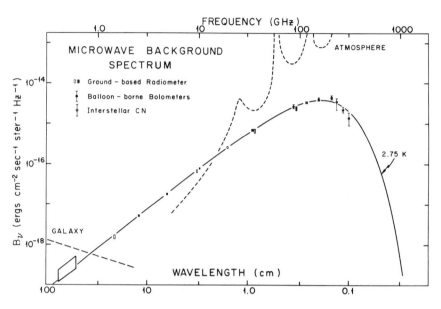

Figure 21.4   Spectrum of the cosmic background radiation, taken from ground-based observations. Most are at centimeter wavelengths, where the Earth's atmosphere is very transparent.

the efforts at Princeton, and the two groups eventually compared notes. It was clear the Penzias and Wilson had found the cosmic background radiation. The confirmation came when the Princeton group carried out their observations at a different wavelength (Fig. 21.4). For their painstaking work in detecting this important signal, Penzias and Wilson were awarded the Nobel Prize in physics in 1977.

The cosmic background radiation is of such significance in cosmology, that its discovery led to extensive efforts to measure its properties as accurately as possible. An immediate consequence of the discovery was the death, for all practical purposes, of the steady-state theories of the universe. However, until it was unambiguously shown that the spectrum is that of a blackbody, there were still some ways for steady-staters to produce something like the background radiation. As Fig. 21.4 shows, the discovery and subsequent confirmations of the background radiation were all on the long-wavelength side of the peak of a 3-K blackbody spectrum. The most convincing observation would be to show that the spectrum does, indeed, turn over at wavelengths shorter than a few millimeters.

However, at wavelengths shorter than about 1 cm, the Earth's atmosphere becomes sufficiently opaque that ground-based radio observations of the required precision are virtually impossible. More recently, observations from high altitudes have helped clarify the situation. Before we discuss those, however, an interesting experiment done in the late 1960s (and still being

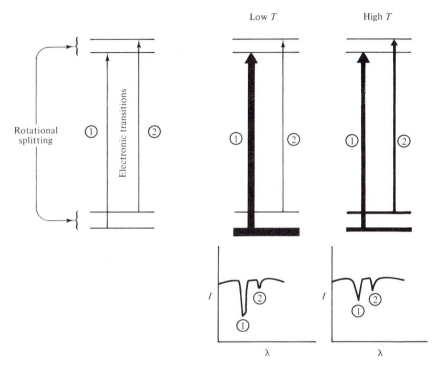

Figure 21.5 The CN experiment to determine the cosmic background temperature at 2.5 mm. On the left the energy levels involved are shown. In each case we are interested in the two lowest rotational states of the two lowest electronic states. The two electronic transitions, 1 and 2, are indicated. The low T figure shows that most of the CN molecules are in the lowest state, making line 1 much stronger than line 2, as shown in the spectrum below. At higher temperature, the next rotational state becomes populated, and line 2 gains in strength relative to line 1.

refined) is worth some mention. The experiment involves optical observations of the interstellar molecule CN, which was originally discovered in 1939.

The basic idea behind the CN experiment is shown in Fig. 21.5. The optical absorption lines are observed when the CN makes an electronic transition from its ground state to the first excited state. In Chapter 14 we saw that for any electronic state, a molecule can be in many different rotational states. Normally, interstellar molecules are studied by radio observations of rotational transitions within the same electronic state. The effect of the rotational energy levels is to split the optical electronic transition into several lines. For example, there are lines corresponding to a transition from a given rotational state of the ground electronic state to the same rotational state in the excited electronic state. The wavelengths of the various transitions are slightly different. In practice, two transitions are observed, one from the ground rotational state and the other from the first excited rotational state.

The relative strengths of the two optical lines is equal to the ratio of the populations in the two rotational states (in the ground electronic state), n_2/n_1. This is given by the Boltzmann equation as

$$n_2/n_1 = (g_2/g_1) \exp(-E_{21}/kT) \qquad (21.6)$$

where g_2 and g_1 are the statistical weights of the two rotational levels, and E_{21} is the energy difference between the two rotational levels. The energy difference between the two levels in CN corresponds to a wavelength of 2.64 mm. If the CN is in a cloud of low density, there will be very few collisions with hydrogen to get molecules from the ground state to the next rotational state. (Collisions with electrons may also be important.) The CN can only get to the higher rotational state by absorbing radiation. If the CN is far from any star, the only radiation available is from the cosmic background. Under these circumstances, the populations will adjust themselves so that the temperature T in equation (21.6) is the temperature of the cosmic background radiation at 2.64 mm.

So, the CN sits out in space, sampling the cosmic background radiation at a wavelength of 2.64 mm. This is a wavelength that we could not study directly from the ground. The CN then modifies the light passing through the cloud. The relative intensity of these optical signals contains the information that the CN has collected on the cosmic background. The optical signals penetrate the Earth's atmosphere. All we have to do is detect them and decode them. To decode them, we take the relative intensities as giving us n_2/n_1. We then solve equation (21.6) for T. When this was done, the temperature was found to be about 3 K. At the time, this was the shortest wavelength at which the blackbody nature of the radiation was confirmed. Some other transitions in CN and CH provided information at shorter wavelengths, but the results were more uncertain.

There is an interesting sidelight to this story. When interstellar CN was

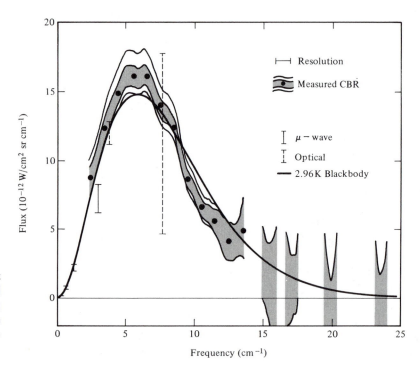

Figure 21.6 Spectrum of the cosmic background radiation, as measured from a high-altitude balloon, allowing measurements on the short wavelength side of the peak, confirming the blackbody nature of the spectrum.

discovered in the late 1930s, the people who observed it also observed both spectral lines. They noted that the relative intensity of the lines corresponded to a temperature of 3 K for the rotational transitions, but did not attach any significance to it. This shows us that, though many important discoveries in astronomy are made unexpectedly, simple luck is not enough. Confronted with the unexpected, the observer still has to be able to recognize that something significant is happening, and then follow up the results.

The most complete picture of the cosmic background radiation spectrum has come from high-altitude balloon observations. The results are shown in Fig. 21.6, and agree quite well with a blackbody spectrum. This best fit to the temperature is about 2.8 K.

21.1.3   Isotropy of the Cosmic Background Radiation

Observational studies of the background radiation have also been directed at seeing the degree to which the radiation is isotropic. In discussing possible deviations from isotropy, *anisotropies*, there are two quantities of importance: (1) the angular scale and (2) the intensity level of the deviation. By "angular scale," we mean how far apart two points have to be on the sky for there to be a difference in intensity. The "intensity level" of the deviation is usually expressed as a fraction of the total intensity. We have already explained why the radiation should appear isotropic. However, there are some possibilities of low-level deviations. Some of these may result from local small-scale irregularities, not of cosmological significance, and others may result from irregularities in the universe of cosmological importance.

One local source of anisotropy is the combined motion of the Earth, Sun, galaxy, local group, and local supercluster with respect to the general Hubble flow. This motion produces a Doppler shift, which varies with direction. If our net motion is with a speed v, in a particular direction, then the radial velocity v_r in a direction θ away from the direction in which we are heading is

$$v_r = v \cos \theta \tag{21.7}$$

In the direction in which we are heading ($\theta = 0°$), we get the maximum blue shift, and the radiation appears slightly hotter. In the opposite direction ($\theta = 180°$), we get the maximum red shift, and the radiation appears slightly cooler. Since this anisotropy is characterized by hot and cold poles, with a smooth variation in between, we call it the *dipole anisotropy*. We should point out that such an anisotropy doesn't violate special relativity by providing a preferred reference frame for the universe. We just happen to be measuring our velocity with respect to the matter that last scattered the background radiation. In fact, if we determined Hubble's constant separately in different

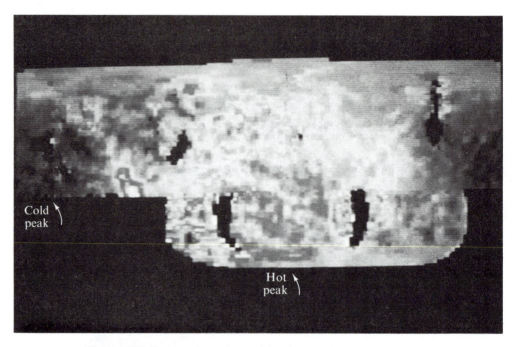

Figure 21.7 Dipole anisotropy in the cosmic background radiation. This is a map of emission over most of the sky, at a wavelength of 1 cm. The projection is on a galactic coordinate system. The lighter areas are slightly hotter than the average temperature, and the darker areas are slightly cooler.

directions, we should measure the same anisotropy. Observations have revealed such an anisotropy. The results are shown in Fig. 21.7. The best data to date indicates that we are moving with a speed of 360 km/s in the direction of the constellation Leo. This includes the Sun's orbital motion about the galactic center, which is actually in the opposite direction. When we correct for this motion, we find that the galaxy is moving with a speed of about 600 km/s in the direction of the galactic coordinates, longitude = 265° and latitude = 25°.

The speed of 600 km/s is too large to be simply our random motion within the Local Group. Instead, it must represent motion of the Local Group. The direction of the motion of the Local Group is within 45° of the direction to the Virgo Cluster and suggests that we are moving under the influence of that cluster. That we are not moving directly toward Virgo suggests that the Virgo Cluster and the Local Group are part of an even larger entity. This general picture has been confirmed in recent studies of the patterns of galaxy red shifts.

There have also been searches for anisotropies on smaller angular scales. The goal of these studies is to learn more about the structure of the universe at the time the temperature was about 3000 K. After all, the background radiation retains an almost perfect record of that time. We think that by that time, the small fluctuations that would eventually grow into gravitationally bound clusters of galaxies already existed. These small clumps may have been slightly hotter than the average material, so the radiation from those directions will appear slightly hotter. By looking for anisotropies on small angular scales, we can learn about the size of and structure of these condensations. This is important in our understanding of galaxy formation. Experiments are still going on, but so far, no anisotropies have been found at the fractional intensity level of about 10^{-5} on an angular scale of a few arc minutes.

In fact, there is an interesting problem in that the background radiation appears to be *too isotropic*. For the background to be very isotropic, the early universe must have been very smooth. However, conditions can only be identical at different locations if they have some way of communicating with each other. Something must have regularized the structure in the early universe. However, such communication can only travel no faster than the speed of light. Two objects separated by a distance greater than that which light can travel in the age of the universe cannot communicate with each other. We see homogeneity on scales that were unable to communicate with each other at the time the radiation departed to reach us now. This is called the *causality problem* in the early universe (Fig. 21.8).

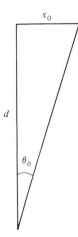

Figure 21.8 Causality and the isotropy of the background radiation.

To see if this is a problem for the cosmic background radiation, we have to relate the current angular separation between two points with their linear separation at the time of decoupling. We now let the two points be separated by an angle θ_0 (where $\theta_0 \ll 1$ rad). If the points are a distance d from us, then their current separation is

$$x_0 = \theta_0 d$$

If we are just seeing the light from these points, it must have been traveling for t_0, the current age of the universe. (We ignore the small difference between the age of the universe and the time since decoupling, since decoupling occurred very early in the history of the universe.) This means that

$$d = ct_0$$

so

$$x_0 = \theta_0 ct_0$$

If we take our reference epoch to be now, so that $R(t_0) = 1$, and we let R be the scale factor at the time of decoupling, then the separation between the two points at the time of decoupling is R times their current separation, or

$$x = \theta_0 ct_0 R$$

The time for light to travel the distance x is x/c, or

$$\Delta t(x) = \theta_0 t_0 R$$

For these two points to communicate, $\Delta t(x)$ must be less than or equal to t, the age of the universe at decoupling. The farthest the two points can be is when $\Delta t(x) = t$, so

$$t = \theta_0 t_0 R$$

We now solve for θ_0, the maximum current angular separation between two points that could have been causally connected at decoupling,

$$\theta_0 = t/Rt_0 \tag{21.8}$$

EXAMPLE 21.1   Causality in the Early Universe

If the age of the universe at decoupling was 10^5 yr, and the current age of the universe is 10^{10} yr, find the maximum current angular separation between two points that can be causally connected.

Solution

To use equation (21.8), we have to find R, the scale factor at the time of decoupling. We know that the temperature of the background radiation at any time t, $T(t)$, is proportional to $1/R(t)$. The background temperature is now 3 K, and was 3000 K at decoupling, so R at decoupling must have been 10^{-3}. This means that

$$\theta_0 = (10^5 \text{ yr})/(10^{-3})(10^{10} \text{ yr})$$

$$= 10^{-2} \text{ rad}$$

This is approximately $1°$.

The above example tells us that points that appear more than one degree apart cannot be causally connected. The background radiation should not appear smooth on scales larger than one degree. However, it clearly does appear smooth on large scales. We will come back to this question in the last section of this chapter.

There is another source of small-scale anisotropy in the background radiation that is not actually cosmological. It has to do with the interaction of the radiation with clusters of galaxies as it passes through. The galaxies in a cluster don't present a very large target, so most of the interaction is with the hot, low-density intracluster gas. This gas is hot enough to be almost completely ionized. The background radiation interacts with the electrons in the gas.

The process by which photons scatter off electrons is called *Compton scattering*. (A. H. Compton worked out the theory of this process and studied it in the laboratory as a demonstration of the photon nature of light.) Generally, in Compton scattering, the photon initially has more energy than the electron, so some energy gets transferred from the photon to the electron. Since the photon loses energy, its wavelength increases. However, it is also possible for the electron to start out with more energy than the photon. In this case, called *inverse Compton scattering* (Fig. 21.9), energy is transferred from the electron to the photon. The wavelength of the photon decreases.

Before	After	
γ 〰️〰️➤ ⓔ	〰️↗ ⓔ 〰️↘	Compton
ⓔ➡️ ⤙〰️	ⓔ➡️ 〰️➤	Inverse Compton

(a)

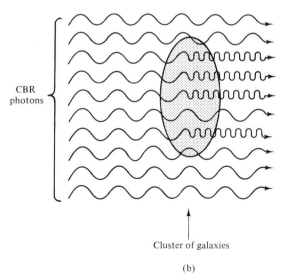

CBR photons

Cluster of galaxies

(b)

Figure 21.9 (a) Compton and inverse Compton scattering. In Compton scattering, a gamma ray strikes a low-energy electron and becomes a lower-frequency photon, with the excess energy going to the electron. In inverse Compton scattering a low-energy photon scatters off a high-energy electron, with the photon gaining energy and the electron losing it. (b) The Sunyaev–Zeldovich effect. As low-energy cosmic background photons strike the hot gas within a cluster of galaxies, inverse compton scattering takes some of the photons from the low-energy side of the 3-K blackbody peak, and transfers them to the high-energy side.

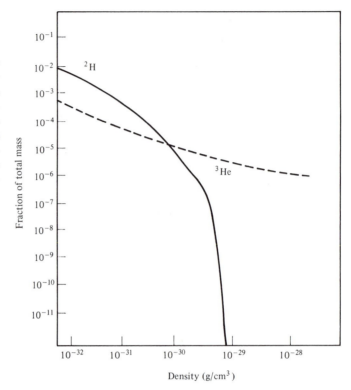

Figure 21.10 Relationship between the abundances of deuterium (^{2}H) and ^{3}H and the density of matter in the universe. Though the actual relationship comes from the density in the first three minutes of the universe, we can relate that to the current density, so the horizontal axis is the current density. It is important to note that this is only the density of matter that participate in nuclear reactions, and doesn't include certain types of particles that have been suggested for dark matter.

The electrons in the intracluster gas have high energies, and the photons of the background radiation have very low energies. When they interact, we have inverse Compton scattering. The photons get a large boost in energy, going in as radio wavelength photons and coming out at shorter wavelengths. This means, however, that if we look in the directions of clusters at radio wavelengths, some of the background radiation photons will have been removed. The background radiation should appear slightly weaker in the directions of clusters of galaxies, since some of the radio photons have been shifted to higher energy. This is called the *Sunyaev–Zeldovich effect* (Fig. 21.9), after the two Russian astrophysicists who proposed it. Many groups have been searching for this effect, but it is a very small one, and the experiments are difficult. After a few false alarms, it now appears that this effect has been detected.

The cosmic background radiation still has a wealth of information to yield about the conditions in the early universe. For this reason, a satellite solely devoted to studying this radiation is being readied for launch. The Cosmic Background Explorer (COBE), scheduled for launch in 1988, should provide answers to many questions about the early universe.

21.2 BIG BANG NUCLEOSYNTHESIS

In the first three minutes of its existence, the universe was hot enough for nuclear reactions to take place. Protons and neutrons combined to form ^{2}H, ^{3}He, and ^{4}He. The deuterium is very reactive, and most of it was used up as quickly as it could be formed. The ^{4}He is very stable, and, in those three minutes, 25% of the mass of the universe was tied up in ^{4}He. The ^{3}He makes up only 10^{-4} to 10^{-5} of the total mass. There is no stable nucleus with mass number, $A = 5$. This served as an effective barrier to the copious formation of heavier elements.

The final mixture of elements produced in the big bang depends on the density of material at $t = 1$ s, when the reactions started. We can relate $\rho(1 \text{ s})$ to ρ_0 (the current density) by knowing $R(1 \text{ s})$. We know R, because it is the ratio of the temperature then to the temperature now. Figure 21.10 shows the results of theoretical calculations of nucleosynthesis for models with different densities, expressed as the current density. Notice that the stability of ^{4}He makes its abundance relatively insensitive to the density. However, the ^{2}H abundance drops off sharply with increasing density. This is because the deuterium is so reactive that a higher density provides it with more opportunities to react and be destroyed. There is a smaller density dependence in the ^{3}He.

The heavier elements (especially ^{7}Li) become more important at higher densities. This is because the high densities provide more reactions capable of building up the heavier elements. For densities greater than ρ_{crit}, C, N, and O become somewhat abundant, but are still many orders of magnitude below their current observed abundances. This means that their current abundances must be explained by production in stars.

The density dependence of the relative abundances of various nuclei provides us with a way of determining the density of the universe. We can measure the relative abundances of certain nuclei, such as D/H. We then use a diagram like Fig. 21.11 to find out what current density this ratio corresponds to. In measuring the current abundance of any nucleus, it is important to account for any modification that has taken place in that particular abundance since the big bang. Most important, the effect of nuclear processing in stars must be taken into account. The abundances that we see now are the net result of big bang and stellar nucleosynthesis. When we want to talk about the results of big bang nucleosynthesis only we speak of *primordial* abundances.

It is important to note that this type of study only gives us the density of nuclear matter (protons and neutrons) in the universe. That is because it is only the nuclear matter that could have participated in the nuclear reactions. If, for example, the dark matter is in the form of neutrinos, then it may be that the nuclear matter is insufficient to close the universe, but the dark matter is sufficient.

Figure 21.11 Particle collision experiments at Fermilab.

Most of the studies of the type outlined above have involved deuterium. The net effect of stellar processing is to destroy deuterium. This means that the current D/H abundance ratio is less than the primordial one. Therefore, the current D/H ratio provides us with an upper limit to the density of nuclear matter. On Earth, the D/H abundance ratio is about 10^{-4}. From Fig. 21.11, we see that this corresponds to a density of about 3×10^{-31} g/cm^3, much less than ρ_{crit}. However, the abundances in the solar system may not be typical of the rest of the galaxy.

The next step is to study interstellar D, in whatever forms it is found. Direct observations of atomic D can be done in the Lyα line (and other Lyman series lines) in the ultraviolet. These lines are observed in absorption against stars. Since the interstellar extinction is so high in the ultraviolet, we cannot study clouds that are very far away. The results for our general area of the galaxy are consistent with the solar system value.

Another possibility is the radio observation of molecules containing deuterium, such as DCN. When we measure the abundance of DCN, we find that it is not much less than that of HCN. However, we now know that this is a result of the fact that chemical reactions involving D and H proceed at different rates for certain molecules. Thus, the abundances of DCN and HCN do not reflect the true D/H abundance ratio. The observation of molecules with D substituted for H tells us more about interstellar chemistry than it does about cosmology.

A final possibility is the observation of the equivalent of the 21-cm line in atomic D. For deuterium, this line is shifted from 21 cm to 90 cm. This makes it difficult to detect against the galactic synchrotron radiation, whose intensity goes up approximately as λ^2. The general conclusion of all of the various D/H experiments is that the density of nuclear matter is probably less than about 10% of the critical density. If this result holds up, it doesn't necessarily mean that the universe is open. It just means that the nuclear matter is not sufficient to close it.

There is an additional note concerning ^{3}He. It also has an abundance that depends on the density, though not as strongly as the D abundance does. If an atom of ^{3}He has one electron removed, a single electron remains. This ion behaves somewhat like a hydrogen atom. It has a transition, analogous to the 21-cm line, at a wavelength of 9 cm. This line has been detected in galactic HII regions in the past few years, but more detailed studies are necessary before the analysis can yield cosmologically significant information.

In this section, we have only discussed what happened between $t = 1$ s and 3 min. Earlier than 1 s, the universe was so hot that the internal structure of the neutrons and protons was important. Before we can understand what happened when the universe was less than 1 s old, we must look at some of the important features of elementary particle physics.

21.3 FUNDAMENTAL PARTICLES AND FORCES

A major goal of physics over the centuries has been the search for the most fundamental building blocks of matter. We have progressed from Earth, Air, Fire, and Water, through the atoms of Democritus, to Mendeleev's realization that the elements showed a regularity that was eventually explained by saying that the elements are not fundamental, but are made up of even smaller structures—nuclei and electrons. We have seen that the nuclei are made up of protons and neutrons.

21.3.1 Fundamental Particles

Since the 1950s, particle physicists have been able to use accelerators to bring particles together at high energy and study their structure (Fig. 21.11). (Before that, cosmic rays were used.) When the energy of the collision is

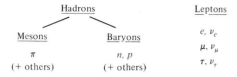

Figure 21.12 Classifications of elementary particles.

just right, particles can be created out of the energy. As accelerators with more energy became available, increasing numbers of "fundamental" particles were found. Physicists began to speculate that these particles were no more fundamental than the 92 elements are.

To follow these results, we divide the various particles into groups, according to their ability to interact via the strong nuclear force (Fig. 21.12). *Fermions* (particles that obey the exclusion principle) that do not interact by the strong force are called *leptons*. Those particles that are strongly interacting are called *hadrons*. We further divide the hadrons into the *baryons*, which are fermions, and *mesons*, which are not fermions. (In general, particles that are not fermions are called *bosons*.) The neutron and proton are baryons, and the pion, discussed in Section 9.2, is a meson.

The most familiar lepton is the electron. The μ and τ particles are heavier versions of the electron, with masses 207 and 3660 times that of the electron, respectively. Neutrinos are also leptons. We think that there are three types of neutrinos, one each to go with the electron, μ, and τ. Only the first two have actually been detected. All of the evidence to date indicates that the leptons are fundamental. They seem to have no internal structure. We still don't know if the six leptons we know of are all that there are. There is some evidence to suggest that there are no others, but we have been surprised before.

The hadrons appear to have internal structure. This can be seen, for example, in experiments that have sufficiently high energy to probe the charge distribution within a proton. We now think that the hadrons are composed of particles, called *quarks*. In the original quark theory, there were three different kinds of quarks. Now, five quarks have definitely been identified, and there is evidence for a sixth. (That would make six quarks and six leptons.) The properties of the quarks are summarized in Table 21.1. Note that they have fractional charges, coming in units of one-third and two-thirds of the fundamental charge e. However, the quarks only combine in ways that produce integral net charges. Each quark has its own antiquark. All the properties of a given antiquark (except mass) are the negative of those for the corresponding quark.

TABLE 21.1   Quark Properties

Name	Symbol	Charge	Mass (GeV)
Up	u	$+\frac{2}{3}e$	0.35
Down	d	$-\frac{1}{3}e$	0.35
Strange	s	$-\frac{1}{3}e$	0.55
Charm	c	$+\frac{2}{3}e$	1.8
Top	t	$+\frac{2}{3}e$	4.5
Bottom	b	$-\frac{1}{3}e$	40.0

In the quark theory, any baryon is a combination of three quarks. For example, a proton is (uud) and a neutron is (udd). A meson is a combination of a quark and an antiquark (not necessarily of the same type as the quark. For example, a positive pion is ($u\bar{d}$), where the bar represents an antiparticle. A negative pion is ($\bar{u}d$), and a neutral pion is ($u\bar{u}$). All known hadrons (and there are over a hundred) can be constructed by these simple rules. Note that though there are six quarks, the only ones we need for everyday life are the u and d. Similarly, the only leptons we need for everyday life are the electron and the electron neutrino.

The quark theory was immediately quite successful, but there were a few important problems left. One was why only the particular combinations mentioned are allowed. The other was that, despite considerable effort, no one has been able to detect a free quark.

21.3.2 Fundamental Forces

Along with the quest for the fundamental particles, physicists are also trying to understand the forces with which the particles interact. The concepts of forces and particles (Table 21.2) are intimately tied together. Without forces, particles would have no meaning since we would have no way of detecting them.

It also appears that particles are necessary as the carriers of forces. For example, *quantum electrodynamics (QED)*, is the theory that describes the electromagnetic force as being carried by photons. These photons may be real or may only exist briefly on energy that can be borrowed because of the uncertainty principle (as discussed in Section 10.3). These photons that live on borrowed energy are called *virtual photons*. The fact that the photon is massless leads to the electromagnetic force being a long-range force (see Problem 21.16). QED has been tested in many ways to very high accuracy, and is a very successful theory.

It is speculated that the gravitational interaction is carried by a massless particle, called the *graviton*. (It is presumed to be massless, because gravity has the same long-range behavior as electricity.) However, no gravitons have

TABLE 21.2 Forces and Particles

Force	Relative strength	Range	Carrier
Strong	1	Short	Pion
Electromagnetic	10^{-2}	Long	Photon
Weak	10^{-13}	Short	W, Z
Gravity	10^{-40}	Long	Graviton (?)

Figure 21.13 Symmetry in an electricity problem. Suppose we wish to calculate the electric field due to the spherical charge distribution in (a). The resulting field must have the same symmetry as the charge distribution (spherical). The result in (b) is clearly wrong, because, if we rotate the page, the charge distribution still looks the same, but the field direction changes. The type of distribution in (c) has the proper symmetry. A consideration of symmetry has allowed us to quickly eliminate unreasonable answers.

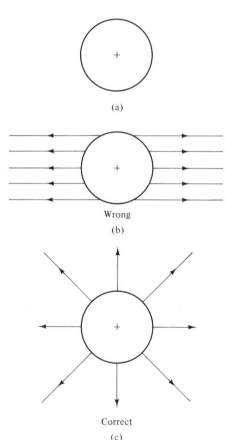

(a)

Wrong

(b)

Correct

(c)

ever been detected, and this theoretical framework is still being developed. We will see that the absence of a quantum-mechanical theory of gravity provides the limitation on how far back we can go in probing the big bang.

In Section 9.2, we saw that the strong nuclear force is carried by the pions. Of course, we now know that pions are not, themselves, fundamental particles, being made up of a quark and an antiquark. Since the pion has a mass, the strong nuclear force is short range. In fact, the mass of the pion was initially inferred from the range of the force (see Problem 21.11).

The weak nuclear force, also a short-range force, is carried by three particles. Two are the positive and negative W (for weak), and the third is a neutral particle, called the Z. These particles are all much more massive than the pion, and the weak force is a short-range force. The W and Z particles have recently been observed with masses 80 and 90 times that of the proton.

21.3.3 The Role of Symmetries

In any study of particles and forces, symmetries play a very important role. Symmetries are important in many areas of physics. When we say that a system has a certain symmetry, we mean that the system doesn't change under some particular transformation. For example, spherical symmetry means that a system doesn't change when we apply a rotation through any angle about any axis through a particular point. Recognizing symmetries can greatly simplify the solving of a problem. If a problem has a certain symmetry, then the solution must have the same symmetry. For example (Fig. 21.13), if we have a spherically symmetric charge distribution, then we must also have a spherically symmetric electric field distribution.

Symmetries also have a deeper importance in physics. Whenever there is a symmetry, there is some quantity that is constant throughout the problem. This means that there is a conservation law. For example, the fact that the laws of physics are not changed by the rotation of a coordinate system leads to the conservation of angular momentum. The fact that the laws of physics are independent of a translation of the origin of a coordinate system leads to conservation of momentum. The fact that the laws of physics are independent of when we start timing leads to conservation of energy.

We can understand the various forces by understanding what symmetries they have, or, equivalently, what conservation laws they obey. In general, a process will take place as long as it doesn't violate some conservation law. If a reaction that you think should take place doesn't, it means that there is some conservation law that you might not be aware of, and this reaction violates that conservation law. For example, before the quark theory had been proposed, there was a group of particles that should have decayed by the strong nuclear force, but did not. Because of this strange behavior, these particles were called "strange" particles. It was proposed that there must be some property of these strange particles which is conserved. The particular decays would then violate this conservation law. When the quark theory was proposed, the *s* (strange) quark was included to incorporate this property.

One interesting property of the weak interaction is that it doesn't obey all of the conservation laws that the other forces do. Before this was realized, it was thought that conservation laws were absolute. The first law found to be broken was that concerning *parity*, which has to do with the symmetry of a system under mirror image reflection. It was realized and demonstrated in 1957 that it is actually possible to set up a beta decay experiment (beta decay taking place via the weak interaction) and its mirror image, and get different results. (For this realization, T. D. Lee and C. N. Yang shared the Nobel Prize in physics in 1959.) Two other symmetries which are violated by the weak interaction involve *charge conjugation* (the interchange of particles and antiparticles) and *time reversal*.

Sometimes we find situations which are inherently symmetric, but somehow lead to an asymmetric result. This is called *spontaneous symmetry breaking* (Fig. 21.14). For example, suppose we toss a coin into the air. As the coin is spinning, it has an equal probability of being heads or tails. As long as the coin stays in the air, the situation is symmetric between heads and tails. However, once the coin falls, the symmetry is broken. Another example is provided by a ferromagnet. If the magnet is heated above its critical temperature, it has spherical symmetry. There is no preferred direction. When we cool the material, it becomes a permanent magnet. Until it cools, all directions are equally probable. Once it cools, one direction is selected and the whole magnet cools pointing in that direction. This is an example of a situation where the situation is symmetric as long as it is hot enough, but cooling the system breaks the symmetry. Nature is symmetric as long as there is enough energy.

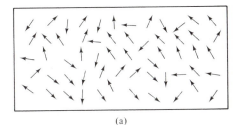

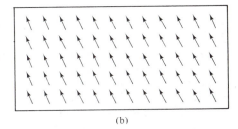

Figure 21.14 Spontaneous symmetry breaking in a magnetic material. (a) The material is above its critical temperature, and the individual dipole moments represented by the arrows have random orientations. (b) If the material is cooled below the critical temperature, all of the arrows will line up, but they could point in any direction. As the material is cooled, one direction emerges as the preferred direction without any influence (such as a magnet) from outside.

21.3.4 Color

We have already mentioned two problems with the quark theory. One is that there was no explanation for why the only allowed combinations are three quarks or one quark and one antiquark. The other is why we have not been able to detect free quarks. There was another problem with the original theory. Quarks have the same spins as electrons. Therefore, they should obey the Pauli exclusion principle. However, some particles are observed which are clearly combinations of three identical quarks, (*uuu*) for example, all in the ground state. This is a violation of the exclusion principle.

The solution was to postulate an additional quark property that could be different for each of the three quarks in a baryon. This property is called *color*. This is just a convenient name, and has nothing to do with real color. The color property is more like electric charge, except that it has three possible values (plus three antivalues) instead of one. (Electric charge is thought of as having one value and one antivalue.) Quark colors can be red (*R*), green

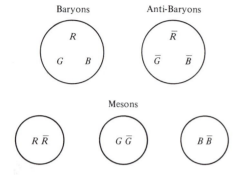

Figure 21.15 Quark color combinations. The top line shows combinations of three quarks to make baryons and antibaryons. For the baryons we need one of each color, and for the antibaryons we need one of each anticolor. The second line shows how to combine a quark and antiquark of a color–anticolor pair to produce a meson. Once the right color combinations are present, flavor combinations are allowed.

(*G*), and blue (*B*). These colors only relate to the ways in which quarks combine, and not to the properties they impart to the particles they make up. To distinguish them from colors, we call the six quark types (*u*, *d*, *s*, *c*, *t*, *b*) *flavors*. Each flavor of quark comes in each of the three colors. The six antiquarks come in corresponding anticolors.

The rule of combining quarks is that we can only have *colorless* combinations like those shown in Fig. 21.15. One way of having a colorless combination is to have a quark of one color and an antiquark with the corresponding anticolor. For example, a pion is ($u\bar{d}$), where the possible color combinations are ($R\bar{R}$), ($B\bar{B}$), or ($G\bar{G}$). This explains the mesons. The other possibility is to have one quark of each color (or one antiquark of each color). So, for a proton, which is (*uud*), the possible colors for three quarks are (*RBG*), (*RGB*), (*BGR*), (*BRG*), (*GRB*), or (*GBR*). We can see from this that a free quark would not be colorless, and is therefore not allowed by the theory. The introduction of color has solved these three major problems.

The properties involving color are more than a set of *ad hoc* rules. They have actually been derived from a mathematical theory, derived only on the assumptions of the types of symmetry that the theory should have. This theory is called *quantum chromodynamics*, or *QCD*, as an analog to QED. In this theory, the strong force between hadrons is no longer a fundamental force. The real force is called the *color force*, acting between quarks. Just as electric charge is a measure of the ability of particles to feel and exert the electromagnetic force, color is a measure of the ability of particles to exert and feel the color force. The strong force between hadrons is a residue of the color force between the quarks within the hadrons. (By analogy, in the nineteenth century physicists thought that the force between neutral molecules, called the *van der Waals force*, was a fundamental force. After the development of the atomic theory in the early twentieth century, it became clear that this was nothing more than the residual force between the electrons and protons within the molecules.).

We have seen that quantum-mechanical theories of forces involve carrier particles. QCD is no different. In fact, the mathematical theory predicts the existence of a group of eight particles carrying the force. These particles are called *gluons*. There is a major difference between QED and QCD. While the photons that carry the electromagnetic force have no electric charge themselves, the gluons that carry the color force have a color charge. This means that gluons can interact with each other and with quarks. They can also change the colors of the quarks they interact with. This makes the theory very difficult mathematically, and the detailed calculations that have characterized the success of QED have not yet been possible for QCD.

Another interesting feature comes out of QCD. The force between two quarks does not fall off with the distance between the quarks. This means that the separation of two quarks requires an effectively infinite amount of energy. Even if we had the energy available to drive two quarks far apart, we could not isolate a quark. As you pulled two quarks apart, you would put so much energy into the system that you would simply create quark–antiquark pairs out of the energy. This new quark and antiquark would bind up with the two quarks you were trying to pull apart (Fig. 21.16). This would lead to *quark confinement*, and tells us we should not see any free quarks. Pulling two quarks apart is like cutting a piece of string. Each piece will always have two ends. You can never get a piece with one end. An interesting feature of the theory is that the quarks behave most freely when they are close together.

21.3.5 Unification of Forces

Following his general theory of relativity, Einstein spent the latter part of his life attempting to ''unify'' the forces of gravity and electromagnetism. By ''unify'' we mean that we would like to explain all of the forces as really being manifestations of one larger force. This search is not new. After all, Maxwell's equations unified the previously distinct forces, electricity and magnetism. Einstein's search for unification was not successful, and some thought that he was wasting his time because the problem has no solution. However, the past decade has seen amazing progress toward this goal. This progress is based on mathematical theories which are, in turn, based on the expectation that nature will have certain symmetries. The progress in unification of the forces is illustrated in Fig. 21.17.

The first of the recent successes was the unification of the electromagnetic and weak forces into one force, called the *electroweak force*. (For this work, S. Weinberg, A. Salaam, and S. Glashow shared the Nobel Prize for physics in 1979.) A major test of this theory was the prediction of the masses of the W and Z particles, the carriers of the weak force, before their discovery. In the electroweak force, the photon and the W (or Z) are essentially the same.

You might wonder how two particles can be the same if one is very massive and the other is massless. The answer to this question tells us what is really meant by unification. In this case, the electromagnetic and weak forces appear to be the same when we deal with particles whose energies are much greater than the photon $-$ W mass difference. At these high energies, the W mass arises from small spontaneous symmetry breaking. However, we don't really notice the difference until the energy is so low that the mass of the W can no longer be ignored. (This is another case of our maxim that nature is symmetric as long as there is enough energy.)

The masses of the W and Z are about 80 GeV (being expressed as mc^2, where the proton mass is about 1 GeV). These masses are barely accessible in today's best particle accelerators. Therefore, we cannot even approach the range where particles will have energies much greater than this so that we

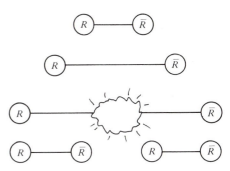

Figure 21.16 One picture of quark confinement. The force between two quarks doesn't fall off with distance. As we pull two quarks apart, we are doing the same amount of work per increase in the separation. Once we have done more work than the rest energy of a quark–antiquark pair, that pair can be created, leaving us with two mesons but no free quarks.

Figure 21.17 Unification of forces. This shows that pairs of forces that, successively, combine together into single forces, as one goes to higher energies from left to right.

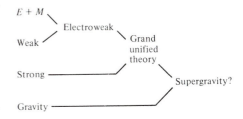

can actually see a world in which the electromagnetic and weak forces are the same. However, there must have been a brief instant in the early universe when the temperature was high enough for this symmetry to be present. This is one of the important connections between cosmology and particle physics, and we will explore it further in the next section.

The next step has been to unify the color (strong) force with the electroweak force, using the same mathematical tools. Theories that do this are called *grand unified theories*, or *GUTs*. These theories are still in the developmental stage. The carriers of this unified force are designated X and Y. Their masses are estimated to be about 10^{14} GeV, or 10^{12} times the mass of the W. This means that the difference between the color force and the electroweak force disappears when energies are much greater than this. This is clearly beyond the range of even accelerators that we could contemplate on Earth. However, these conditions existed, even more briefly than the conditions for the electroweak unification, in the big bang.

One prediction of GUTs is that the proton is not stable. Our normal understanding of the apparent stability of the proton is that it is the lowest-mass baryon. The electroweak and color forces all obey a *conservation of baryons*. Since a decay must always be to a lower-mass particle, if a proton decayed, the result could not be a baryon. Hence the stability of the proton. However, GUTs say that at high enough energies, there is no difference between the electroweak force and the color force. This means that the difference between hadrons and leptons could be lost.

According to these theories, the proton should decay into a positron and a pion, with a half-life of roughly 10^{31} yr. This doesn't mean that half of the protons wait this long and then all decay at once. There is a steady stream of decays, and half will be gone after that time. The probability of any given proton decaying this year is small, but there are a lot of protons around. This is now being tested by looking for evidence for proton decays in large tanks of water. The signature would be a flash of light as the relativistic pion passes through the water. Experiments now show that if the proton decays, its lifetime must be at least a factor of 10 longer than predicted, meaning that some modification of the theory is necessary.

Another prediction of the GUTs is the existence of *magnetic monopoles*, particles that serve as a source of magnetic fields that are not due to the motion of charges. The magnetic field around a magnetic monopole would look like the electric field around a proton. The simplest GUT theory predicts that there should be almost as many magnetic monopoles as there are baryons, but we clearly don't see them.

Even though there are still problems with GUTs, some theoreticians are forging ahead with the final step, the inclusion of gravity in the unified forces. These theories have been called *super GUTs*, or *supergravity*. Needless to say, they are still in the very speculative stages.

21.4 MERGING OF PHYSICS OF THE BIG AND SMALL

21.4.1 Back to the Earliest Times

One of the most fascinating developments in the past decade has been the interaction between two frontier areas in physics, elementary particle physics and cosmology. In the last section we saw that the big bang provides a unique opportunity for testing the predictions of the various unified theories. It is as if an experiment was done for us, some 15 billion years ago, and the data were left around in coded form. We only have to decode the results. For example, the current abundance of helium in the universe tells us that the three neutrino types that we have so far (e, μ, and τ) are likely to be the only ones there are.

Some of the recent developments in particle physics may be able to help solve some of the outstanding problems in cosmology. We have already discussed some of these. The problems that we would like to be able to address include:

1. Is the universe open or closed? If it is closed, what form does the dark matter take?
2. Why is the universe so close to being flat? As we discussed in Chapter 20, Ω evolves away from 1 on a timescale of the age of the universe. For it to be so close to 1, it had to be even closer to 1 near the beginning.
3. Where did the monopoles go?
4. How can the background radiation be isotropic on scales larger than those that correspond to regions that were causally related?
5. How did galaxies form?
6. For every baryon in the universe, there are approximately 10^{10} photons. Why is this?

To see how particle physics can help answer these questions, we go through a scenario of the early universe, taking into account the aspects of particle physics that we discussed in the previous section. Many of these points are still quite speculative, but they give some idea of the scope of the phenomena. The general ideas are summarized in Fig. 21.18. As we go through the scenario, notice the extremely short timescales.

We cannot say anything about the first 10^{-43} s. This time is called the *Planck time*. To describe phenomena on this timescale we need a quantum theory of gravity. It is even possible that on this short timescale the continuous fabric of space-time breaks down.

From 10^{-43} s onward, the temperature was so high that GUTs were

Figure 21.18   Time line of the early universe. The left column gives the age in seconds (and any other appropriate units). The middle column gives the composition of the universe. The next column shows when various forces cease to become important as the temperature gets too low. The last columns give the temperature, both in Kelvins, and expressed as the average kinetic energy (in Gev) per particle at that temperature.

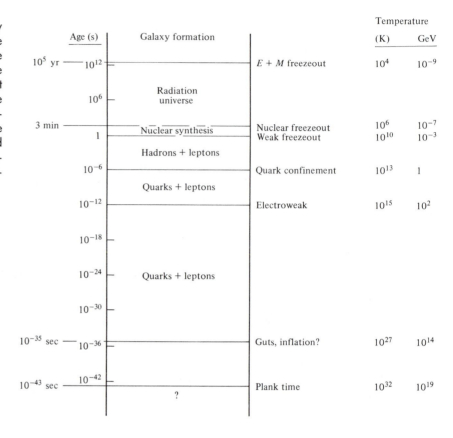

important. Since the electroweak and color forces are the same at these high energies, the differences between quarks and leptons disappear. (This is one of the reasons why we think that the number of quark types should be equal to the number of lepton types.) During this period, baryon number is not conserved. The breaking of the symmetries involving parity (mirror reflection) and charge conjugation (particle–antiparticle interchange) results in a slight excess in the amount of matter over the amount of antimatter. This excess is one particle extra for every 10^{10} particle–antiparticle pairs produced.

This excess explains the observation that there are 10^{10} photons for every baryon in the universe. As the universe cooled, particle–antiparticle pairs would annihilate, each producing a photon. For every 10^{10} particles that could find an antiparticle and produce a photon, there was one particle left over. This slight imbalance is very important. If there was a perfect balance between matter and antimatter, the universe would now be all radiation, and we wouldn't be here.

At the end of this period, the universe has cooled to the point where the average energy is no longer much greater than the mass of the X and Y particles. This means that the symmetry between the color and electroweak forces is broken. From that point on, baryon number is conserved. By 10^{-12} s, the universe has cooled to the point where the electromagnetic and weak forces separate. At this point the weak force is as strong as the electromagnetic. From this point on it gets weaker. It is still hot enough for the quarks to move in a fluid. By 10^{-6} s, the quarks become confined in hadrons.

At 1 s, the temperature is low enough for the weak force to have weakened to the point where neutrinos are decoupled from the rest of the matter. The neutrinos should now be at a temperature of 2 K. They are cooler than the radiation because the photons acquired extra energy from later electron–positron recombinations. These low-energy neutrinos are very hard to detect. However, if they turn out to have a rest mass of about 20 eV, then the total mass of these neutrinos will exceed the mass of nuclear matter in the universe by a factor of 10. That could be enough to close the universe. It may also be that clumps of neutrinos formed the centers of attraction for the formation of clusters of galaxies.

Nuclear reactions continue for the first 5 min. By 10^5 yr, the matter has cooled enough to become neutral, and the radiation decouples, remaining to be detected as the cosmic background radiation.

21.4.2   Galaxy Formation

We now look at galaxy formation. As mentioned in Section 21.1, we believe that galaxies grew out of small fluctuations in the density that were present before the radiation and matter decoupled. (This is why anisotropies in the cosmic background radiation may tell us about the sizes of these fluctuations.) There are two basic galaxy formation scenarios. (1) The galaxies may have formed first, and then gathered into clusters. (2) The initial condensations may have been larger, forming the protoclusters first, with the galaxies then forming out of density fluctuations within the forming cluster. In this latter scenario, calculations show that the original large-scale structures would be flattened, and these are referred to as pancakes.

In deciding between these scenarios, it is important to understand the nature of the dark matter in the universe. Even if there is not enough dark matter to close the universe, there must at least be enough to account for the virial masses of clusters. Even this amount would place the density of dark matter as being significantly greater than the density of luminous matter. This means that the dynamics of the universe and galaxy formation is dominated by the dark matter. The result of theoretical calculations with massive neutrinos as dark matter are shown in Fig. 21.19.

The exact nature of the dark matter also influences how the forming galaxies or condensations can cool. This is an important part of the formation process, since collapse will proceed more efficiently if there is a way to get rid of excess energy. For example, calculations show that if the dark matter is in the form of massive neutrinos, then the formation of large superclusters

Figure 21.19 Computer model for evolution of gravitational concentrations in the early universe. In this model, massive neutrinos provide the dominant source of gravitation. To get a 3-dimensional feel, successive frames should be stacked over each other. This gives a feel for the concentration of material into "pancakes."

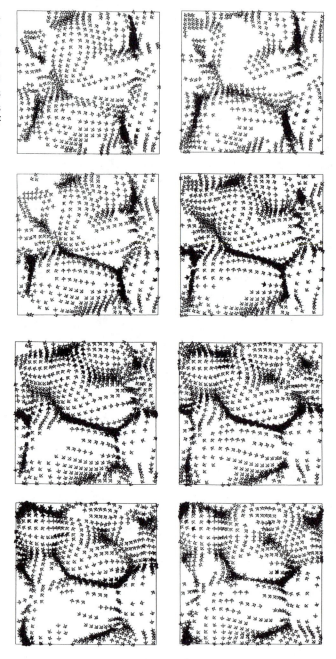

and voids can be explained, but galaxy formation is not efficient. If the dark matter takes some other, colder form, such as more massive subatomic particles that have not yet been detected in the laboratory, then galaxies can form more efficiently. However, with these, it may be difficult to explain superclusters. Some cosmologists think that, if we can decide which formed first, clusters or galaxies, then we can decide what type of particles make up the dark matter.

21.4.3 Inflation

Particle physics has one more surprise for us, relating to the nature of the vacuum during the GUT era. We have already seen, in Chapter 7, that the quantum-mechanical vacuum is very different from the classical vacuum. It is not simply "nothing." It is just the lowest energy state, but it can have a lot of activity. In particular, we have the continuous creation and destruction of particle–antiparticle pairs. It has been proposed that during the GUT era, the nature of the vacuum changed, and the universe underwent a phase change, like changing from ice to water. According to the theory, the energy released in this phase change drove an extremely rapid expansion of the universe. In this expansion, the scale factor changed by a factor of about 10^{35} almost instantaneously!! This rapid growth is called the *inflation* of the universe. It is illustrated in Fig. 21.20.

Though it is still a speculative theory, inflation might explain some of our cosmological puzzles:

1. It solves the monopole problem by saying that all of the monopoles were produced before inflation, and were carried far apart by the inflation. Therefore, their density is very low now.

2. To see how it solves the flatness problem, we go back to our expanding-sphere analogy. Remember, we can only see within our horizon, whose radius is ct_0, where t_0 is the current age of the universe. The inflation caused the radius of curvature of the sphere to be so large that the section contained within our horizon has a very small fraction of the surface. Since we can only see a small fraction of the surface, it appears to be quite flat.

3. It solves the causality problem by saying that everything within our horizon was so close together before inflation that it could all be causally connected.

4. It may even solve the galaxy formation problem by saying that galaxies (or the matter to form them) condensed around irregularities in the distribution of the phases as the inflation ended.

It must be remembered that these ideas are quite speculative. However, even if the specifics are wrong, they may have opened many new possibilities to us in our understanding of the universe.

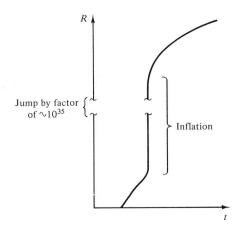

Figure 21.20 Inflationary universe. The normal expansion at the lower left is interrupted by a sharp increase in R, and then the normal expansion resumes.

CHAPTER SUMMARY

We looked in this chapter at the conditions in the very early stages of the universe. We looked at how those conditions affect the appearance of the universe today, and at observations that we can do today to tell us about the early universe.

An important relic of the early universe is the cosmic background radiation. These are the photons that last scattered off the matter just before the matter became neutral, at a temperature of about 3000 K. The photons have been red-shifted by the expansion of the universe, and are now seen with the spectrum of a 3-K blackbody.

We looked at how the background radiaton was discovered by Penzias and Wilson, and at how we observe various properties of the radiation. We saw how anisotropies in the radiation can tell us about fluctuations in the density of the early universe. We also discussed the problem of causality, relating to the very smooth appearance of the radiation.

We looked at the formation of elements in the big bang, and saw how the deuterium abundance can be a test of the amount of matter in the universe.

We looked at the developments in particle physics that have had a major impact on cosmology. In particular, we have seen that when temperatures are very high, as they were briefly in the early big bang, the differences among the various forces disappears, and processes can take place that don't normally take place today. We saw, for example, how this can account for the ratio of photons to baryons in the universe.

Finally, we looked at another prediction of particle physics—inflation. In an inflationary universe, there was a brief period of rapid growth. If the inflation really took place, it could solve nagging cosmological problems, such as the flatness of the universe, the isotropy of the background radiation, and the problem of our finding no magnetic monopoles.

QUESTIONS AND PROBLEMS

21.1 What do we mean by "recombination" and "decoupling" in connection with the cosmic background radiation?

21.2 (a) What is the present energy density in the cosmic background radiation, and (b) what was it when it was emitted at a temperature of 3000 K?

21.3 (a) Suppose the scale factors at two times, t_1 and t_2, are R_1 and R_2, respectively, and the background radiation temperatures are T_1 and T_2. What is the relationship between T_1 and T_2, in terms of R_1 and R_2? (b) If we know that the background temperature is now 3 K, and we let $R = 1$, write an expression for the background temperature at any time when the scale factor is $R(t)$.

21.4 Assume that the energy density, integrated over all wavelengths, in the background radiation is given by the Stefan–Boltzmann law. Assume that we know the cosmological red shift, and show that these also lead to the conclusion that the temperature must scale as $1/R$.

21.5 Use the Wien displacement law to find the peak wavelength of a blackbody for a temperature of (a) 3K; (b) 3000 K.

21.6 In the CN experiment to study the cosmic background radiation, what must the population ratio be for the temperature to be 3 K? (*Hint:* Take g_2/g_1 to be 3.)

21.7 What are the difficulties in studying the cosmic background radiation from the ground?

21.8 What is the significance of the dipole anisotropy in the cosmic background radiation?

21.9 What is the purpose of searching for small-scale anisotropy in the background radiation?

21.10 For the dipole anisotropy, the fractional increase in temperature in the direction in which we are heading, $\Delta T/T$, is given by $-v/c$, where v is our speed. Justify this relation, using Wien's displacement law and the Doppler shift relation.

21.11 Use the result of Problem 21.10 to predict $\Delta T/T$ resulting from the Earth's orbital motion. This has already been detected.

21.12 Why do we expect the pattern of the dipole anisotropy in the CBR to be matched by an anisotropy in the Hubble flow?

21.13 What can the deuterium abundance tell us about the density of the universe?

21.14 When we look at the DCN/HCN abundance ratio, why don't we measure the true D/H abundance ratio?

21.15 (a) Assume that a force is carried by a virtual particle of mass m. Assume that this particle can exist for a time, $\hbar/mc^2$, and that it travels close to c. What is the approximate range of the force? (b) If the range of the strong force is 10^{-13} cm, what mass particle should carry the force?

21.16 If the electromagnetic force is carried by virtual photons, which can live for a time $\hbar/mc^2$, explain how the force can be felt at any range. (*Hint:* Think about photons of different wavelength and energy.)

21.17 What is the temperature at which the average kinetic energy is equal to the mass of the W?

21.18 (a) What is the temperature at whch the average kinetic energy is equal to the mass of the X? (b) What is the mass of the X in grams?

21.19 If the average lifetime of the proton is 10^{31} yr, how much water would you have to watch to detect 100 proton decays in one year?

21.20 Work out the different combinations of u and d quarks that can make allowed particles. Give their electronic charge and approximate mass. [*Hint:* Note that the order doesn't matter; that is, (uud) and (udu) are the same particle.]

21.21 Why do we now not expect to see a free quark?

21.22 What cosmological problems would inflation help solve?

PART V

THE SOLAR SYSTEM

In studying the solar system, we find an important exception to our concept of astronomical objects being so remote that we cannot hope to visit them in the foreseeable future. Men have already visited our nearest neighbor in the solar system, the Moon, and brought pieces back to be studied in a normal earthbound laboratory. Unmanned probes have landed on Venus and Mars, and have flown close to all of the other planets except Neptune and Pluto (and Voyager 2 will reach Neptune in 1989). Clearly, the opportunity for even limited close-up viewing has had a major impact on our understanding of the solar system.

However, the study of the solar system is not simply devoted to sending probes when we feel like it. The spacecraft have followed literally centuries of study by more traditional astronomical methods. By the time the first probe was launched to any planet, astronomers had already developed a picture of what they expected to find. Most of these pictures did not survive the planetary encounters. But they did provide a framework for asking questions, and for deciding what instruments were important to place on the various probes.

We have also had the advantage of having the Earth as an example of a planet to study. It has been possible to develop ideas about planetary interiors, surfaces, atmospheres, and magnetospheres by studying the Earth. For that reason, we have devoted one chapter of this part to the Earth, viewed, not as our home base, but as just one planet. In studying the Earth we will generate ideas which we will extend to the other planets.

We will study the other planets in two groupings of similar planets, the inner and outer planets. Within each grouping, we don't study all aspects of a given planet before going on to the next planet. Instead, we study a given aspect, atmospheres, for example, of all the planets in the group. This allows us to extend common ideas to similar objects, looking for similarities and differences.

We will also see that much of the physics we have used in other astronomical problems—orbits, energy transport, hydrostatic equilibrium, tidal effects, and using spectroscopy to study remote objects, to mention a few examples— fit very naturally into our study of the solar system. Therefore, rather than trying to give a complete list of all the facts revealed by various probes, we will emphasize the underlying physics.

Overview of the Solar System

22.1 MOTIONS OF THE PLANETS

To the observer who views the sky almost every night, as ancient man must have, it is clear that most of the objects maintain their relative positions. These are the stars. However, apart from the Sun and Moon, a small number of objects move with respect to the stars. These are the planets. The study of the motions of the planets has occupied astronomers for centuries. These motions do not appear to be simple. The planets occasionally seem to double back along their paths, as shown in Fig. 22.1. This doubling back is known as *retrograde motion*. Historically, any explanation of the motions of the planets had to include an explanation of this retrograde motion.

The earliest models for our planetary system placed the Earth at the center of the system. This idea was supported by Aristotle in approximately 350 B.C. His view was that the planets, and the Sun and Moon, moved in circular orbits about the Earth. The picture of Hipparchus was refined by Claudius Ptolemy, in Alexandria, Egypt, around A.D. 140. In order to explain retrograde motion, he added addiitonal circles, called *epicycles*. As shown in Fig. 22.2, each planet moves around its epicycle as the center of the epicycle orbits the Earth. To obtain a closer fit to the observed motion, higher-order epicycles can be added.

An opposing picture was supported by the sixteenth-century Polish astronomer Nicholas Copernicus. In the Copernican system, the Sun is at the center of the planetary system. This picture is therefore called *heliocentric*. Copernicus showed that the retrograde motion could be an artifact caused by the motion of the Earth, as shown in Fig. 22.3. The Copernican system seemed more natural to some. However, since Copernicus used only circular orbits (as opposed to ellipses), his detailed predictions of planetary positions had small errors. (To get around these errors, epicycles were used.) The debate over the two pictures of our planetary system continued.

Figure 22.1 Retrograde motion of Mars, as it changes its direction against the background of stars.

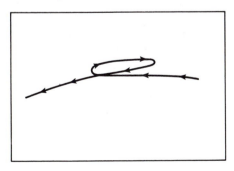

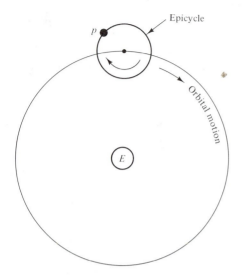

Figure 22.2 Epicycles. In this picture, the Earth E is at the center. The planet P doesn't simply orbit the Earth, it goes around a circle, which, in turn, orbits the Earth. If the planet's motion along the epicycle is faster than the epicycles's motion around the Earth, then the planet can appear to go backward for parts of each orbit. More epicycles can be added to this picutre.

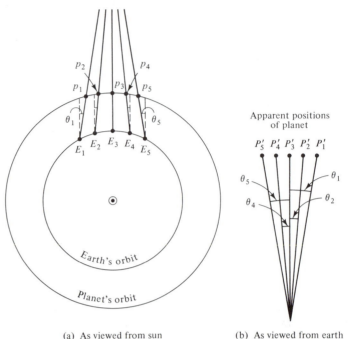

(a) As viewed from sun   (b) As viewed from earth

Figure 22.3 Retrograde motion in the heliocentric system. The Sun is at the center. We consider the Earth at five positions E_1 through E_5 with the planet at P_1 through P_5 at the same times. We use the line of sight from the Sun, through E_3 and P_3, as a reference direction. The dashed lines are all parallel to that direction, and the angles θ_1 through θ_5 keep track of the differences between the line of sight from the Earth to the planet and the reference direction. We see that, since the Earth is moving faster than the planet, the line of sight goes from being ahead of the dashed line to being behind the dashed line. The view from Earth is plotted at the upper right, and the apparent position of the planet on the sky is indicated by P_1' through P_5'. During this part of their orbits, the planet appears to move backward on the sky.

When Galileo Galilei turned his newly invented telescope to the planets, he found that Venus does not appear as a perfect circle. It goes through a series of phases, similar to those of the Moon. The size of its disc also changes as the phases change. These observations can fit more easily into the heliocentric picture but not by the Earth-centered one (because Venus would not always be the same distance from Earth). Though Galileo was persecuted for holding that the heliocentric picture is the true one, his work had influence on later scientific thought. Work switched from trying to understand what was at the center of the planetary system to trying to understand how the planets, the Earth included, move around the Sun.

Extensive observations were carried out by Tycho Brahe, at Uraniborg,

Denmark, late in the sixteenth century. Brahe moved to Prague in 1597, and died four years later. His results were taken over by an assistant, Johannes Kepler. Based on his analysis, Kepler published two laws of planetary motion in 1609, and a third law nine years later. Together these are known as *Kepler's laws*. It is important to remember that these laws were based on observation, and were not derived from any particular theory.

Before discussing Kepler's laws, we look briefly at how we survey the solar system, measuring the period and sizes of orbits. We take advantage of certain geometric arrangements. These are shown in Fig. 22.4.

When we talk about the orbital period of a planet, we mean the period with respect to a fixed reference frame, such as that provided by the stars. This period is called the *sidereal period* of the planet. However, we most easily measure the time it takes for the planet, Earth and Sun to come back to a particular configuration. This is called the *synodic period*. For example, the synodic period might be the time from one opposition to the next. How do we determine the sidereal period from the synodic period?

Suppose we have two planets, with planet 1 being closer to the Sun than planet 2. (For simplicity, we assume circular orbits.) The angular speed ω_1 of planet 1 is therefore greater than that of planet 2, ω_2. The relative angular speed is given by

$$\omega_{rel} = \omega_1 - \omega_2$$

Since $\omega = 2\pi/P$, where P is the period of the planet, the period of the relative motion of the two planets P_{rel} is related to P_1 and P_2 by

$$(1/P_{rel}) = (1/P_1) - (1/P_2) \tag{22.1}$$

Now we let one of the planets be the Earth, and express the periods in years. First we look at the Earth plus an inner planet. This means that P_1 is the period of the planet, and P_2 is 1 yr. Equation (22.1) then becomes

$$(1/P_{rel}) = (1/P) - 1 \quad \text{(inner planet)}$$

Similarly for the Earth and an outer planet, equation (22.1) becomes

$$(1/P_{rel}) = 1 - (1/P) \quad \text{(outer planet)}$$

In each case, P_{rel} is the synodic period, and P is the sidereal period.

We now look at how the sizes of various planetary orbits are determined. The technique is different for planets closer to the Sun than Earth and farther from the Sun than Earth. Figure 22.5 shows the situation for a planet closer to the Sun. When the planet is at its greatest elongation, it appears farthest from the Sun. The planet is then at the vertex of a right angle, as shown by

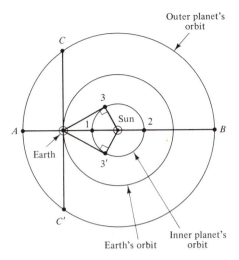

Figure 22.4 Configurations of the Earth and the inner and outer planets. Positions of the inner planets are indicated by numbers: (1) inferior conjunction, (2) superior conjunction, (3 and 3') greatest elongation. Positions of the outer planets are indicated by letters: (A) opposition, (B) conjunction, (C and C') quadrature.

Figure 22.5 Diagram for finding distance to inner planet.

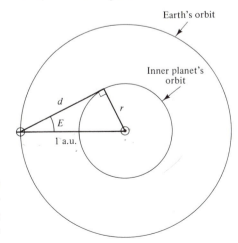

Fig. 22.5. Since we can measure the angle E between the Sun and the planet, we can use the right triangle to write

$$\sin E = r/1 \text{ AU}$$

where r is the distance from the planet to the Sun. This equation can be solved for r to give us the distance to the planet, measured in astronomical units.

Methods like this give us distances in terms of the astronomical unit. Even if we don't know how large an astronomical unit is, we still have all of the distances on the same scale, so we can talk about the relative separations of the planets. The current best measurement of the astronomical unit comes from situations like that in Fig. 22.5. We can now bounce radar signals off planets, such as Venus. By measuring the roundtrip time for the radar signals (which travel at the speed of light), we know how far the planet is from the Earth. The right triangle in equation (22.5) gives us

$$\cos E = d/1 \text{ AU}$$

Since E is measured and d is known from the radar measurements, the value of the astronomical unit can be found.

It is more complicated to find the distance to an outer planet. There are two different methods. The easier one was derived by Copernicus, but it is not good for tracing out a full orbit. It just gives the distance of the planet from the Sun at one point in the orbit. Kepler's method of tracing the whole orbit is shown in Fig. 22.6. We make two observations of the planet, one

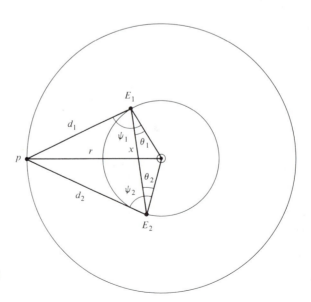

Figure 22.6   Diagram for finding distance to outer planet.

sidereal period of the outer planet apart. The earth is at E_1 and E_2 when these are made. The angles ψ_1 and ψ_2 are directly determined. The angles θ_1 and θ_2 are known, as well as the distance x. (If the Earth's orbit is circular, then $\theta_2 = \theta_2$.) We therefore know $\psi_1 - \theta_1$ and $\psi_2 - \theta_2$, and can find d_1 and d_2, and then, finally, r. The advantage of this method is that each point in the planet's orbit can be traced, with the observations overlapping in time.

We have already encountered Kepler's laws (Fig. 22.7.) when we discussed the orbits of binary stars. After all, orbiting planets and orbiting stars must obey the same laws of physics. The first law states:

1.   The planets move in elliptical orbits with the Sun at one focus.

The second law states:

2.   A line from the Sun to the planet sweeps out equal areas in equal times.

This is a result of conservation of angular momentum. The third law states:

3.   The square of the period is proportional to the cube of the semimajor axis, a.

This follows from the inverse square law for gravity. (Actually, Newton deduced the inverse square law by seeing what gravitational law was necessary to give Kepler's third law.)

There is, of course, nothing in Kepler's laws that tells us how far each planet should be from the Sun. However, for over 200 years it has been known that there is a simple relationship, giving the radii of planetary axes in astronomical units. To generate the law, we take the series (0, 3, 6, 12, 24, . . .), in which each term after the second is the double of the previous term. If we add 4 to each number and divide by 10, we get the approximate distances to the planets out to Uranus, with an extra term for the asteroid belt. This is known as *Bode's law*. It is not really a law, in the sense that there is any physical basis for it. It may just be a mathematical curiosity. (It also holds for the major moons of Jupiter and Saturn.)

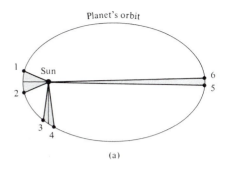

(a)

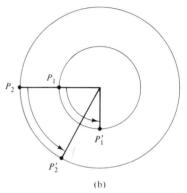

(b)

Figure 22.7   (a) Kepler's second law. Each of the shaded triangles has the same area. (b) Kepler's third law. In the time the inner planet moves from P_1 to P_1', the outer planet moves from P_2 to P_2'.

22.2   THE MOTION OF THE MOON

As the Moon orbits the Earth, it goes through its cycle of phases, as shown in Fig. 22.8. The Moon's sidereal period is 27.3 days. However, the motion of the Earth around the Sun causes the phases to cycle in 29.5 days. (This number actually varies by up to 13 hours since the Earth doesn't move at a constant rate about the Sun.) The Moon rotates with the same period, so we always see almost the same face. The Moon's axis of rotation is inclined by

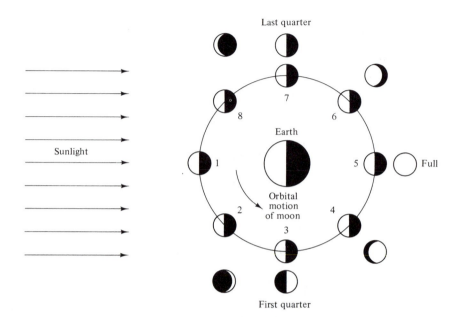

Figure 22.8 Lunar phases. The numbered images on the circle show the actual position and illumination of the Moon. The outer images show the appearance of the Moon, as seen from Earth. (1) New moon; (2) waxing crescent; (3) first quarter; (4) gibbous; (5) full moon; (6) gibbous; (7) last quarter; (8) waning crescent.

$1.5°$ with respect to the plane of its orbit. This inclination contributes to an effect, known as *libration*, which allows us to see more than 50% of the surface of the Moon. Another contribution to libration comes from the fact that the Moon is so close to the Earth that observers on opposite sides of the Earth see the Moon to be rotated through approximately $2°$. These effects, as shown in Fig. 22.9, allow us to see 59% of the Moon's surface.

EXAMPLE 22.1   Forces on the Moon

Calculate the relative strength of the forces exerted on the Moon by the Earth and by the Sun.

Solution

The force exerted by each body is proportional to its mass, and inversely proportional to its distance from the Moon. Therefore,

$$F_{SM}/F_{EM} = \frac{GM_{\odot}/r_{SM}^{2}}{GM_{E}/r_{EM}^{2}}$$

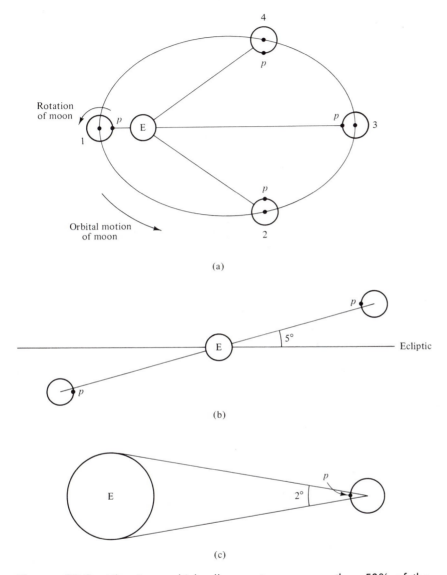

(a)

(b)

(c)

Figure 22.9   Librations, which allow us to see more than 50% of the Moon's surface. (a) The effect of the eccentricity of the Moon's orbit. The Moon rotates on its axis at a constant rate, but orbits the earth at a variable rate, so at various times the rotation gets ahead or behind the orbital motion, allowing us to see a total of an extra 6° around. In the figure, the point *P* keeps track of the steady rotation of the Moon. We can see that the line from the earth to the Moon passes through *P* at positions 1 and 3, but not at 2 and 4. (b) The effect of the inclination of the Moon's orbit, relative to the ecliptic, by 5°, allowing us to under and over the poles. (c) The effect of the finite size of the earth, allowing us to look from different directions, providing a 2° effect.

$$= \left(\frac{M_\odot}{M_E}\right)\left(\frac{r_{EM}}{r_{SM}}\right)^2$$

$$= \left(\frac{2 \times 10^{33} \text{ g}}{6 \times 10^{27} \text{ g}}\right)\left(\frac{3.85 \times 10^5 \text{ km}}{1.50 \times 10^8 \text{ km}}\right)^2$$

$$= 2.2$$

This means that the Sun exerts twice as great a force on the Moon as the Earth does. It is therefore not really proper to talk about the Moon orbiting the Earth. The Moon actually orbits the Sun, with the Earth causing the curvature of the Moon's orbit to change. This is shown in Fig. 22.9. Notice that the Moon's path is always concave toward the Sun. This is because the net force on the Moon is always inward, even when it is between the Earth and Sun.

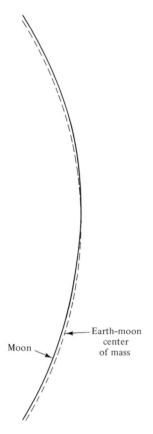

Figure 22.10 Orbit of the Moon, relative to the Sun. The Moon's orbit must always be concave towards the Sun.

The plane of the Moon's orbit is also inclined by 5° relative to the plane of the Earth's orbit (Fig. 22.10). If it were not for this inclination, there would be a total eclipse of the Sun and Moon every month. The plane of the Earth's orbit intersects that of the Moon's orbit in a straight line. This is called a *line of nodes*. We can only have an eclipse when the Moon is close to a node. Also, we can only have an eclipse when the Moon is new or full. Putting these together, we can only have an eclipse when the new or full Moon is near or at a node. The geometry is shown in Fig. 22.11.

The two times a year when the Sun is near a node are called *eclipse seasons*. Each season is approximately 38 days long. Each new or full Moon during the eclipse season results in at least a partial eclipse. The eclipse season is sufficiently long that each total eclipse has a corresponding partial eclipse of the other object either two weeks before or after. The plane of the Moon's orbit precesses with respect to the plane of the Earth's orbit with a period of 18.6 years. The tilt of the orbit doesn't change, but the line of nodes moves around the sky. This results in an eclipse year that is actually 346.6 days long. Therefore, there may be up to seven (lunar plus solar) eclipses a year, of which two to five will be solar.

In a total lunar eclipse, the whole Moon passes through the darkest part of the Earth's shadow, the *umbra*. In a partial eclipse, only part of the Moon passes through the umbra. In some eclipses, the Moon only passes through the lighter (outer) part of the Earth's shadow, the *penumbra*. Penumbral eclipses are hardly noticeable. The Moon is never completely dark, even during a total eclipse. Some sunlight is refracted by the Earth's atmosphere, and illuminates the Moon. Since the atmosphere filters out blue light better than red, the Moon appears red. Particulates in the Earth's atmosphere sometimes block sunlight more than at other times, and different eclipses have

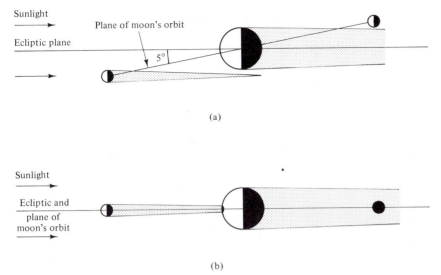

Sunlight

Plane of moon's orbit

Ecliptic plane

5°

(a)

Sunlight

Ecliptic and
plane of
moon's orbit

(b)

Figure 22.11 Eclipse geometry. Since the Moon's orbit is inclined by 5°, whether or not there is an eclipse depends on whether the Earth is in a part of its orbit where the new and full moon are in the ecliptic. (a) The Moon is out of the ecliptic at new and full moons, and no eclipse takes place. (b) The Moon is in the ecliptic at new and full moons, and eclipses can take place. The times of year when eclipses are possible are called eclipse seasons.

different amounts of light reaching the Moon. Eclipses just after major volcanic eruptions are particularly dark.

We have already discussed the value of total solar eclipses in studying Lunar eclipses can be seen from any point on the nighttime side of the Earth. They are currently of limited scientific value. They used to provide information on the thermal properties of the lunar soil. Astronomers could use radio and infrared observations to see how fast the soil cooled when the sunlight was removed. However, we now have samples of lunar material in the laboratory. As the Earth's shadow passes across the Moon, its round shape is a constant reminder of the Earth's roundness.

We have already discussed the value of total solar eclipses in studying the solar corona (Chapter 8). Since the Sun and Moon subtend almost the same angle as seen from the Earth, total eclipses are very brief, as the Moon's shadow sweeps quickly along the Earth. Also, totality can only be seen from a very narrow band on the Earth. On some occasions, the Moon is far enough from the Earth that it can't cover the whole solar disc. We see a bright ring, or annulus, around the Sun. From this appearance, we get the name *annular eclipse*.

22.3 STUDYING THE SOLAR SYSTEM

In the subsequent chapters in this part of the book, we will be discussing the individual members of the solar system. For any object, we will be interested in such things as the surface, the interior, the atmosphere, and the formation and evolution. Comparisons with the Earth are also interesting. For example,

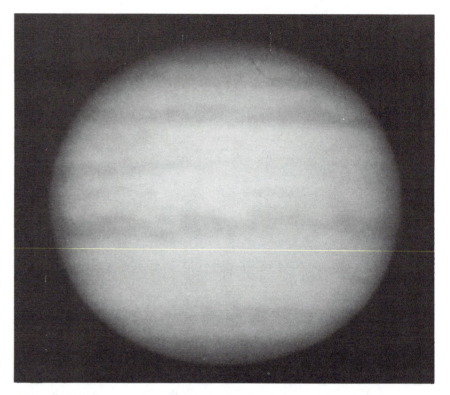

Figure 22.12   Jupiter as seen from Earth.

we think that any theory of atmospheric structure that can explain the Earth's atmosphere should, with the input of the right parameter, be able to explain that of Mars or Jupiter.

In studying the solar system, theoretical tools now include sophisticated computer modeling of various processes. The processes include accretion of interstellar material to form the various planets, or evolution of an atmosphere. These models are a great aid in interpreting complex observations.

We have undergone a great revolution in our observations of the solar system. Ground-based observations, yielding images and spectra, as well as masses and spin rates, have been quite useful. However, the opportunity afforded by space probes (Figs. 22.12 and 22.13) to take a close-up look has greatly added to our knowledge. Flybys have have given us detailed pictures, as well as allowing us to study objects in spectral bands that do not penetrate the Earth's atmosphere. They can also directly sample the environment of the object being studied. Landings can provide even more information, allowing us to study the conditions at a site. In the case of the Moon, we have had the additional luxury of bringing samples back for study in our laboratories. This has brought the solar system into the realm of the geologist.

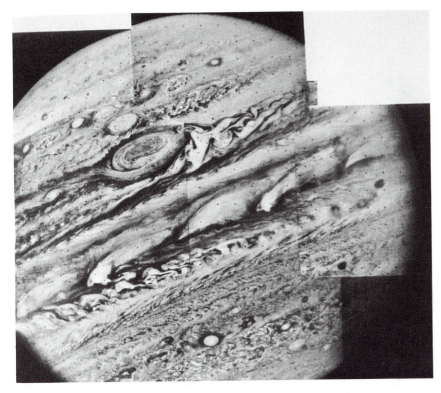

Figure 22.13 Jupiter as seen from Voyager 1.

22.4. TRAVELING THROUGH THE SOLAR SYSTEM

In this section we look at the mechanics relating to space probes to other planets. It is important to remember that during most of the flight of a planetary probe, it is unpowered. This means that it is simply in an orbit about the Sun. When we talk about orbiting space probes, the mass of the probe is much less than that of the Sun or any planet. We therefore do not have to consider the recoil of the Sun or any planet, making the problem a little easier.

EXAMPLE 22.2 Orbit and Escape from the Earth

What is the speed for a circular orbit just above the Earth's atmosphere? How does this compare with the escape velocity?

Solution

For an object of mass m, in a circular orbit of radius r, around a planet of mass M, the gravitational force must provide the acceleration for circular motion, so

$$mv_{\text{orb}}^2/r = GmM/r^2$$

where v_{orb} is the orbital speed. Solving for v_{orb} gives

$$v_{\text{orb}} = (GM/r)^{1/2}$$

$$= \left[\frac{(6.67 \times 10^{-8}\ \text{dyn-cm}^2/\text{s}^2)\ (6 \times 10^{27}\ \text{g})}{(6.4 \times 10^8\ \text{cm})}\right]^{1/2}$$

$$= 7.8 \times 10^5\ \text{cm/s}$$

$$= 7.8\ \text{km/s}$$

$$= 28{,}000\ \text{km/hr}$$

The escape speed is that with which we must launch an object from the surface, such that it can get infinitely far away with zero kinetic energy. This means that if we define potential energies to be zero when the objects are infinitely far apart, the total energy of an object with the escape speed is zero. If v_{esc} is the escape speed, then

$$\tfrac{1}{2}mv_{\text{esc}}^2 - GMm/r = 0$$

Solving for v_{esc} gives

$$v_{\text{esc}} = (2GM/r)^{1/2}$$

Note that comparing v_{orb} and v_{esc} gives

$$v_{\text{esc}} = \sqrt{2}\,v_{\text{orb}}$$

Therefore, the escape speed for the Earth is 11 km/s.

In considering travel to other planets, we use elliptical orbits. After all, an object whose orbit touches both the Earth and Mars cannot be in a circular orbit about the Sun. (We should note that parabolic or hyperbolic orbits can be used for one-time flybys, such as the Pioneer and Voyager.) When we studied binary stars, we found that the velocity of a particle in an elliptical orbit of semimajor axis a, when the object is a distance r, from the mass M at the focus, is

$$v^2 = GM[(2/r) - (1/a)] \qquad (22.2)$$

We can use this to find the total energy of the orbit,

$$E = \tfrac{1}{2}mv^2 - GMm/r$$
$$= \tfrac{1}{2}GMm[(2/r) - (1/a)] - GMm/r \qquad (22.3)$$
$$= -GMm/a$$

Note that, if you are in an orbit of a given semimajor axis a, and fire a rocket such that your energy increases, you must go to a higher orbit (making your energy less negative). In the higher orbit, your speed will actually be slower. Thus, firing your engine to "accelerate" the rocket, has the interesting effect of reducing its tangential speed.

Another consequence of equation (22.3) is that the energy of an orbit depends only on the semimajor axis, and not on the eccentricity. If we launch a rocket, once its fuel is used up, its energy is determined by its location and its speed. How, then, can we determine the eccentricity of the resulting orbit? The direction of motion when the fuel is used up determines the eccentricity. It also determines the total angular momentum L. Referring to Fig. 22.14, the angular momentum is given by

$$L = mvr \sin \theta$$

The least eccentric orbit has the highest angular momentum for a given semimajor axis. It should be noted that $\theta = 90°$ doesn't give a circular orbit unless the speed is the right speed for a circular orbit at that radius. However, $\theta = 90°$ means that the point where the fuel is spent will be the aphelion or perihelion of the orbit.

. The trick in launching a planetary probe is to place it in an orbit that intersects the planet's orbit when the planet is at this point of intersection. In choosing the direction of launch we can take advantage of the Earth's motion. There are many different orbits that can be chosen. However, the minimum energy orbit is one that has the Earth at aphelion and the planet at perihelion (if the planet is closer to the Sun) or one that has the Earth at perihelion and the planet at aphelion (if the planet is farther from the Sun).

EXAMPLE 22.3 Minimum Energy Orbit to Venus

Find the semimajor axis and eccentricity of a minimum-energy orbit from the Earth to Venus. Find the necessary launch speed and the time for the trip.

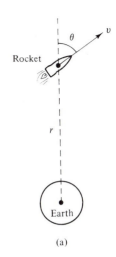

(a)

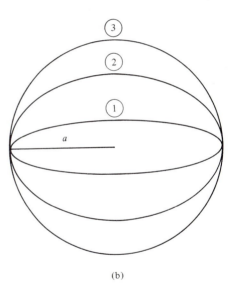

(b)

Figure 22.14 Angular momentum and orbital eccentricity. (a) If a rocket stops its powered flight at a distance r from the Earth while moving with velocity **v**, its energy is determined (as the sum of kinetic plus potential energies at that point), independent of the direction of motion. This means that the semimajor axis of the ellipse is determined. However, the direction of motion determines the eccentricity. (b) Ellipses with the same semi-major axis, but with different eccentricities.

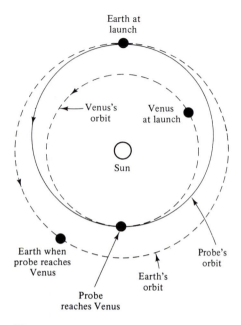

Earth at launch

Venus's orbit

Venus at launch

Sun

Earth when probe reaches Venus

Probe's orbit

Earth's orbit

Probe reaches Venus

Figure 22.15   Orbit of space probe to Venus.

Solution

In this case the Earth is at aphelion and Venus is at perihelion. The situation is shown in Fig. 22.15. The major axis of the orbit is the sum of the distance of the Earth from the Sun and the distance of Venus from the Sun. Therefore, the semimajor axis is

$$a = \tfrac{1}{2}(R_e + R_V)$$

$$= 0.86 \text{ AU}$$

The eccentricity is defined as the separation between the foci, divided by $2a$. The Sun is at one focus, R_V from the perihelion. Therefore, the symmetry of the orbit requires the other focus to be R_V from the aphelion. The distance between the foci is therefore

$$d = 2a - 2R_V$$

$$= R_e + R_V - 2R_V$$

$$= R_e - R_V$$

The eccentricity is therefore

$$e = \frac{R_e - R_V}{2a}$$

$$= \frac{R_e - R_V}{R_e + R_V}$$

$$= 0.16$$

At launch, $r = R_e$, so the launch speed is given by

$$v^2 = GM_\odot \left(\frac{2}{R_e} - \frac{2}{R_e + R_V} \right)$$

This gives $v = 27.3$ km/s. Note that the Earth's orbital speed is 30 km/s, so the probe must be going 2.7 km/s slower than the Earth. This means that we must launch the probe in the opposite direction from the Earth's motion.

We can find the length of the trip by using Kepler's third law. It tells us that the period of the orbit P in years is given by

$$P^2 = a^3$$

For $a = 0.86$ AU, we find $P = 0.80$ yr. Therefore, the trip to Venus takes half of this time, or 0.40 yr (approximately 5 months).

Once we know how long the trip will take, we must launch when Venus is at the place in its orbit that will take it to the rendezvous place in the travel time of the probe. The short period of time when a launch is possible is called a *launch window*. The size of the window depends on how far we are willing to deviate from a minimum-energy orbit. In particular, orbits with higher energy might be used since they have shorter travel times. Since the launch requires the correct relative position of the Earth and Venus, the launch window opens once per synodic period of Venus. This explains the spacing of launches to a given planet.

There is a trick that can be used to minimize the energy needed to visit a planet beyond Jupiter. In this, we take advantage of an elastic gravitational encounter between a space probe and a massive planet. This is known as a *gravity assist*.

To understand how this works, lets look at an analogous situation, a ball bounced off the front of an oncoming train. This is illustrated in Fig. 22.16. We assume that the mass of the ball is much less than that of the train and that the collision is completely elastic. The train is moving with speed u, and the ball is moving toward the train with speed v. As viewed from the train, the ball is coming toward the train at a speed of $u + v$. In the elastic collision, the train is so massive that it doesn't recoil, so the ball simply reverses its speed relative to the train. It is now moving, relative to the train, at a speed of $u + v$. As viewed from the ground, its speed is now $v + 2u$. In the collision, the ball has taken a little energy from the train (very small relative to the total energy of the train), and speed is now greater than its initial speed by twice the speed of the train.

With a space probe, we cannot have the probe bounce directly off Jupiter, for example. However, we can arrange the trajectory to have the same effect (Fig. 22.17). We generally don't get the full advantage of a head-on collision, but the speed of the probe can be increased by an amount of order the speed of Jupiter. This technique has been used for some of the Pioneer missions (Fig. 22.18) and for Voyagers 1 and 2.

22.5.   FORMATION OF THE SOLAR SYSTEM

One of our goals in studying the solar system is understanding how it formed. As we study the planets we will see that they provide many clues to the solar system's history. In this section, we briefly outline some of the ideas that have been proposed. Any theory of the formation of the solar system should

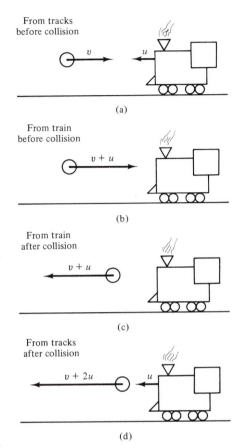

Figure 22.16 Analogy for gravity assist. (a) A low-mass ball with speed v approaches a high-mass train with speed u, for a head-on collision. (b) The same situation, as viewed from the train. The train is at rest, and the ball has a speed v + u. (c) Still looking from the train, the elastic collision simply reverses the direction of the ball, so it is now moving away from the train with a speed v + u. (d) We revert to the view from the ground. If the ball is moving ahead of the train at a speed v + u, then the speed of the ball, relative to the ground, must be this speed, plus the speed of the train u giving a speed of v + 2u.

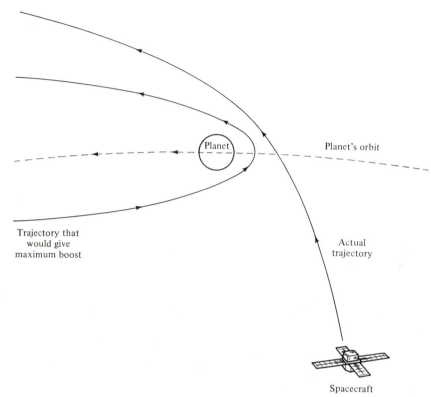

Figure 22.17 Trajectories for gravity assist. The space probe doesn't have to collide with the planet; it only has to make a close gravitational encounter. The maximum boost would come from the trajectory that comes closest to a head-on collision. However, since the spacecraft is coming from well inside the planet's orbit, this is not practical. The actual trajectory still gives almost half of the maximum boost in speed.

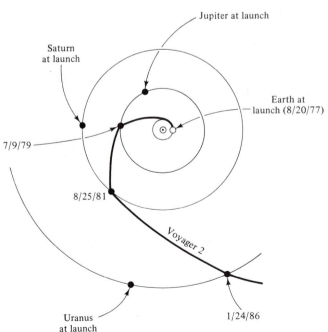

Figure 22.18 Orbit of Voyager 2 shows the effects of gravitational assists in going past each planet.

be able to explain such things as the fact that the planets' orbits are approximately in the same plane, and the fact that the planets orbit in the same direction. In addition, it must be able to explain the distribution of angular momentum in the solar system. Also, the different compositions and appearances of the planets must be explained.

Historically, two basic scenarios have been discussed. In one, the solar system formed as a by-product of the Sun's formation. The material left over from the Sun's formation is the material out of which the planets formed. This idea was discussed by René Descartes in 1644, and was elaborated upon by Immanuel Kant, and further by Pierre Simon de Laplace, who was the first to take the effects of angular momentum into account. In the other theory, originally proposed by Georges Leclerc de Buffon, the material to form the planets was ripped from the Sun by the effects of a passing object, possibly a comet, as suggest by Buffon.

Because of our present understanding of star formation, we now think that the solar system is the remnant of material that collapsed to form the Sun (Fig. 22.19). The original cloud might have been spherical. However, it must have been rotating, since we know that the solar system has angular momentum. The result of the rotation is that collapse perpendicular to the axis of rotation is retarded, while that parallel to the axis of rotation continues. This means that the spherical cloud can flatten to a disk. It is this disk out of which the planets probably formed.

Attempts have been made to calculate the minimum amount of material in the solar nebula. This is the amount of material in the Sun and planets, plus that which escaped during the formation. Estimates put the mass of the nebula in the range of a few $M_\odot$, though there are some theories in which it is much smaller. In understanding how this nebula produced the planets, there is a problem involving the angular momentum distribution. The Sun has only 2% of the angular momentum of the solar system (see Problem 22.16), but it would be expected that most of the angular momentum is in the central condensation. To explain this, it has been proposed that the material to form the planets fell slowly into the cloud around the already forming Sun.

In following the evolution of the solar nebula, we must keep track of three types of materials: gases, ices, and rocks. Most of the mass was in the gas (as most of the mass of the interstellar medium is in gas). However, gas can only be held to a growing planet by gravity, so it escapes from all but the largest objects. The ices are water (H_2O), carbon dioxide (CO_2), nitrogen (N_2), along with some ammonia (NH_3) and methane (CH_4). These make up 1.4% of the mass of the solar system. The rocks are iron oxides, and silicates of magnesium, aluminum, and calcium. Some of the iron was metallic and some was in iron sulfide (FeS). They can only be destroyed at high temperatures, in excess of 2000 K. They make up 0.44% of the mass (not including the Sun) in the solar system. They are particularly prominent in the inner planets, while the ices are prominent in the outer planets.

Figure 22.19 Formation of the solar system. (a) A rotating interstellar cloud. (b) The cloud begins to contract. Since angular momentum is conserved, the rotation gets faster. (c) The rotation is fast enough to slow collapse perpendicular to the direction of motion, so a disk forms. The center is collapsing fastest, forming a denser concentration that will eventually become the Sun. (d) When the rotation prevents further collapse of the disk, it breaks up into smaller clumps, so that some of the angular momentum is taken up by the orbital motion of the clumps. The clumps can then collapse. (e) Clumps of material gather together, forming planets, as the proto-Sun begins to radiate, and generate a large wind. (f) The wind clears the debris from the solar system.

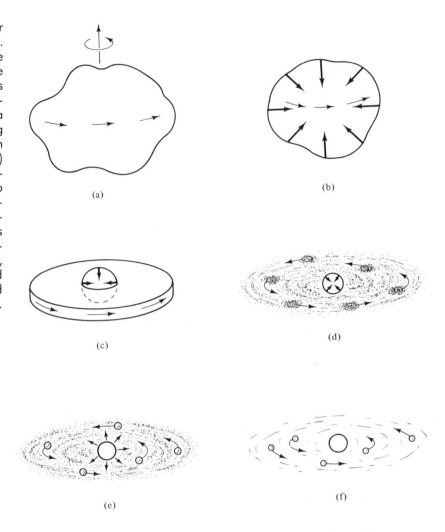

The collapsing nebula had a higher temperature in the center (near the forming Sun) than at the edge. Therefore, some different materials condensed at different distances from the center. When the temperature was about 3000 K near the center, it was a few hundred K in the regions of planetary formation. This was when the matter had already become opaque.

The accretion of the nebula probably took place over 10^4 to 10^5 years. The first step in the process was for small grains to clump together. The grains collided, sometimes making larger ones, and sometimes breaking into smaller ones. The process produced many grains about 1 cm in size. These grains were large enough to settle through the gas to the midplane of the nebula. This brought the clumps closer together, and allowed for even more collisions. Calculations indicate that the thin sheet of grains could then clump into objects with sizes of a few kilometers (essentially asteroid-sized objects). About 1000 of these could then form a gravitationally bound group. The high angular momentum of these groups prevented their total collapse. However, collisions eventually led to a dynamical relaxation, and these groups served as the cores for further condensation of bodies orbiting at the same distance from the Sun.

We have mentioned that there was a temperature variation with distance from the Sun. Near the Sun, volatile materials in the grains vaporized. The grain composition near the Sun was therefore different from that far from the Sun. The inner planets are essentially rock. Simple collisions around the already formed cores may have been sufficient to finish the formation of the planet. The masses of these inner planets were so small that gases could not be held, and escaped.

Farther out, the temperature was low enough for the ices to become important. The planets formed out of a combination of rock and ices of water and ammonia. Jupiter and Saturn built up collapsing cores massive enough to hold gas. They are therefore mostly hydrogen and helium. The satellite systems around Jupiter and Saturn probably grew from the gas disk forming around the planet. This process repeated the formation of the planets on a smaller scale.

It is interesting that Jupiter is at the distance from the Sun where the temperature is close to the freezing point of water. It is the innermost planet that could have formed from large amounts of ice. Since the density of material in the solar nebula decreased with distance from the Sun, Jupiter is the icy planet that had the most gas available to attract. It is probably no accident that the largest planet formed where it did. Uranus and Neptune have less hydrogen and helium. They probably retained these gases from their cores, but did not gather any more out of the surrounding material.

The scenario that we have just discussed probably left a large amount of debris around the planets. However, the solar system is now relatively clean. Where did all of the leftover material go? We think the Sun went through a stage when its wind was much stronger than it is now, much like the winds seen in T Tauri stars. The peak mass loss rate may have been $1 \, M_\odot/10^6$ yr. This wind carried sufficient energy and momentum to sweep out the debris and to stop the infall into the solar nebula.

Collisions among particles also helped clear the solar system. Some of the debris crashed directly into planets, leaving the craters that we still see on Mercury, Mars, and the Moon. Other collisions led to ejection of bodies from the solar system. Finally, the momentum carried by the sunlight itself may have helped clear the solar system. The momentum carried by light is its energy, divided by the speed of light. That is,

$$p = E/c$$

$$= h/\lambda$$

The radiation would have been able to sweep out small dust particles, and also carry away the orbital angular momentum of the dust. Finally, some collisions could also reduce the orbital energy of the dust, allowing it to fall into the Sun. This is known as the *Poynting–Robertson effect*. In these collisions, aberration of light, discussed in Chapter 12, makes it appear to the dust particle that the light from the Sun is coming from slightly ahead in the orbit. Thus, the particles lose energy, and eventually spiral into the Sun.

CHAPTER SUMMARY

In this chapter we began our study of the solar system by looking at the arrangements of the planets and at the clues we have to the formation of the solar system.

We first looked at the evidence for a Sun-centered system, and discussed the observational basis for Kepler's laws. We also discussed the relationship between Newton's laws of motion and gravitation and Kepler's laws.

We looked at the advantages of space probes for studying the planets, and discussed orbital considerations in a planetary mission. We saw how a gravity assist could reduce the energy needed for a particular trip.

Finally, we discussed the formation of the solar system. We discussed the accretion of material around the forming Sun, and the importance of temperature variations with distance from the proto-Sun. Finally, we discussed various ways by which the debris could have been cleared from the early solar system.

QUESTIONS AND PROBLEMS

22.1   Use the tabulated sidereal periods of the planets to find their synodic periods.

22.2   Why is the frequency of launch windows for a space probe to a planet dependent on the synodic period of the planet?

22.3   What are the angles of greatest elongation for Mercury and Venus?

22.4   Fill in the trigonometric steps in finding the distance to an outer planet, using Kepler's method.

22.5   How does the orbital angular momentum per unit mass vary with distance from the Sun?

22.6   Show how the inverse square law force can be derived from Kepler's third law for circular orbits.

22.7   Show that viewers on opposite sides of the Earth see the Moon rotated through 2°.

22.8   What is the escape velocity from our point in the solar system?

22.9   For a minimum-energy orbit to Mars, find the semimajor axis, the eccentricity, and the time of flight.

22.10   Find the speed and radius of a satellite in synchronous orbit above the Earth.

22.11   Express equation (22.2) in a form that gives v in terms of the escape velocity.

22.12   Find the escape velocities for all the planets.

22.13.   How does the energy of an object with the escape speed compare with the energy of an object in a circular orbit?

22.14   For the minimum-energy orbit to Venus, draw a diagram, showing the relative positions of the Earth and Venus at launch.

22.15   If a ball of mass m bounces elastically off the front of a train of mass M, where $M >> m$, with the ball moving at speed v and the train at speed u, what fraction of the kinetic energy of the train is lost in the collision?

22.16   Find the angular momentum of the Sun, and compare it with the angular momentum, orbital plus rotational, of all the planets.

22.17   Why is the location of Jupiter the place where we would expect the most massive planet?

22.18   Estimate the amount of material that can be cleared out of the solar system in 1 yr by a solar wind with a mass loss of $10^{-6}\,M_{\odot}$/yr and a speed of 200 km/s. Assume that the material to be cleared out is at the position of the Earth.

22.19   Estimate the rate at which material can be swept out from the position of the Earth by the radiation pressure from the Sun.

23

The Earth and Moon

We start our discussion of the planets with the one with which we are most familiar, the Earth. In understanding the processes that are important on the Earth, both now and in the past, we are setting a framework for our understanding of the other planets. Therefore, in this chapter, we will develop many of the ideas that should apply to all planets, both in terms of what properties are important and how we measure them.

23.1 HISTORY OF THE EARTH

23.1.1 Early History

We try to understand the formation of the Earth in terms of the basic scenarios, discussed Chapter 22, for the formation of the solar system. Somehow the Earth accreted from the material in the original solar nebula (Fig. 23.1). As material fell toward a central core, gravitational potential energy was converted into kinetic energy. This kinetic energy was then available to heat the forming planet, as particles moving at high speeds collided. In addition, heat was provided by the radioactive decay of isotopes of potassium, thorium, and uranium. Such decays led to heating, because the energetic particles—alpha, beta, and gamma—were absorbed by the surrounding rock. The kinetic energy of the particles then went to heating the rock. The relatively massive alpha particles are particularly effective in this heating.

This heating resulted in a molten (liquid) interior. Since materials are free to move in a liquid, heavier elements, such as iron and nickel, sank to the center, while lighter elements, such as aluminum, silicon, sodium, and potassium, floated to the surface. This process is called *differentiation*. The iron and nickel form the current core of the Earth. So much heat was trapped by the Earth that its core is still molten. The thorium and uranium were squeezed out of the core, and carried along in crystals to the surface. These radioactive materials provide a heating close to the surface.

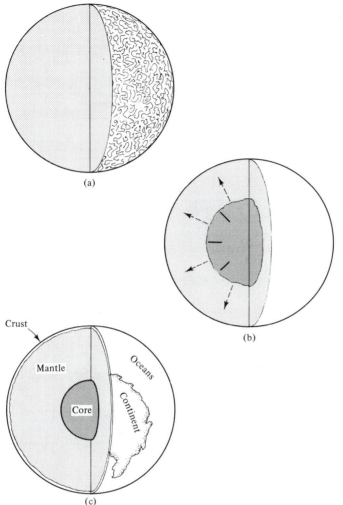

Figure 23.1 Steps in the formation and development of the Earth.

(a)

(b)

Crust

Mantle

Oceans

Continent

Core

(c)

The molten core is responsible for the Earth's magnetic field. The rapid rotation of the Earth leads to a dynamo process. In this process, a small magnetic field plus convection creates electrical currents flowing through the fluid. These currents produce a larger magnetic field, and so on. This process takes energy from the rotation of the Earth. The Earth's magnetic field is not fixed. The magnetic pole wanders irregularly. In addition, geological records indicate that the direction of the magnetic field has reversed every few hundred thousand years. The actual reversals take about ten thousand years, and there is a period when the field is very weak. The causes of these reversals are not well understood.

The upper part of the Earth's crust is composed of igneous, sedimentary, and metamorphic rocks. (Igneous rocks are volcanic in origin, sedimentary rocks are deposited, and metamorphic are either of the two that have been altered by high pressures.) The igneous rocks include granite. They have been enriched in silicon, aluminum, and potassium. The lower part of the crust is composed of *gabbro* and *basalts*. These are dark rocks containing silicates of sodium, magnesium, and iron. They result from partial melting of materials in the crust. (Gabbros have coarser crystals than do basalts.)

An important clue to the Earth's interior comes from volcanic activity. This activity results from heating of the crust, producing the molten material. Volcanic activity is very important in mountain building. It also provides a means of venting certain materials from the interior to the surface. In particular, we think that this is the source of the carbon dioxide (CO_2), methane (CH_4), and water (H_2O), as well as sulfur-containing gases in our atmosphere.

As the water vapor was ejected into the atmosphere, the temperature was low enough for it to condense. Other gases dissolved in the water, and combined with calcium and magnesium leached from surface rocks. This had the important effect of removing most of the carbon dioxide from the Earth's atmosphere.

23.1.2 Radioactive Dating

Much of what we know about the ages of rocks in the Earth's crust comes from *radioactive dating*. In this process, we take advantage of the fact that if we start with some number N_0 of nuclei of a radioactive isotope, the number left after a certain time t is given by

$$N(t) = N_0 \exp(-t/\tau_e)$$

The quantity τ_e is the time for the number in the sample to fall to $1/e$ of its original value. We can also write this expression in terms of the *half-life*, $\tau_{1/2}$, which is the time for the number in the sample to fall to one-half of its original value,

$$N(t) = N_0(\tfrac{1}{2})^{-t/\tau_{1/2}} \tag{23.1}$$

Comparing these two expressions (see Problem 23.3), tells us that

$$\tau_{1/2} = 0.693\tau_e$$

The behavior of $N(t)$ vs. t is shown in Fig. 23.2.

The half-lives of various nuclei can be measured in the laboratory (see Problem 23.4). Therefore, if we known N_0 and $N(t)$, we can solve for t, the time that has elapsed since the sample had N_0 nuclei of the particular isotope. It is important to choose an isotope whose half-life is comparable to the time period you are trying to measure. If the time period is much longer than the

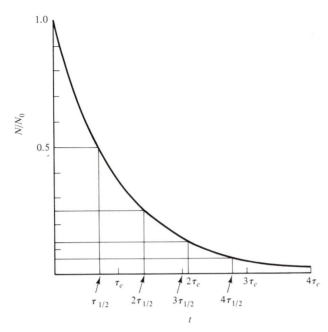

Figure 23.2 Radioactive decay. The vertical axis is the number of remaining nuclei N divided by the original number of nuclei N_o. Half-lives are indicated by $\tau_{1/2}$ and lifetimes to fall by $1/e$ are indicated by τ_e.

half-life, there will be very few nuclei left to measure. If the time period is much less than the half-life, very few decays will have taken place, and it will be hard to detect a difference between N_0 and $N(t)$. From this consideration, ^{238}U (uranium-238), is ideal for studying the Earth's crust, which we believe to be billions of years old. The half-life of ^{238}U is 4.6 billion years.

A practical problem is that we generally don't know how much of a given isotope the Earth started out with. This means that we must employ indirect methods, generally involving a knowledge of the end-products of the decay. For example, the alpha decay of ^{238}U is followed by a series of alpha decays, eventually leading to the stable isotope, ^{206}Pb (lead-206). If we can assume that all of the ^{206}Pb on Earth came from such decays, we could use the amount of ^{206}Pb as an indicator of the original amount of ^{238}U.

However, the Earth may have formed with ^{206}Pb, so we have to use even more involved techniques to correct for that effect. These techniques often involve measuring the relative abundances of certain isotopes, such as ^{204}Pb and ^{206}Pb. Such measurements require a means of separating the isotopes. This separation is much easier for gases than for solids. For this reason, we actually use a different age tracer. For example, ^{40}K (postassium-40) beta decays with a half-life of 1.3 billion years. The decay product is ^{40}Ar (argon-40). Since argon is a noble gas, it doesn't stay bound to the rocks. We can collect the argon and study the relative abundances of various isotopes.

In considering the history of the Earth, we must remember that the atmoshere has eroded the surface, altering its characteristics. (As we will see below, the motions of the continents also alter the surface.) The oldest rocks we see are dated at 3.7 billion years. (The oldest fossil cells are dated at 3.4 billion years). We think that the surface underwent significant alteration 2.2 to 2.8 billion years ago. This was a period of volcanic mountain building with the Earth having a very thin crust. The crust was probably broken into smaller platelets.

At that time the atmosphere was mostly water, carbon dioxide, carbon monoxide, and nitrogen. It is thought that any ammonia and methane could not have lasted very long. There was little oxygen, since oxygen is a product of plant life. Since there was no oxygen, there was also no ozone (O_3) layer to shield the Earth from the solar ultraviolet radiation. We think that this ultraviolet radiation must have stimulated the chemical reactions to synthesize the simplest organic compounds. It has been shown in the laboratory that such reactions are greatly enhanced in the presence of ultraviolet radiation.

23.1.3 Plate Tectonics

The layer below the thin crust is kept heated by radioactive decay. The amount of heat is not sufficient to melt the material completely, but it keeps it from being completely solid. It has the consistency of a plastic. It doesn't deform very quickly, as a liquid would. However, under a steady pressure, it will flow slowly. The solid layer above the plastic region is called the *lithosphere*. The Earth's lithosphere is broken into *plates*. The name is meant to suggest objects that are much larger in their dimensions parallel to the Earth's surface than they are in the vertical direction. The plates float on top of the plastic layer.

As they float, the plates move slowly. Since they carry the continents with them as they move, we refer to this motion as *continental drift*. The general term for any process involved in the movement or deformation of planetary surfaces is *tectonics,* so continental drift is also called *plate tectonics*. Figure 23.3 shows how the Earth's continents have moved in recent geological history, over periods of hundreds of millions of years. The actual motion is being driven by convection forcing material up from below into some of the narrow gaps between the plates. One such region is called the *mid-Atlantic ridge*. Throughout the ridge, fresh material is appearing on the sea floor, as the plates move away from the ridge.

The regions where the plates meet are characterized by a high level of geological activity. When one plate is forced under another, the resulting upward pressure can build great mountain ranges, such as the Himalayas. These plate boundaries also show a high frequency of volcanoes and earthquakes. The volcanoes result from warm material being pushed upward. The earthquakes result from the fact that the slippage of the plates past each other is not smooth. For long periods the plates might not move as the pressure increases. Eventually, the pressure becomes too great, and there is a sudden movement along the boundary. The line along which this slippage takes place is called a *fault line*. One is shown in Fig. 23.4.

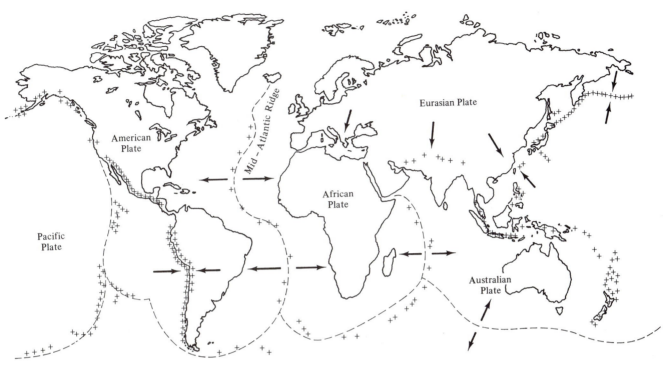

Figure 23.3 The main tectonic plates are outlined by areas of geological activity, earthquakes and volcanos, marked by plus signs (+). Arrows indicate direction of motion of plates.

Figure 23.4 Aerial view of San Andreas fault line in California.

23.2   TEMPERATURE OF A PLANET

The temperature of the Earth is determined by a balance between the energy absorbed from the Sun and the energy given off by the planet. For planets like the Earth, heat from inside doesn't have much effect on the surface temperature. The planetary temperature for which these balance is called the *equilibrium temperature* of the planet. The actual energy transport might be complicated by the presence of an atmoshere, but we will first calculate the equilibrium temperature, ignoring atmospheric effects. In the absence of an atmosphere, this calculation is essentially the same as that for an interstellar grain (see Chapter 14).

We start by calculating the energy received per second. The energy per second given off by the Sun is its luminosity, $L_\odot$. (We could leave this in terms of the luminosity, or write the luminosity as $4\pi R_\odot^2 \sigma T_\odot^4$.) At the distance d of the planet from the Sun, the solar luminosity is spread out over a surface area of $4\pi d^2$. This means that the power per unit surface area is $L_\odot/4\pi d^2$. As seen from the Sun, the projected area of the planet is πR_p^2. The planet therefore intercepts a power equal to this area multiplied by the power per surface area. Finally, not all of the sunlight is absorbed by the planet. A fraction a, the albedo, is reflected. The amount absorbed is equal to $(1 - a)$ multiplied by the amount that actually strikes. In this calculation, we are assuming that the albedo is the same at all wavelengths. Therefore, the power absorbed by the planet is

$$\text{Power absorbed} = \frac{L_\odot R_p^2 (1 - a)}{4d^2} \tag{23.2}$$

We now look at the power radiated. We assume that the planet rotates fast enough so that there is no great difference between the day and night temperatures, so we treat the temperature of the planet as a constant. (This is a good approximation for the Earth, but not for the Moon.) The power radiated per unit surface area is $e\sigma T_p^4$, where e is the *emissivity*. The emissivity can range from zero to one, and is one for a perfectly blackbody. Multiplying this by the planet's surface area $4\pi R_p^2$, we get the total power radiated

$$\text{Power radiated} = 4\pi R_p^2 e \sigma T_p^4 \tag{23.3}$$

If we equate the power absorbed with the power radiated, we can solve for the equilibrium temperature of the planet, giving

$$T_p = \left[\frac{L_\odot (1 - a)}{16\pi d^2 \sigma e} \right]^{1/4} \tag{23.4}$$

This calculation doesn't account for the fact that the albedo and emissivity vary with wavelength. We must integrate the energy received over the spectral energy distribution of the Sun, and integrate the energy radiated over the spectral energy distribution of the Earth. In each case we must incorporate a frequency-dependent albedo and emissivity inside the integral. For example, remembering the luminosity of the Sun,

$$L_\odot = 4\pi R_\odot^2 \int_0^\infty B_\nu(T_\odot)d\nu$$

where B_ν is the Planck function. The power absorbed by the Earth then is

$$\text{Power absorbed} = [\pi R_\odot^2 R_p^2/d^2] \int_0^\infty (1 - a_\nu)B_\nu(T_\odot)d\nu \qquad (23.5)$$

Similarly, the power radiated is

$$\text{Power radiated} = 4\pi R_p^2 \int_0^\infty e_\nu B_\nu(T_p)d\nu \qquad (23.6)$$

Equating these gives

$$4\int_0^\infty e_\nu B_\nu(T_p)d\nu = (R_0^2/d^2) \int_0^\infty (1 - a_\nu)B_\nu(T_\odot)d\nu \qquad (23.7)$$

This equation cannot be solved directly for T_p, even if we know a_ν and e_ν. But, since we know T_0, we can evaluate the right-hand side of the equation. We then have to try different values of T_p in the left-hand side, probably using a computer to evaluate the integral, until we find the value of T_p which makes the left-hand side equal to the right-hand side. To be even more rigorous, we should also account for the fact that the temperature is not constant across the surface of a planet.

   We can take advantage of the fact that most of the Sun's energy is in the visible and most of the energy given off by the Earth is in the infrared. We can assume that there is a constant albedo a_ν in the visible, and a constant emmisivity e_{ir} in the infrared. Since e_ν and a_ν are constant over the region of each integral where B_ν is significant, we can factor them out of the integral. The result is similar to equation (23.4):

$$T_p = \{[L_\odot/16\pi d^2\sigma][(1 - a_\nu)/e_{ir}]\}^{1/4} \qquad (23.8)$$

   On the Earth, the albedos are different for the oceans and the land. They are also different for cloud cover. When we take this into account, the equilibrium temperature is 246 K. However, this is still not the temperature that we measure at the ground. We have not yet considered the important effects of radiative transfer in the atmosphere.

23.3 THE ATMOSPHERE

The Earth's atmosphere provides us with a multitude of phenomena whose complexity may make them seem beyond understanding. However, we can apply many of the basic ideas that we have already discussed for stars, such as energy balance and hydrostatic equilibrium, to get a reasonable understanding of the phenomena. With the aid of fast computers, these ideas can be applied to study the atmosphere in great detail. In this section, we will look at some of the concepts common to all planetary atmospheres, and see how they apply to the Earth's.

To relate the basic variables that describe a gas—temperature, density, and pressure—we must have an equation of state. We can treat planetary atmospheres as ideal gases, so the equation of state is simply

$$P = (\rho/m)kT \tag{23.9}$$

In this expression, m is the average mass per molecule. For the Earth's atmosphere, this is approximately 29 times the mass of the proton, reflecting the fact that the atmosphere is mostly N_2 (molecular weight 28) and O_2 (molecular weight 32). The pressure at the surface is 10^6 dyn/cm^3, also called *one atmosphere,* or *one bar*. For a surface temperature of 300 K, the density is 1.1×10^{-3} g/cm^3.

23.3.1 Pressure Distribution

The vertical distribution of pressure and density is governed by the condition of hydrostatic equilibrium, just as in stars. The weight of each layer is supported by the pressure difference between the bottom and the top of that layer. For stars, we treated the layers as spherical shells. We could do that for the Earth, also. However, the Earth's atmosphere is so thin that we can treat it as a plane parallel layer (see Problem 23.7).

The condition of hydrostatic equilibrium then becomes

$$dp/dz = -\rho g \tag{23.10}$$

We now have two equations, the equation of state and the equation of hydrostatic equilibrium. However, we have three unknowns, T, P, and ρ. When we look at the temperature distribution (discussed later in this section), we find that, especially in the lower atmosphere, the temperature doesn't deviate by large amounts from its value at the ground T_0. We therefore make the approximation that the temperature is constant. Knowing T, we can use equation (23.9) to substitute for the density in equation (23.10), so that the only remaining variable is the pressure. This gives

$$dP/dz = -(mg/kT_0)P$$

To integrate this, we want all of the P dependence on one side and all of the z dependence on the other side. This gives

$$dP/P = -(mg/kT_0)\,dz$$

We now integrate this, with the limits on pressure being P_0, the surface pressure, and P, the pressure at the altitude of interest, and the limits on z being 0 and z, the altitude of interest. That is,

$$\int_{P_0}^{P}(dP'/P') = -(mg/kT_0)\int_0^z dz' \qquad (23.11)$$

Integrating and substituting the limits gives

$$\ln(P/P_0) = -(mg/kT_0)z$$

where we have used the fact that $\ln(P/P_0) = \ln(P) - \ln(P_0)$. If we raise e to the value on the left-hand side, it should equal e raised to the value on the right-hand side. Remembering that $\exp[\ln(x)] = x$, this gives

$$P/P_0 = \exp[-(mg/kT_0)z)]$$

The quantity kT_0/mg has dimensions of length, and is the distance over which the pressure falls to $1/e$ of its original value. We call this quantity the *scale height*,

$$H = kT_0/mg \qquad (23.12)$$

In terms of the scale height, the pressure distribution becomes

$$P = P_0\exp(-z/H) \qquad (23.13)$$

This variation of pressure with altitude is shown in Fig. 23.5.

EXAMPLE 23.1 Scale Heights

Compute the scale height for the Earth's atmosphere as well as for an atmosphere of pure oxygen (O_2) and one of pure hydrogen (H_2).

Solution

For a gas of molecular mass Am_p (where m_p is the proton mass), the scale height is

Figure 23.5 Pressure vs. altitude in the Earth's atmosphere. In plotting quantities about the atmosphere, we usually plot altitude on the vertical axis, even though it is the independent variable. The scale height is indicated by *H*.

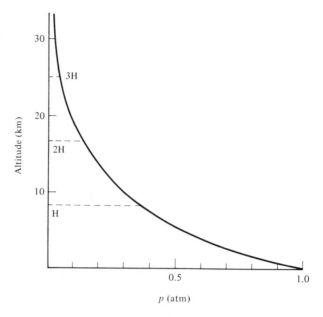

$$H = (300 \text{ K})(1.38 \times 10^{-16} \text{ erg/K})/(1.67 \times 10^{-24} \text{ g})(980 \text{ cm/s}^2)A$$

$$= 2.53 \times 10^7 \text{ cm}/A$$

$$= 2.53 \times 10^2 \text{ km}/A$$

For $A = 29$, $H = 8.7$ km. For oxygen, $A = 32$, so $H = 7.9$ km; and for hydrogen, $A = 2$, so $H = 125$ km.

Another important constituent of the Earth's atmosphere is water vapor. Equation (23.13) is not a good description of its distribution. This is because the atmospheric temperature is close to that at which the water condenses. The water vapor may have a normal distribution at low altitudes. However, it may be almost totally absent at the cooler, higher altitudes.

23.3.2 Retention of an Atmosphere

Even if a planet is formed with an atmosphere, that atmosphere will not necessarily be retained. Any molecules moving faster than the escape speed can escape from the top of the atmosphere. If the molecules are not replaced from below, by gases escaping from the planet's interior, the atmosphere will eventually be lost.

At any temperature, the molecules with a lower molecular mass will be moving faster, so these are the most likely to escape. We can calculate the value of A for which the mean molecular speed $(3kT/m)^{1/2}$ is equal to the escape speed for the planet $(2GM/R)^{1/2}$. Equating these, writing m as Am_p, and solving for A, gives

$$A = 3kTR/2GMm_p$$

For the Earth, $A = 0.1$. This tells us that there is no molecule for which the average speed is sufficient for escape. However, we have seen that because there is a distribution of speeds, many molecules move significantly faster than the average.

The speed distribution, called a *Maxwell–Boltzmann distribution*, is shown for both hydrogen and oxygen in Fig. 23.6. The distribution is such that the number of molecules with speeds between v and $v + dv$ is given by

$$N(v)dv = (\text{constant})v^2 \exp(-mv^2/2kT)dv \qquad (23.14)$$

Note the $\exp(-\text{energy}/kT)$, similar to the Boltzmann equation for the populations of energy levels.

Note from Fig. 23.6 that hydrogen has a greater fraction of its molecules moving faster than the escape speed than does oxygen. The molecules moving faster than the escape speed will be able to escape. At first you might think that this is the end of the process, with the rest of the molecules being trapped. However, the remaining molecules will collide with each other, reestablishing a Maxwell–Boltzmann distribution. Since the escaping molecules took away more than the average energy per molecule, if there is no replacement of the energy, the new distribution will be at a lower temperature. The gas will be cooled by the escape of the faster molecules (just as a liquid is cooled by the evaporation of the faster molecules). However, if there is a source of energy, such as sunlight or heat from the ground, equilibrium can be established at

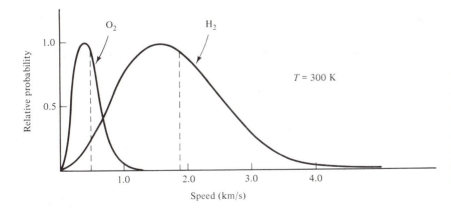

Figure 23.6 Maxwell–Boltzmann distribution for oxygen and hydrogen molecules at $T = 300$ K. For each curve, the vertical axis is a relative probability of finding a molecule at a given speed. In each case, the vertical dashed line is for a molecule whose energy is ³⁄₂ kT. Notice the large number of molecules with greater than this average speed.

the same temperature as before, and the same fraction of the remaining molecules will be moving faster than the escape speed. It is by this process that the atmosphere escapes.

This escape can only take place from the highest levels of the atmosphere. A molecule lower down in the atmosphere might start out going faster than the escape speed, but it will collide with other molecules, losing energy before it can escape. The layer of the atmosphere from which molecules can escape is called the *exosphere*. The thickness of the exosphere is taken to be equal to the average distance between collisions in the gas at high altitudes.

23.3.3   Temperature Distribution

The temperature distribution with altitude, shown in Fig. 23.7, reflects the variety of mechanisms by which energy enters the atmosphere. Even though the Sun is the ultimate source of energy for the atmosphere, it is not the direct heat source over most of the atmosphere. Most of the Sun's visible energy passes through directly to the ground, and is not absorbed by the atmosphere. The heated ground then gives off radiation characteristic of its temperature, so the radiation from the ground is mostly in the infrared. Infrared radiation from the ground is then trapped in the lower atmosphere. Thus, the ground is actually the immediate source of energy for the lower atmosphere, explaining why the temperature in the lower atmosphere decreases as one gets farther from the ground. As the ground heats the air just above it, that air expands and rises. This convection is another means of energy transport from the ground to the lower atmosphere. (The above discussion is very simplified. In certain situations, ''temperature inversions'' are present in which warmer air is on top of the cooler air, and there is little convection. With little convection, pollution can build up.)

Figure 23.7 Temperature vs. altitude in the Earth's atmosphere. Layers are divided according to the important energy balance mechanisms, so temperature behavior differs from layer to layer.

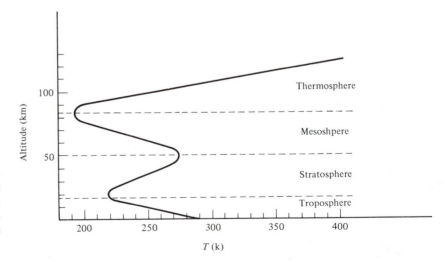

One place where solar energy is directly absorbed is the *ozone layer*. The ozone directly absorbs the solar ultraviolet radiation. This protects us from the solar ultraviolet. It also makes the ozone layer warmer than the layers below it.

The lower part of the atmosphere, heated by radiation from the ground, is called the *troposphere*. It is the part of the atmosphere to which our daily existence is confined. We now look at radiative energy transport from the ground into the troposphere, as outlined in Fig. 23.8. The ground is at temperature T_e. We assume that the troposophere has a height z_t, and we are interested in the energy reaching and leaving some intermediate height z.

The power per unit area reaching z from the ground is σT_e^4. The power per unit area radiated at z is σT^4, where T is the temperature at z. However, some of the radiation emitted at z does not escape. It is absorbed at a higher altitude, to be reemitted. The energy density in the layer between z and z_t, due to energy emitted at z is $4\sigma T^4/c$. Therefore, the energy per unit area in this layer is the energy per unit volume, multiplied by the thickness $z_t - z$, so

$$\text{Energy/area} = (4\sigma T^4/c)(z_t - z)$$

If all of this energy left z is in a time, t, then the power per unit area is the energy per area, divided by t,

$$\text{Power/area} = (4\sigma T^4/ct)(z_t - z)$$

Each time a photon is absorbed and reemitted, its direction is changed in a random way. We say that the photon does a *random walk*. If a walk is not random, all of the steps are in the same direction. If there are N steps of length L, the distance from the original point will be NL. However, in a random walk, a lot of time is spent backtracking. It can be shown (see Problem 23.12) that for a random walk in one dimension (steps back and forth along a line), the distance from the original position after N steps is $N^{1/2}L$, and for a three-dimensional walk, the distance is $(N/3)^{1/2}L$. (In each case, the total distance actually traveled by photons is NL. In the random walk, some of this is lost in doubling back.)

Using this, the number of steps required for a photon to do a random walk of length $z_t - z$ is

$$N = 3[(z_t - z)/L]^2$$

where L is the photon mean free path. The length of each step is L, so the time for this number of steps is

$$t = NL/c$$

$$= (3/Lc)(z_t - z)^2$$

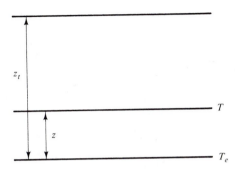

Figure 23.8 Radiative transport in the troposphere. We are considering material a height z above the ground, while the top of the troposphere is at a height z_t.

Using this, the power per area leaving z becomes

$$\text{Power/area} = \frac{4\sigma T^4(z_t - z)}{c(3L/c)(z_t - z)^2}$$

$$= \tfrac{4}{3}\sigma T^4 \frac{L}{z_t - z}$$

However, the path length $z_t - z$, divided by the photon mean free path L, is the optical depth τ (as discussed in Chapter 6), so

$$\text{Power/area} = 4\sigma T^4/\tau$$

Equating this to the power received from the ground σT_e^4, and solving for T, we have

$$T^4 = \tfrac{3}{4}T_e^4\,\tau$$

A more accurate treatment of the radiative transfer, originally worked out by Sir Arthur Eddington for the atmosphere of stars, modifies our result to

$$T^4 = \tfrac{3}{4}T_e^4(\tau + \tfrac{2}{3}) \qquad (23.15)$$

At the top of the troposphere, $\tau = 0$. (All photons escape upward.) This means that $T = 0.84\,T_e$. If we use $T_e = 246$ K, then $T = 207$ K. The infrared optical depth from the bottom to the top of the troposphere is about 2, giving $T = 1.2\,T_e$, or about 293 K.

Note that the atmosphere near the ground is warmer than if there were no trapping of infrared radiation near the ground. The heating of the lower atmosphere by trapped infrared radiation near the ground is called the *greenhouse effect*. (The idea is that the windows of greenhouses are transparent to the visible incoming solar radiation but opaque to the outgoing infrared radiation from the Earth, trapping heat in the greenhouse. Real greenhouses also are aided by the fact that the walls eliminate breezes.) On the Earth, the greenhouse effect is relatively small. As we will see, it is much more significant on Venus.

Convection also plays an important role in energy transport in the lower atmosphere. As the air near the ground is heated, it expands. The buoyant air will continue to rise until it reaches air of its own density. The more rapidly the temperature falls off with altitude, the more rapidly the pressure will fall off with altitude. A faster falloff in pressure means a larger pressure difference between the top and bottom of any parcel of air. The larger pressure difference provides a larger buoyant force, and the air rises more quickly, making convection more important. Therefore, convection becomes more important when the falloff in temperature with altitude is larger.

As the air rises, it encounters lower pressure and expands. The gas does work in the expansion. It takes no energy in from its surroundings, so it cools. A process in which no energy is exchanged is called *adiabatic*. The convection process itself will modify the *temperature gradient dT/dz* to the value appropriate for an adiabatic process. If the temperature gradient established by radiation is less than the adiabatic gradient, the temperature gradient is not enough to drive convection. However, if the temperature gradient is greater than the adiabatic gradient, convection will set in. Remembering that dT/dz is negative, we can say that the condition for significant convection is that

$$|dT/dz|_{\text{rad}} > |dT/dz|_{\text{rad}}$$

EXAMPLE 23.2 Adiabatic Temperature Gradient

Find the value of dT/dz when the energy flow is dominated by convection near the ground.

Solution

We don't have to consider the whole atmosphere to do this. We only have to look at a small volume, or parcel of air, and see how it behaves. Suppose we have a parcel of air, of volume V_0, rising from just above the ground, where the temperature is T_0 and the pressure is P_0. As the parcel rises, its temperature, pressure, and volume are T, P, and V. For an adiabatic process, these quantities are related by

$$PV^{\gamma} = P_0 V_0^{\gamma}$$

where γ is the ratio of the specific heats at constant pressure and constant volume. For a monatomic gas, $\gamma = \frac{5}{3}$. For a diatomic gas, $\gamma = \frac{7}{5}$. Since the Earth's atmosphere is mostly N_2 and O_2, we use $\frac{7}{5}$. (If there is a lot of water vapor present, γ will be different.)

We can eliminate the volume, using the ideal-gas law

$$V = NkT/P$$

where N is the number of molecules in the parcel of air. This gives

$$P^{1-\gamma}T^{\gamma} = P_0^{1-\gamma}T_0^{\gamma}$$

For convenience, we will call this constant quantity C. Solving for T gives

$$T^\gamma = CP^{\gamma-1}$$

To get dT/dz, we differentiate both sides with respect to z, giving

$$\gamma T^{\gamma-1}(dT/dz) = C(\gamma - 1)P^{\gamma-2}(dP/dz)$$

$$= C(\gamma - 1)P^{\gamma-2}(-\rho g)$$

where we have used the hydrostatic law $dP/dz = -\rho g$. Solving for dT/dz, and substituting for C, we have

$$dT/dz = -(T_0\rho_0/P_0)g(\gamma - 1)/\gamma$$

From the ideal-gas law, $T\rho/P = m/k$, so

$$dT/dz = -(mg/k)(\gamma - 1)/\gamma$$

$$= 9.8 \text{ K/km}$$

As the warmer air rises from the ground, it must be replaced by cooler air from above. This leads to a general flow pattern like that shown in Fig. 23.9. We find *convection cells* with hot air rising alternating with cooler air falling. (We see this same pattern in a boiling pot of water, or in the Sun's atmosphere, causing the granular appearance.)

The situation becomes more complicated when there is water vapor in the air. As the air rises and cools, it may pass through the temperature at which the water condenses into liquid droplets. When this happens, clouds form. Usually, this takes place at a particular altitude where the temperature is right, so the cloud bottoms form a reasonably flat layer, like that in Fig. 23.10. If the convection is weak, thin cloud layers will form. However, if the convection is strong, it can continue the cloud building upward, forming cumulus clouds. The convection can even be strengthened by the energy release when water goes from gas to liquid (the heat of vaporization). In cases of particularly strong convection, thunderheads, like those in Fig. 23.11, are formed. Thunderheads tend to form on summer afternoons, when the solar heating of the ground creates very large temperature gradients, driving a very strong convection.

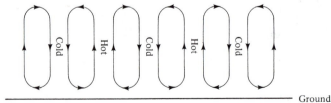

Figure 23.9 Convection cells. This motion carries hot air up and cold air down.

Figure 23.10   Cloud layers result from convection. The clouds layers are at the tops of convection cells.

Figure 23.11   Thunderheads form under conditions producing very strong convection.

23.3.4 General Circulation

Just as local air flows are in response to temperature differences which cause pressure differences, global air flows are subject to the same processes. The general tendency is for air to be heated at the equator, rise, and then flow toward the poles, where it cools, falls, and returns to the equator.

This simple pattern is disturbed by the effects of the Earth's rotation. Since the rotating Earth is an accelerating reference frame, we observe pseudo-forces. One of these is a centrifugal force. It doesn't play an important role. However, another, the *Coriolis force,* does.

The origin of this pseudoforce is shown in Fig. 23.12. First we look at the flow from the pole to the equator, as viewed from above the pole. The horizontal component of the velocity for air leaving the pole is less than that for air at the equator. The air from the pole will therefore lag behind the air at the equator. For air flowing from the equator to the pole, the opposite situation applies, and the air from the equator gets ahead of that at the pole.

In general, air flows from high- to low-pressure areas in the atmosphere (23.13). Air flowing from a high to a low, going toward the equator, will lag behind the low. Air flowing in the other direction will get ahead. This results in a counterclockwise circulation around lows and clockwise circulation around highs in the northern hemisphere. The opposite situation pre-

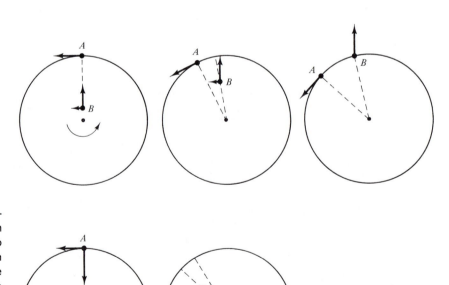

Figure 23.12 The coriolis pseudoforce. In the upper series, we look at an object thrown from a high latitude to the equator. *A* is the point aimed at on the equator, and *B* is the point where the object is launched. However, the horizontal motion at *B* is less than that at *A*, so the object reaches the equator behind *A*. We look at an object thrown from *A* to *B* in the lower series. The object has a greater horizontal motion than does *B* and, therefore, gets ahead of *B*.

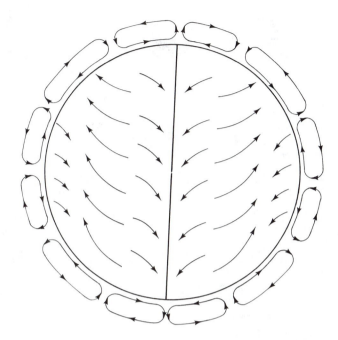

Figure 23.13 Circulation patterns on the Earth. The closed loops represent the major cells within the atmosphere, while the open arrows represent the general wind directions at the surface.

vails in the southern hemisphere. This pattern is particularly evident in the circulation around hurricanes, which have very low pressure centers (Fig. 23.14).

In the past few years, major improvements have been made in our understanding of the large-scale atmospheric circulation. Computers are used to model the Earth's atmosphere in considerable detail. The equations governing the fluid flow and energy balance are solved, taking such details as terrain into account. For such a model to be successful in predicting the movement of weather systems over periods of a few days, or even possibly weeks, it is important to have detailed information on conditions all over the Earth (Fig. 23.15). Data are continuously collected from a variety of ground stations, from ships and weather buoys at sea, and from balloons sent up from various places on a regular schedule. Satellite observations are also important. The observations over the ocean are particularly important, since much of the energy that passes into the atmosphere is stored in the water.

As the models are refined and the data gathering is improved, the predictions get better and better. We have now even reached the point where we can use the same approach to trying to understand the global atmospheric properties of other planets, as we will see in the next few chapters.

Figure 23.14 In this photograph, the coriolis force results in the spiraling motion around the large hurricane (Allen).

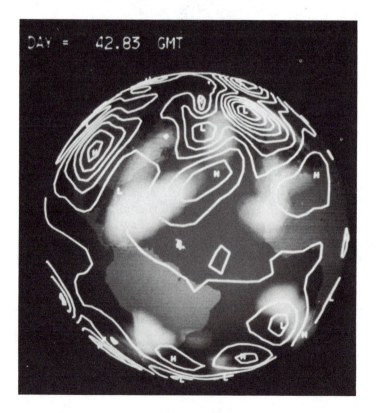

Figure 23.15 Computer-generated global weather simulation.

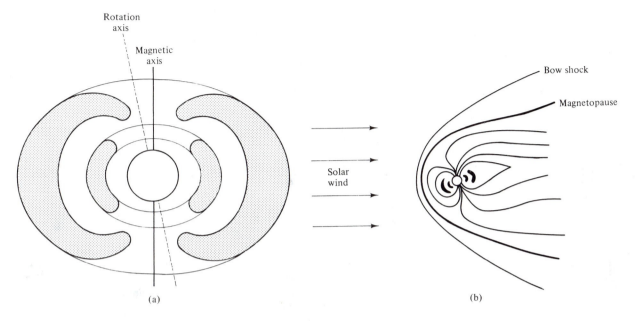

Figure 23.16   Van Allen radiation belts and the Earth's magnetosphere.
(a) The radiation belts are groups of trapped particles, concentrated into
two bands, each with the shape of a doughnut that has been hollowed along
its inner rim. (b) The Earth's magnetic field deflects the charged particles of
the solar wind. The protected region is the magnetosphere, and the boundary is called the *magnetopause*. Just outside the magnetopause is a shock
wave that looks like the bow wave made when a ship plows through water.

23.4   THE MAGNETOSPHERE

One of the early surprises of our unmanned space program was the discovery
of belts of charged particles high above the Earth's surface. These particles
are a source of low-frequency synchrotron radiation. This radiation over-
whelmed the detectors on early spacecraft, making it appear as if no particles
were present. However, James van Allen correctly interpreted the strange
result as indicating the presence of large numbers of charged particles. We
call these belts of charged particles the *van Allen radiation belts* (Fig. 23.16).

These particles are trapped in the Earth's magnetic field. We call this
part of the atmosphere the *magnetosphere*. When we discussed solar activity
(Chapter 8), we said that charged particles will follow helical paths around
magnetic field lines. This is because the force on the particles is perpendicular
to both the field lines and the velocity of the particle. This means that there
can be no force along the field lines. The component of the velocity along
the field stays fixed, as the particles execute circular motion perpendicular to
the field lines.

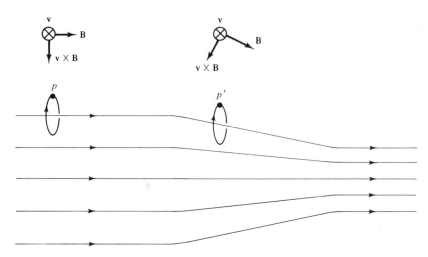

Figure 23.17 Magnetic mirror. Particles spiral magnetic field lines. For simplicity, we look only at positively charged particles, though the same argument can be carried out for negative particles. In the region where the magnetic field lines are parallel, on the left side, the magnetic force on a particle at *P* is downward. This just keeps the particle orbiting. However, when the field lines get closer, meaning that the field is getting stronger, the magnetic force of the particle at *P'* has a slightly rearward component. This is the force that slows and reverses the component of the motion parallel to the field lines.

The situation is different if the particles are moving from a region of a weaker magnetic field into one of a stronger magnetic field. This is illustrated in Fig. 23.17. The stronger field is represented by the field lines getting closer together. We divide the magnetic field into two components, $\mathbf{B}_{\parallel}$, parallel to the z-axis, along which the particles are drifting, and $\mathbf{B}_{\perp}$, perpendicular to this axis. If the field were constant, $\mathbf{B}_{\perp}$, would be zero.

We look at the force on an electron moving in the indicated spiral path. Note that the sense of the spiral is determined by the charge of the particle. $\mathbf{B}_{\parallel}$ doesn't change the z-component of the electron's speed. However, $\mathbf{v} \times \mathbf{B}_{\perp}$ points to the right. Therefore, the force on the electron points to the left. This is opposite to the direction of drift of the electron. The drift is therefore slowed down, and the motion reversed. It is as if the electron bounced off a mirror. We call it a *magnetic mirror*. Note that the same thing happens for a positively charged particle (see Problem 23.16). The charged particles therefore spiral back and forth, trapped in the Earth's magnetic field.

The particles come from interplanetary space, mostly from the solar wind. In fact, the Earth's magnetic field shields the Earth from most of the solar wind by trapping the particles. Because of the dipole nature of the Earth's magnetic field, these charged particles get closer to the Earth's surface near the magnetic poles. When they get into the denser parts of the atmosphere, the energy causes the air to glow, causing the aurorae (Fig. 23.18).

Figure 23.18 When solar wind particles penetrate the Earth's atmosphere, they cause the glowing aurorae.

23.5 TIDES

A number of phenomena on Earth depend on the fact that the gravitational force exerted by the Moon (or the Sun) on the Earth is slightly different at different parts of the Earth. As we have already discussed (Chapter 11), any effect which depends on the difference between the gravitational forces on opposite sides of an object is called a *tidal effect*.

If we have an object of mass m, a distance r from an object of mass M, the force on the object of mass m is

$$F = GMm/r^2$$

If we move the object, the force changes. The rate of change of the force is

$$dF/dr = -2GMm/r^3$$

The change in the force ΔF in going from r to $r + \Delta r$, is

$$F = (dF/dr)\Delta r$$

$$= (-2GMm/r^3)\Delta r$$

The change in acceleration, Δa, is $\Delta F/m$, or

$$\Delta a = (-2GM/r^3)\Delta r \qquad (23.16)$$

Note that the tidal effects fall off as $1/r^3$, faster than the $1/r^2$ falloff of the gravitational force itself.

EXAMPLE 23.3   Tidal Effects on Earth

Compare the strength of the tidal effects exerted on the Earth by the Sun and the Moon.

Solution

We can express the result as a ratio, using equation (23.16).

$$\left(\frac{\Delta a_{SE}}{\Delta a_{ME}}\right) = \left(\frac{M_\odot}{M_M}\right)\left(\frac{r_{ME}}{r_{SE}}\right)^3$$

$$= \left(\frac{2 \times 10^{33} \text{ g}}{7.3 \times 10^{25} \text{ g}}\right)\left(\frac{385 \times 10^3 \text{ km}}{150 \times 10^6 \text{ km}}\right)^3$$

$$= 0.46$$

Even though the Sun exerts a greater gravitational force on the Earth than the Moon does, the closeness of the Moon makes its tidal effects greater.

These tidal effects of the Sun and Moon are responsible for the ocean tides on the Earth (Fig. 23.19). To see what happens, let's only consider the effects of the Moon. We look at three points: (1) the point closest to the Moon; (2) the center of the Earth; and (3) the point farthest from the Moon. We see that $a_1 > a_2 > a_3$, as viewed from the rest frame of the Moon. However, as viewed from the Earth, all accelerations must be relative to a_2. In this frame, the acceleration of (1) is $a_1 - a_2$ toward the Moon; that of (2) is zero; and that of (3) is $a_3 - a_2$, and is therefore directed away from the Moon. This means that there will be a high tide on the side nearest the Moon and also on the side farthest from the Moon. (We can think of the tide on the near side as the water on that side being pulled away from the Earth, and we can think of the tide on the far side as the Earth being pulled away from the water.

The Sun has a similar effect, but only half as great in size. When the Moon and Sun pull along the same line, the difference between high and low

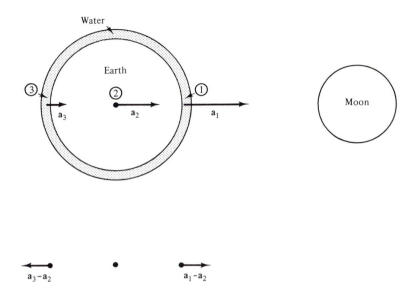

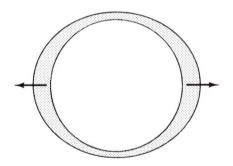

Figure 23.19 Tides on the Earth. Above, we look at the different accelerations caused by the gravitational attraction of the Moon. In the center, we express these accelerations relative to that of the center of the Earth. The result is that the water is pulled away from the Earth on the side facing the Moon, and the Earth is pulled away from the water on the far side.

tides is greatest. When they pull at right angles to each other, the difference between high and low tides is the least.

The height of the water tide is actually not found to be as large as one would calculate. This is because the Earth is not completely covered by water, and is also not solid, It is therefore distorted by these tidal forces. This constant reshaping of the Earth helps to heat the Earth's interior and also dissipates the Moon's orbital energy. The Earth doesn't respond instantaneously to these tidal effects. Its rotation causes the bulge on the near side facing the Moon to get ahead of the Earth–Moon line. This means that the Earth exerts a torque on the Moon, increasing its orbital speed, making the Moon move farther from the Earth. The torque that the Moon exerts on the Earth causes its rotation to slow, keeping the total angular momentum of the Earth–Moon system fixed. Similarly, the Earth has exerted torques on a

Figure 23.20 Precession caused by the Sun. We treat the Earth as a sphere with an extra bulge around the equator, and look at the torque exerted by the Sun on that bulge. We see that F_1 is greater in magnitude than F_2, so there is a net torque on the Earth, whose direction, given by the right-hand rule, is out of the page. This torque would rotate the Earth, but the Earth already has some angular momentum, so the change in angular momentum must be added to the existing angular momentum, as shown in the upper right. The result is a new angular momentum whose magnitude is the same as before, but whose direction is different. Below, we show the torques at different points in the Earth's orbit. The torque can sometimes be zero, but when it is not zero, it always points in the same direction.

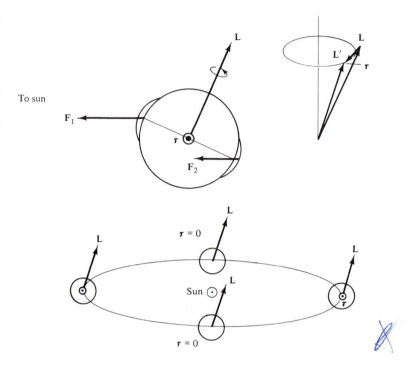

distorted Moon, slowing its rotation, producing a rotation period equal to the Moon's orbital period.

The tidal effects of the Sun and Moon on the Earth produce a torque that results in a continuous change in the direction of the axis of rotation. This change is called *precession*, and is illustrated in Fig. 23.20. To see how the effect comes about, we just consider the effect of the Sun on the Earth. The Earth is not a perfect sphere, but has a larger diameter across the equator than between the poles. This *oblate*, or flattened, appearance is due to the Earth's rotation. We can think of the nonrigid Earth as deforming under the effects of the centrifugal force. Since the Earth is not spherical, the Sun can exert a torque on the Earth.

To see the direction of this torque, we idealize the Earth as a sphere with an extra band around the equator. The side of the band closest to the Sun feels a slightly larger force than the side farther from the Sun. When the Earth is at position 1 (first day of winter in the northern hemisphere), the greater force on the near side of the band produces a torque that points out of the page. Three months later, at position 2, the force on the near and far sides of the band both pass through the center of the Earth, so the torque is zero. Three months later, at position 3, the torque is again outward. Three month's later, at position 4, it is again zero. The torque is sometimes zero, but it is never in the opposite direction. Therefore, averaged over the year, there is a nonzero torque. The torque goes through two full cycles in its value over the course of the year. The effect of the Moon is essentially the same, with the torque going through a full cycle in its value once each lunar orbit.

We now see the effects of this average torque $\langle\boldsymbol{\tau}\rangle$. If the Earth's angular momentum is originally $\mathbf{L}$, then after a short time dt the new torque $\mathbf{L}'$ is given by

$$\mathbf{L}' = \mathbf{L} + d\mathbf{L}$$

where

$$d\mathbf{L} = \langle\boldsymbol{\tau}\rangle dt$$

Note that $d\mathbf{L}$ is much smaller in magnitude than $\mathbf{L}$, and is perpendicular to $\mathbf{L}$, so it can only change the direction of $\mathbf{L}$, not the magnitude. To change the direction of $\mathbf{L}$, the direction of the Earth's axis must change. The axis sweeps out a circle on the sky, returning to its current position in 26,000 years. Since the torque on the Earth is not constant in time, the precession doesn't occur smoothly. The small variations in the rate of precession are called *nutation*.

Precession has important observational consequences. Since the Earth's axis is changing its orientation, the Earth's equatorial plane is doing the same. Therefore, the point of intersection of the celestial equator and the ecliptic, the vernal equinox, changes its position on the sky. It has recently moved into the constellation Aquarius (the "Age of Aquarius"). Remember, for this average precession, the 23.5° tilt doesn't change, just the direction in which the axis points. Since the vernal equinox is the starting point for the astronomical coordinate system of right ascension and declination, the positions of all stars appear to change. It may seem like the change should be small, since it takes 26,000 years to go through a full cycle, but this amounts to approximately 50″ per year.

23.6   THE MOON

The Moon provides us with the most spectacular example of a body that has been taken from the realm of remote sensing by traditional astronomical techniques to that of up-close study, including the luxury of bringing samples back for studies in the laboratory (Fig. 23.21).

Even from the Earth we can see a variety of lunar features. The *highlands* are lighter colored. They contain mountains and valleys as well as long canyons, or *rilles*. The *maria*, once thought to be oceans because of their smooth, dark appearance, are more level. We can also see many different types of *craters*. Some have bright rays of ejected material. In some areas, we see several layers of cratering, with younger craters appearing to cross

(a)

Figure 23.21   Views of the Moon. (a) The near side. (b) Apollo 8 photograph of Galenius crater, 52 miles across. (c) A relatively new crater, the Flansteed ring, 24 miles across.

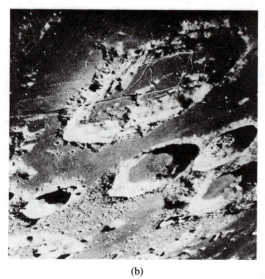

(b)

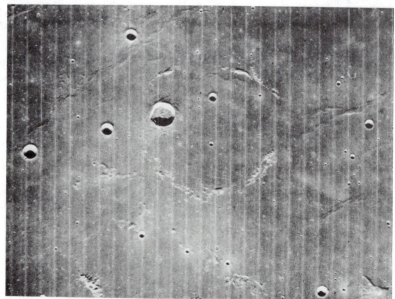

(c)

the walls of older craters. One way of determining the relative age of surface features on the Moon is to look at the relative density of craters.

As we have said, lunar exploration has taken the Moon to the realm of the geologist. (In fact, the last manned lunar landing included a geologist, Harrison ''Jack'' Schmitt.) There were six manned lunar landings from 1969 through 1972, Apollos 11 through 17, with the exception of Apollo 13, which aborted its mission after an explosion en route. These missions returned 382 kg of lunar rock. A variety of regions were visited, as shown in Fig. 23.21, and in the later flights, a vehicle allowed a study of an extended region around each landing site. In addition to bringing back lunar samples, the astronauts left a variety of monitoring equipment on the Moon. This equipment included seismometers to detect moonquakes and meteor impacts and x-ray and radio-activity detectors. There were also three unmanned Russian Luna flights (16, 20, and 24), which, from 1970 through 1976, returned 310 g of rock from different sites.

23.6.1 The Lunar Surface

The lunar soil is a layer ranging from 1 to 20 m in depth. It is called *regolith*, a combination of powder and broken rocky rubble. It is the result of meteoritic bombardment. We are talking about a large number of small meteorites that would have burned up in the Earth's atmosphere, so we see no corresponding material on the Earth. Larger meteor impacts have spread material around and mixed the material in a given region. The soil contains no water. It does have a high proportion of *refractory*, or high melting point, materials, such as calcium, aluminum, and titanium.

The rocks are all basalts, meaning that they resulted from the cooling of lava. (Remember, the Earth has igneous rocks, but also has sedimentary rocks, such as limestone and shale.) The rocks contain some silicates (py-roxene, olivine, feldspar, tridymite, ilmenite) as well as oxides of iron and titanium. It also contains three new minerals—named tranquilityite, armal-colite (after the three Apollo 11 astronauts, Armstrong, Aldrin, and Collins), and pyrroxferroite.

There are some differences between the maria and the highlands. The maria are basalts, like the lavas from the Hawaiian volcanos. The highlands rocks contain anorthosites, gabbro, and norrite, which were probably cooled slowly deep within the Moon. They also contain more feldspar than do the maria rocks. There are also variations in element composition. The maria rocks have more titanium, magnesium, and iron, while the highlands rocks are rich in calcium and aluminum. The highlands rocks are also older. The maria rocks have ages in the range 3.1–3.8 billion years. These youngest rocks on the Moon are as old as the oldest rocks found on Earth.) The highlands rocks have ages ranging from 3.9 to 4.48 billion years. This latter

Figure 23.22   Far side of the Moon.

number occurs in several places. There are also differences between the near and far sides of the Moon. The far side has more craters and fewer maria (Fig. 23.22). It also has a higher average altitude above the Moon's average radius.

Based on the properties of the rocks, and their distribution, it has been possible to deduce a scenario for the evolution of the lunar surface. Some of the parts of this scenario had been speculated on before the lunar explorations, but the dating of the lunar rocks gives exact times since certain important events. We think that the Moon formed 4.6 billion years ago. (Some small green rock fragments in the Apollo 17 samples are that old.) The surface underwent extensive melting and chemical separation for the next 200 million (0.2 billion) years. This melting resulted from the heat generated in extensive meteoritic bombardment, as well as some radioactivity. This bombardment created the large basins, hundreds of kilometers across. Prominent among these basins are Imbrium and Orientale. The bombardment eased approximately four billion years ago. Residual stored heat and some radioactivity resulted in melting down to 200 km below the surface.

From 3.1 to 3.8 billion years ago (the ages of the rocks in the maria), lava rose and filled the basins, creating the maria. Since then the Moon has been quiet, except for occasional meteor impact. The footprints left on the Moon will remain sharp for millions of years.

23.6.2 The Lunar Interior

Much of what we know about the lunar interior is deduced from lunar seismology. There are still small quakes, with about one-billionth the energy of a typical quake on Earth. Some of the quakes are the results of meteor impacts. By watching the propagation of seismic disturbances around and through the Moon, we can draw some conclusions about the interior. We can also learn about the lunar interior from the heat flow rate through the surface. According to instruments in the Apollo landers, this flow is about 2 microwatts/cm², or about one-third that on Earth.

The basic lunar structure is depicted in Fig. 23.23. The core has a radius of some 700 km. At this point we are unsure of its composition or detailed structure. The absence of a magnetic field suggests that it is probably not molten. (However, some rocks show evidence for a magnetic field in the past.) Above the core is a partly molten zone 400 km thick. The deep quakes are thought to originate in this zone. Above this is a mantle, and above the mantle is a thin crust. The crust is variable in thickness, ranging from 0 to 65 km thick.

Anomalies in the orbits of early lunar orbiters indicated that the mass of the Moon is not spherically symmetrically distributed. Mass concentrations, or *mascons*, have been detected beneath the circular maria, such as Imbrium, Serenitatis, Crisium, and Nectaris. We think that these resulted from volcanic lava filling the basins. The fact that these concentrations have survived the 3 billion years since the basin filling suggests that the lunar interior is quite rigid. If it were not rigid, the mascons would have sunk a long time ago.

23.6.3 Lunar Origin

Prior to the Apollo landings, three basic theories of the origin of the Moon had been proposed:

Fission. In this theory, the Moon broke away from the earth. This theory was particularly popular when it was realized that the Moon is the right size to have left behind the Pacific Ocean basin when it broke away. However, we now know that the plate tectonics on the Earth mean that the Pacific Ocean has been in its current arrangement for a relatively short time, and that its current appearance is unrelated to its appearance some 4 billion years ago. The fission theory is supported by the fact that, when we allow for the effects of erosion on Earth, the surfaces of the Earth and Moon have some similarities. However, there are also some major composition differences.

Capture. In this theory, the Moon passed close to the Earth, and was captured into its current orbit. One problem with this theory is that capture would more likely result in a much more eccentric orbit. It is also unlikely that the orbit would lie so close to the plane of the ecliptic. Theoreticians have not been able to work out a reasonable dynamical scenario.

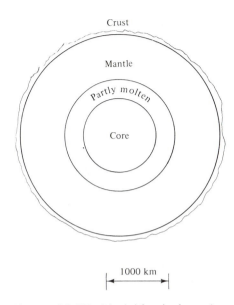

Figure 23.23 Model for the lunar interior.

Condensation. In this theory, the Moon formed as part of that section of the solar nebula that formed the Earth. Perhaps it gathered after the Earth was well under way in formation. Possibly, the Earth simply formed as a double planet.

This is one question about the Moon that was not resolved by the Apollo landings, and all three theories are still under consideration.

CHAPTER SUMMARY

In this chapter we looked at the properties of the Earth and Moon. We also looked at the physical processes responsible for their current properties.

We began by looking at the formation of the Earth from material collected from the original solar nebula. We looked at sources of heating, gravitational potential energy released in the collapse of the planet, and radioactivity. We also saw how radioactive materials can be used do date surface features. We saw how the heat sources below the Earth's surface lead to plate tectonics, or continental drift.

We saw how a balance of energy in and out determines the equilibrium temperature of a planet.

We saw how some of the basic ideas developed for discussing stellar structure, hydrostatic equilibrium, and energy transport, can be used to describe a planet's atmosphere. We saw how the pressure must vary so that each layer of the atmosphere is supported by the pressure difference between the air just below and above it. We also discussed how molecules with greater than average energy can escape from an atmosphere.

We also saw how energy transport is related to the temperature distribution in the atmosphere. In particular, we saw how the Sun heats the ground, and the ground then heats the lower atmosphere. The trapping of infrared radiation near the ground leads to a greenhouse effect. We also saw how the magnitude of the temperature gradient, relative to the adiabatic gradient, determines whether convection will take place.

We saw how the Earth's magnetic field shields us from charged particles. It also traps those particles, producing radiation belts.

We also looked at tidal effects on the Earth. Tidal effects result from the change in the gravitational force across an object. These tidal effects lead to our ocean tides and to the precession of the Earth.

Finally, we discussed the Moon. We saw how the Apollo missions greatly added to our knowledge of the Moon, through the samples brought back and the instruments left behind. The important feature of the lunar surface is that it has not been eroded in the past 4.4 billion years. It therefore preserves a record of the early solar system.

QUESTIONS AND PROBLEMS

23.1   Show that if the number of nuclei in a sample is given by equation (23.1), then, following any arbitrary starting time, the half-life is the time for the number to be reduced to half of its value at that arbitrary time.

23.2   If the number of nuclei in a sample is given by equation (23.1), show that the rate of radioactive decay (number of decays per second) is proportional to $N(t)$.

23.3   Show that $\tau_{1/2} = 0.693\tau_e$.

23.4   (a) How would you measure the half-life of an isotope whose half-life is long compared to your lifetime? (b) How would you determine the number of nuclei of that isotope in a particular sample (assuming you know the half-life)?

23.5   How would the equilibrium temperature of a planet be modified if the planet always kept the same side toward the Sun?

23.6 Express the equilibrium temperature of a planet as a function of that of the Earth and the distance of the planet in astronomical units. Use your equation to construct a table of equilibrium temperatures for the nine planets.

23.7 If we are interested in getting distances through the atmosphere to within 1%, down to what angle from the zenith can we approximate the Earth's atmosphere as a plane parallel layer?

23.8 Assuming that the pressure in the Earth's atmosphere varies according to equation (23.13), (a) what is the pressure, relative to that at sea level, at an altitude of 2 km? (b) How high do you have to go for the pressure to drop to half its value at sea level?

23.9 If the pressure in the Earth's atmosphere varies according to equation (23.13), show that the pressure at sea level is equal to the weight of the air in a unit area column from the bottom of the atmosphere to the top.

23.10 If the pressure in the Earth's atmosphere varies according to equation (23.13), find an expression for the column density of air $N(z)$ as a function of z. The column density is the number of molecules in a cylinder of unit area, from height z to the top of the atmosphere.

23.11 Why does it take longer to boil an egg at high altitude than at sea level?

23.12 If after N steps of length L in a random walk in one dimension, the distance from the origin is $N^{1/2}L$, show that this distance is $(N/2)^{1/2}L$ and $(N/3)^{1/2}L$ for two-dimensional and three-dimensional random walks, respectively.

23.13 Find the pressure distribution in an atmosphere where the temperature distribution is

$$T(z) = T_0 - bz$$

23.14 Using the fact that precession goes through a full cycle in 26,000 years, calculate the average torque on the Earth. (*Hint*: It is necessary to calculate the angular momentum of the Earth.)

23.15 When we applied hydrostatic equilibrium to the Earth's atmosphere, we found that the pressure falls exponentially. However, when we applied it to the ocean, the pressure varied only linearly with depth. How can you account for this difference?

23.16 Show that positively and negatively charged particles are reflected in the same way by a magnetic mirror. (*Hint*: Remember that oppositely charged particles spiral magnetic field lines in opposite directions.)

24

The Inner Planets

The solar system naturally divides into two groups of planets, separated by the asteroid belt. The four inner planets have many things in common with the Earth, while the next four planets present worlds of an entirely different type. (Pluto is an additional enigma.) In this chapter we look at Mercury, Venus, and Mars, comparing their properties with each other, and with the Earth.

24.1 BASIC FEATURES

24.1.1 Mercury

Mercury is the closest to the Sun, and is not much larger than our Moon. There is an interesting story concerning its rotation period. Since it is so close to the Sun, we never get a really good view of Mercury, and surface features are hard to recognize. By noting the positions of large surface features, it appeared that the rotation period was 88 days, the same as the planet's orbital period. This would have meant that Mercury always kept the same side toward the Sun. Since Mercury is so close to the Sun, it seemed plausible that some tidal effect could keep its rotation period synchronized with its orbital period.

However, the situation was corrected following radar observations. Radio waves were bounced off Mercury and then detected back on Earth. The planet's rotation causes a spread in the Doppler shifts of the reflected waves. From the amount of the Doppler shift, we can tell how fast the planet is rotating. Astronomers were surprised to find that Mercury's rotation period is 59 days, or two-thirds of the 88-day orbital period. This relationship means that when Mercury is favorably placed for observations, it often has the same side toward the Earth, making it seem that it always keeps the same side toward the Sun. This simple ratio is probably no accident. It may result from the varying tidal effects, as Mercury has a rather eccentric orbit.

Our only closeups of Mercury have come from the Mariner 10 spacecraft, which actually made three fly-bys of Mercury. The orbit of Mariner 10 was arranged to bring it back close to Mercury periodically. The closest fly-by was within 300 km, and allowed a very detailed study of surface features.

24.1.2 Venus

Venus has been referred to as our "sister planet." Its size is very close to that of the Earth. It should also encounter somewhat similar solar heating conditions. It actually receives twice as much solar energy as does the Earth. For some time it was thought that Venus might be a good candidate for finding life. At the very least, it is a good candidate for testing theories which explain various aspects of the Earth. Study of the surface is hindered by thick clouds.

The planet's rotation period is 243 days. The sense of the rotation is opposite to that of the orbital motion. This is called *retrograde rotation*. If the planets simply condensed out of a rotating nebula, conservation of angular momentum tells us that the planets should all be rotating on their axes in the same sense as they are orbiting. Therefore, the retrograde rotation needs some specific explanation. It has been suggested that the drag of a heavy atmosphere and tidal effects of the Sun could be responsible.

Exploration of Venus by spacecraft has been quite extensive. Especially notable are the Soviet Venera landers, which also sampled the atmosphere on the way down to the surface. In addition, the Pioneer Venus probes (in 1978) provided a wealth of information. One spacecraft went into orbit and used radar to map the surface features. The other sent probes through the atmosphere.

24.1.3 Mars

Mars has long held our fascination. Except during dust storms, there is no thick cloud cover, so we have had a comparatively good view of the surface. Particularly intriguing were color changes with season that suggested some form of vegatation. The axis of Mars is tilted by 25°, so we would expect the seasons to be like those on Earth. The white polar caps also change size with the season, raising the possibility that they hold water ice. In addition, the rotation period is very close to that of the Earth.

Mars has also been the subject of extensive exploration. The Mariner spacecraft flew by, sending back photos of a barren, crater-marked surface. Two Viking spacecraft left orbiters around Mars and also sent landers to the surface. These landers sampled the soil, and even carried out a search for life. Some chemical effects were found that mimic some simple aspects of certain living things, but no evidence for life was found.

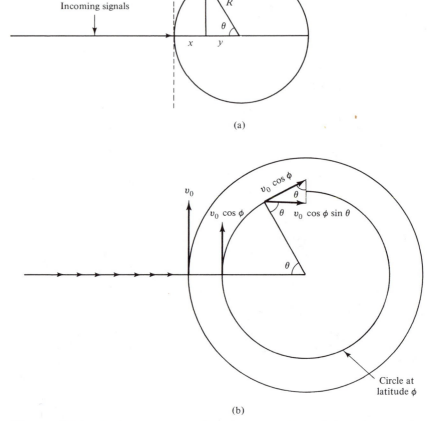

(a)

(b)

Figure 24.1   Radar mapping of a planet. (a) Time delay. (b) Doppler shift.

24.1.4   Radar Mapping of Planets

Since we gain so much information from radar mapping of planetary surfaces, it is useful to go over the basic ideas behind it. A raised surface feature may show up as a stronger than average radio echo. However, the problem is to determine where on the surface this feature is. There are two effects that help us locate a feature. One is the time delay for a signal returning to Earth, and the other is the Doppler shift. These are both illustrated in Fig. 24.1.

We first look at the time delay. Since the surface is round, different parts of the surface are different distances from our radio telescope. These different distances mean that the light (or radio wave) travel times will be different for waves bouncing off different parts of the surface. We can express time delays relative to that of a wave bouncing off the closest point. According to the figure, the extra distance that the signal has to travel is $2x$, so the time delay is

$$\Delta t = 2x/c$$

We see that

$$x = R - y$$

$$= R - R \cos \theta$$

$$= R(1 - \cos \theta)$$

This means that the time delay is

$$\Delta t = (2R/c)(1 - \cos \theta) \tag{24.1}$$

If we measure the time delay, and we know the planet's radius R, then we can solve equation (24.1) for θ. This does not give us the point uniquely. There is a whole ring of points that all have the same θ. As viewed from the radio telescope, lines of constant time delay appear as concentric rings about the closest point.

We now look at the Doppler shift. If a point on the equator moves with speed v_0, then a point a latitude φ moves with speed (see Problem 24.13),

$$v(\varphi) = v_0 \cos \varphi$$

New we view this point from above the pole, assuming that the line from the point to the pole makes an angle θ with the line from the pole to the telescope. The Doppler shift depends on the radial velocity $v_r(\theta, \varphi)$, which is given by

$$v_r(\theta, \varphi) = v(\varphi) \sin \theta$$

$$= v_0 \cos \varphi \sin \theta \tag{24.2}$$

As seen from the telescope, lines of constant Doppler shift form concentric rings about the point on the equator that is just appearing from the back side, and the point on the equator that is just about to disappear.

By combining time delay and Doppler shift data, we limit the source of the echo to two possible points. These are the two points where the time delay circle intersects the Doppler shift circle. The remaining ambiguity in the location of a feature can be removed by observing at a different time when that feature's location and Doppler shift have changed.

Figure 24.2 (a) Mosaic of Mercury, taken by Mariner 10 during its approach to the planet. (b) Bright rays from a crater 60 miles across.

(a)

(b)

24.2 SURFACES

24.2.1 Mercury

The Mariner 10 spacecraft has mapped 50% of the planet's surface (Fig. 24.2), with the third and closest encounter showing features 50 m across. An obvious feature is the extensive cratering. The surface is reminiscent of the Moon, but there are also significant differences. The craters are flatter than on the Moon. The larger surface gravity on Mercury means that there is more slumping in the sides of the craters. The surface may also be more plastic than that of the Moon. There is evidence for some erosion, most likely by micrometeorites.

There are a number of other unusual features. For example, there are long fractures, called *scarps*. These indicate some large-scale compression of the surface. There is also a series of irregular features called *weird terrain*, which was probably caused by shocks from impacts of large objects.

Infrared observations (also from Mariner 10) indicate that the surface has a layer of fine dust, several centimeters thick. These observations also indicate surface temperatures ranging from about 700 K on the day side to 100 K on the night side. This large temperature difference means that there is very little heat flow, via soil or atmosphere, from the warm side to the cool side. The extreme temperature changes from day to night can introduce stresses that cause weakening of the surface features.

Figure 24.3 Venus.

24.2.2 Venus

The total cloud cover on Venus (Fig. 24.3) means that we must rely on radar mapping to see the large-scale surface features. Maps have now been made of 93% of the surface, and are shown in Fig. 24.4. Original maps were made from ground-based telescopes, particularly the large dish at Aricebo. The more recent data are from orbiting radar.

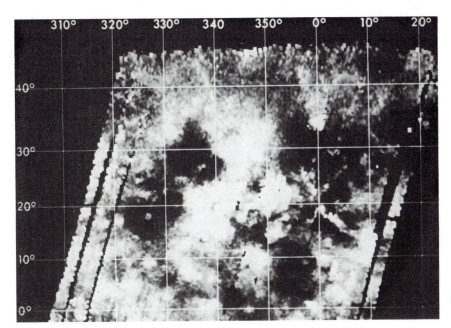

Figure 24.4 Radar maps of lowlands on Venus.

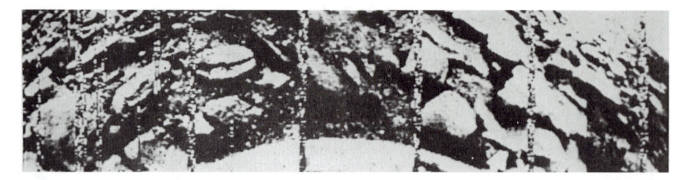

Figure 24.5 Surface of Venus from Russian Venera lander.

Venus is relatively flat. Approximately 60% of the surface lies within 500 m of the most common radius, and only 5% of the surface is outside 2 km of this radius. However, the total range of elevation, about 13 km, is comparable to that on Earth. About 20% of the planet's surface is covered by lowland plains, and about 10% by true highlands. The remaining 70% is described as rolling uplands. These uplands have a variety of features, some of which may be large impact craters. There is evidence for volcanoes, some of which may still be active.

The maps show two large "continents." By using the term "continent" we do not mean to imply that there is evidence of plate tectonics on Venus. One continent is called *Terra Ishta*. It is in the northern hemisphere. It is approximately the size of the United States, and is several kilometers above the mean radius of the planet. It has a high mountain, *Maxwell Montes*, which is 12 km above the mean radius (as compared with 9 km for Mt. Everest on earth). The western part is a plateau some 2,500 km across. The other continent, *Aphrodite Terra*, is twice as large as Terra Ishta, and has a rougher terrain. It has a large highland region with deep (3 km) depressions, hundreds of kilometers wide and 1000 km long.

Photographs from the surface, such as in Fig. 24.5, show angular rocks. This was surprising, since it was expected by some that the thick atmosphere would lead to considerable erosion, smoothing the rocks. However, a very low wind speed has been found on the surface, so the erosion is not as great as originally thought. The soil is basalt, providing additional evidence for vulcanism. The surface temperature is a roughly constant 750 K.

24.2.3 Mars

Even before Viking, many of the early myths about Mars had been dispelled. Mariner 4 showed a cratered surface. Mariners 6 and 7 showed some signs of erosion. Mariner 9 arrived during a planetwide dust storm. We now know that it is the dust storms that we have been interpreting as seasonal changes in vegetation. When the dust cleared, the Mariner 9 cameras showed a variety of features, including volcanoes, canyons, craters, terraced areas, and channels (Figs. 24.6, 24.7).

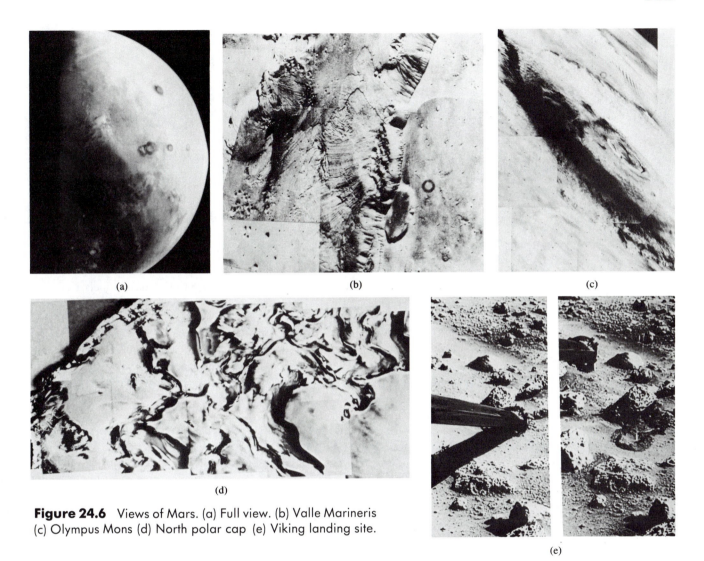

Figure 24.6 Views of Mars. (a) Full view. (b) Valle Marineris (c) Olympus Mons (d) North polar cap (e) Viking landing site.

The largest volcano, *Olympus Mons*, rises 25 km and is some 600 km across at the base. This means that it has the very shallow slopes typical of shield volcanos on Earth. (An example of such volcanos is the Hawaiian chain.) However, Olympus Mons is much higher than Mauna Kea, an impressive statistic in view of the fact that the Earth is larger than Mars. We interpret this as meaning that there are no plate tectonics on Mars. On Earth, the moving plates keep the sites of volcanic eruptions moving. New lava goes into making new mountains, not making old mountains larger. Therefore, we see a chain in Hawaii, but only one large mountain on Mars.

The channels are interesting, because they may have held water long ago. If they did, then it must have been a long time ago, since the channels

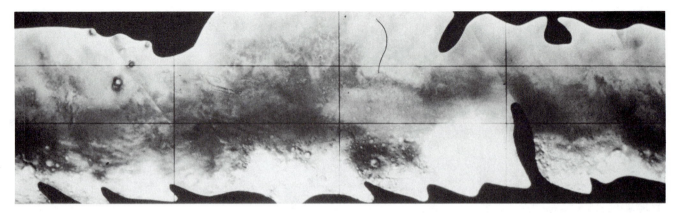

Figure 24.7   Mosaic of Mars.

are cratered. This means that water has not eroded the channels since the heavy cratering in the solar system, possibly some four billion years ago.

Mars shows a hemispheric asymmetry. The southern hemisphere has more craters, and it is 1 to 3 km above the mean radius of the planet. It also seems to have an old part, with many craters, and a less old part. The northern hemisphere has volcanic plains around large volcanos.

An interesting large feature is the *Tharsis ridge*, which is a large bulge. It is the largest area of volcanic activity, and has many young craters. It also has a number of rilles and fractures. A large equatorial canyon, *Valles Marineris*, extends away from Tharsis. There is evidence for wall collapse and channel formation. The whole ridge may be uplifted crust.

The seasonal changes in the polar caps have also been of considerable interest. It has long been speculated that water is stored there in the winter and released in the summer. The release is by sublimation, a phase change directly from the solid to the gas phase. However, the caps never sublime totally, with some part remaining through the summer. It now appears that the part that never sublimes is composed of water. The growth of the caps in the winter is due to the freezing of CO_2. (On Earth we call frozen CO_2 dry ice; it is used to keep things cold because it has a lower freezing point than water.)

24.3   INTERIORS

24.3.1   Basic Considerations

Since we cannot directly observe the interior of a planet, we must come up with indirect methods of determining the interior structure. We briefly go over some of the types of evidence that we can use in studying planetary interiors.

The average density of a planet can give us information. For example, consider the simple structure shown in Fig. 24.8. The planet has a core with density ρ_c and radius R_c, while the mantle has density ρ_m, with outer radius R_m, the radius of the planet. In this case, the mass of the planet is

$$M = (4\pi/3)[R_c^3 \rho_c + (R_p^3 - R_c^3)\rho_m]$$

The average density of the planet is its mass, divided by its volume,

$$\bar{\rho} = M_p/[(4\pi/3)R_p^3]$$
$$= \rho_c(R_c/R_p)^3 + \rho_m[1 - (R_c/R_p)^3] \tag{24.3}$$

If we know the material that is likely to make up the mantle and the core, we can estimate ρ_m and ρ_c. The average density is easily determined, so we can find (R_c/R_p).

Additional information can come from the moment of inertia, since it depends on how the mass is distributed. We can measure the moment of inertia I by seeing the effects that perturbing torques have on the planet. The moment of inertia for the case shown in Fig. 24.8 is

$$I = \tfrac{2}{5}(4\pi/3)(R_c^3 \rho_c R_c^2 + R_p^3 \rho_m R_p^2 - R_c^3 \rho_m R_c^2)$$

$$= (8\pi/15)(\rho_c R_c^5 + \rho_m R_p^5 - \rho_m R_c^5)$$

$$= (8\pi/15)R_p^5\{\rho_c(R_c/R_p)^5 + \rho_m[1 - (R_c/R_p)^5]\} \tag{24.4}$$

For planets that we can place seismometers on, we can learn about the interior activity. We can also learn about the propagation of seismic waves through the interior. (For example, one Apollo experiment involved allowing a spent Lunar Module to crash into the Moon, and then measuring the seismic waves with seismometers that had been left behind.) Seismic waves travel at different speeds through different materials. They will also be reflected from boundaries between different materials. Analyzing these waves can tell us about the composition and size of various interior sections.

Additional information comes from magnetic field, which should require a molten iron core, and some rotation, for stirring up the core, creating a dynamo. Heat flow measurements can tell us about how close radioactive material is to the surface. This tells us whether the mantle is well-mixed or differentiated.

We can also carry out theoretical modeling of planetary interiors (Fig. 24.9), just as we do for stellar interiors (Section 9.3). Just as in stars, planetary interiors must be in hydrostatic equilibrium, meaning that

$$dP/dr = -\rho GM(r)/r^2$$

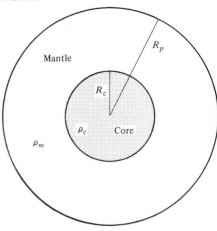

Figure 24.8 Planet with core and mantle.

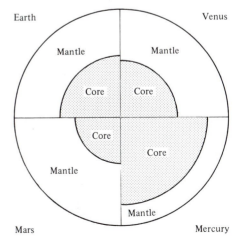

Figure 24.9 Model interiors for the four inner planets. In each case, the sizes are relative to the planet's radius, so the total radius of each planet is the same in this figure.

where $M(r)$ is the mass interior to radius r. If the density can be approximated at being constant, then

$$M(r) = (4\pi/3)\rho r^3 \tag{24.5}$$

so the equation of hydrostatic equilibrium takes the form

$$dP/dr = -(4\pi/3)G\rho^2 r \tag{24.6}$$

We can use equation (24.6) to estimate the central pressure of a planet, by taking the pressure at the surface to be zero, and integrating,

$$\int_0^{P(r)} dP' = -(4\pi/3)G\rho^2 \int_{R_p}^{r} r' \ dr'$$

to get

$$P(r) = (2\pi/3)G\rho^2(R_p^2 - r^2)$$

In comparing this result with more sophisticated models of the Earth, we find that it underestimates the central pressure by a factor of about 2. (Among other things, our assumption of constant density is an oversimplification.)

24.3.2 Mercury

Even though Mercury's surface looks like that of the Moon, Mercury's density is higher. We estimate that an iron core makes up about 60% of Mercury's mass. This is a greater percentage than for either the Earth or Moon. This is because Mercury condensed under higher temperature conditions than did the Earth and Moon. The radius of the core is 1800 km, leaving a 700 km mantle. The core radius is 72% that of the planet.

A surprising result of the Mariner 10 studies was the discovery of a weak magnetic field around Mercury. This field is 1% the strength of the Earth's field. It is surprising since Mercury's low mass would suggest that the core is not hot enough to be molten. Also, the rotation period is so slow that it is hard to see how convection can be started in the core. There is also radioactivity coming from near the surface, suggesting a differentiated interior. This differentiation would have also required some period of melting in the interior.

The surface plains may have resulted from volcanic flooding, suggesting an active past. Mercury also has scarps, unlike the Moon. These scarps may have resulted from a contraction of the surface, something which would have also required a molten history for the planet.

24.3.3 Venus

The interior structure is believed to be very similar to that of the Earth. There are some composition differences. We think that the core of Venus formed later than that of the Earth. The lithosphere is also about twice the thickness of that of the Earth. Venus also has no measurable magnetic field. This may be an effect of the planet's slower rotation.

24.3.4 Mars

The density of Mars, 3 g/cm^3, is low enough to suggest that the core cannot be very large. It is about 1200 km in radius, meaning that it is only about 40% of the planetary radius. The core is probably a combination of iron and iron sulfide. If it is all iron, it is probably even smaller than currently estimated. We have already said that the existence of large volcanos indicates that there are no plate tectonics, also arguing for a cooler interior than the Earth.

24.4 ATMOSPHERES

24.4.1 Mercury

Not much of an atmosphere was expected on Mercury (Fig. 24.10), and only a small amount of gas was found. The surface pressure is 10^{-15} atm. The gas has been detected by ultraviolet spectroscopy. This small amount of gas is 98% helium. This light atom should have escaped long ago. This suggests that it is being replaced continuously. Two possible sources are the solar wind and the alpha decay of uranium or thorium. Remember, alpha particles are ^{4}He nuclei.

Most of the remaining 2% of the gas is hydrogen. This also may be from the solar wind. In addition, small traces of oxygen, carbon, argon, nitrogen and xenon have been found.

24.4.2 Venus

The atmosphere of Venus (Fig. 24.10) is very interesting, especially in comparison with that of Earth. It was first observed in 1761 by the Russian, Lomonosov, who saw the backlighted atmosphere during a transit of Venus across the Sun.

The surface pressure is much higher than on Earth, 90 atm. The atmosphere is 96% CO_2 and 3.5% N_2. The total amount of N_2 on Venus is comparable to that in the Earth's atmosphere (see Problem 24.5). However, on the Earth the nitrogen is the primary constituent. Venus also has very small

Figure 24.10 Atmospheric compositions of Venus, Earth, and Mars.

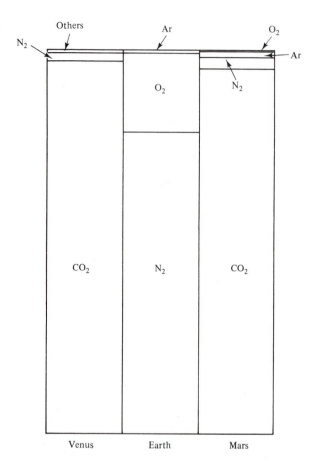

quantities of water, sulfur dioxide, argon, carbon monoxide, neon, hydrogen chloride, and hydrogen fluoride.

The large amount of CO_2 produces a very strong greenhouse effect on Venus. The 750 K is some 400 K higher than the temperature would be without an atmosphere (Fig. 24.11). It is intriguing that two planets could start out so close in conditions and end up so different. The crucial difference between Venus and the Earth seems to be the extra sunlight, making Venus initially somewhat warmer than the Earth. On the Earth, water condensed, while on Venus, it remained as a gas, and escaped. On Earth, the water kept the CO_2 bound up in the rocks, in the form of various carbonates. Without the water on Venus, this couldn't happen, and the CO_2 stayed in the atmosphere. (The amount of CO_2 in the rocks on Earth is comparable to the amount of CO_2 in the atmosphere of Venus.)

Once Venus had more CO_2 in its atmosphere than the Earth did, the greenhouse effect heated the lower atmosphere. This heating released more

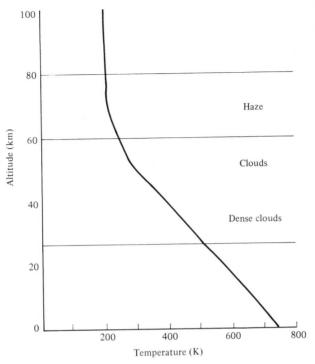

Figure 24.11 Temperature distribution in the atmosphere of Venus. The region below the dense clouds is clear.

CO_2 into the atmosphere, increasing the greenhouse effect. This situation is called a *runaway greenhouse effect*. A small difference in initial conditions ends up with a large difference in final conditions. Above the clouds on Venus (about 50 km above the surface), the greenhouse effect is no longer very strong, and the temperature drops to about 300 K.

The clouds are made up of sulfuric acid (H_2SO_4). (There is a lot of sulfur in the atmosphere, because there is no water to remove it.) The haze above these clouds enhances the greenhouse effect. There are three cloud layers between 48 and 80 km above the surface. The sulfuric acid droplets are about 2 μm in size. The presence of some water at the lower altitudes has washed the sulfuric acid away, and the lower atmosphere is clear. Many were surprised by the clarity of the Venera pictures from the surface. There is also some lightning below the clouds.

There is a general westward circulation of the winds (Fig. 24.12). The wind speed is a modest 1 m/s near the ground, growing to a substantial 100 m/s near the cloud tops. There is a general pole-to-equator circulation, but it is not broken into cells, as it is on the Earth. This circulation minimizes the temperature difference between the equator and the poles to a few degrees Kelvin.

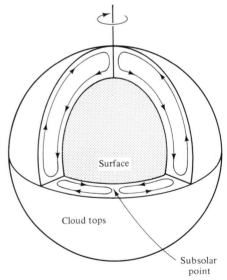

Figure 24.12 Atmospheric circulation on Venus. There is only one cell from the equator to the pole. There is a break in the east–west flow, with the air rising at the point receiving maximum solar energy.

24.4.3 Mars

The surface pressure on Mars (Fig. 24.10) is 7×10^{-3} atm. This is sufficient for the atmosphere to be of some importance. It is 95% CO_2, leading to a small greenhouse effect. The temperatures are raised by about 5 K above the value they would have without an atmosphere (Fig. 24.11). The atmosphere is 2.7% nitrogen and 1.6% argon. There are traces of oxygen, carbon monoxide, water vapor, neon, krypton, xenon, and ozone. There are blue/white clouds made up of carbon dioxide and water. (Most of the water, however, is tied up permanently in the northern polar cap.) We think that Mars had a more plentiful atmosphere in the past, possibly with a pressure of 1 atm. At that time there was probably more water. This earlier atmosphere probably came from an era of volcanism.

Strong winds sweep over the planet, lifting up large amounts of dust into the air. There are a number of small dust storms and a few global storms per Martian year. (Some atmospheric scientists have pointed to the dust storms on Mars as providing a test of the theories that say that the aftermath of a nuclear war on Earth would produce enough dust in the atmosphere to cool our planet, the so-called "nuclear winter.")

The density in the atmosphere is so low that the winds are not efficient at energy transport. Therefore, large temperature variations can exist across the planet. For example, in the winter the poles get as cold as 150 K, which is cold enough to freeze CO_2. At its maximum extent, the CO_2 cap extends down to a latitude of about 50°. The similarity between the rotation period of Mars and the Earth produces similarities in the circulation patterns. However, since there are no oceans on Mars, the surface responds more quickly to heating changes. Therefore, the hottest point on Mars is always the point closest to the Sun, the *subsolar point*. This results in a circulation pattern in which one large cell spans the equator. There is also a large day/night temperature difference, a few tens of degrees K.

The Viking landers also carried out a search for microscopic life on Mars. Three different experiments were carried out to try to detect changes in small samples of material as a result of the metabolism of microscopic life. Evidence for some unusual chemical reactions was found, but there was no evidence for living organisms.

24.5 MOONS

No moons have been found for Mercury and Venus. Mars has two moons that are much smaller than the Earth's. Phobos is 22 km across, and orbits Mars in 7 hr 40 min. Demos is 15 km across and orbits in 30 hr. Since one moon orbits faster than Mars's rotation period and the other is slower, from

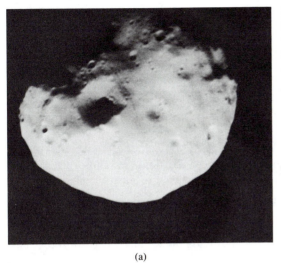

(a)

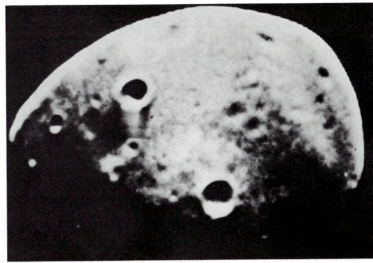

(b)

Figure 24.13 Moons of Mars. (a) Phobos. (b) Deimos.

the surface they appear to move across the Martian sky in opposite directions. Since the orbital period of Demos is close to Mars's rotation period, its synodic period is 137 hr. It takes almost 3 days to go from rising to setting, and another 3 days to reappear.

Each moon has an irregular shape, as shown in Fig. 24.13. They appear dark, like certain asteroids. This has led to the conjecture that these two satellites were asteroids that were captured by Mars.

CHAPTER SUMMARY

In this chapter we looked at the properties of the inner planets of the solar system.

We saw what could be learned from ground-based observations. We saw how some observations could lead to misinterpretations, such as with the rotation period of Mercury. We saw how radar mapping is a useful tool, both from Earth and from spacecraft.

We discussed the wealth of information that has been obtained through fly-bys, orbiters, and landers. Venus and Mars have been visited extensively. In the case of Venus, the spacecraft allowed us to probe the dense cloud layer that hides the surface.

We have seen certain similarities among the inner planets. For example, there is extensive cratering on Mercury, Mars, and our Moon. However, there are also many differences. For example, the Earth is the only planet to have an interior structure that produces plate tectonics. We see very thin atmospheres on Mars and Mercury, while Venus has a much thicker atmosphere than does the Earth. Of course, there is also the large runaway greenhouse effect on Venus, probably initiated by the fact that with a slightly higher temperature than the Earth, the CO_2 was not bound into the rocks.

QUESTIONS AND PROBLEMS

24.1 Compare the tidal effects of the Sun on Venus with the tidal effects of the Sun and Moon on the Earth.

24.2 For radar mapping of a planet, what time and frequency resolution are needed in measuring the time delays and Doppler shifts to provide a resolution of 1 km on the surface?

24.3 Approximate the central pressures of Mercury, Venus, Earth, and Mars, and compare them.

24.4 How far under water do we have to go on Earth to get a pressure of 90 atm?

24.5 Estimate the total amount of nitrogen in the atmosphere of Venus, and compare it with that of Earth.

24.6 How do the surfaces of Mercury and the Moon compare?

24.7 Why is it surprising that Venus has retrograde rotation?

24.8 What is the evidence for the existence of water on Mars?

24.9 How does the interior of Mars compare with that of Earth? What are the clues that allow us to make this comparison?

24.10 Using the mass, radius, and core radius of the Earth (given in Chapter 23), find the ratio of the density of core material to the density of mantle material.

24.11 Estimate the adiabatic temperature gradients near the surfaces of Venus and Mars.

24.12 To what altitude do we have to go in the Venus atmosphere to reach a pressure of 1 atm?

24.13 Show that if a point on the equator of a planet moves with a speed v_0 due to rotation, that a point at latitude φ moves at $v_0 \cos \varphi$.

The Outer Planets

In the outer planets, we find a considerable contrast with the four inner planets. We therefore study them as a group, comparing surfaces, interiors, and atmospheres (Fig. 25.1).

25.1 BASIC INFORMATION

25.1.1 Jupiter

Jupiter is by far the most massive planet in the solar system. It is 318 times as massive as the Earth, and is a respectable 0.1% of the mass of the Sun. (The rest of the planets together only have 129 Earth masses.) Jupiter's density is much lower than that of the inner planets, 1.3 g/cm³, vs. 5.4, 5.3, 5.5, and 3.9 g/cm³ for Mercury, Venus, Earth, and Mars, respectively. This suggests that the composition of Jupiter is basically different from that of the inner planets. This is due, in part, to the larger gravity on Jupiter, 2.54 g at the cloud tops. The larger gravity means that lighter gases have been retained.

The atmosphere is 85% hydrogen and 15% helium, with a variety of trace constituents. This composition is much closer to that of the Sun than it is to the inner planets. The motions in the atmosphere are affected by the planet's rapid rotation. The period is 9.92 hr at the equator. The rotation period is greater at the poles, since the planet rotates differentially. The rapid rotation produces large shear and Coriolis effects. These are manifested in the appearance of bands and spots such as the *Great Red Spot*, which is 14,000 by 30,000 km, and has persisted for hundreds of years.

The energy output of Jupiter is interesting. It seems to radiate over 50% more energy than it receives from the Sun. In addition, it is also a strong nonthermal radio source.

Space exploration has greatly improved our view of Jupiter. Pioneers 10 and 11 (with Pioneer 11 giving a good view of the pole) and Voyager 1 and 2 have provided us with spectacular images as well as a variety of other observations. These closeup observations have also added to the already known extensive moon system. They also revealed a ring around the planet.

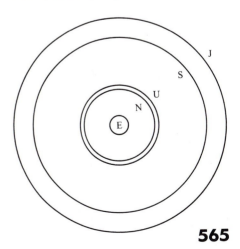

Figure 25.1 Relative sizes of Jupiter, Saturn, Uranus, Neptune and Earth.

Figure 25.2   Saturn, image from Voyager 1, at a distance of 31 million km.

25.1.2 Saturn

Saturn's mass is 0.3 times that of Jupiter, or 95.2 Earth masses. It has the lowest density of the planets, 0.7 g/cm^3. With this low density, its acceleration of gravity at the cloud tops is only slightly greater than that on Earth, 1.07 g. Its composition is similar to that of Jupiter.

Like Jupiter, it rotates rapidly, in 10.7 hr at the equator, and differentially. Through a telescope we can see bands in the cloud structure. However, the bands do not show as much contrast as those of Jupiter.

Of course, Saturn (Fig. 25.2) is best known for its prominent ring. From Earth, we see the rings as three main structures, and can deduce that they are very thin. Like Jupiter, Saturn has an extensive moon system, with 17 moons having been identified to date. Pioneers 10 and 11 as well as Voyagers 1 and 2 have revealed a great complexity in the structures of the rings in addition to surprising views of the moons.

25.1.3 Uranus

Uranus is the closest planet that has not been known since ancient times (Fig. 25.3). It was discovered in 1781 by William Herschel. Its mass is considerably less than that of Jupiter and Saturn, but at 14.6 Earth masses, it is definitely not in the class of the inner planets. Its density, 1.2 g/cm³, is similar to that of Jupiter and Saturn. Though its density is similar to Jupiter's, their different sizes may reflect differences in composition. Its surface gravity is slightly less than that of the Earth, at 0.87 g.

From ground-based telescopes we can tell that Uranus has a high albedo, suggesting a coating of clouds. It is difficult to identify features for the purpose of measuring a rotation period. Values from 12 to 24 hr have been proposed, but the most recently accepted value is 16 hr. The axis is tilted by 98°. The large angle means that the axis is close to the plane of the ecliptic. The value greater than 90° means that the rotation is retrograde. It is in the opposite sense from the planet's orbital motion. Infrared observations of the clouds suggest a temperature of 58 K.

Uranus has five moons. In addition, it has a ring system, discovered accidentally during an occultation of a star by Uranus. We had our first closeup from space when Voyager 2 flew by in January 1986.

Figure 25.3 Uranus, taken from Voyager 2, from a distance of 9.1 million km. The left image shows how Uranus would appear to the eye. The picture on the right is enhanced to show the cloud structure.

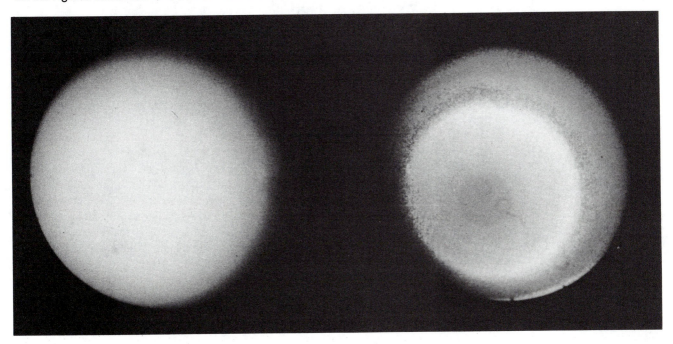

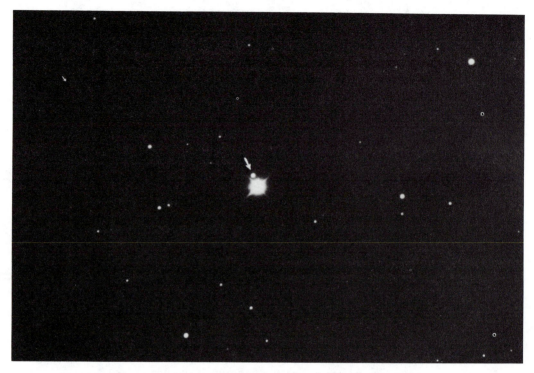

Figure 25.4 Neptune. The arrow points to one of the moons.

25.1.4 Neptune

The discovery of Neptune (Fig. 25.4) makes an interesting story. After following the orbit of Uranus, astronomers found that it did not move exactly in its predicted path. The perturbation of a planet beyond Uranus was suspected. Calculations of the possible location of the planet were carried out independently in 1845 by John C. Adams in England, and Urbain Leverier in France. Adams presented his calculations to the Astronomer Royal, who was not impressed, and did not carry out the easy observations that would have been necessary to test the idea. Leverier had no better luck in France. Finally, based on Leverier's calculations, Johannes Galles, in Germany, carried out the observations, and found Neptune in 1846. We credit both Adams and Leverier with the successful prediction. As an interesting sidelight, recent readings of Galileo's notes indicate that he may have actually observed Neptune, noting its change in position relative to the stars, but not having enough observations to identify it as a new planet.

Neptune's mass is 25.2 times that of the Earth, similar to that of Uranus. From occultations we can tell that its radius is 3.88 Earth radii. Using these numbers, its density is 1.6 g/cm^3. The acceleration of gravity on its surface is 1.14 g. Its rotation period is also hard to determine, with published values ranging from 17 to 26 hr. The currently accepted value is 25.8 hr. We can deduce the presence of an atmosphere by the rate at which starlight dims during an occultation.

Neptune has two confirmed moons, with a third being suspected.

25.2 ATMOSPHERES

25.2.1 Jupiter

Jupiter's atmosphere contains hydrogen and helium in the same proportion as in the Sun's atmosphere. This suggests that Jupiter has its original atmosphere. A number of minor constituents have been identified as well. Ammonia (NH_3) and methane (CH_4) are the most prominent. In addition, C_2H_6 (ethane), C_2H_2 (acetylene), H_2O, PH_3, HCN (hydrogen cyanide), and CO (carbon monoxide) have been identified.

The temperature is 125 K at the cloud tops. It increases downward from there at the rate of about 2 K/km. This is approximately the adiabatic temperature gradient (Fig. 25.5) (see Problem 25.11). Above the cloud tops, the temperature rises with altitude. The emitted radiation is approximately inde-

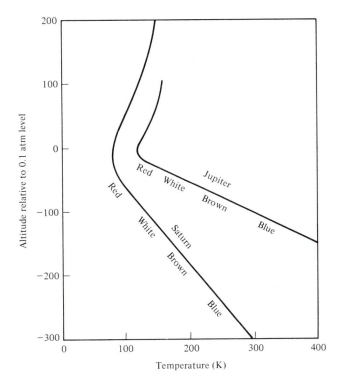

Figure 25.5 Temperature distributions in the atmospheres of Jupiter and Saturn. Since we don't know where the surfaces are, we use as our reference point the place where the pressure is 0.1 atm. The colors written along each curve indicate the dominant color material at that temperature.

pendent of latitude. This is true despite the fact that the solar heating is greatest at the equator. This may mean that winds are effective at distributing heat from the equator to the pole. Such large winds would require large temperature differences to drive them. Since we don't see these differences in the upper atmosphere, the transport must take place in the lower atmosphere. Another possibility is that Jupiter has an internal heat source that supplies more heat to the poles than to the equator.

EXAMPLE 25.1   Energy from Jupiter

Compare the energy given off by Jupiter with the energy it receives from the Sun. Assume that Jupiter radiates like a blackbody at 125 K, the temperature of the cloud tops.

Solution

Jupiter's luminosity is given by its surface area $4\pi R_J^2$, multiplied by the total power per unit area σT_J^4, or

$$L_J = (4\pi R_J^2)(\sigma T_J^4)$$

The power received by the Sun is the solar luminosity $(4\pi R_\odot^2)(\sigma T_\odot^2)$, divided by the area of a sphere at the distance of Jupiter from the Sun $4\pi d^2$, where d is the distance of Jupiter from the Sun, and multiplied by the projected area of Jupiter πR_J^2, or

$$P_{\text{rec}} = (4\pi R_\odot^2)(\sigma T_\odot^4)(\pi R_J^2)/(4\pi d^2)$$

Taking the ratio of these gives

$$\frac{L_J}{P_{\text{rec}}} = 4\left(\frac{T_J}{T_\odot}\right)^4\left(\frac{d}{R_\odot}\right)^2$$

$$= (4)\left(\frac{125\text{ K}}{5500\text{ K}}\right)^4\left[\frac{(5.2)(1.5\times10^6\text{ km})}{7\times10^5\text{ km}}\right]$$

$$= 1.3$$

The situation is even worse than this, because Jupiter doesn't absorb all of the energy that strikes it. This indicates that Jupiter gives off more radiation than it receives from the Sun. Jupiter must have some internal heat source.

The fact that the temperature falloff is close to the adiabatic rate suggests that convection is the important form of energy transport below the cloud tops. Convection is probably more effective than radiation, because the high opacity in the lower atmosphere limits the rate of radiative energy transport. Because of the temperature distribution, we think that there are three major cloud layers, resulting from the fact that different constituents condense at different temperatures and pressures. We think that the highest cloud layer is ammonia ice, the middle layer is ammonium hydrosulfide (NH_4SH), in the form of crystals, and the lowest layer is water, in a mixture of liquid and ice.

The varying conditions mean that the chemistry is different at different altitudes. We think that the different cloud colors reflect different compositions, a result of the varying chemistry. The different colors come from different temperature ranges, and therefore from different levels. For example, the blues correspond to the warmest regions, and are therefore closest to the surface. They are probably only seen through holes in the higher layers. Brown, white, and red come from progressively lower temperatures, meaning they come from progressively higher levels. We have still not identified all of the compounds responsible for the various colorations. Other factors besides temperature also affect the chemistry. For example, some regions have more lightning than others, and the energy from the lightning can help certain reactions go in one place while they don't go in other places.

The east–west winds are quite substantial. They flow at about 100 m/s near the equator, and about 25 m/s at the higher latitudes. These speeds are with respect to the average planetary rotation period at a given latitude. The winds flow in alternating east–west and west–east bands. These alternating wind patterns correspond to the alternating color bands. On Jupiter there are five or six pairs of alternating bands in each hemisphere. For comparison, on the Earth there are only the westward (in the northern hemisphere) trade winds at low latitudes and the eastward jet stream at high latitudes.

The brighter-colored bands are called *zones*. They appear to have gas rising. The dark-colored bands are called *belts*. (Fig. 25.6) They appear to have the gas falling. The tops of the belts are about 20 km lower than the

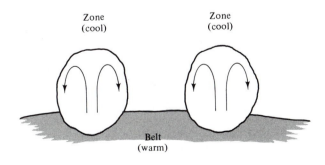

Figure 25.6 Convection on Jupiter produces the belts and zones.

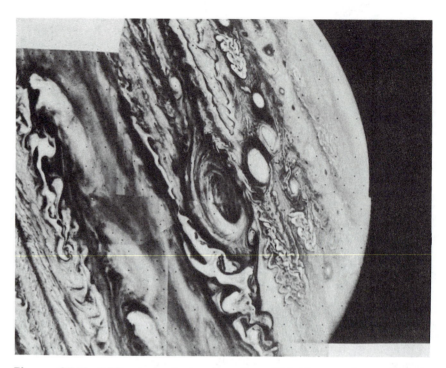

Figure 25.7   Eddies in Jupiter's atmosphere. This Voyager 1 mosaic was taken from a distance of 6.5 million km.

tops of the zones. The belts are brighter in the infrared, indicating a higher temperature. The stability of the large-scale cloud patterns has not been understood. The locations of the bands are constant, even though their colors occasionally change. Close-up pictures have shown small circular regions, called *eddies* (Fig. 25.7). We generally associate eddies with the dissipation of energy. We would therefore expect the patterns to wash out. The fact that the east–west bands are stable indicates that their flow must continue deeper into the atmosphere than the eddies.

The most famous cloud feature on Jupiter is the Great Red Spot (Fig. 25.8). It has been observed for more than 300 years. It covers more than 10° in latitude. It is also surrounded by white ovals, which have flows that would dissipate energy very quickly. In trying to explain its stability, there are two questions that must be answered: (1) How can it be maintained for so long as a stable fluid flow? (2) What is the energy source to replace the energy lost in the eddy flow around the spot? The most successful model has been to say that the spot is analogous to a hurricane on Earth. Large vertical convection currents allow it to draw energy from the latent heat of condensing materials below. On the Earth, hurricanes draw energy from that released when water vapor is condensed. This is why hurricanes intensify when they

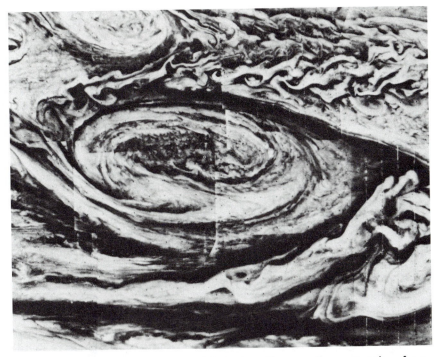

Figure 25.8 Jupiter's Great Red Spot, in a Voyager 1 mosaic taken from 1.8 million km.

pass over large bodies of water. There are several different models for the details of how the spot works, but it does seem that, with the conditions in Jupiter's atmosphere, such a storm should be stable for hundreds of years!

25.2.2 Saturn

It should not be surprising that we find many similarities between the atmosphere of Saturn and that of Jupiter. However, there are certain differences. These differences arise from the fact that Saturn is farther from the Sun than is Jupiter. Saturn's lower gravity and lower rotational speeds are also important.

The temperature is 95 K at the cloud tops. Saturn gives off approximately twice as much energy as it receives from the Sun (see Problem 25.4). The temperature rises as one gets deeper into the atmosphere, just as on Jupiter, but the rate of temperature change is about half that on Jupiter. Figure 25.5 shows the temperature profile of both planets, for comparison.

The winds on Saturn are much greater than on Jupiter. They are about 450 m/s at the equator and about 100 m/s at higher latitudes. There is also an alternating pattern, with the speeds alternating between 0 m/s and about 100 m/s. Saturn does not have as many bands as Jupiter does. The lower temperature means that there are chemical differences. This is evidenced by the fact that the bands don't show as much contrast as those of Jupiter.

Voyager showed the equivalent of the Great Red Spot on Saturn. It had not been seen from Earth because of the low color contrast.

25.2.3   Uranus and Neptune

The atmospheres of Uranus and Neptune are hydrogen rich, like those of Jupiter and Saturn. However, Uranus and Neptune contain a higher proportion of heavy materials. This is because these lower-mass planets probably retained less of their original hydrogen than did Jupiter and Saturn.

The temperatures of Uranus and Neptune are almost the same, about 57 K. One would expect Neptune to be cooler than Uranus, but it appears to give off more heat than it receives from the Sun. It is thought that this may be related to the way in which the atmosphere traps the heat.

Both atmospheres contain methane. This molecule has been identified from its infrared spectrum. Both planets have a greenish color, and it is believed that this color comes from the methane.

There is a difference in the cloud content of the two atmospheres. Uranus has an atmosphere that is cold and clear to great depths. There appear to be no clouds or haze in the lower atmosphere. There may be some higher up. Neptune has a variable haze. We think that the haze is composed of aerosol particles or methane ice crystals. A comparative study of these two atmospheres, especially as more data become available, should provide a good test of our ability to model planetary atmospheres with enough detail to predict or explain the differences.

25.3   INTERIORS

25.3.1   Jupiter

We cannot directly sample the interior of Jupiter. However, we can use physical laws to construct computer models of the interior. We can then see which models produce results that agree with observations. The basic structure that emerges from these models is shown in Fig. 25.9.

The outermost layer is a hydrogen–helium envelope. Within that, extending from a radius 10,000 to 54,000 km, is a liquid region. This liquid is hydrogen. The pressure in this region rises to 40 million times the atmospheric pressure at the Earth's surface. Under these conditions, the hydrogen forms into a metal. This metallic hydrogen region contains 73% of the plan-

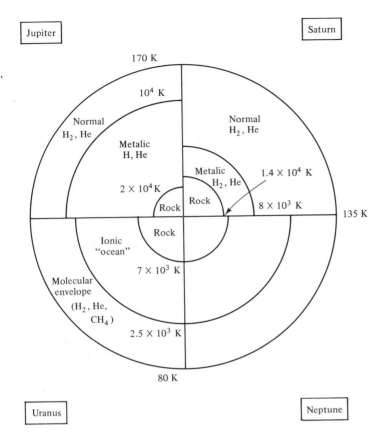

Figure 25.9 Interiors of Jupiter, Saturn, Uranus and Neptune. These are from model calculations, and are expressed relative to the planet's radius. The numbers at each boundary are estimated temperatures.

et's mass. Within the liquid region is a core. The core may be made of rock and ice materials, though it has been suggested that the hydrogen continues right into the center. The pressure is 80 million bar; the temperature is about 25,000 K. The core contains about 4% of Jupiter's mass.

The transition from the gaseous region to the liquid region is probably a gradual one. The transition from the normal liquid hydrogen to the metallic hydrogen probably takes place over a small range of radius. We think that the interior has excess energy stored from the time of the collapse of the planet. The energy is so large that is has not all escaped yet. This is probably the source of the excess energy that Jupiter gives off.

25.3.2 Saturn

The general structure of Saturn is probably very similar to that of Jupiter, as shown in Fig. 25.9. There are some differences, however. Saturn has a larger core, containing about 26% of the mass. The central pressure is 50 million bar, and the central temperature is about 20,000 K.

There is a smaller metallic hydrogen zone. The range of radii is from 16,000 to 28,000 km. The temperature in this zone ranges from 9,000 to 12,000 K. This zone contains about 17% of Saturn's mass.

Our understanding of Saturn's excess heat is not as good as that for Jupiter's. Some other explanation is needed. It has been suggested that some energy comes from helium condensing and sinking through the less dense material toward the core.

25.3.3 Uranus and Neptune

Uranus and Neptune have higher densities than Jupiter and Saturn. This suggests a different composition. The interior structure is deduced from models. This structure is shown in Fig. 25.9.

The cores are rock. The rock is mostly silicon and iron. Over the core is a mantle. This mantle probably contains liquid water, ammonia, and methane. Over the mantle is a crust of hydrogen and helium. It may be in the form of a high-density gas. The central pressure is the same for both planets, about 20 million bar. The central temperature is about 7,000 K.

Voyager 2 provided a good view of Uranus from a distance of 5 planetary radii. There is very little visible structure in the atmosphere. There is, however, a polar haze, probably composed of methane or acetylene. There is also an extended neutral hydrogen (H and H_2) corona. Ultraviolet emission from this corona had previously been detected from Earth-orbiting satellites. The rotation period of the atmosphere appears to be just under 17 hours, and wind speeds seem to be in the range of tens of kilometers per hour.

Voyager 2 also revealed an interesting temperature distribution. The equator is cooler than either of the poles. In addition, the pole facing the Sun is cooler than the shadowed pole. This is obviously an interesting problem in energy transport.

25.4 RINGS

Jupiter, Saturn, and Uranus have ring systems. Saturn's has been known since Galileo, while those of Jupiter and Uranus are recent discoveries. We first review the basic properties of the rings, and then we consider the dynamical effects that are important in shaping the rings.

25.4.1 Basic Properties

A telescopic view of Saturn's rings is shown in Fig. 25.10. The rings shine by reflected sunlight. Therefore, the brighter areas are those of higher optical depth, meaning a greater amount of reflected sunlight.

Three main rings are apparent. The A ring is the farthest from the planet.

Figure 25.10 Saturn's rings, in an enhanced Voyager 1 image.

It has a width of 20,000 km, ranging from 2.02 to 2.27 times the planetary radius. Its thickness is less than 200 m. The optical depths in the A ring are in the range 0.3 to 0.5, so approximately half the light striking them passes through. The B ring is in the middle, and is the brightest, because it has a large optical depth, slightly greater than one. The B ring extends from 1.52 to 1.95 times the planetary radius. The A and B rings are separated by a gap, called *Cassini's division*. The innermost ring is the C ring, and it is the darkest. This means that is has the lowest optical depth, about 0.1. It extends from 1.23 to 1.52 times the planetary radius.

Additional rings have been found, both from the ground and from space-craft. The D ring is a faint ring, inside the C ring (1.11 to 1.23 radii). The E ring is a very faint ring, outside the A ring (3 to 8 radii). It has optical depths of less than 10^{-6}. The F ring is just beyond the A ring (2.37 radii), and was discovered by Pioneer 11. Finally, the G ring, is a faint ring, 2.8 radii from the planet.

The rings are composed of individual particles, rather than being solid structures. We can tell this from the spectra of the rings. The Doppler shifts vary across the rings, as shown in Fig. 25.11. These variations are those to be expected for individual particles orbiting Saturn, with orbital speeds being given by Kepler's third law. The particle sizes probably range from a few

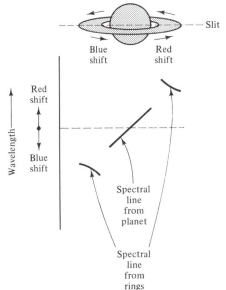

Figure 25.11 Spectra of Saturn's rings, showing change in Doppler shift as one goes farther out in the ring. In the upper diagram the placement of a slit is indicated. Below the schematic spectrum shows the Doppler shift of a particular line as one moves across the slit.

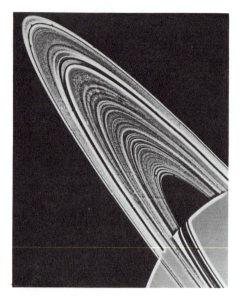

Figure 25.12   Closeup of Saturn's rings from Voyager 1 image.

centimeters to 10 m. The total mass of the ring particles is 10^{17}–10^{19} kg. (This is only about 10^{-6} of the Moon's mass.) The albedo of the particles is 0.5. Infrared observations suggest that the particles are ices of water and other molecules.

The Voyager fly-bys produced spectacular pictures of the rings. We find that there is an extensive pattern of detailed structure. Each ring is divided into many smaller rings, as shown in Fig. 25.12.

The first evidence for rings around Jupiter (Fig. 25.13) came from Pioneer 11 in 1974. The discovery was confirmed by Voyager 2. The rings are different from those of Saturn. They fall within a narrow range of radii, 1.72 to 1.81 planetary radii. Their width is 6000 km. Their thickness is less than 30 km. The rings are very faint, because their optical depth is very low, less than 10^{-5}. The rings appear smooth, except for a 600-km-wide enhancement at 1.79 R_J.

The particles are much smaller than in Saturn's ring, being a few microns across. They have a low albedo, suggesting that they are silicates rather than ice. Such small particles are not expected to stay in the ring very long, so it seems that they must be replaced continuously. It is possible that some ring material is being thrown off the closest Galilean moon, Io, and from two embedded smaller moons.

Figure 25.13   Jupiter's ring, in a Voyager 2 image from a distance of almost a million miles.

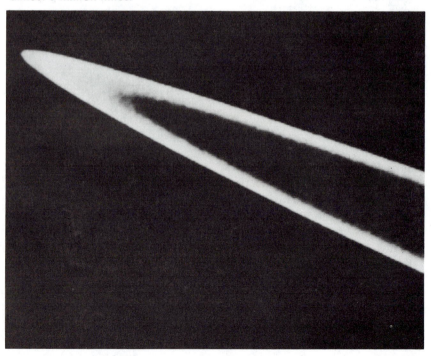

Figure 25.14 Uranus rings.

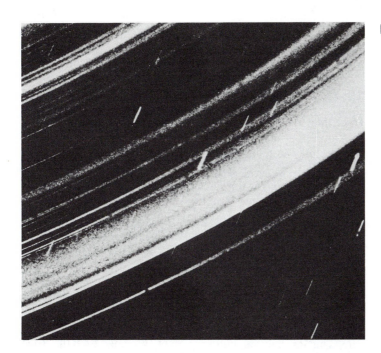

The discovery of a ring system around Uranus (Fig. 25.14), in March 1977, was a great surprise. An occultation of a star by Uranus was being observed. Some of the observations were being carried out from the air, in the Kuiper Airborne Observatory. Normally, this observatory is used for infrared observations. However, in the case of the occultation, it was used for optical observations, with the airplane providing a means of getting the telescope to a favorable viewing point. Shortly before the occultation, a number of brief dips were noted in the light coming from the star. The dips repeated after the occultation. These dips were the result of the ring passing in front of the star.

This observation, and many follow-up observations, have revealed a system of nine narrow rings. The rings extend from 1.60 to 1.95 planetary radii. The designations of the rings, from nearest to farthest, are 6, 5, 4, α, β, η, δ, and ϵ. Each ring is about 10 km wide, except for ϵ, which is about 100 km wide. The ring particles orbit with eccentricities ranging from 0.001 to 0.01. There are also some changes in width. For example, the ϵ ring varies in distance from the planet by about 800 km, and in width from 20 to 100 km. The gaps between the rings are quite clear. The rings themselves have a low albedo, making them difficult to see.

Voyager 2 provided closeup information on the rings. New rings were discovered. Radio observations of the ϵ ring showed the same opacity at 3.6 and 13 cm wavelengths. This means that the particles must be larger than both wavelengths, probably larger than 30 cm. Backlighted views show the regions away from the known rings to be filled with very thin rings. The particles in these rings may be only a micron across.

25.4.2 Ring Dynamics

These three ring systems pose a number of interesting questions. One obvious question is that of their origin. Other questions center around their structure. Why are they so thin? Why are there multiple rings? Why is the appearance so different from planet to planet? All of these questions relate to the dynamics of the particles in the rings. We must understand the forces that perturb the motion of the ring particles, and the response of those particles to the forces.

When we study rings, our first dynamical consideration is the role of tidal effects. The result is to cause different parts of the object to accelerate differently. If some particles are very close to a massive object, the tidal effects due to the massive object can prevent the particles from staying together under the influence of their mutual gravitational attraction. We define the *Roche limit* (Fig. 25.15) of a planet as the minimum distance that an object can be from a planet and still be held together by *gravitational forces only*. This last point is quite important. We are within the Roche limit of the Earth, but we are not torn apart. This is because we are held together by electromagnetic forces, not gravitational forces.

We can estimate the Roche limit by considering the simple situation in Fig. 25.16. We have an object, represented by two particles, each of mass m, a distance x apart. One particle is a distance R from a planet of mass M, and the other particle is a distance $R + x$ from the planet. We assume that $x << R$. The attractive force between the two particles is

$$F_{att} = Gm^2/x^2$$

The tidal force F_{tid} is the difference between the forces exerted by the planet on the two objects. We can write it as

$$F_{tid} = (dF/dR)\Delta R$$

Since the force on a particle of mass m, a distance R from the planet, is

$$F = GMm/R^2$$

we have

$$dF/dR = -2GMm/R^3$$

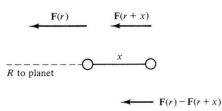

Figure 25.15 Roche limit. We compare the attractive forces between the two spheres with the difference between the forces that the planet exerts on each sphere.

This makes the tidal force

$$F_{\text{tid}} = -2GMmx/R^3$$

where we have taken $\Delta R = x$.

We find the limiting value of R, the Roche limit, by equating the magnitudes of the tidal force and the attractive force between the particles. This gives

$$2GMmx/R^3 = Gm^2/x^2$$

Simplifying, we have

$$2M/R^3 = m/x^3$$

We further note that the mass of the two particles is $2m$, and the volume occupied is approximately x^3, so m/x^3 is approximately twice the density ρ of the object. Making this substitution and solving for R gives the Roche limit as

$$R_{\text{Roche}} = (4M/\rho)^{1/3} \qquad\qquad (25.1)$$

In calculating the Roche limit for a planet, we have to enter a value for the density of the material that is trying to hold itself together. Obviously, the greater the density, the closer it can venture to the planet. At a safe limit, we often take the density of the planet itself. That is, we find the Roche limit for an object with the same density as the planet. In general, an object just forming around a planet will have a lower density than the planet, so it must be farther away than objects as dense as the planet.

We find that all of the ring systems lie within the Roche limits for the three planets. This leads to the idea that the rings are made up of particles that would have formed moons. However, the particles were too close to the planet for the gravitational attraction to allow the moons to form. The arrangements of the various rings, relative to the Roche limit, are shown in Fig. 25.16.

This begins to give us some picture of how rings may have evolved. The scenario is illustrated in Fig. 25.17. As each planet started to form, the material around it formed into a disk. For the material far enough from the forming planet, the collection into a moon was possible. However, for material inside the Roche limit, a moon could not form. Therefore, the material continued to orbit in a disk. Not all of the particles were originally in the thin disk. However, those that weren't had to pass through the disk twice per orbit. During these passages, collisions with particles in the disk changed their orbits. Eventually, the orbits were changed sufficiently that the strag-

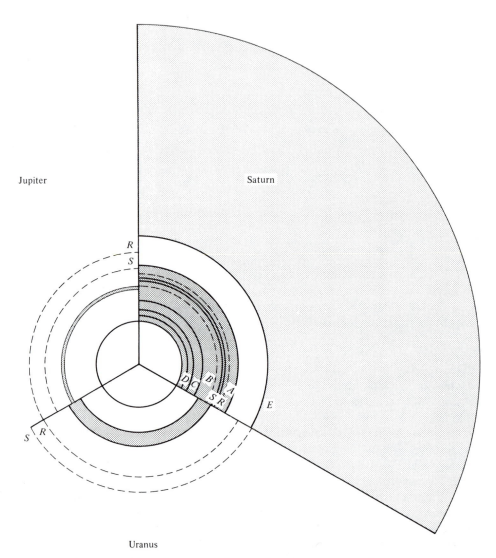

Figure 25.16 Rings and Roche limits for Jupiter, Saturn and Uranus. Again, each diagram is given in units of that planet's diameter. The shaded areas are the actual rings. (For Saturn, the rings are indicated by their letter designations.) The dashed line labeled R is the Roche limit for material with a density of 1 g/cm³. The dashed line labelled S is the radius for a synchronous orbit about the planet. Certain electromagnetic effects change sign at this radius.

glers joined the disk. The flattening was not total, since collisions in three dimensions always leave some residual random motion in the direction perpendicular to the plane.

Within the plane of the disk, the ring could spread out over a range of

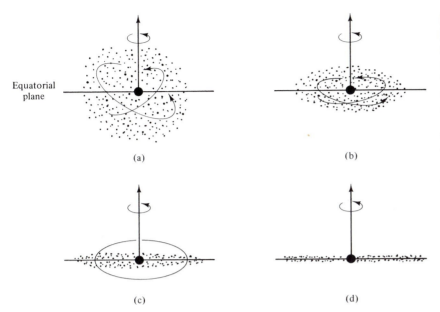

(a)

(b)

(c)

(d)

Figure 25.17 Ring evolution. (a) Particles start in a rotating cloud. Some sample orbits are shown. Particles in such orbits collide, and the collisions tend to reduce the motions perpendicular to the equatorial plane. (b) The cloud flattens by these collisions. (c) Eventually most of the cloud is in a disk, with a few particles having large motions perpendicular to the equatorial plane. However, these particles pass through the plane twice per orbit, and collisions eventually bring them into the plane, producing the thin distribution shown in (d).

radii. One mechanism for this spreading involved collisions between particles in which one was closer to the planet than the other. The closer one would overtake the farther one in their orbits. A gravitational encounter would then slow the inner one and speed up the outer one. This would cause the inner one to fall farther in and the outer one to move farther out, meaning a spreading of the ring. Another effect, Poynting–Robertson drag, discussed in Chapter 22, causes smaller particles to move inward after losing energy under the impact of photons.

The final appearance of rings around a given planet depends greatly on the position of the ring relative to the orbits of the planet's satellites. The gravitational effects of the satellites sculpt the rings by perturbing the orbits of the ring particles. This may have the effect of producing detailed structure, as in Saturn's rings, or of confining the ring to some range of radii, as happened in the case of Uranus. Satellites embedded within the rings can serve as sources for particles in the ring, replacing particles removed by other effects. The moons can also remove particles from the ring.

The effects of multiple moons in sculpting the rings are particularly important when the moons have *resonant orbits*. By this we mean that the ratios of the orbital periods are equal to the ratios of small integers, all illustrated in Fig. 25.18. This means that certain configurations of the satellites repeat with regularity, on a relatively short timescale. Thus, the perturbing effects of those repeating arrangements are strongly reinforced. (Similarly, if we push a swing at its resonant frequency, the effects of the pushing are reinforced.) An example of this is found in the asteroid belt between

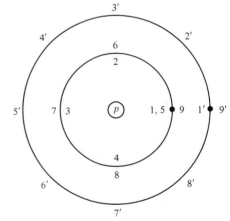

Figure 25.18 Resonant orbits. In this case, the inner satellite has half the period of the outer satellite. Positions of the inner satellite are marked 1 through 9, while the positions of the outer satellite at the same times are marked 1' through 9'. The important point is that every orbit of the outer satellite, the two are close together, so any perturbations can be amplified, in much the way as pushing a swing at its resonant frequency can amplify its motion.

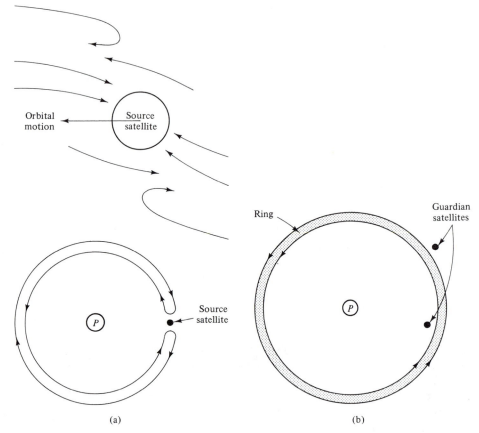

Figure 25.19 Satellites and rings. (a) Source satellites. The upper figure shows the motions of particles, relative to the satellite. The lower figure shows the full orbit of the particles that get turned around near the satellites. (b) Guardian satellites. Satellites just inside or outside the rings can keep the rings confined.

Mars and Jupiter. There are gaps, called *Kirkwood gaps*, in which few asteroids are found. They correspond to orbits whose period would be related to Jupiter's by the ratios of small integers. In Jupiter's moon system, the periods of the orbits of Io, Europa, and Ganymede have the ratio 1:2:4. These are important in influencing the appearance of Jupiter's ring. (It also has important effects on Io, as we will see in the next section.)

In Saturn's rings the Cassini division is at a radius where the orbital period would be half of that of Mimas. It has been suggested that various resonances are important in producing the many gaps seen in the A ring as well as establishing the boundaries of the A and B rings. Shepherding satellites may confine rings, such as the F rings, but other rings are confined without any such satellites. It has also been suggested that satellites, called *guardian satellites* produce a spiral density wave, just like the density waves in spiral galaxies, and that these density waves clear the regions of resonant orbits. In addition, a satellite near a ring can push it away. This accounts for the narrowing of the ring of Uranus.

Another interesting effect results when two small moons have orbits very close to each other, as shown in Fig. 25.19. The inner moon overtakes the outer moon. In a gravitational encounter, the inner moon is pulled out, and the outer moon is pulled in. The moons exchange orbits. This process then repeats when they overtake again. As viewed from the rotating system orbiting with the more massive moon, the less massive moon executes a *horseshoe orbit*.

A moon within a ring has a similar effect on particles. As seen from the moon, particles approach from the outer leading side and the inner trailing side. Their paths are altered by tidal effects of the planet, since the rings are inside the Roche limit (Fig. 25.20). This causes some of the particles to follow looped paths, similar to the horseshoe orbits. Since the tidal effects are directed toward the planet, particles have a hard time sticking to the moon on the side of the moon closest and farthest from the planet, but can stick to the leading and trailing edges.

Figure 25.20 Saturn's F ring, with shepherding satellite.

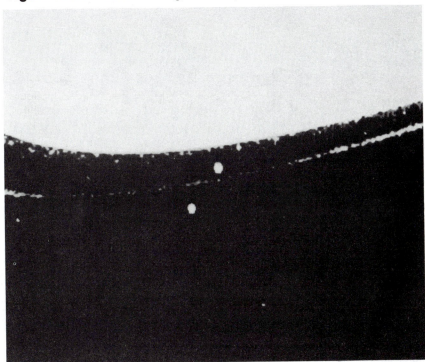

25.5   MOONS

The planets we have discussed in this chapter provide us with an interesting variety of moons. In this section we discuss the most important of those moons, planet by planet.

25.5.1   Jupiter

The four largest moons of Jupiter were discovered by Galileo, and are therefore called the *Galilean satellites*. They are, in order of distance from Jupiter, Io, Europa, Ganymede, and Callisto. The smallest, Europa, has a radius of 1563 km, and the largest, Ganymede, has a radius of 2638 km. Their densities are 3.55, 3.04, 1.93, and 1.8 g/cm^3, respectively. This means that the density falls with distance from Jupiter. The densities of Io and Europa are comparable to those of the Moon and Mars. These moons are close enough to Jupiter to be shielded from the solar wind by Jupiter's magnetic field.

They are massive enough to affect each other's orbits. We have already mentioned the orbital resonance among the inner three (with the periods being in the ratio 1:2:4). This resonance keeps Io's orbit elliptical. It also produces large tidal disruptions of Io, which dissipate energy in Io's interior, keeping it hot.

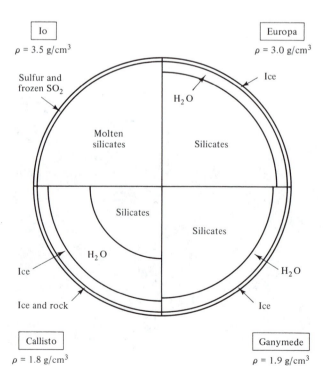

Figure 25.21   Structures of the Galilean satellites, from theoretical models. Again, each is plotted relative to the radius of the particular satellite.

The higher densities of Io and Europa suggest that they contain a significant amount of silicate rock. The lower densities of Ganymede and Callisto suggest that they contain much less rock. Jupiter condensed in the part of the solar nebula where the temperature was low enough for water ice to form. We therefore expect the nonrock part to be water ice. The composition differences among the four satellites could have occurred because Jupiter itself was an important heat source as it collapsed. Thus, the nearer satellites were heated by Jupiter more than the farther ones. This would explain why the nearer ones do not have any water ice.

In Fig. 25.21 we compare the basic internal structure of these four moons. Callisto's density is the least well known, of the four. Models suggest that it is about 40%–60% (by mass) silicate. This silicate is concentrated in a core. Over the core is a 850-km-thick section of liquid water, with a 250-km-thick crust of water ice. Ganymede has 60%–80% silicates, but has a basically similar structure. Its high albedo (0.4) also points to an ice surface. Europa is thought to have a silicate core, with a water layer and a thin (150 km thick) crust. Its high albedo (0.6) is also indicative of an ice surface. Io has a molten silicate interior and a frozen sulfur dioxide (SO_2) and sulfur surface with active volcanoes. We now look at these four moons in some more detail.

One obvious feature of Io (Fig. 25.22) is its yellow color. We think that this results from the sulfur in the surface material. There are no impact features. The surface is volcanic. We have even obtained pictures of volcanic eruptions in progress. The volcanic flows might also contain sulfur. They may be basalts colored with sulfur. There are also many calderas, over 200 of which are more than 20 km across. A few large shield volcanic mountains (like Olympus Mons, on Mars) are evident. There are also nonvolcanic mountains. One consequence of this volcanic activity is that the surface is recycled on short timescales.

This extensive volcanic activity is unexpected for such a small object. We would not expect such a small object to have a molten interior. However, we have already seen that the gravitational effects of the other moons, enhanced by orbital resonances, distort Io (Fig. 25.23). This distortion changes its orientation as the moon moves around in its orbit. The effect is to cause internal friction, heating the interior. This results in a very large heat flow, about 2 W/m^2, through the surface. For comparison, the average heat flow on earth is 0.06 W/m^2, while some geologically active areas on Earth have heat flows as high as 1.7 W/m^2.

Io is also strongly affected by Jupiter's magnetosphere. As Io orbits the planet, Jupiter's magnetic field sweeps by at 57 km/s. According to Faraday's law, this induces a potential difference across Io (see Problem 25.8). The potential difference is about 600 kV. The ionized gas between Io and Jupiter is a good conductor, so currents flow parallel to the magnetic field. These currents are as high as 10^6 amp! This interaction between Io and Jupiter may explain why nonthermal radio bursts from Jupiter are more frequent when Io is in certain positions.

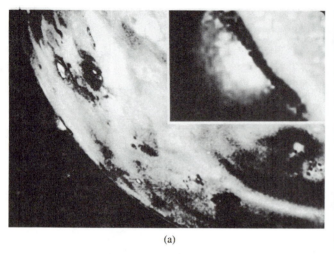

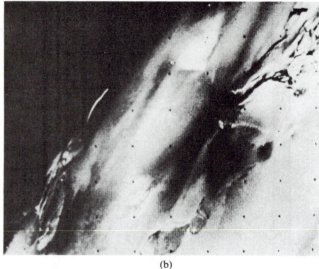

(a)

(b)

Figure 25.22   Io. (a) Volcanic plume. (b) Closeup of surface.

Figure 25.23   Orbital resonances and tidal heating for Io. An orbital resonance with Europe keeps Io's orbit elliptical. This means that, even though Io rotates at a constant rate, equal to its orbital period, it doesn't go around its orbit at a constant rate. If we think of Io as an ellipsoid, with an inner and outer section, the effect of the torques caused by Jupiter is for the inner layer to be out of phase with the outer layer. The two layers move against each other, and generate heat.

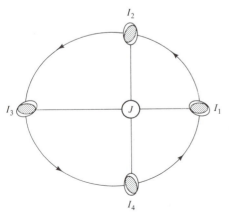

Io also has an atmosphere. It is quite irregularly distributed, with more gas being over the warmer areas. This suggests that the atmosphere is being replaced by the volcanic activity. A major constituent is sulfur dioxide (SO_2). Io also has a dense ionosphere, with a density of about 10^4 to 10^5 cm^{-3}. A large cloud has also been detected around Io. This cloud was first detected from Earth by observations of sodium, and is referred to as the *sodium cloud*. However, sodium was just one of the easiest element to observe, and other constituents are present. The cloud extends tens of thousands of kilometers along the orbit.

Our knowledge of Europa (Fig. 25.24) is less detailed. The closest flybys have produced pictures with only 4-km resolution. However, we can tell that the surface is relatively flat. There is a complicated pattern of lines. There is also a lack of large craters, suggesting a young surface. The surface is probably made of ice. We think that the lines are tension patterns in the ice. This tension may have resulted from an expansion of the surface, probably by about 5%. There are some dark patches. These are probably composed of silicates.

Ganymede (Fig. 25.25) also has an ice surface. It is covered with irregular light and dark regions. The dark regions are heavily cratered, indicating that they are the older part of the surface. The craters in these regions are relatively flat. The lighter regions are grooved. The grooves appear to be alternating ridges and troughs. The pattern suggests tension, as on Europa.

Callisto (Fig. 25.26) has a heavily cratered surface. These craters are also flat, typical of an icy surface. An unusual feature are large ring structures. These are probably the result of violent impacts in the past.

Figure 25.24 Europa.

Figure 25.25 Ganymede, in a Voyager 1 image from 246,000 km.

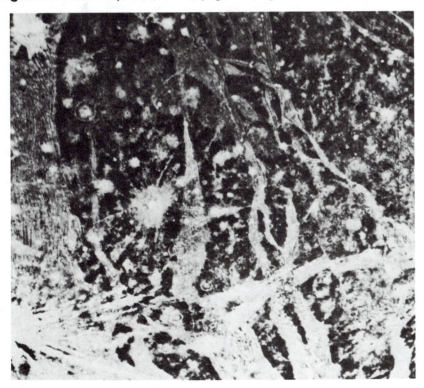

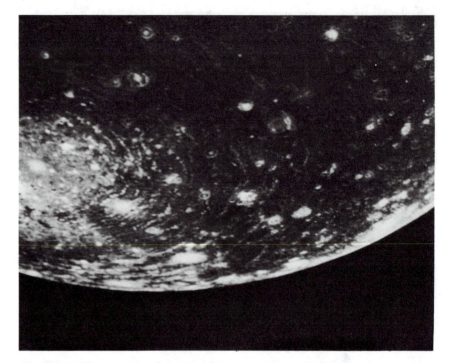

Figure 25.26   Callisto, in a Voyager 1 image from 350,000 km.

25.5.2 Saturn's Moons

The moon of Saturn we know best is its largest, Titan (Fig. 25.27). It has the most significant atmosphere of all the moons in the solar system. Voyager 1 passed within 5000 km of Titan, giving us a good opportunity to study it.

Its density is less than 2 g/cm^3, suggesting a mixture of rock and ice. Its composition and structure may have been affected by heat given off by Saturn during its formation. The model of the interior, shown in Fig. 25.28, has 55% of the mass in a rock core. It is possible that internal heating has taken place. Tidal distortion should not be as important for Titan and Saturn as for Io and Jupiter (see Problem 25.9). There may be some radioactivity, but it seems most likely that energy has been stored from the time of accretion. Beyond the core, there are probably layers of different crystal structures of ice.

Titan's atmosphere contains methane (CH_4). This was originally detected from the Earth. However, closer observations have shown that nitrogen (N_2) makes up 80–95% of the atmosphere. It has also been suggested that there is some argon. Ultraviolet radiation triggers a chemistry that produces traces of ethane (C_2H_6), acetylene (C_2H_2), and ethylene (C_2H_4). These are found lower in the atmosphere. Higher up, hydrogen cyanide (HCN) is formed.

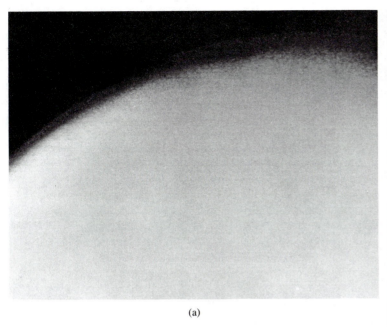

(a)

(b)

Figure 25.27 Two of Saturn's moons. (a) Titan, showing the atmosphere. (b) Dione.

There is no oxygen, since it is tied up in frozen water. The chemistry produces a smog, possibly with some seasonal variation.

The surface pressure is 1.6 bar at the surface. The surface temperature is 93 K. The highest temperature in the atmosphere is about 150 K, in the upper atmosphere. The lowest temperature is 70 K, about 40 km above the ground. An interesting feature of this temperature range is that it includes the temperature of the triple point of methane, 90.7 K. The triple point is that at which the three phases—gas, liquid, and solid—can coexist. We might therefore expect to find a frozen methane surface, with methane oceans and methane clouds in the atmosphere. Others have speculated on oceans of ethane rather than methane.

25.5.3 Moons of Uranus

Voyager 2 provided our first good look at the five known moons, and discovered a number of smaller moons. As with Jupiter and Saturn, the moons of Uranus show a variety of features. Oberon has large albedo variations over its surface. It is cratered, and a small mountain was photographed on the limb. Titania has craters, ridges, rilles, and scarps. Its cratering is sparse compared to that of other moons. Ariel has many bright reflected regions as

Figure 25.28 Internal structure of Titan, as deduced from theoretical models.

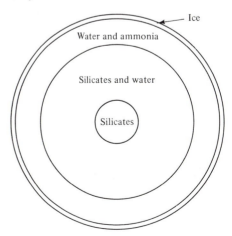

Ice

Water and ammonia

Silicates and water

Silicates

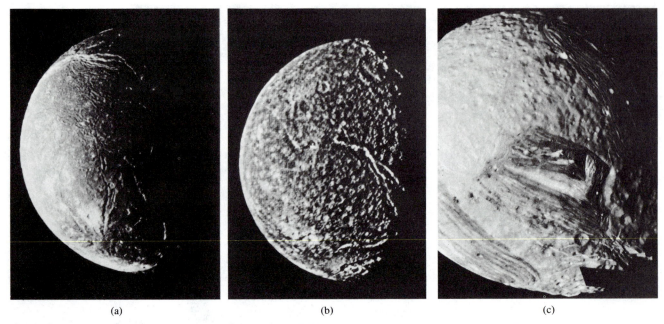

(a)　　　　　　　　　　(b)　　　　　　　　　　(c)

Figure 25.29   Moons of Uranus. (a) Ariel; (b) Titania; (c) Mirande.

well as deep scarps and valleys. There are many filled-in features, indicating an active surface. Some of the most intriguing structures are found on Miranda. It has three distinct types of surface: an old surface, a scored region, and a region with very complex structures (Fig. 25.29).

25.6   PLUTO

Pluto was discovered in 1930, following an extensive search, by Clyde Tombaugh. The search was initiated by Percival Lowell after it was thought that a planet beyond Neptune was perturbing Neptune's orbit. Calculations narrowed the range of possible locations and the search was carried out. As Fig. 25.30 shows, Pluto doesn't stand out very well against the background of stars. It is detectable as a planet only by its very slow motion with respect to the stars.

For Pluto to have a perturbing effect on other planets, its mass must be greater than that of the Earth. For this reason, since its discovery, Pluto's mass has been overestimated. We now know that its mass is much less than previously thought. This means that Pluto has no measurable effect on other planets. In a sense, Pluto's discovery was accidental. It was the result of an extensive search of a particular region of the sky. For this reason, other searches have been carried out for a ''tenth planet,'' none with success.

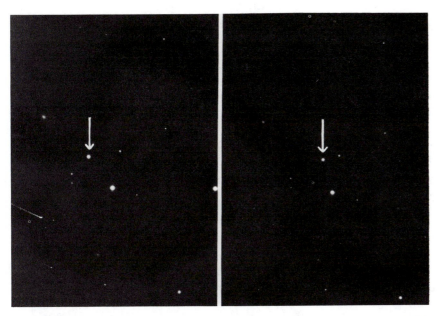

Figure 25.30 Pluto, indicated by the arrow, showing its motion in 24 hours.

Pluto's mass is now known reasonably accurately. This is because a moon was discovered orbiting Pluto in 1978. This moon is named Charon. By studying the orbital period and size, we can determine Pluto's mass. It is about $1/500\ M_E$. Charon's mass is about 5–10% of that of Pluto. Charon was originally found as a bump on one side of the planet, as seen in Fig. 25.31.

Figure 25.31 Pluto, with Charon showing up as a bulge on one side of the planet.

Pluto's size has been estimated from its failure to occult certain stars. However, our best measurements of its size now come from speckle interferometry. Using the derived size and the mass given above, Pluto's density is 0.5–1.0 g/cm^3. This density suggests that its composition is similar to that of the moons of the giant planets. It has been suggested that Pluto's surface is frozen methane and that its atmosphere is also composed of methane.

Pluto's size, density, and orbit raise questions about its status as a planet. Its orbit is the most eccentric of the planets. It even spends part of its orbit closer to the Sun than Neptune. It is closer than Neptune now, reaching perihelion in 1989. It has been suggested that Pluto is really an escaped moon of Neptune. This would explain its small size, low density, and crossing of Neptune's orbit. However, when we trace back the orbits of Neptune and Pluto, we find no time when they were actually close together. Thus, if Pluto did escape from Neptune, its orbit must have been perturbed since then. Thus, Pluto's origin still remains a mystery.

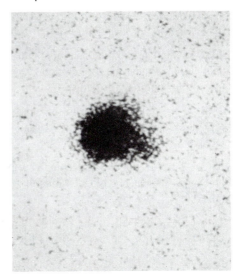

CHAPTER SUMMARY

In this chapter we looked at the outer planets in the solar system. Four of them, Jupiter, Saturn, Uranus, and Neptune, are much larger than the inner planets, and have structures that are very different from those of the inner planets. Pluto is much more like one of the inner planets.

Because of their large masses, these planets, especially Jupiter and Saturn have been able to retain much of their initial supply of hydrogen, giving them compositions that are dominated by the hydrogen. The minor constituents play important roles in the atmospheric structure, energy transfer, and appearance.

Both Jupiter and Saturn give off more energy than they receive from the Sun. It is speculated that this comes from gravitational potential energy liberated during the collapse of the planets.

The general circulation on Jupiter and Saturn is quite complicated, producing bands with varying color and wind flow. Large, hurricanelike systems, such as the Great Red Spot, are found. The atmospheres of Uranus and Neptune are not as dense or as active.

The interiors of Jupiter and Saturn are mostly liquid hydrogen, with the possibility of metallic hydrogen in the center. Uranus and Neptune are thought to have rocky cores.

We looked at the properties of the rings around Jupiter, Saturn, and Uranus. We saw how tidal effects may be important in ring formation. We also looked at the dynamical effects that provide the intricate structure of the rings and gaps.

Finally, we looked at the major moons of these planets. These moons provide a diversity of surface and atmospheric phenomena.

QUESTIONS AND PROBLEMS

25.1 Find the gravitational potential energy of Jupiter. Assume that this amount of energy has been released over a period of four billion years. How does the average rate of energy release compare with that in the sunlight received by Jupiter?

25.2 Confirm Kepler's third law for the moons of Jupiter. Use the information to derive Jupiter's mass.

25.3 Calculate the rate of temperature falloff for an adiabatic process near the Jupiter cloud tops. Compare it with the similar number for Saturn.

25.4 From the data given in the chapter, show that Saturn gives off approximately twice as much power as it receives from the Sun.

25.5 What is the ratio of solar energy per second per unit surface area reaching Uranus to that reaching Neptune?

25.6 Show that all of the rings of Jupiter, Saturn, and Uranus lie within the Roche limits for these planets.

25.7 In deriving the Roche limit, we ignored the fact that particles are orbiting the main planet. This introduces pseudoforces in the rest frame of the orbiting material. How do these pseudoforces affect the Roche limit calculation?

25.8 Use Faraday's law to derive an expression for the potential difference induced across a planet by a magnetic field B sweeping across the planet's surface at a speed v. Take the planet's radius to be R.

25.9 Compare the magnitude of the tidal effects that Jupiter exerts on Io with those that Saturn exerts on Titan.

25.10 List the moons of the solar system in order of decreasing size. Where do the smallest planets fit in?

25.11 Estimate the adiabatic temperature gradients at the altitude of minimum temperature for both Jupiter and Saturn.

25.12 Compare the solar energy received by Jupiter, Saturn, Uranus, and Neptune.

Minor Bodies in the Solar System

There are a vast number of smaller objects in the solar system, not as substantial as our moon, but which provide important clues on the history of the solar system. These are the *asteroids*, *comets*, and *meteoroids*.

26.1 ASTEROIDS

Most of the asteroids lie in a band between the orbits of Mars and Jupiter. This band is called the *asteroid belt* (Fig. 26.1). Over 3000 have been catalogued to date, but there are many more. The ones that have been catalogued are the brightest, and presumably the largest. We would expect there to be many more small ones. The total mass of the asteroids is less than the mass of the Moon.

The sizes of asteroids are determined from stellar occultations (Fig. 26.2). The orbits of many asteroids are well known. When we talk about the orbit of an asteroid (or any other object), we are talking about the path of its center of mass. When the center of mass comes close to passing in front of a star, the asteroid can only occult the star if the asteroid is large enough. When an occultation is expected, astronomers from various points on the Earth watch. If they all see the occultation, the asteroid is larger than some size. If none sees the occultation, then the asteroid is smaller than some size. If some see it and others don't, the size of the asteroid can be determined quite accurately.

Of the asteroids that have been studied in this way, only six are larger than 300 km, 200 are larger than 100 km, and there are many that are smaller than 1 km. If we know the size and apparent brightness of an asteroid, we can determine its albedo (see Problem 26.9). Albedos for asteroids fall in the range 3%–50%.

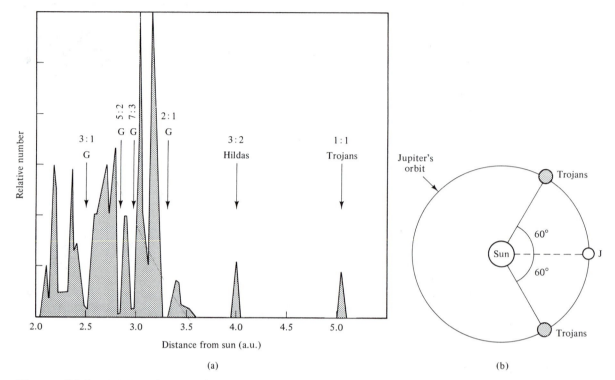

(a)   (b)

Figure 26.1 (a) Distribution of asteroids. The places marked G are gaps, with the orbital period relation to Jupiter given above. The two groups, the Hildas and Trojans are also indicated. (b) Locations of the Trojans, at two Lagrangian points of the Jupiter–Sun system.

Not all of the asteroids are in the asteroid belt. Some, the *Apollos*, cross the Earth's orbit. Another group, known as the *Trojans*, are grouped around the Lagrangian points of Jupiter. (Remember, the Lagrangian points are the points where the gravitational forces of the Sun and Jupiter, and the pseudo-force, the centrifugal force, sum to zero.) It is thought that Phoebe, Saturn's outermost moon, which is only 200 km across and quite dark (low albedo), is a captured asteroid. It has also been suggested that the smaller moons of Jupiter, as well as the moons of Mars, are captured asteroids. Astronomers

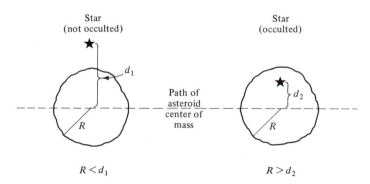

Figure 26.2 Occultations of stars by asteroids tell us about the asteroid sizes.

think that some of the larger asteroids would have a similar appearance to Phobos and Deimos.

Asteroids are classified according to their surface type. E types have a very high albedo. They are rare and are found at the inner edge of the belt. S types have lower albedos. They are more abundant and are found from the inner part to the center of the belt. M types are also abundant, being found in the middle of the belt. They have moderate albedos. The most abundant are the C types, which are found in the outer parts of the belt. They are characterized by a low albedo. We will see in Section 26.3 that there is some correspondence between asteroid types and meteorite types.

The brightnesses of asteroids vary with time. Studies of their light curves reveal periodic variations, with up to a factor of 3 in brightness. This is interpreted as indicating either an irregular shape or an irregular surface coverage. In the latter case, the asteroid would have a dark and a light side. In some cases, elongated shapes have been detected.

The origin of the asteroids is still not clear. For a long time, it had been speculated that the asteroid belt was the remainder of a planet that was destroyed. The total mass of the asteroids, about 2% that of the Earth, would not provide a very massive planet. However, it does seem likely that the large tidal effects of Jupiter kept a planet from forming too close to Jupiter. The asteroids may therefore be debris that would have formed into a planet, if Jupiter had not prevented it.

26.2 COMETS

Every now and then, a spectacular comet (Fig. 26.3) is visible for a few weeks, and we are reminded of this phenomenon. However, most comets are faint, and do not attract much general attention. They are still very important in our understanding of the history of the solar system. Over 600 comets are known. Their masses are less than 10^{-9} times the mass of the Earth.

Figure 26.4 shows the basic structure of a comet. The smallest part is called the *nucleus*. It is only a few kilometers across. One way of determining sizes is by bouncing radio waves off the surface and measuring the strength of the reflected signal. Since the early 1950's, the conventional picture, advanced by Fred L. Whipple, is that the nucleus is a "dirty snowball." It contains dust, plus ices of water, carbon dioxide, ammonia, and methane. When a comet gets close to the Sun, material is ejected from the nucleus. This material acts like the exhaust of a rocket, and provides thrust for the comet. This may actually alter the orbit of comet. It is one of the reasons why the exact prediction of comet orbits is difficult.

Outside the nucleus is an extended region of gas and dust, called the *coma*. It is 10^5 to 10^6 km in extent, and shines by reflected sunlight. The material in the coma flows outward at about 0.5 km/s. The outflow of gas

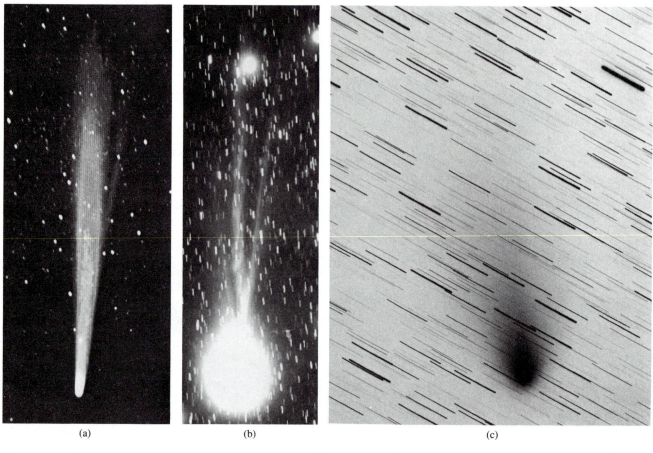

(a)   (b)   (c)

Figure 26.3 Comets. (a) Comet Halley in 1910. This is a computer-reconstructed image. (b) Comet Halley on December 8, 1985. (c) Comet Giacobin–Zinner, on September 12, 1985, just after it was visited by the International Cometary Explorer.

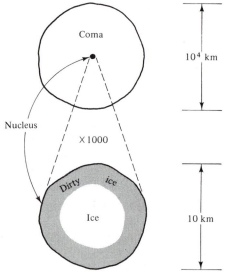

Figure 26.4 Structure of a comet.

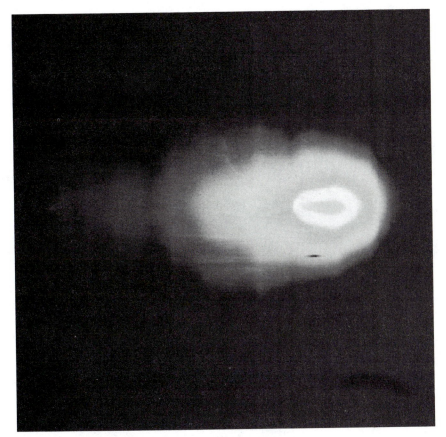

Figure 26.5 IRAS image of comet IRAS–Araki–Alcock.

drags dust away from the nucleus. The sunlight reflects off both gas and dust in the coma. The coma can also emit radiation from excited gas. Spacecraft observations have shown Lyman alpha emission in the ultraviolet. These indicate a hydrogen cloud up to a factor of 10 larger than the coma itself. It is thought that this hydrogen comes from the breakup of water molecules and OH radicals by solar ultraviolet radiation. Spectra (Fig. 26.6) of the coma have indicated the presence of a number of simple molecules, such as NH, NH_3, OH, and H_2O.

When a comet gets relatively close to the Sun, it may develop a large *tail*. This tail can be up to 1 AU in extent, but is of such low density that it doesn't contain an appreciable fraction of the mass of the comet. Only about 1/500 of the mass of the comet is in the tail.

There are actually two tails (Fig. 26.7). The *gas tail* is blown straight out by the interaction of the solar wind with the comet. The gas tail always points away from the Sun. Variations in the solar wind produce a varied appearance along the length of the tail. A number of molecular ions have

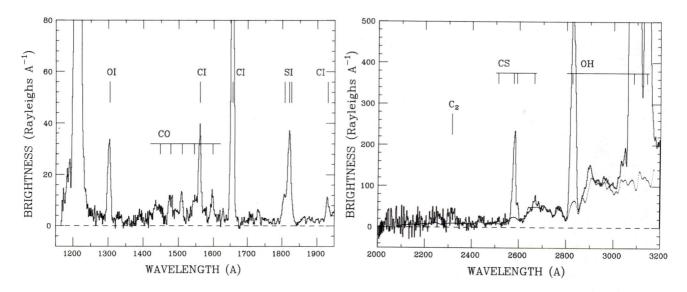

Figure 26.6   Ultraviolet spectra of Comet Halley.

been detected in the gas tail: CO^+, CO_2^+, CH^+, CN^+, N_2^+, OH^+, and H_2O^+. Since it contains ions, it is also called the *ion tail*. Emission from the CO^+ is responsible for the blue color of the tail. The gas tail can be up to 10^8 km long. The *dust tail* is material that is left behind in the orbit. We see it as a smooth curve, tracing out the comet's orbit. It is ejected by radiation pressure from sunlight. The dust tail can be up to 10^7 km long. Sometimes, a tail appears to be pointing toward the Sun. This is an illusion, caused by the appearance of the tail pointing away from the Sun, as viewed from the Earth in particular positions with respect to the comet and Sun.

We can estimate the effect of radiation pressure on dust grains. The

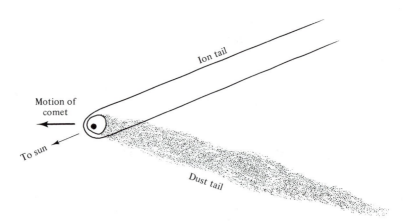

Figure 26.7   Dust and ion tails.

source of radiation pressure is the momentum carried by photons. For a photon of energy E the momentum is

$$p = E/c \qquad (26.1)$$

Suppose we want to calculate the pressure a distance r from the Sun. Imagine a spherical shell at this radius. The force F on the shell is just the momentum per second carried by the photons reaching the shell. That is,

$$F = dp/dt$$

The pressure P is the force, divided by the area of the shell $4\pi r^2$, so

$$P = \frac{dp/dt}{4\pi r^2}$$

By equation (26.1),

$$dp/dt = \frac{dE/dt}{c}$$

but dE/dt is just the energy per second emitted by the Sun, or the solar luminosity $L_\odot$. Therefore,

$$P = L_\odot/4\pi r^2 c \qquad (26.2)$$

At a distance of 1 AU from the Sun, the pressure is 5×10^{-5} dyn/cm^2, which is 5×10^{-11} atm.

Despite this small pressure, if a grain is small enough, the force on the grain due to radiation pressure can exceed the gravitational force that the Sun exerts on the grain. Small grains are important, because the gravitational force depends on the mass of the grain, which is proportional to d^3 (where d is the grain size), and the radiation pressure force is pressure, multiplied by area, so it depends on d^2. Therefore, the ratio of the forces

$$F_{rad}/F_{grav} \propto (d^2/d^3)$$
$$\propto (1/d)$$

Since the ratio of the forces is proportional to $1/d$, the radiation pressure is more effective for the smaller grains. Note that equation (26.2) tells us that the radiation pressure force is proportional to $1/r^2$ (where r is the distance from the Sun). This is the same dependence as the gravitational force on r. Therefore, the ratio of the forces is independent of the distance from the Sun.

EXAMPLE 26.1   Radiation Pressure

For grains of a given density 1 g/cm^3, find the grain size for which gravity and radiation pressure are equal.

Solution

The gravitational force on a grain of mass m is

$$F_g = GM_\odot m/r^2$$

$$= GM_\odot(4\pi/3)d^3\rho/r^2$$

The radiation pressure force is

$$F_{\rm rad} = (L_\odot/4\pi r^2 c)(\pi d^2)$$

Equating these and solving for d gives

$$d = 3L/16\pi GM_\odot \rho c$$

$$= \frac{\frac{3}{16}(4 \times 10^{33}~{\rm erg/s})}{(6.7 \times 10^{-8}~{\rm dyn\text{-}cm^2/s^2})(2 \times 10^{33}~{\rm g})(1~{\rm g/cm^3})(3 \times 10^{10}~{\rm cm/s})}$$

$$= 6 \times 10^{-3}~{\rm cm}$$

For grains smaller than this, radiation pressure will push out more strongly than gravity pulls in.

Our current picture of the origin of comets places them in a cloud, called the *Oort cloud*, some 50,000 AU from the Sun. This cloud is like a spherical shell around the solar system. The existence of such a cloud is suggested by the fact that comets appear to be bound to the solar system (no hyperbolic orbits), and that they come from all directions, orbiting with equal frequency in either direction relative to the orbital motions of the planets. The cloud may contain 10^{12}–10^{13} comets, giving it a total mass in the range 1–10 Earth masses. From time to time, one of the objects in the cloud has its orbit severely perturbed, and starts to head for the inner solar system. At this point, the comet is only the material that will be seen as a nucleus when the comet is in the inner solar system. There is no coma or tail. As the comet gets closer to the Sun, it develops the coma, and then the tail, as discussed above. The coma begins to appear when the comet is about 3 AU from the Sun. At this point, the temperature is about 215 K, which is right for the sublimation of water ice to form water vapor.

The comet is brightest at perihelion. The tail has its longest physical length then. However, the apparent length of the tail depends on the viewing angle from the Earth. Therefore, the tail may not appear to be longest at perihelion. Once the comet has passed through the inner solar system, it continued outward in its orbit. Most of the orbits are closed (ellipses), so the comets will return, unless the orbit is perturned when it passes close to a planet. Orbital eccentricities and periods vary considerably. For example, comet Encke has a period of 3.3 yr and comet Kohoutek has a period of 80,000 yr. It is thought that most short-term comets have had their orbits severely perturbed by Jupiter.

The appearance of a particular comet near perihelion is often hard to predict in advance. For example, it was predicted that comet Kohoutek (1974) would be very bright. This was based on the fact that it appeared to be bright when it was far away. The comet was not as bright as predicted. It is now speculated that the comet was making its first pass by the Sun, and therefore behaved differently than a comet making a return visit. On the other hand, comet West (1975) put on a better show than expected. People who remember the spectacular view of Halley's comet in 1910 were disappointed by the unfavorable viewing angle in 1986. The closest approach of Halley's comet to Earth in 1986 was much farther than that in 1910. In addition, the view in 1986 was marred for many observers by the spread of ''light pollution'' in the past 76 years.

Some comets pass quite close to the Earth. For example, IRAS–Araki–Alcock (1983) passed within 4.6 million km. This comet was of interest because of its simultaneous discovery by the IRAS satellite, in the infrared, and two more traditional ground-based optical observers.

Comets are important in our understanding of the solar system. We think that the Oort cloud is left over from the material that condensed to form the solar system. The cloud is far enough out to be unaffected by the heating in the solar system itself. Therefore, the composition of comets should reflect the composition of the original solar nebula. This is one reason that there was considerable activity in flying spacecraft close to Halley's comet in 1986. The United States was not one of the countries sending a Halley's comet probe. However, NASA was able to redirect a satellite intended for solar wind studies to fly through the tail of comet Giacobini–Zinner, in 1985. An initial analysis indicates that the tail is wider than originally thought.

26.3 METEOROIDS

Meteoroids are small chunks of matter left in space. They are up to tens of meters in diameter. When the Earth encounters a meteoroid, the meteoroid may fall through the Earth's atmosphere, being heated by the friction between the air and the meteoroid. It glows brightly as it streaks across the sky (Fig.

Figure 26.8 Earth moving through meteoroids. From the Figure, you can see that the activity is greatest when the observer is on the leading edge of the earth, which occurs after midnight. The maximum effect is actually at dawn, so the best time to see meteors is just before it begins to get light in the morning.

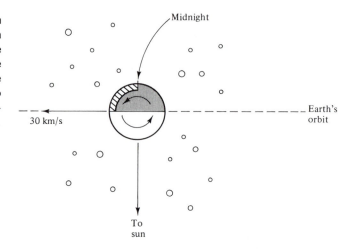

26.8). At this point, we refer to it is a *meteor*. Most meteors burn up as they pass through the atmosphere. However, some do reach the ground. The ones that reach the ground are called *meteorites*.

Some very small meteoroids, much less than 1 mm across, may settle into the upper atmosphere, and can actually be collected by balloons. These are of interest, because there are many more small meteoroids than large ones (Fig. 26.9), and it is these small ones that are responsible for the erosion on the Moon. (The Moon has no atmosphere to protect it.)

Most meteoroids that produce meteors are probably the debris of comet tails. They are therefore left behind in the orbit of the comet (Fig. 26.10). When the Earth crosses the comet's orbit, we see a large number of meteors— a meteor shower. These showers occur at the same times each year, since they represent the passage of the Earth through the orbit of a comet. This scenario also explains why we see most meteors after midnight. After midnight, an observer is on the side of the Earth facing in the direction of the orbital motion of the Earth.

Since we think that comets are left over from the formation of the solar system, meteorites give us a chance to examine that material directly. Two different compositions have been identified. One group, called *irons*, are mostly nickel and iron. The other type, called *stones*, have an appearance similar to ordinary rocks and are hard to find on the ground unless the fall was witnessed. There is also a type, *stony irons*, which is a combination of the two types of material. Most of the meteorites are stones.

Most of the stones contain small rounded glasslike particles, called *chondrules*. Meteorites containing chondrules are called *chondrites*. The others are called *achondrites*. Chondrites with large amounts of carbon are called *carbonaceous chondrites*. It is believed that these are the oldest meteorites. An example of a famous carbonaceous chondrite is the *Murchison meteorite*, which fell on Australia in 1969. This meteorite contains amino acids of a

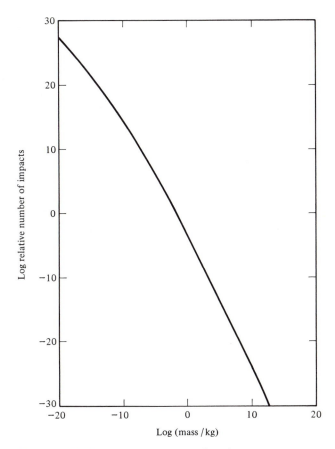

Figure 26.9 Meteoroid size distribution.

type (left-handed vs. right-handed) not found on Earth. The largest carbonaceous chondrite is the Allende meteorite which fell on Mexico in 1969. We think that carbonaceous chrondrites were never strongly heated after formation, and therefore preserve the original material out of which the solar system formed. The Allende meteorite has centimeter-sized inclusions of minerals rich in calcium and aluminum.

It has been suggested that there is some relationship between asteroid types and meteorite types. In this picture, the C-type asteroids are like the carbonaceous chondrite meteorites. The S-type asteroids are like the ordinary chondrites or the stones.

Radioactive dating tells us that meteorites are 4.6 billion years old. This confirms our idea that they are part of the debris left over from the formation of the solar system. Studies are now being done of the relative abundances of various elements and various isotopes of those elements to understand the composition of the material out of which the solar system formed. So far,

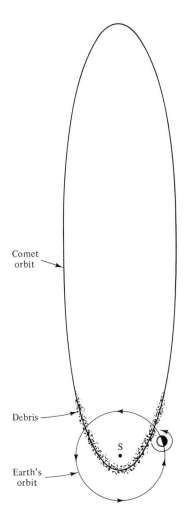

Figure 26.10 Meteoroids and comets. The meteoroids are debris left by a comet along its orbit. We see a meteor shower as the Earth passes through this debris.

some unusual abundances have been found. For example, xenon-129 is formed by the beta decay of iodine-129, with a half-life of 17 million years. A lot of xenon-129 has been found, suggesting a large production of iodine-129 just prior to the formation of the meteoroids. It has been suggested that these unusual abundances imply a supernova explosion in our neighborhood around the time of formation of the solar system. However, this is still quite speculative, and some claim that they can explain the abundances without invoking any unusual events.

CHAPTER SUMMARY

In this chapter we looked at asteroids, comets, and meteoroids. These small objects give us important clues to the history of the solar system.

We saw that most asteroids are between Mars and Jupiter, but there are notable exceptions. Even within the asteroid belt, there are differences in composition as one gets farther from the Sun.

We think that comets may provide us with the best record of the material out of which the solar system formed.

This is because they reside well beyond the orbit of Pluto, and only make brief visits to the inner solar system. We saw how the appearance of a comet changes as it approaches the Sun. We saw the processes that are important in the development of a tail.

For the most part, meteoroids are the debris left behind by comets. When they fall to Earth, they provide us with information on the composition of material in the forming solar system.

QUESTIONS AND PROBLEMS

26.1 What is the force on the Earth due to the Sun's radiation pressure? How does this compare with the gravitational force on the Earth?

26.2 It has been suggested that radiation pressure from the Sun could be used to propel a large spacecraft toward the outer solar system. How large a solar sail would be needed to provide an acceleration of 1 m/s^2 for a 10^6 kg spacecraft?

26.3 Suppose an asteroid is a distance d from Earth. Its center of mass is going to pass within an angle θ of a star. How large does the asteroid have to be to occult the star?

26.4 How far can the center of mass of a 100-km-radius asteroid pass from the direct line between the Earth and a star and still have the asteroid occult the star, if the asteroid is 3 AU from Earth?

26.5 Use the mass of a typical comet and the mass of the Oort cloud to estimate the total number of comets in the cloud.

26.6 Draw a diagram to show how a tail pointing away from the Sun could be viewed from Earth as pointing toward the Sun.

26.7 Estimate the kinetic energy of a 100-m-diameter object, of density 5 g/cm^3, striking the Earth with a speed equal to the escape speed from the Earth. Find one phenomenon on Earth that has a comparable energy associated with it.

26.8 Explain why the presence of xenon-129 in large quantities in a meteoroid means that large quantities of iodine-129 were produced just before the formation of the meteoroid.

26.9 For an asteroid of radius r and albedo a a distance d from the Earth, and a distance R from the Sun, find an expression for the amount of reflected sunlight reaching the Earth. You may treat the asteroid as a disk, oriented so that sunlight will be reflected toward the Earth.

Glossary of Symbols

a	Semimajor axis of ellipse
	Albedo
a	Acceleration
A	Area
	Atomic mass number
	Extinction
	Oort galactic rotation constant
	Angle to convergent point of moving cluster
b	Semiminor axis of ellipse
	Distance that a trajectory would pass an object if there were no deflection
	Bottom quark
B	Blue filter
	Oort galactic rotation constant
$B_\nu(T)$	Planck function at temperature T
B	Magnetic field
BC	Bolometric correction
c	Speed of light
	Charm quark
d	Distance from Earth to astronomical object
	Telescope diameter
	Deuteron
	Down quark
dM	Mass of thin shell
d_0	Reference distance for absolute magnitudes
e	Eccentricity of ellipse
	Electron
	Charge of proton
	Emissivity
e^+	Positron
e^-	Electron
E	Energy
E	Electric field
f	Energy flux
	Focal length
	Fraction of radiation absorbed

$\mathbf{F}$	Force
g	Acceleration of gravity
g_i	Statistical weight of ith state
G	Universal gravitation constant
h	Planck's constant
$\hbar$	Planck's constant$/2\pi$
H	Hydrogen
	Hubble constant
H_0	Current value of Hubble constant
HI	Atomic hydrogen
HII	Ionized hydrogen
Hα	First Balmer line
H$_2$	Molecular hydrogen
i	Inclination of orbit
I	Moment of inertia
	Intensity
J	Angular momentum quantum number
	Angular momentum
k	Boltzmann constant
	Constant, giving curvature of universe
K	Kinetic energy
KE	Kinetic energy
l	Galactic longitude
	Path length for absorption
L	Luminosity
	Mean free path
$\mathbf{L}$	Angular momentum
$L_\odot$	Solar luminosity.
m	Mass
	Magnitude
	Order of interference maximum
m_e	Electron mass
m_g	Gravitational mass
m_i	Inertial mass
M_J	Jeans mass
m_n	Neutron mass
m_P	Proton mass
m_r	Reduced mass
M	Mass
	Absolute magnitude
$M(r)$	Mass interior to radius r
$M_\odot$	Solar mass
n	Principle quantum number
	Neutron
	Density
	Index of refraction
n_i	Level population (i can be any index)

N	Column density
	Number of lines on a grating
	Number of neutrons in a nucleus
p	Parallax angle
	Proton
$\mathbf{p}$	Momentum
P	Power
	Period of orbit
	Probability
	Pressure
PE	Potential energy
q	Electric charge
	Deceleration parameter
q_0	Current value of deceleration parameter
r	Radius
R	Radius
	Rydberg constant
	Resolving power of prism or grating
	Ratio of total-to-selective extinction
	Distance from center of galaxy
	Scale factor in cosmology
	Rate
R_J	Jeans length
$R_\odot$	Solar radius
R_S	Schwarzschild radius
s	Strange quark
t	Time
	Top quark
t_{ff}	Free-fall time
t_{rel}	Relaxation time
T	Temperature
T_i	Ionization temperature
T_k	Kinetic temperature
T_x	Excitation temperature
u	Up quark
	Speed
U	Potential energy
	Ultraviolet filter
v	Speed
v_a	Orbital speed at aphelion or apastron
v_{esc}	Escape speed
v_p	Orbital speed at perihelion or periastron
v_r	Radial velocity
v_T	Tangential velocity
v_{rms}	Root-mean-square speed
$\mathbf{v}$	Velocity
V	Visible filter
	Gravitational potential
	Volume

w	w particle
$\mathbf{x}$	Position
X_r	rth ionization state of element X
z	Zenith distance (angle away from zenith)
	Red shift $(\Delta\lambda/\lambda_0)$
Z	Number of protons in a nucleus (atomic number)
α	Alpha particle, or helium nucleus
β	v/c
	Beta particle (electron or positron)
γ	Special relativistic factor $(1 - \beta^2)^{-1/2}$
	Photon
	Gamma ray
	Ratio of specific heats
δ	Increment
Δ	Increment
Δs	Space-time interval
Δv	Velocity spread
$\Delta\lambda$	Wavelength shift or spread
$\Delta\nu$	Frequency shift or spread
$\Delta\theta$	Telescope angular resolution due to diffraction
ϵ	Rate of energy generation, per unit mass
Φ_B	Magnetic flux
κ	Opacity
λ	Wavelength
$\lambda_{\max}$	Wavelength at which spectrum peaks
λ_0	Rest wavelength
Λ	Cosmological constant
μ	Proper motion
ν	Frequency
	Neutrino
$\bar{\nu}$	Antineutrino
ν_0	Rest frequency
ρ	Density
ρ_{crit}	Density to close the universe
σ	Cross section
	Stefan–Boltzmann constant
τ	Optical depth
τ_e	Lifetime to fall to $1/e$ of initial value
$\tau_{1/2}$	Half-life
ω	Angular speed
Ω	Angular speed
	Ratio, ρ/ρ_{crit}, for the universe

Physical and Astronomical Constants

Physical Constants

Speed of light	$c = 2.99792456 \times 10^{10}$ cm/s
	$= 2.99792456 \times 10^5$ km/s
Gravitation constant	$G = 6.6732 \times 10^{-8}$ cm^3/g-s^2
Boltzmann constant	$k = 1.3806 \times 10^{-16}$ erg/K
Planck's constant	$h = 6.6262 \times 10^{-27}$ erg/s
Stefan–Boltzmann constant	$\sigma = 5.6696 \times 10^{-5}$ erg/cm^2-K^4-s
Wien displacement constant	$\lambda_{max}T = 2.89789 \times 10^{-1}$ cm/K
Rydberg constant	$R_\infty = 1.097373 \times 10^5$/cm
Avogadro's number	$N_A = 6.022169 \times 10^{23}$/mol
Atomic mass unit	$u = 1.66053 \times 10^{-24}$ g
Mass of proton	$m_p = 1.6726 \times 10^{-24}$ g
Mass of neutron	$m_n = 1.6749 \times 10^{-24}$ g
Mass of electron	$m_e = 9.1096 \times 10^{-28}$ g
Mass of hydrogen atom	$m_H = 1.6735 \times 10^{-24}$ g
Charge of proton	$e = 4.8033 \times 10^{-10}$ esu
Bohr radius	$a_0 = 5.29177 \times 10^{-9}$ cm

Astronomical Constants

Astronomical unit	$AU = 1.4959789 \times 10^{13}$ cm
	$= 1.4959789 \times 10^8$ km
Parsec	$pc = 3.0856 \times 10^{18}$ cm
	$= 3.0856 \times 10^{13}$ km
	$= 3.2615$ ly
Light year	$ly = 9.4605 \times 10^{17}$ cm
Solar mass	$M_\odot = 1.9891 \times 10^{33}$ g
Solar radius	$R_\odot = 6.9598 \times 10^{10}$ cm
	$= 6.9598 \times 10^5$ km

Solar luminosity	$L_\odot = 3.83 \times 10^{33}$ erg/s
Earth mass	$M_\oplus = 5.977 \times 10^{27}$ g
Earth radius (equatorial)	$R_\oplus = 6.37817 \times 10^8$ cm
	$= 6.37817 \times 10^3$ km
Earth–Moon distance	$= 3.84403 \times 10^{10}$ cm
	$= 3.84403 \times 10^5$ km
Moon mass	$M_M = 7.35 \times 10^{25}$ g
Moon radius	$R_M = 1.738 \times 10^8$ cm
	$= 1.738 \times 10^3$ km
Direction of galactic center	$\alpha(1950) = 17^h42.4^m$
	$\delta(1950) = -28°55'$
Galactic center–Sun distance	$R_0 = 8.5$ kpc
Orbital speed of Sun about the galactic center	$v_0 = 220$ km/s

Units and Conversions

Prefixes and Symbols for Powers of 10

	Symbol	Prefix
10^{-18}	a	atto
10^{-15}	f	femto
10^{-12}	p	pico
10^{-9}	n	nano
10^{-6}	μ	micro
10^{-3}	m	milli
10^{-2}	c	centi
10^{-1}	d	deci
10^{1}	da	deca
10^{2}	h	hecto
10^{3}	k	kilo
10^{6}	M	mega
10^{9}	G	giga
10^{12}	T	tera

Length

1 in. = 2.54 cm (exact)
1 m = 1.0936 yd
1 km = 0.6214 mi
$1\ \text{Å} = 10^{-8}$ cm

(See Appendix A for astronomical length units.)

Energy

1 Joule = 10^{7} erg
1 erg = 6.242×10^{11} eV
$m_p c^2 = 938.3$ MeV
$m_e c^2 = 0.511$ Mev
1 Jansky (Jy) = 10^{-26} W/m^2 Hz

613

D

Planet and Satellite Properties

TABLE D.1 Planet Properties

Planet	a (A.U.)	Orbital period (yr)	i (deg)	e	Rotation period	Axis tilt (deg)	Mass (Earth)	Radius (Earth)
Mercury	0.387	0.241	7.00	0.206	58.7^d	0.	0.055	0.382
Venus	0.723	0.616	3.39	0.007	$243.^d$	$-2.$	0.815	0.949
Earth	1.000	1.000	0.00	0.017	23.9^h	23.5	1.000	1.000
Mars	1.52	1.88	1.85	0.093	24.6^h	24.0	0.107	0.533
Jupiter	5.20	11.9	1.30	0.049	9.92^h	3.1	318.	11.2
Saturn	9.54	29.5	2.49	0.056	10.7^h	29.	95.2	9.45
Uranus	19.2	84.0	0.77	0.047	23.9^h	-82.1	14.6	4.10
Neptune	30.1	165.	1.77	0.009	17.8^h	28.8	17.2	3.88
Pluto	39.4	248.	17.2	0.249	6.39^d	$\geq 50.$	0.002	0.24

Notes: a = semi-major axis of orbit. i = inclination of orbit. e = eccentricity of orbit. Rotation period is sidereal, so $23^h56^m04.1^s$. 1 A.U.. = 149,600,000 km. Mass and radius are given as fraction of Earth's. Mass of Earth = 5.997×10^{27} g. Radius of Earth (equatorial) = 6378 km.

TABLE D.2   Satellite Properties

	a (10^3 km)	Orbital period (days)	e	i (deg)	Mass (Moon)	Radius (km)
Earth						
Moon	385	27.3	0.055	18.3–28.6 V	1.00	1738
Mars						
Phobos	9.38	0.319	0.018	1.0	1.3×10^{-7}	13
Deimos	23.5	1.26	0.002	2.0	2.7×10^{-8}	7.5
Jupiter						
Adrastea	127	0.295	0.	0.		20
Metis	128	0.297	0.	0.		20
Amalthea	181	0.489	0.003	0.4		135
Thebe	221	0.670	0.	0.		40
Io	422	1.77	0.000 V	0.0	1.2	1816
Europa	671	3.55	0.000 V	0.0	0.66	1563
Ganymede	1070	7.16	0.001 V	0.2	2.0	2638
Callisto	1880	16.7	0.01	0.2	1.5	2410
Leda	11100	240.	0.146 V	26.7		5
Himalia	11500	251.	0.158 V	27.6		90
Lysithea	11700	260.	0.130 V	29.0		10
Elara	11700	260.	0.207 V	24.8		40
Ananke	20700	617 R	0.17 V	33.		10
Carme	22400	692 R	0.21 V	16.		15
Pasiphae	23300	735 R	0.38 V	35		20
Sinope	23700	758 R	0.28 V	27		15
Saturn						
Atlas	138	0.602	0.002	0.3		20
Inner F	139	0.613	0.004	0.0		70
Outer F	142	0.629	0.004	0.1		55
Epimetheus	151	0.694	0.009	0.3		70
Janus	152	0.695	0.007	0.1		110
Mimas	186	0.942	0.020	1.5	0.0005	196
Enceladus	238	1.37	0.004	0.0	0.001	255
Tethys	295	1.89	0.000	1.1	0.01	530
Telesto	295	1.89				17
Calypso	295					17
Dione	377	2.74	0.002	0.0	0.014	560
S XII	377	2.74	0.005	0.2		18
Rhea	527	4.52	0.001	0.4	0.034	765

TABLE D.2   Satellite Properties (Continued)

	a (10^3 km)	Orbital period (days)	e	i (deg)	Mass (Moon)	Radius (km)
Titan	1220	16.0	0.029	0.3	1.8	2575
Hyperion	1480	21.3	0.104	0.4		205
Iapetus	3560	79.3	0.028	14.7 V	0.026	730
Phoebe	13000	550. R	0.163	30.		110
Uranus						
1986U7	49	0.33				15
1986U8	53	0.37				10
1986U9	59	0.44				30
1986U3	62	0.47				35
1986U6	63	0.48				25
1986U2	64	0.50				35
1986U1	66	0.52				45
1986U4	70	0.56				25
1986U5	75	0.62				25
1985U1	86	0.77				85
Miranda	130	1.42	0.000	0.0	0.0005	240
Ariel	191	2.54	0.003	0.0	0.03	585
Umbriel	266	4.16	0.004	0.0	0.01	590
Titania	436	8.74	0.002	0.0	0.04	795
Oberon	583	13.5	0.001	0.0	0.04	775
Neptune						
Triton	355	5.88 R	0.000	20.	0.8	1600
Nereid	5560	360.	0.75	28	2×10^{-8}	470
Pluto						
Charon	17	6.39	0.0		0.02	400

Notes: a = semi-major axis of orbit. e = eccentricity of orbit. i = inclination of orbit. V = variable. R = retrograde. Masses are given as fraction of Moon's. Mass of Moon = 7.32×10^{25} g. Tables D.1 and D.2 were compiled from material appearing in the "Solar System Data Bank," © 1986, Astronomical Society of the Pacific.

Properties of Main Sequence Stars

Sp.T.	M	B–V	T(K)	M	M/M☉	R/R☉	L/L☉
O5	−6	−0.45	35,000	−10.6	39.8	17.8	3.2×10^5
B0	−3.7	−0.31	21,000	−6.7	17.0	7.6	1.3×10^4
B5	−0.9	−0.17	13,500	−2.5	7.1	4.0	6.3×10^2
A0	+0.7	0.00	9,700	0.0	3.6	2.6	7.9×10^1
A5	+2.0	+0.16	8,100	+1.7	2.2	1.8	2.0×10^1
F0	+2.8	+0.30	7,200	+2.7	1.8	1.4	6.3
F5	+3.8	+0.45	6,500	+3.8	1.4	1.2	2.5
G0	+4.6	+0.57	6,000	+4.6	1.1	1.05	1.3
G5	+5.2	+0.70	5,400	+5.1	0.9	0.93	7.9×10^{-1}
K0	+6.0	+0.84	4,700	+5.8	0.8	0.85	4.0×10^{-1}
K5	+7.4	+1.11	4,000	+6.8	0.7	0.74	1.6×10^{-1}
M0	+8.9	+1.39	3,300	+7.6	0.5	0.63	6.3×10^{-2}
M5	+12.0	+1.61	2,600	+9.8	0.2	0.32	7.9×10^{-3}

Astronomical Coordinates and Timekeeping

Coordinate Systems

When we want to locate a star, or any other astronomical object, we only need to specify its direction. We don't care about its distance. We therefore need only two coordinates, two angles, to locate an astronomical object. Sometimes, it is convenient to think (as the ancients did) of the stars as being painted on the inside of a sphere, the celestial sphere. Just as we can locate any place on the surface of the Earth with two coordinates, latitude and longitude, we need two coordinates to locate an object on the celestial sphere.

We choose coordinate systems for convenience in a particular application. In general, to set up a coordinate system we first identify an equator and then choose coordinates that correspond to latitude and longitude.

A convenient system for any particular observer is the *horizon system*. The horizon becomes the equivalent of the equator in that system. The angle around the horizon, measured from north, through east, south, and west, is the *azimuth*. The angle above the horizon is called the *elevation*. Instead of elevation, we can use the *zenith distance*, which is the angle from the zenith (overhead) to the object. From their definitions, we can see that the sum of the zenith distance and the elevation is always 90°. The azimuth ranges from 0° to 360°, and the elevation from −90° to 90° (with negative elevations being below the horizon). The problem with this system is that, as the Earth rotates, the azimuths and elevations of the stars change in a complicated way. It would not be very useful to prepare a catalog of stars, just giving their azimuths and elevations.

One solution is to use a coordinate system based on the projection of the Earth's equator onto the celestial sphere, the *celestial equator*. Such a system is called an *equatorial coordinate system*. The angle above or below the celestial equator is called the *declination*, and is designated with the symbol δ. It ranges fom −90° to 90°. As the Earth rotates, the declinations of objects don't change. Also, the declination doesn't change as observers move around the Earth. There are two ways to measure the other coordinate:

(1) We can measure the angle, going westward, from the observer's meridian. This is called the *hour angle H*. We can measure it from 0° to 360°, but it is convenient to express it in units of time, from 0 to 24 hr. As the Earth rotates, the hour angle of each object changes, but in a simple way, increasing by one hour for each hour that the Earth rotates. In addition, at any instant, observers at different longitudes will measure different hour angles for the same object. (The hour angles differ by the difference in longitudes.)

(2) To get a coordinate that is the same for all observers and doesn't change with time, we must fix that coordinate with respect to a location on the celestial sphere. We choose as our reference point one of the two intersections of the celestial equator and the ecliptic (the Sun's path around the sky). These two points are called the *equinoxes*. (When the Sun is at either equinox, all observers on Earth have a 12-hr day and a 12-hr night. This occurs on the first day of spring and the first day of fall.) We choose the *vernal equinox*, the point where the Sun is on the first day of spring, as our starting point for the coordinate, *right ascension*, designated by the symbol α. The right ascension is measured from 0 to 24 hr, increasing from west to east. That is, objects with higher right ascensions cross an observer's meridian later than objects with lower right ascensions.

One effect of the Earth's precession is to move the equinoxes by about 50 arc sec per year. Thus, the origin of our coordinate system is drifting. Therefore, in compiling a catalog of objects, the time, or *epoch*, at which the coordinates apply must be specified. The epoch is usually put in parentheses after the α and δ, for example, α(1950) and δ(1950). An observer can then calculate where the objects will be on the date they are to be observed (a relatively simple calculation). To keep things simple, we generally agree on standard epochs, and keep our catalogues on a common standard. For example, many current catalogs use the standard epoch 1950. Catalogs that are just coming out are beginning to use the standard epoch 2000. By changing the standard epoch every 50 years, we don't have to change too often, but keep the catalog coordinates reasonably close to the actual coordinates. (In 50 years, the origin will have moved by less than one degree.)

Other coordinate systems are useful for studying particular sets of objects. For example, in studying the solar system, *ecliptic coordinates* are useful. In this system, the *ecliptic latitude* β is measured above or below the ecliptic (from −90 to 90°), and the *ecliptic longitude* λ is measured around the ecliptic, starting at the vernal equinox and increasing eastward (from 0° to 360°).

A coordinate system that is useful for studying galactic structure is the *galactic coordinate system*. (We touched on this briefly in Chapter 16.) The *galactic latitude b* is measured above and below the galactic plane. The *galactic longitude l* is measured from the galactic center, increasing in the same direction as the right ascension.

Once an object is located in one coordinate system, those coordinates can be transformed into any of the other systems. The equations for those transformations are beyond the scope of this brief summary, but the calculations can be done on a hand calculator, and certainly by a computer that would be involved in pointing a large telescope.

Timekeeping

It is natural to use the Earth's rotation as a basis for timekeeping. We can keep track of the Earth's rotation by noting the motion of the stars. For each rotation of the Earth the stars make a full circle in the sky. We can therefore measure time by choosing a star, or other point in the sky, and seeing the fraction of its daily circle that it has made.

We choose the reference point to be the vernal equinox. We measure a time, called *local sidereal time (LST)*, by the progress of the vernal equinox. When the vernal equinox is at an observer's meridian, the LST is zero for that observer. One hour later, the LST is 1 hr; in addition, the hour angle of the vernal equinox is 1 hr. This means that the *LST is simply the hour angle of the vernal equinox*. When the LST time is 1 hr, objects with a right ascension of 1 hr will be on the meridian. This means that the *LST is also equal to the right ascension of the object that happens to be on the meridian*. Observers at two different points on Earth will have different LSTs. The LSTs will differ by the longitude difference between the two observers.

Most of our civil timekeeping is referenced to the Sun. We could define a local solar time, based on the hour angle of the Sun. This is what a sundial measures, but this is not very useful for civil systems. For uniformity, civil systems utilize time zones. This means that the Sun can be as much as a half an hour ahead or behind your local time, even more for some very wide time zones.

As the Earth moves around the Sun, the Sun appears projected against a changing background of stars (progressing through the constellations of the Zodiac). This means that the right ascension of the Sun is continuously changing. The right ascension of the Sun increases by an average of $3^m 56.56^s$ per day. The time for the Sun to go from one passage of your meridian to the next is this much longer than the time for a star to go from one passage to the next. Therefore a day by the Sun, a *solar day*, is longer, by this amount, than a day according to the stars, a *sidereal day*.

Another problem with solar time is that the right ascension of the Sun does not change smoothly. This is the result of two effects: (1) The Earth's orbit is elliptical, so the Earth moves faster when it is closer to the Sun, and slower when it is farther. This variable speed is mirrored in the apparent motion of the Sun against the background of stars. (2) The Sun moves along the ecliptic, which makes a $23\frac{1}{2}°$ angle with the celestial equator. Therefore, the right ascension and declination of the Sun are both changing. Even if the Sun were to move along the ecliptic at a constant rate, its right ascension would change at a variable rate. Because of these two effects, we define a fictitious object, called the *mean Sun*, which moves along the celestial equator at a uniform rate. Time kept by the mean Sun is called *mean solar time*, and is time that would be kept by a clock. The relationship between the mean Sun and true Sun, and thus the relationship between sundial and clock time, is given by a quantity called the *equation of time*. This quantity is depicted graphically by the distorted figure ''8'' that appears in the empty areas of some globes. This is called an *analemma*.

To the extent that solar time is used in astronomical timekeeping, it is usually *Universal Time (UT)*, which is the mean solar time at Greenwich, England. It is often useful to convert from UT to LST for any observer. This is done with the aid of a publication, like the *Astronomical Almanac*, published by the U.S. Naval Observatory. The Almanac gives, for each date, the LST at Greenwich at 0^h UT. To this, we add the UT of interest, multiplied by a factor to account for the difference between sidereal and solar time. This gives the LST at Greenwich at the UT of interest. We then subtract the longitude of the observer L. We can write this as

$$\text{LST}(L, \text{UT}) = \text{LST}(0, 0) - L + \text{UT}(1 + 2.738 \times 10^{-3})$$

where LST(0,0) is the LST at Greenwich at 0^h UT.

All of this is complicated by the effects of precession, and the wobble of the Earth, known as *nutation*. While *true sidereal time* is the actual hour angle of the vernal equinox, our sidereal clocks really keep a *mean sidereal time*. There is also a problem in the definition of a year. A *sidereal year* is the time for the Sun to return to the same place with respect to the fixed background of stars. We could use this definition, but after a long time, the precession will cause the seasons to occur in the wrong months. For this reason, we use a *tropical year*, defined as the time it takes for the Sun to travel from vernal equinox to vernal equinox, remembering that the equinox moves while this is happening. A sidereal year has 365.2564 mean solar days, while a tropical year has 365.2422 mean solar days.

Our definition of the year also brings us back to a definition for a Universal Time whose rate is really constant. We base *Ephemeris time* on the rate of the mean Sun at the beginning of the year 1900. Ephemeris time is a certain fraction of the tropical year 1900 (ephemeris year), which contains 365.242199 mean solar days, so the ephemeris second is defined as 1/31,556,925,974 of an ephemeris year.

Abundances of the Elements

Atomic no.	Symbol	Element	Atomic weight	Relative abundance by number[a]
1	H	Hydrogen	1.008	1.00
2	He	Helium	4.003	1.45×10^{-1}
3	Li	Lithium	6.939	1.00×10^{-9}
4	Be	Beryllium	9.013	2.51×10^{-10}
5	B	Boron	10.812	6.31×10^{-10}
6	C	Carbon	12.012	3.02×10^{-4}
7	N	Nitrogen	14.007	9.12×10^{-5}
8	O	Oxygen	16.000	6.76×10^{-4}
9	F	Fluorine	18.999	2.51×10^{-7}
10	Ne	Neon	20.184	2.75×10^{-4}
11	Na	Sodium	22.991	1.66×10^{-6}
12	Mg	Magnesium	24.313	2.88×10^{-5}
13	Al	Aluminum	26.982	1.91×10^{-6}
14	Si	Silicon	28.09	2.95×10^{-5}
15	P	Phosphorus	30.975	3.39×10^{-7}
16	S	Sulphur	32.066	1.66×10^{-5}
17	Cl	Chlorine	35.454	2.51×10^{-7}
18	Ar	Argon	39.949	4.17×10^{-6}
19	K	Potassium	39.103	7.59×10^{-8}
20	Ca	Calcium	40.08	1.66×10^{-6}
21	Sc	Scandium	44.958	8.13×10^{-10}
22	Ti	Titanium	47.90	6.61×10^{-8}
23	V	Vanadium	50.944	6.03×10^{-9}
24	Cr	Chromium	52.00	2.40×10^{-7}
25	Mn	Manganese	54.940	1.26×10^{-7}
26	Fe	Iron	55.849	7.94×10^{-6}

Atomic no.	Symbol	Element	Atomic weight	Relative abundance by number[a]
27	Co	Cobalt	58.936	5.25×10^{-8}
28	Ni	Nickel	58.71	8.51×10^{-7}
29	Cu	Copper	63.55	4.47×10^{-8}
30	Zn	Zinc	65.37	1.91×10^{-8}
31	Ga	Galium	69.72	2.82×10^{-10}
32	Ge	Germanium	72.60	1.51×10^{-9}
33	As	Arsenic	74.924	2.0×10^{-10}
34	Se	Selenium	78.96	1.6×10^{-9}
35	Br	Bromine	79.912	4.0×10^{-10}
36	Kr	Krypton	83.80	1.6×10^{-9}
37	Rb	Rubidium	85.48	2.24×10^{-10}
38	Sr	Strontium	87.63	5.62×10^{-10}
39	Y	Yttrium	88.908	2.51×10^{-10}
40	Zr	Zirconium	91.22	2.5×10^{-10}
41	Nb	Niobium	92.91	5.0×10^{-11}
42	Mo	Molybdenum	95.95	8.32×10^{-11}
43	Tc	Technetium	99.0	
44	Ru	Ruthenium	101.07	3.31×10^{-11}
45	Rh	Rhodium	102.91	6.03×10^{-12}
46	Pd	Palladium	106.4	1.78×10^{-11}
47	Ag	Silver	107.874	5.0×10^{-12}
48	Cd	Cadmium	112.41	3.16×10^{-11}
49	In	Indium	114.82	7.9×10^{-12}
50	Sn	Tin	118.70	3.55×10^{-11}
51	Sb	Antimony	121.78	4.0×10^{-11}
52	Te	Tellurium	127.61	1.0×10^{-10}
53	I	Iodine	126.909	2.5×10^{-11}
54	Xe	Xenon	131.30	1.0×10^{-10}
55	Cs	Caesium	132.91	1.3×10^{-11}
56	Ba	Barium	137.35	1.29×10^{-10}
57	La	Lanthanum	138.92	2.5×10^{-11}
58	Ce	Cerium	140.13	4.0×10^{-11}
59	Pr	Praseodymium	140.913	6.3×10^{-12}
60	Nd	Neodymium	144.25	3.2×10^{-11}
61	Pm	Promethium	147.0	
62	Sm	Samarium	150.36	1.0×10^{-11}
63	Eu	Europium	151.96	5.0×10^{-12}
64	Gd	Gadolinium	157.25	1.3×10^{-11}
65	Tb	Terbium	158.930	2.5×10^{-12}
66	Dy	Dysprosium	162.50	1.6×10^{-11}
67	Ho	Holmium	164.937	3.2×10^{-12}

Atomic no.	Symbol	Element	Atomic weight	Relative abundance by number[a]
68	Er	Erbium	167.27	7.9×10^{-12}
69	Tm	Thulium	168.941	1.3×10^{-12}
70	Yb	Ytterbium	173.04	1.3×10^{-11}
71	Lu	Lutecium	174.98	2.0×10^{-12}
72	Hf	Hafnium	178.50	4.0×10^{-12}
73	Ta	Tantalum	180.955	2.0×10^{-12}
74	W	Tungsten	183.86	1.3×10^{-11}
75	Re	Rhenium	186.3	4.0×10^{-12}
76	Os	Osmium	190.2	2.0×10^{-11}
77	Ir	Iridium	192.2	1.6×10^{-11}
78	Pt	Platinum	195.10	4.0×10^{-11}
79	Au	Gold	196.977	5.0×10^{-12}
80	Hg	Mercury	200.60	7.9×10^{-12}
81	Tl	Thallium	204.38	3.2×10^{-12}
82	Pb	Lead	207.20	4.0×10^{-11}
83	Bi	Bismuth	208.988	5.0×10^{-12}
84	Po	Polonium	210.0	
85	At	Astatine	211.0	
86	Rn	Radon	222.0	
87	Fr	Francium	223.0	
88	Ra	Radium	226.05	
89	Ac	Actinium	227.	
90	Th	Thorium	232.047	2.0×10^{-12}
91	Pa	Protactinium	231.	
92	U	Uranium	238.03	1.0×10^{-12}
93	Np	Neptunium	237.05	
94	Pu	Plutonium	242.0	
95	Am	Americium	242.0	
96	Cm	Curium	245.0	
97	Bk	Berkelium	248.0	
98	Cf	Californium	252.0	
99	Es	Einsteinium	253.0	
100	Fm	Fermium	257.0	
101	Md	Mendelevium	257.0	
102	No	Nobelium	255.0	
103	Lr	Lawrencium	256.0	
104	Rf	Rutherfordium	261.0	
105	Ha	Hahnium	262.0	

[a]Abundances are by number, relative to hydrogen. These represent the best determinations of solar or solar system abundances. No entry means that the abundance is not well determined.

Photo Credits

Chapter 3
3.2	Deutches Museum, Munich.
3.4	George Herbig, Lick Observatory, with permission of *Astrophysical Journal*.
3.5, 3.9	Hale Observatory.

Chapter 4
4.6	Lick Observatory.
4.7a	Hale Observatory.
4.7b	NOAO.
4.9a	Hale Observatory.
4.9b	NOAO.
4.10	Hale Observatory.
4.11a	Harvard Smithsonian Center for Astrophysics.
4.11b	NOAO.
4.12a,b	NOAO.
4.13	Harvard Smithsonian Center for Astrophysics.
4.16	NOAO.
4.17	Institute for Astronomy, University of Hawaii.
4.18	Space Telescope Science Institute.
4.19a,b	Hale Observatory.
4.19c	Harvard Smithsonian Center for Astrophysics.
4.19d	George Carruthers, Naval Research Laboratory.
4.20	Giovanni Fazio, CFA.
4.21	NASA.
4.22	NASA/JPL/Infrared Processing & Analysis Center—Caltech.
4.23	NASA/JPL/Infrared Processing & Analysis Center—Caltech.
4.24a	Courtesy NRAO/AUI.
4.24b	Marc Kutner.
4.25	Thomas Pauls, NRL.
4.28, 4.29	Courtesy NRAO/AUI.
4.30, 4.31, 4.32	Harvard Smithsonian Center for Astrophysics

Chapter 5
5.6	Hale Observatory.
5.16	David Evans, University of Texas, Austin.
5.17	Simon Worden, with permission of *Astrophysical Journal*.

Chapter 6
6.1	Hale Observatory.
6.6	NOAO.
6.7, 6.11	Hale Observatory.
6.15	David Lynch, Aerospace Corporation.
6.16, 6.17, 6.20	Hale Observatory.
6.24	William Livingston, NOAO.
6.25, 6.28, 6.29	Hale Observatory.

Chapter 7
7.13	Courtesy Calspan Corporation, Buffalo, NY.
7.15	NASA.
7.16a	Marc Kutner.
7.16b	NASA.
7.21	William Hamilton, Louisiana State University.

Chapter 8
8.6	Hale Observatory, supplied by F. Vrba, USNO.
8.7	William Herbst, Van Vleck Observatory.
8.8, 8.9, 8.10	George Herbig, Lick Observatory.
8.12	NASA/JPL/Infrared Processing & Analysis Center—Caltech.

Chapter 9
9.12, 9.13	Raymond Davis, Brookhaven National Laboratory.

Chapter 10
10.8, 10.10	NOAO.

Chapter 11
11.1	Lick Observatory.
11.3a	Hale Observatory.
11.3b	Courtesy NRAO/AUI.
11.3c	Harvard Smithsonian Center for Astrophysics.
11.5	Hale Observatory.

625

11.8	Joseph Taylor, Princeton University.
11.9	Joseph Taylor, Princeton University with permission of *Astrophysical Journal*.
11.12	NOAO.
11.13, 11.14	Joseph Taylor, Princeton University.

Chapter 12

12.5	Gary Emerson, E. E. Barnard Observatory.
12.6, 12.7a	P. Joss, Massachusetts Institute of Technology.
12.7b	Courtesy Walter H. G. Lewin, Massachusetts Institute of Technology, from the *Annals of the NY Academy of Sciences,* Vol. 302, p. 218 (1977).
12.10	E. A. Boldt, NASA/GSFC with permission of *Astrophysical Journal*.
12.11	Dr. Jerome Kristian, Mount Wilson and Las Campanos Observatory. (Palomar Observatory photo.)
12.12	Bruce Margon, University of Washington.

Chapter 13

13.1, 13.2	NOAO.
13.9	William Harris, McMaster University.

Chapter 14

14.6	F. Gillette, NOAO.
14.11	C. Heiles, University of California, Berkeley.
14.13	Courtesy NRAO/AUI.

Chapter 15

15.2	Harvard–Smithsonian Center for Astrophysics and *Scientific American*.
15.3	NOAO.
15.5	John Bally, AT&T Bell Laboratories, with permission of *Astrophysical Journal*.
15.9, 15.10	Ronald Snell, FCRAO, with permission of *Astrophysical Journal*.
15.11, 15.12	Stuart Vogel, University of California, Berkeley, with permission of *Astrophysical Journal*.

Chapter 16

16.1	Hale Observatory.
16.2	William Harris, McMaster University.
16.6	Leo Blitz, University of Maryland, with permission of *Astrophysical Journal*.
16.8	Frank Kerr, University of Maryland, with permission of *Astrophysical Journal*.
16.9	Plot made from data presented by Georgelin and Georgelin, *Astronomy and Astrophysics,* 1976, Vol. 49, p. 57.
16.10	Thomas Dame, CFA.
16.11	Kathryn Mead, Naval Research Laboratories.
16.13a	George Rieke, University of Arizona, with permission of *Astrophysical Journal*.

16.13b	K. Y. Low, Hale Observatory.
16.14	John Lacey, University of Texas, Austin.

Chapter 17

17.2	NOAO.
17.3a	Hale Observatory.
17.3b,c,d	NOAO.
17.4a,c	Hale Observatory.
17.4b, 17.5, 17.7a	NOAO.
17.7b	Hale Observatory.
17.8	NOAO.
17.10	Robert Kennicut, University of Minnesota and Paul Hodge, University of Washington.
17.11	Philip Seiden, IBM.
17.12, 17.13, 17.14	E. Brinke, Sterrewacht & Leiden.
17.14	Judith Young, FCRAO.
17.15	Antony Starke, AT&T Bell Laboratories.
17.16	NASA/JPL/Infrared Processing & Analysis Center—Caltech.
17.20	Alar Toomre, Massachusetts Institute of Technology.
17.21a	NOAO.
17.21b	Robert Kennicut, University of Minnesota.
17.21c	Michelle Kaufman, Ohio State University, Frank Bash and Butler Hine, University of Texas, Austin, and Arnold Rots, NRAO.
17.22	Vera Rubin, Department of Terrestrial Magnetism.

Chapter 18

18.1	P. J. E. Peebles, Princeton University.
18.2	NOAO.
18.3	Harvard Smithsonian Center for Astrophysics.
18.4a	NOAO.
18.4b	Harvard Smithsonian Center for Astrophysics.
18.5a	Alar Toomre, Massachusetts Institute of Technology.
18.5b	Hale Observatory.
18.6	Hale Observatory.
18.7	J. B. Oke, Hale Observatory, with permission of *Astrophysical Journal*.
18.10	R. B. Tully, Institute for Astronomy, University of Hawaii.
18.11	Margaret Geller, CFA, with permission of *Astrophysical Journal*.

Chapter 19

19.3	Courtesy NRAO/AUI.
19.6a	S. Unwin, OVRO.
19.8	Yerkes Observatory.
19.11	NOAO.
19.12	Maarten Schmidt, Hale Observatory.

Index

AAVSO, 233
Aberration, 54–55
  of starlight, 140
Absolute magnitude, 23–24
Absorption coefficient, 114
Absorption lines. *See also* Fraunhofer lines;
    Hydrogen, stellar; Quasars
  interstellar, 319–322
  interstellar cyanogen, 464–466
  interstellar hydrogen, 320–321
  solar, 120
  stellar, 27–43
  21-cm, 320–321
Abundances, 622–624
  cosmological, 473
  solar system, 214
Accelerators, 474–475
Active galaxies, 419–439
Albedo, 312, 317–318, 595. *See also* individual objects for their albedos
Allende meteorite. *See* Meteorite
Allowed states. *See* Quantum mechanics
Alpha decay, 204–205
Alpha particle, 204–205
Alpher, Ralph, 460, 463
Aluminzing, 59
Analemma, 629
Andromeda galaxy. *See* M31
Ångstrom, 9
Angular momentum, 98–99, 505, 509
Angular resolution, 50–52, 77, 79–81
Angular velocity, 89–90
Annular eclipse. *See* Sun, eclipses

Antenna, 77
Anti-matter, 464
Apastron. *See* Orbits, elliptical
Aperture synthesis, 80–81
Apollo asteroids, 596
Apollo program, 543
Apparent magnitude, 9–11
Argon, 225, 560, 563
Ariel. *See* Uranus, moons
Arizona, University of, 61
Associations. *See* OB Associations
A stars, 42
Asteroid belt, 595–596
Asteroids, 595–597
Astigmatism, 54–55
Astrometric binary, 86
Astrometry, 86
Astronomical unit (AU), 22
Astrophysics, 5
Atmosphere. *See* individual planets and objects, atmosphere
Atom, 30–37
Atomic number, 201
AU. *See* Astronomical unit
Aurora, 535–536
Autocorrelator, 78
Azimuth, 618

Background radiation. *See* Cosmic background radiation
Balloon, 74
Balmer, J. J., 30
Balmer series, 30–31, 35

Basalt, 543
Bell Burnell, Jocelyn, 258
Bell Laboratories, 75–76, 463
Bending of light. *See* General theory of relativity
Beta decay, 205
Beta particle, 205
Big bang. *See* Cosmology
Binary stars, 85–106
  black holes in, 280–283
  close, 272–285
  novae, 275–277
  pulsar, 279–280
  stellar sizes, 105
  white dwarfs, 275–277
  x-ray sources, 277–280
Binding energy, 35
Bipolar flows, 349–352
Black body, 12–17. *See also* Cosmic background radiation
Black hole
  in active galactic nuclei, 422–424
  basic properties, 166–173
  stellar, 269–270
BL Lacertae objects, 438
Blueshift. *See* Doppler shift
Bode's law, 497
Bohr, Neils, 31
Bohr atom, 31–37
Bok, Bart, 76
Bok globules, 305
Bolometer, 24, 71–72
Bolometric magnitude, 24

Boltzmann constant, 15
Boltzmann distribution, 38–39
Boltzmann equation, 39
Bound-free transition, 117
Bound states, 35
Brahe, Tycho, 495
Bremsstrahlung. *See* Free–free radiation
Brightness, 7–9
B stars, 43
Butterfly diagram. *See* Sun, sunspots
B − V. *See* Color

California Institute of Technology. *See* Hale
    Observatories
Callisto. *See* Jupiter, moons
Camera, 52–55
Cannon, Annie Jump, 29
Canyon. *See* Mars
Carbonaceous chondrite. *See* Meteorites
Carbon cycle. *See* CNO cycle
Carbon dioxide, greenhouse effect and. *See*
    Venus, greenhouse effect
Carruthers, George, 70
Cas A, 251
Cassegrain telescope. *See* Telescopes, optical
Cassini's division, *See* Saturn, rings
Cassiopeia A, 251
Celestial equator, 619
Celestial poles, 619
Celestial sphere, 619
Center of mass, 91
Cepheid variable, 232–239, 375
Charge, electric, 33
Charm, 476
Chondrite, carbonaceous. *See* Meteorites
Chromatic aberration, 54–55
Chromosphere. *See* Sun, chromosphere
Closed universe. *See* Cosmology
Clusters, 287–300. *See also* Galactic clus-
    ters; Globular clusters
    distances to, 289–291
    dynamics, 291–297, 403–409
    of galaxies, 401–409
Cluster variables. *See* RR Lyrae stars
CNO cycle, 211–212
Collapsing clouds. *See* Star formation
Color
    and filter systems, 19
    stellar, 18–19
    wavelength, 13–14
Color force, 479–480
Color-magnitude diagram. *See* Hertz-
    sprung–Russell diagram
Color of quark, 479–480
Column density, 323
Coma. *See* Comets
Coma cluster of galaxies, 402
Comets
    coma, 579–599

composition, 599–600
Halley's, 598–603
head, 597
mass, 597
nucleus, 597–599
origin, 602–603
snowball model of, 597
spectra, 600
tail, 599–603
Composite spectrum, 85
Compton scattering, 17
    inverse, 471
Cone, exit. *See* General theory of relativity,
    black holes
Conjunction, 495
Conservation laws
    angular momentum, 98–99
    energy, 99
    momentum, 92
    symmetries, and, 478–479
Continental drift. *See* Earth, plate tectonics
Continuous spectrum, 7–17
Continuum. *See* Blackbody radiation; Con-
    tinuous spectrum
Convection, 530–531
Convection zone, 118
Convergent point, 290–291
Coordinate systems, 618–620
Copernicus, Nicholas, 493–494
Core. *See* individual objects
Corona, solar, 111, 124–127
Coronagraph, 124
Cosmic background radiation
    discovery, 463–464
    isotropy, 467–471
    nucleosynthesis and, 472
    origin, 460–463
    spectrum, 464–467
Cosmological constant, 450
Cosmology, 440–488. *See also* Cosmic
    background radiation
    expansion of the universe, 409–416
    inflation, 486–487
    models, 441–442, 455–458
    Newtonian, 444–450
    particle physics and, 475–488
    relativity and, 450–454
Coude focus, 59
Crab nebula, 252–253
Crab pulsar, 261–262
Craters. *See* individual objects
Cross section, 113–114
Curtis, Heber D., 375
Curvature of space. *See* General theory of
    relativity
Cygnus A, 420
Cygnus loop, 251
Cygnus X-1, 273, 281–283
Cytherian. *See* Venus

Dark matter, 396
Davis, Raymond, Jr., 224–226
Deceleration parameter, 448–449
Declination, 619
Degeneracy
    electron, 240–244
    neutron, 254–256
Deimos, *See* Mars, moons
δ Cephei, 233–234
Density. *See* individual objects
Density wave, 389–390
Deuterium, 472–475
Differential forces. *See* Tides
Differential rotation
    Jupiter, 565
    Milky Way galaxy, 356–367
    Sun, 131
Differentiation, 514
Diffraction, 50–52, 77
Diffraction grating, 65
Dione, *See* Saturn, moons
Dispersion
    of light, 64
    of pulsar pulses, 268–269
Distance ambiguity, 366–367
Distance modulus, 23
D lines. *See* Sodium
Doppler broadening, 121–123
Doppler effect, 86–89
    anisotropy of CBR, 467–468
    binary pulsar, 279–280
    binary stars, 86–91
    Cepheid variables, 235–237
    clusters of galaxies, 291–297
    Hubble's law, 401–416
    line broadening, 121–122
    planetary nebulae, 238–239
    quasars, 431
    radar mapping, 550–552
    radial velocities, 86–89
    relativistic, 148–151
    rotation of galaxies, 356–367, 396
    Saturn's rings, 577
Doppler shift. *See* Doppler effect
Dust. *See* Interstellar dust
Duty cycle. *See* Pulsars
Dwarf stars. *See* Hertzsprung–Russell dia-
    gram, main sequence

Earth, 514–541
    age, 517–518
    atmosphere, 522–535
    continental drift. *See* Plate tectonics
    history, 514
    interior, 514
    magnetic field, 535–537
    magnetosphere, 535–537
    plate tectonics, 518–519
    precession, 540–541

Earth (*Continued*)
　radiation belts, 535–537
　tides, 537–541
Eccentricity. *See* Ellipse
Eclipse
　geometry of, 500–501
　lunar, 501
　solar, 124–125
Eclipsing binary, 85–86, 105
Ecliptic, 619
Einstein, Albert, 451
Einstein Observatory, 82
Electromagnetic force, 477
Electromagnetic radiation. *See* Electromag-
　netic spectrum
Electromagnetic spectrum, 9–11
Electromagnetic waves, 9–10
Electron, 30–36
　degeneracy, 240–244
Electron volt (eV), 34–35
Elements, abundances, 622–624
Ellipse, 97–102
Elongation, 495
Emission lines, 27–28, 329
　in quasars, 430
　in stars, 28
　in T Tauri stars, 195
　21-cm, 321
Energy
　conservation, 99
　transport, 220–221
Energy levels, 31–36
　hydrogen, 34–35
　molecules, 328–329
Ephemeris time, 621
Epicycles, 493
Epoch, 619
Equation of state, 219
Equation of time, 629
Equations of stellar structure. *See* Stellar
　structure
Equivalence, principle of. *See* General the-
　ory of relativity
Ergosphere. *See* General theory of relativity,
　black holes
Escape velocity, 503–504
Eta Carina nebula, 343
Euclidean geometry. *See* General theory of
　relativity, foundations
Event horizon. *See* General theory of relativ-
　ity, black holes
Evolution, stellar, *See* Stars, evolution
Evolutionary tracks. *See* Hertzsprung–Russell
　diagram
Excitation temperature, 39
Exit cone. *See* General theory of relativity,
　black holes
Expansion of the universe. *See* Cosmology
Extinction. *See* Interstellar extinction
Eyepiece, 55–57

Field, magnetic. *See* individual objects
Filter, 12
Flare, solar. *See* Sun
Flash, helium, 231
Flash spectrum, 124
Flows, energetic, 349–353
Flux, 7–8, 20
Focal length, 53–57
Focal ratio, 54
Focus, 53–57. *See also* Ellipse
Force
　color, 479–481
　electromagnetic, 477
　gravitational, 92–95
　magnetic lines of, 129–130
　nuclear, 201–215
　strong, 477
　unification, 481–487
　weak, 477
Formaldehyde. *See* Molecules, interstellar
Fourier transform spectroscopy, 66
Fraunhofer, Joseph, 27
Fraunhofer spectrum, 27–28
Free-fall time, 181–184
Free–free radiation, 345
Frequency, 9–10
F stars, 42
F-stop, 54
FU Orionis, 195–196
Fusion, 206–208

Galactic cannibalism, 409
Galactic center, 371–373
Galactic clusters, 287–288
　age of, 298
　H–R diagram, 298
Galactic coordinates, 358, 619–620
Galaxies
　active, 382, 419–439
　barred spiral, 379
　clusters of, 401–409
　dark matter in, 396
　density waves, 389–394
　ellipticals, 376–377, 381
　formation of, 485–486
　halos, 396
　Hubble type, 375–376
　interacting, 408–409
　irregular, 381
　Local Group, 403
　Local Supercluster, 416
　Milky Way, 355–373
　N, 427–429
　peculiar, 382
　quasars, 429–437
　rotation, 394–396
　Seyfert, 382, 427–429
　spiral, 378, 382–388
　spiral structure of, 388–394
　starburst, 388

superclusters of, 416–417
　SO, 380–381
Galileo Galilei, 494
Gamma rays, 369
　in nuclear reactions, 205
Ganymede. *See* Jupiter, moons
Gas. *See* Interstellar gas; individual objects
General theory of relativity, 155–173,
　450–454
　bending of electromagnetic radiation,
　　162–163
　black holes, 166–173
　curved spacetime, 155–156
　gravitational radiation, 166, 279–280
　gravitational redshift, 163–165
　orbiting bodies, 161–162
　principle of equivalence, 157–158
　tests of, 159–166, 279–280
Geodesic, 156
Giant planets, 565–592
Giant stars. *See* Hertzsprung–Russell diagram;
　Red giants
Glitch. *See* Pulsar
Globular clusters, 287, 289, 291–299
　Cepheids in, 233
　dynamics, 291–297
　H–R diagram, 297–299
　RR Lyrae stars in, 237
Globlule, 305
Gluons, 479–481
Gravitation, universal, 92–95, 177–178,
　444–450, 504–505. *See also* General
　theory of relativity
Gravitational lenses, 437–439
Gravitational potential energy, 99, 177–178,
　444–450
Gravitational radiation. *See* General theory
　of relativity
Gravitational redshift. *See* General theory of
　relativity
Gravitational waves. *See* General theory of
　relativity
Gravitons, 477
Gravity. *See* Gravitation, universal
Gravity assist, 507–508
Grazing incidence, 81–82
Great Red Spot. *See* Jupiter
Green Bank. *See* National Radio Astronomy
　Observatory
Greenhouse effect. *See* Venus, greenhouse
　effect
Ground state. *See* Energy levels
G stars, 41

H. *See* Hydrogen
H_2. *See* Molecules, interstellar
HI regions, 320–324
HII regions, 342–346
Hale, George Ellery, 130

Hale Observatories, Palomar Mountain, 58
Half-life, 516–517
Halley's Comet. *See* Comets
H alpha, 31, 35, 123
Haro, Guillermo, 195
Hawking, Stephen, 172
HCN. *See* Molecules, interstellar
H₂CO. *See* Molecules, interstellar
HDE 226868, 281–283
HEAO (High Energy Astronomy Observatory), 82
Heliocentric theory, 493
Helium, and cosmology, 472–475
Helium flash, 231
Henry Draper Catalogue, 29
Herbig, George, 195
Herbig-Haro objects, 195–196
Hercules cluster, 402
Hercules X-1, 277–279
Herman, Robert, 460, 463
Herschel, William, 567
Hertz, 9
Hertzsprung, Enjar, 43
Hertzsprung–Russell diagram, 43–46
  ages of clusters, 297–298
  evolution on, 230
  galactic clusters, 297–298
  globular clusters, 297–299
  luminosity class, 44–46
  main sequence, 44–46
  red giant, 44–45
  red supergiant, 44–45
  Sun, position of, 45
  white dwarfs, 44–46
  zero-age main sequence, 230
Hewish, Antony, 258
High-Energy Astronomy Observatory. *See* HEAO
High-energy astrophysics, 81–82
H₂O. *See* Water
Horizontal branch. *See* Hertzsprung–Russell diagram
Horsehead Nebula, Color plate 1
Hour angle, 619
H–R diagram. See Hertzsprung–Russell diagram
Hubble, Edwin P., 409
Hubble classification. *See* Galaxies
Hubble's constant, 409–416
Hubble's law, 409–416
Hubble time, 412
Hyades, 291
Hydrogen. *See also* HI; HII; Molecules, interstellar
  atom, 30–37
  Balmer series, 30, 35
  energy levels, 35
  fusion, 209–211
  interstellar, 320–324
  isotopes, 201

negative ion, 117
  planetary atmospheres, 525
  recombination lines, 345
  stellar, 41–42
Hydrostatic equilibrium, 216–219
Hydroxyl radical. *See* Molecules, interstellar
HZ Herculis, 277–279

IAU. *See* International Astronomical Union
Ideal gas, 219
Inclination, 91, 95
Index of refraction, 52, 55, 268–269
Inertia, 158
Inertial frames. *See* Special theory of relativity
Infrared, telescopes, 71–75. *See also* individual objects
Infrared Astronomy Satellite (IRAS), 74–75, 197
Infrared Telescope Facility (IRTF), 73
Initial mass function (IMF), 335
Interferometer
  radio, 79–81
  VLA, 79–80
  VLBI, 80–81
International Ultraviolet Explorer (IUE), 71
Interstellar dust
  composition, 314
  electric charge, 314–316
  evolution, 318–319
  size, 313–314
  temperature, 317–318
Interstellar extinction, 306–313
Interstellar gas, 319–331
Interstellar hydrogen, molecular. *See* Molecules, interstellar
Interstellar magnetic fields, 321, 336–338
Interstellar medium, 305–332
Interstellar molecules. *See* Molecules, interstellar
Interstellar reddening. *See* Interstellar extinction
Inverse Compton scattering, 471
Io. *See* Jupiter, moons
Ion, 39–40
Ionization, 39–40
Ion–molecule reaction *See* Molecules, interstellar
Iron
  abundance, 622
  solar, 123
  supernovae and, 248–249
Irregular galaxies. *See* Galaxies, classification
Isotopes, 201
  cosmological, 472–475
  dating, 516–517
Isotropy. *See* Cosmic background radiation

Jansky, Karl, 75–76
Jeans length, 178–180, 339
Jovian planets. *See* Jupiter; Neptune; Saturn; Uranus
Julian days, 194
Jupiter
  atmosphere, 569–573
  bands, 571–572
  belts, 571–572
  composition, 556, 565
  density, 565
  diameter, 565
  differential rotation, 565
  energy source, 570
  Great Red Spot, 565, 572–573
  interior, 574–575
  mass, 565
  moons, 586–590
  rings, 578, 582
  rotation, 565
  spacecraft observations, 502–503, 565
  temperature, 569
  zones, 571–572

K. *See* K stars
K-corona. *See* Sun, corona
Kepler, Johannes, 494–495
Kepler's laws
  binary stars, 94–99
  solar system, 495–497
Kinematic distance, 366–367
Kinetic energy, 33
Kinetic temperature, 38–45
Kirchhoff, Gustav, 27
Kitt Peak, 66–67. *See also* National Optical Astronomy Observatory
Kitt Peak National Observatory. *See* National Optical Astronomy Observatory
K stars, 41
Kuiper Airborne Observatory (KAO), 73–74

Lagrangian points, 273
Large Magellanic Cloud. *See* Magellanic Clouds
Lasers, 160. *See also* Masers
Leverrier, Urbain, 568
Libration, 498–499
Light
  bending of. *See* General theory of relativity
  speed of, 9
  waves. *See* Electromagnetic radiation
Light cone. *See* General theory of relativity, black holes
Lighthouse model. *See* Pulsars
Limb darkening, 118–119
Line profile, 121–122
Lobes. *See* Radio galaxies
Local group, 403

Local standard of rest (LSR), 358
Local supercluster, 416
Lorentz contraction. *See* Special theory of relativity
Lorentz transformation. *See* Special theory of relativity
Lowell, Percival, 592
Luminosity, 14–15, 103–104, 199
Luminosity class, 45
Lunar. *See* Moon
Lunar occultation. *See* Occultation
Lyman series, 5

M1. *See* Crab Nebula
M31, 387
M42. *See* Orion Nebula
M51, 393
M81, 393
M82, 383, 386
M87, 377
M92, 289
McDonald Observatory, 160
Magellanic Clouds, 381–382
Magnetic field. *See also* individual planets
    interstellar, 336–339
    neutron stars, 257–258
    radio galaxies, 421–422
    solar, 129–131
    synchrotron emission, 250–252
Magnetic mirror, 535–536
Magnetic monopole, 482–483
Magnification, 56–57
Magnitude
    absolute, 23–24
    apparent, 9–11
    bolometric, 24
Main sequence, 199–227
Major axis. *See* Ellipse
Mantle, 514–515
Mare. *See* Moon
Mariner spacecraft. *See* individual planets
Mars
    axis, 549
    atmosphere, 562–563
    canyons, 555, 557
    channels, 557
    climate, 557
    craters, 549, 555
    Deimos, 563–564
    density, 559
    dust storms, 555, 562
    interior, 559–560
    life on, 549, 563
    Mariner spacecraft, 549, 554–555
    moons, 563–564
    Olympus Mons, 555–556
    Phobos, 563–564
    polar caps, 555, 557
    retrograde motion, 493

sand dunes, 555
spacecraft, 549, 554–556
surface, 554–557
temperature, 563
Valles Marineris, 555, 557
Viking spacecraft, 549–556
volcanoes, 555
water, 557
Mascons. *See* Moon
Masers, 346–349
Mass. *See also* individual objects
    gravitational, 158
    inertial, 158
    rest, 154
    stellar, 102–104
Mass function, 104
Mass loss, 237–239, 249–252
Mass–luminosity relation, 102–104
Mass number, 201
Mauna Kea, Hawaii, 67
Maunder butterfly diagram. *See* Sun, sunspots
Maunder minimum. *See* Sun, sunspots
Max Planck Institute fur Radioastronomie, 77
Maxwell, James Clerk, 10
Maxwell's equations, 10
Mayall telescope, 67
Mean free path, 114
Mean Sun, 621
Mercury
    advance of perhelion, 161
    atmosphere, 560–561
    craters, 552–553
    days on, 548
    interior, 559
    magnetic field, 559
    orbit, 548
    rotation, 548
    surface, 552–553
    temperature, 553
Mesons, 476
Metals and populations, 299
Meteorites
    abundances in, 605
    ages of, 605
    Allende, 605
    chondrites, 604
Meteoroids, 603–606
Meteors, 604
Meteor showers, 605
Methane, 569, 574
Michelson, A. A., 141
Michelson–Morley experiment, 141
Microwaves. *See* Radio astronomy
Milky Way, 355
Milky Way Galaxy
    center, 371
    disk, 367–371

globular cluster distribution, 355
mass, 357–358
molecular studies, 367–369
populations, 299–300
rotation, 356–367
spiral structure, 369–371
21-cm studies, 364, 367–369
Mini black holes. *See* General theory of relativity, black holes
Minor axis. *See* Ellipse
Minor planets. *See* Asteroids
Mira variables, 233
Mirror. *See* Telescopes, optical
Missing mass problem. *See* Dark matter
MMT. *See* Multiple mirror telescope
Molecules
    in comets. *See* Comets
    in planets. *See* individual planets
    interstellar, 324–331, 339–342, 386–388
Momentum
    angular, 98–99, 328–329
    conservation of, 92
    linear, 92
Monopoles, magnetic, 482–483
Moon, 541–546
    age of rocks, 543–544
    craters, 541–543
    distance from Earth, 498–500
    eclipses, 501
    exploration, 543
    gravity, 545
    highlands, 541
    interior, 545
    librations, 498–499
    maria, 541
    mascons, 545
    meteorites, 544
    orbit, 497–501
    origin of, 545–546
    phases of, 497–498
    rilles, 541
    surface, 543–544
Moons. *See* individual planets
Mossbauer effect, 165
Mount Palomar, 58, 60. *See also* Hale Observatories
Mount Wilson. *See* Hale Observatories
Moving clusters, 289–291
M stars, 41
Multiple mirror telescope (MMT), 61

NASA. *See also* specific satellites
    Apollo program. *See* Moon
    planetary exploration. *See* individual planets
National Aeronautic and Space Administration. *See* NASA
National Optical Astronomy Observatory, 66–67

National Radio Astronomy Observatory, 76–79, 325
Naval Research Laboratory, 70
Nebulae. *See also* individual nebulae
  absorption, 306–313
  dark, 305
  emission, 342–346
  planetary, 237–239
  reflection, 306
  solar, 509
Neptune
  atmosphere, 569, 574
  density, 568
  diameter, 568
  discovery, 568
  interior, 575–576
  rotation, 569
  temperature, 574
Neutrino, 205, 249
  possible mass, 396
  solar, 223–227
Neutron, 205, 249
  decay, 205
  nuclear energy levels, 206
Neutron degeneracy, 254–256
Neutron star, 252–258
  density, 254–255
  magnetic field, 257–258
  pulsar, 258–268
  rotation, 256–257
Newtonian reflector, 59
Newton's law of gravitation, 92
N galaxies. *See* Galaxies, N
NH_3, 509. *See also* Molecules, interstellar
Nonthermal radiation. *See* Synchrotron radiation
Novae
  binary stars, 275–277
  Nova Cygni 1975, 276
NRAO. *See* National Radio Astronomy Observatory
NRL. *See* Naval Research Laboratory
Nuclear reactions, 204–207
  binding energy, 202–204
  CNO cycle, 211–212
  decays, 204–206
  fission, 206
  fusion, 206–208
  proton–proton chain, 210
Nucleosynthesis, 213–214
Nucleus
  atomic, 30, 201–215
  galactic, 371–373
Nutation, 621

OAO-3. *See* Copernicus satellite
OB associations, 352
Objective lens, 55–57
Objective prism, 64

Oblateness, 540
Occultation
  asteroid, 596
  lunar, 106–107
  quasar, 429
Occulting disk, 124–125
OH. *See* Molecules, interstellar
Olbers's paradox, 443–444
Olympus Mons. *See* Mars
Oort, Jan H., 360
Oort cloud, 602–603
Oort rotation constants, 360–363
Opacity, 112–116, 220–221
Open clusters. *See* Galactic clusters
Open universe. *See* Cosmology
Opposition, 495
Optical depth, 113–116
Optical doubles, 85
Optical observatories, 66–69. *See also* individual observatories
Orbits
  binary stars, 89–102
  elliptical, 97–102, 505–507
  planets, 493–497
  space probes, 505–507
Orion, 352–353
Orion molecular cloud, 350–353
Orion Nebula, Color Plate 2
Oscillating universe. *See* Cosmology
O stars, 43
Outer planet, 495–497, 565–593
Ozone, 518

Palomar Observatory. *See* Hale Observatories
Parabola, mirror. *See* Telescopes, optical
Parallax
  spectroscopic, 46–47
  trigonometric, 20–22
Parsec, 22
Pauli exclusion principle, 240
Peculiar galaxies. *See* Galaxies, peculiar
Penzias, A. A., 463
Perihelion. *See* Binary pulsar; Comet; Mercury
Period
  Cepheid, 232–237
  orbit, 91, 94–95
  pulsar, 258–269
Period–luminosity relation, 235–237
Perseus cluster, 402
Phase, 497–498
Phobos. *See* Mars, moons
Photocell, 63
Photography, 62–63
Photometry, 62
Photomultiplier, 63
Photon, 17–18
Photon sphere. *See* Black holes

Photosphere, 116–123. *See also* Sun
pion, 476
Pioneer spacecraft. *See* Jupiter; Saturn; Venus
Planck, Max, 16
Planck's constant, 16
Planck's law, 15–16. *See also* Black body
Planetary nebulae, 237–239
Planets. *See also* individual planets
  atmospheres, 522–535
  formation, 507–512
  interiors, 557–559
  Kepler's laws, 495–497
  orbits, 493–497, 506
  temperatures, 520–521
Plate tectonics. *See* Earth, plate tectonics
Pleiades, 288
Pluto
  discovery, 592
  moon, 593
  orbit, 593
Polar caps. *See* individual planets
Polaris, 233
Polarization
  starlight, 312
  synchrotron radiation, 250–253
Population of energy level, 38–39
Population I, 299–300
Population II, 299–300
Positron, 205
Potential energy
  electrical, 33
  gravitational, 177–178
p–p chain. *See* Proton–proton chain
Precession, 540–541
Pressure, 216–219. *See also* individual planets, atmospheres
Prime focus, 59
Primeval solar nebula, 507–512
Primordial abundances, 473
Principal quantum number, 31–35
Prism, 64
Prominences. *See* Sun
Proper motion, 289
Proton, 205, 209
Proton–proton chain, 210
Protostars, 176–198
  collapse, 178–192
  H–R diagram, 190–192
  infrared radiation, 187–190
Ptolemaic theory, 493
Ptolemy, Claudius, 493
Pulsars, 258–269
  binary, 280
  Crab, 261–262
  discovery, 258–260
  dispersion, 268–269
  distances, 269
  explanations, 260–263

Pulsars (*Continued*)
  glitches, 266–267
  lighthouse model, 261
  magnetic field, 257–258
  optical, 262–263
  slowdown, 263–265
Pumping. *See* Masers

q_0. *See* Deceleration parameter
QSO. *See* Quasars
QSR. *See* Quasars
Quadrature, 495
Quantization. *See* Quantum mechanics
Quantum mechanics, 37–38
  Bohr atom, 31–37
  Compton effect, 17
  de Broglie waves, 37
  history, 15–18
  particle–wave duality, 37–38
  probability, 37–38
  quantum electrodynamics, 477
  uncertainty principle, 241–244
  wave function, 37
Quarks, 475–483
Quasars, 429–437
  active galaxies, 436
  detection, 429–433
  discovery, 429–432
  Doppler shift, 431
  energy source, 436
  lensed, 437–439
  redshift explanations, 434–436
  spectra, 430, 433
Quasi-stellar objects. *See* Quasars

Radar astronomy, 550–552
Radial velocities, 87–91
Radian, 20
Radiation. *See* Black body; Cosmic background radiation; Electromagnetic radiation; Light; Planck's law
Radiation pressure, 511, 600–602
Radiative transfer, 112–117
Radioactive isotopes, dating with, 516–518
Radio astronomy. *See also* Radar astronomy; Telescopes, radio
  history, 75–76
  interstellar medium, 345. *See also* HI regions; HII regions; Molecules, interstellar
  planets. *See* individual planets
Radio waves. *See* Electromagnetic waves; Radar astronomy; Radio astronomy
Rayleigh–Jeans law, 15–16
Reber, Grote, 76
Reciprocity failure, 63
Recombination lines. *See* HII regions
Reddening, 309–311
Red giants, 44–45, 229–232

Redshift. *See* Black holes; Cosmology; Doppler effect; General theory of relativity; Hubble's law; Quasars
Red supergiants, 44–45
Reflecting telescope. *See* Telescopes, optical
Refracting telescope. *See* Telescopes, optical
Refraction, 53
Relativity. *See* General theory of relativity; Special theory of relativity
Relaxation, 294–296
Rest wavelength, 88
Retrograde motion, 493–494
Retrograde rotation, 549
Right ascension, 619
Ritchey–Cretien telescopes. *See* Telescopes, optical
Ring Nebula, 239
Roche limit, 580–581
Roche lobe, 273–274
Root-mean-square velocity, 122
Rotation. *See also* individual objects
  retrograde, 549
  synchronous, 272, 497, 548–549, 588
Rotation curve, 358–360
RR Lyrae stars, 237
Russell, Henry Norris, 43
Rutherford, Ernest, 30
Rydberg constant, 30

Sagittarius A. *See* Galactic center
Saha equation, 40
SAS-1 (Uhuru), 81
Satellites, artificial, 503–507. *See also* individual probes; Orbits
Saturn, 566, 573–576, 582, 585
  atmosphere, 573–574
  density, 566
  diameter, 56
  interior, 575–576
  internal heating, 576
  mass, 566
  moons, 590, 591
  rings, 576–580, 582, 585
  rotation, 566
  spacecraft to, 566
Scale height, 23
Scattering, 306
Schmidt, Maarten, 430
Schmidt camera. *See* Telescopes, optical
Schrödinger, Erwin, 37
Schrödinger's equation, 37
Schwarszchild radius. *See* Black holes
Seeing, 52
Seyfert galaxies. *See* Active galaxies; Galaxies, Seyfert
Shapley, Harlow, 375
Sidereal day, 621
Sidereal period, 495
Sidereal time, 621

Silicates, 314
Simultaneity. *See* Special theory of relativity
Singularity. *See* Black holes
Small astronomy satellite. *See* SAS-1
Small Magellanic Cloud. *See* Magellanic Clouds
Smithsonian Astrophysical Observatory. *See* Center for Astrophysics
Snowball model of comets. *See* Comets
SNU, 223–227
Solar nebula, 507–512
Solar system. *See also* individual objects
  age, 517–519
  exploration, 503–507
  formation, 507–512
  Kepler's laws, 495–497
Solar telescopes. *See* Telescopes, optical
Solar time, 621
Solar wind. *See* Sun
Source counting. 457
Space telescope, 68–69
Spacetime. *See* Special theory of relativity
Space velocity, 289
Special theory of relativity, 139–155
  Doppler shift, 148–158
  energy and mass, 153–155
  foundations of, 139–142
  four-vectors, 151–155
  light cone, 153
  Lorentz contraction, 145–147
  Lorentz transformation, 151–155
  Maxwell's equations and, 139–140
  Michelson–Morley experiment, 141
  principle of relativity, 139
  simultaneity, 141–142
  space–time, 151–155
  speed of light, 140–143
  tachyons, 155
  time dilation, 142–155
Speckle interferometry, 107
Spectral classes. *See* Spectral types
Spectral lines, 12, 27–28. *See also* individual objects
  absorption, 27–28
  broadening, 121–122
  emission, 27–28
  formation of, 30–43
  profile, 121–122
Spectral types, 29–30, 41–44, 102. *See also* Hertzsprung–Russell diagram
Spectrograph, 64–66
Spectrometer, 64–66
Spectroscopic binaries, 86
Spectroscopic parallax, 46–47
Spectroscopy, 64–66
Spectrum. *See* individual objects
Speed of light, 9. *See also* Special theory of relativity
Spherical aberration, 54–55

Spin, 240
Spin–flip, 320
Spiral galaxies. *See* Galaxies, spiral
Spiral structure. *See* Galaxies, spiral
SS433, 283–285
ST. *See* Hubble Space Telescope
Stars. *See also* individual types; Galactic
    clusters; Globular clusters; Hertz-
    sprung–Russell diagram; Spectral types;
    Variable stars
  absolute magnitude, 23–24
  binary, 85–106
  brightness, 7–9
  close binaries, 272–285
  clusters, 287–300
  colors, 11–12, 18–19, 24
  distances, 19–24
  energy generation, 199–215, 223–227
  evolution, 229–231
  evolutionary tracks, 229–231
  formation of, 176–184, 335–353
  interiors, 215–222
  luminosity, 14–15
  magnitudes, 8
  main-sequence, 199–226, 229
  mass, 102–104, 229–232
  mass loss, 237–239, 249–252
  mass–luminosity relation, 103–104
  models, 222
  neutrinos, 223–226
  oscillations, 234–236
  proper motion, 289
  radial velocity, 87–91
  sizes, 104–108
  space velocity, 289
  spectra, 27–47
  spectral types, 41–43
  structure, 215–222
  temperatures, 41–43
  variable, 232–237
Statistical weights, 39
Steady state theory. *See* Cosmology
Stefan–Boltzmann constant, 14–15
Stefan–Boltzmann law, 14–15
Stellar atmospheres. *See* Stars, atmospheres
Stellar populations, 299–300
Stimulated emission, 329
Streamers, coronal. *See* Sun
Strong force, 477
Sun, 110–135
  absorption lines, 120
  active, 128–133
  atmosphere, 111, 116–127
  Balmer lines, 120
  chromosphere, 123–124
  composition, 111
  continuous spectrum, 117
  convection zone, 118
  core, 111

  corona, 111, 124–128
  coronal holes, 127
  diameter, 111
  Doppler broadening, 121–123
  eclipses, 124–125
  flares, 132–133
  flash spectrum, 123–124
  Fraunhofer lines, 119–120
  granulation, 118
  interior, 111
  limb, 110, 119
  luminosity, 15, 111
  magnetic field, 129–131
  mass, 111
  neutrinos, 223–226
  opacity, 112–116
  oscillation, 118
  photosphere, 111, 116–123
  prominences, 133
  rotation, 120, 131
  sunspots, 120–132
  supergranulation, 123
  temperature, 111, 121–122, 126
Sunspots. *See* Sun
Supercluster, 416–417
Supergiant. *See* Hertzsprung–Russell diagram
Superluminal expansion, 424–427
Supernovae, 248–250, 276
  distances to galaxies, 414
  types, 276
Supernova remnants, 250–253
Synchronization of clocks. *See* Special theory of relativity
Synchronous rotation, 272
Synchrotron radiation, 250–252, 345,
  419–421
Synodic period, 495

Tachyons. *See* Special theory of relativity
Tangential velocity, 289
Taylor, Joseph H., 279
Telescopes, 49–82
  gravity wave, 166
  infrared, 71–75
  neutrino, 223–227
  optical, 49–62, 66–69, 117. *See also* Mul-
    tiple Mirror Telescope; individual obser-
    vatories
  radio, 75–81
  space. *See* individual telescopes and satel-
    lites
  ultraviolet, 69–71
  x-ray, 81–82
Temperature. *See also* individual objects
  cosmic background radiation. *See* Cosmol-
    ogy
  excitation, 39
  ionization, 40
Terrestrial planets. *See* individual planets

Thermal radiation, 345
3C 273, 424, 429–433
Tidal forces, 256, 537–540, 580–582
Time, 620–621
Time dilation. *See* Special theory of
  relativity
Titan. *See* Saturn, moons
Titus–Bode law, 497
Tombaugh, Clyde, 592
Total eclipses. *See* Moon, eclipses; Sun,
  eclipses
Transit, Venus, 561
Transition, 31–37
Transverse velocity, 289
Trigonometric parallax. *See* Parallax, trigo-
  nometric
Triple-alpha process, 210–211
Tritium. *See* Hydrogen
Trojan asteroids, 596
T Tauri, 192–194
T Tauri stars, 192–197
21-cm radiation, 320–324
Twinkling, 52
Twin paradox. *See* Special theory of
  relativity
Two-phase model, 323
Tycho Brahe, 495

UBV filters, 19
Uhuru satellite, 81
Ultraviolet astronomy, 69–71. *See also* Telescopes, ultraviolet
Umbra. *See* Sun, eclipse; Sun, sunspots
Uncertainty principle, 241
Universal time (UT), 621
Universe. *See* Cosmology
Uranus
  atmosphere, 574
  axis, 567
  density, 567
  interior, 575–576
  moons, 591–592
  rings, 579–580, 582
  rotation, 567
  spacecraft studies, 567
  temperature, 567, 574

Valles Marineris. *See* Mars
Van Allen, James, 535
Van Allen radiation belts, 535–537
Variable stars, 232–236. *See also* Binary
  stars; Cepheids; RR Lyrae stars
Veil Nebula, 251
Vela pulsar, 259
Venera spacecraft, 554
Venus
  atmosphere, 561–562
  cloud cover, 561
  craters, 553–554

Venus (*Continued*)
  greenhouse effect, 561–562
  interior, 559
  phases, 494
  radar maps, 553–554
  rotation, 549
  spacecraft observations, 549, 553–554
  surface, 553–554
  temperature, 561–562
  transits, 561
  Venera landings, 549, 553–554
Vernal equinox. *See* Equinoxes
Very Large Array (VLA), 79–81
Viking missions. *See* Mars
Virgo cluster, 402
Virial theorem, 184–187
Visible light, 9–11
Visual binaries. *See* Binaries, visual
VLA. *See* Very Large Array

Voids, 416–417
Volcanoes. *See* individual objects
Voyager, 508

Water
  Earth, 516, 530–531
  interstellar, 346–347
  Mars, 557
Wave front, 79
Wavelength, 9–10
Wave–particle duality, 37–38
Weak force, 477–481
Weird terrain, 553
Whipple, Fred, 597
Whirlpool galaxy, 393
White dwarfs, 45–46, 240–246
  Chandrasekhar limit, 246
  density, 243
  electron degeneracy, 240–244

  H–R diagram location, 44
  mass, 243
  novae, 275–276
  relativity test, 165
  size, 45–46
Wien's displacement law, 13–14
Wilson, Robert, 463
Windows, atmospheric, 11
W particles, 477
W Virginis stars. *See* Cepheids

X-ray astronomy, 81–82. *See also* Telescopes, x-ray

Young, Thomas, 9

Zeeman effect, 320–321
Zero-age main sequence (ZAMS). *See* Hertzsprung–Russell diagram